Operator Theory: Advances and Applications
Vol. 142

Editor:
I. Gohberg

Singular Integral Operators, Factorization and Applications

International Workshop on Operator Theory and Applications IWOTA 2000, Portugal

Albrecht Böttcher
Marinus A. Kaashoek
Amarino Brites Lebre
António Ferreira dos Santos
Frank-Olme Speck
Editors

Springer Basel AG

Editors:

Albrecht Böttcher
Faculty of Mathematics
Technical University Chemnitz
09107 Chemnitz
Germany

Marinus A. Kaashoek
Department of Mathematics and Computer Science
Vrije Universiteit Amsterdam
De Boelelaan
NL-1081 HV Amsterdam
The Netherlands

Amarino Brites Lebre
António Ferreira dos Santos
Frank-Olme Speck
Departamento de Matemática
Instituto Superior Técnico
Ave. Rovisco Pais
1049-001 Lisboa
Portugal

Managing Editor:
Antonio.Ferreira.Santos@math.ist.utl.pt

2000 Mathematics Subject Classification: 47-xx; 46-xx; 45-xx

A CIP catalogue record for this book is available from the
Library of Congress, Washington D.C., USA

Bibliographic information published by Die Deutsche Bibliothek
Die Deutsche Bibliothek lists this publication in the Deutsche Nationalbibliografie; detailed bibliographic data is available in the Internet at <http://dnb.ddb.de>.

ISBN 978-3-0348-9401-2 ISBN 978-3-0348-8007-7 (eBook)
DOI 10.1007/978-3-0348-8007-7

Originally published by Birkhäuser Verlag, Basel – Boston – Berlin in 2003
Softcover reprint of the hardcover 1st edition 2003
Printed on acid-free paper produced from chlorine-free pulp. TCF ∞
Cover design: Heinz Hiltbrunner, Basel

ISBN 978-3-0348-9401-2

9 8 7 6 5 4 3 2 1

www.birkhauser-science.com

Table of Contents

Editorial Introduction

This volume contains the proceedings of the International Workshop on Operator Theory and Applications held at the University of Algarve in Faro, Portugal, September 12–15, in the year 2000.

The main topics of the conference were

- ▷ Factorization Theory;
- ▷ Factorization and Integrable Systems;
- ▷ Operator Theoretical Methods in Diffraction Theory;
- ▷ Algebraic Techniques in Operator Theory;
- ▷ Applications to Mathematical Physics and Related Topics.

A total of 94 colleagues from 21 countries participated in the conference. The major part of participants came from Portugal (32), Germany (17), Israel (6), Mexico (6), the Netherlands (5), USA (4) and Austria (4). The others were from Ukraine, Venezuela (3 each), Spain, Sweden (2 each), Algeria, Australia, Belorussia, France, Georgia, Italy, Japan, Kuwait, Russia and Turkey (one of each country).

It was the 12th meeting in the framework of the IWOTA conferences which started in 1981 on an initiative of Professors I. Gohberg (Tel Aviv) and J.W. Helton (San Diego). Up to now, it was the largest conference in the field of Operator Theory in Portugal. The previous IWOTA meetings were held in

Santa Monica, California, USA	(1981)
Rehovot, Israel	(1983)
Amsterdam, The Netherlands	(1985)
Mesa, Arizona, USA	(1987)
Rotterdam, The Netherlands	(1988)
Sapporo, Japan	(1991)
Vienna, Austria	(1993)
Regensburg, Germany	(1995)
Bloomington, Indiana, USA	(1996)
Groningen, The Netherlands	(1998)
Bordeaux, France	(2000)

Thus, during the World Mathematical Year 2000, there were, as a unique event up to now, two conferences of the IWOTA series in the same year, both with a high number of participants.

The IWOTA-Portugal 2000 had a duration of four days and was organized in 25 invited and 49 contributed talks.

This volume contains 20 contributions. The majority of articles is centered on the analysis of various classes of singular operators, the factorization of operators and matrix functions, algebraic methods in approximation theory, and applications to diffraction theory. The volume contains some further papers with related topics from fractional calculus, complex analysis, operator algebras and partial differential equations.

The organizers gratefully acknowledge the support of the following institutions:

- Fundao para a Cincia e a Tecnologia
  (through Programa Praxis and Fundo de Apoio à Comunidade Científica),
- Fundação Calouste Gulbenkian,
- Fundação Luso-Americana para o Desenvolvimento,
- Centro de Matemática Aplicada (financed by FCT).

The organizers are also greatly indebted to the Instituto Superior Técnico, from U.T.L., and the University of Algarve at Faro for their general support. Last but not least, the University of Algarve at Faro provided excellent facilities for the realization of the

Lisbon, May 2002 The Editorial Board

Note on the Memory of Professor Erhard Meister (1930–2001)

On July 5th 2001 the invited speaker Erhard Meister died at the age of 71. IWOTA-Portugal 2000 was the last conference that he attended, but he was busy with scientific work until the very last days of his life. The present proceedings contain his last publication.

Erhard Meister was born on February 12th 1930 in Bernburg, studied mathematics and physics in Heidelberg and finished his PhD with Heinz Söhngen at Saarbrücken in 1958. He was full professor at the Technical University of Berlin from 1966, at the University of Tübingen from 1970 and at the Darmstadt University of Technology from 1974 to 1998. In 1974, he met Israel Gohberg at Darmstadt – the beginning of a long period of fruitful international collaboration and friendship. Convinced of the importance of international contacts, he initiated and sponsored plenty of visits of foreign specialists to Germany, especially from Eastern European countries. The benefits for the scientific community and for individual progress are gratefully remembered by his colleagues and pupils.

The scientific work of Professor Meister belongs on the whole to applications of mathematics, in a wide sense, to mathematical physics. His principal areas of research included aero and fluid dynamics, complex analysis, integral transformations, integral equations and operator theory with a specialization in the mathematical diffraction and scattering theory for microwaves in acoustics, electromagnetics and elasto-dynamics, aperiodic wave propagation and, finally, quaternionic analysis. Among more than one hundred publications we can find various survey articles and three textbooks. His research activities resulted in the supervision of twenty-five PhD theses.

Professor Meister was a member of the editorial boards of seven mathematical serials of international reputation. He was chairman of the GAMM committee on applied analysis and mathematical physics, worked as main referee for applied mathematics for the DFG and for the DFG committee "Sonderforschungsbereiche", and he organised several international conferences.

With the death of Erhard Meister we lost a scientist devoted to mathematics in all its aspects and to science in a broad sense. His friends and colleagues remember him as an enthusiastic mathematician, mentor and impressive person who promoted mathematics in all fields of applied science and the progress of the studies of everybody in his sphere.

The Editorial Board

Program of IWOTA-Portugal 2000

Tuesday, 12 September 2000

Morning Sessions

09:30–09:45		*Opening Session*	Room: Anf. Azul
Chairman	F. dos Santos	*Session 1*	Room: Anf. Azul
09:50–10:35	Harold Widom	*Toeplitz determinants, random matrices and random permutations*	
10:40–11:00		*Coffee*	
Chairman:	M. Kaashoek	*Session 2*	Room: CP-Anf.A
11:00–11:45	Israel S. Kac	*Spectral theory of a string and its applications in the probability theory*	
11:50–12:35	Albrecht Böttcher	*Randomly perturbed Toeplitz and Laurent matrices*	
Chairman:	V. Maz'ya	*Session 3*	Room: CP-Anf.B
11:00–11:45	Reinhard Mennicken	*A factorization theorem for the transfer function associated with a 2×2 operator matrix having unbounded couplings*	
11:50–12:35	Harm Bart	*Logarithmic residues of Fredholm operator valued functions and sums of finite rank projection operators*	
12:40–14:00		*Lunch*	

Afternoon Sessions

Chairman:	I. Spitkovsky	*Session 4* Factorization Theory	Room: CP-Anf.A
14:00–14:25	Georgii Litvinchuk	*On solvability theory of generalized Hilbert and Carleman boundary value problems*	
15:00–15:25	Torsten Ehrhardt	*Invertibility of Toeplitz + Hankel operators and singular integral operators with flip*	
15:30–15:55	Juan Rodrguez	*Factorization of singular integral operators with a linear-fractional Carleman shift*	
Chairman:	M. Gorbachuk	*Session 5* Topics in Operator Theory	Room: CP-Anf.B
14:00–14:25	L. von Wolfersdorf	*A regularization of the auto-correlation equation*	
14:30–14:55	Elias Wegert	*On a class of nonlinear singular integral equations*	

15:00–15:25	Harald Woracek	*An indefinite generalization of a canonical system of differential equations*
15:30–15:55	Matthias Langer	*A characterization of intermediate Weyl coefficients*
Chairman:	G. Hsiao	*Session 6* Room: CP-2.21 Applications to Mathematical Physics
14:00–14:25	Christiane Tretter	*The essential spectrum of a system of singular differential operators occurring in magnetohydrodynamics*
14:30–14:55	Héctor Oviedo	*Applications of quaternionic analysis to the solutions of boundary values problems for Maxwell's equations in chiral media*
15:00–15:25	Vladislav Kravchenko	*Integral representations for spinor fields and the radiation condition for the Dirac operator*
15:30–15:55	Angela Hommel	*Fundamental solution of a discrete Dirac operator*
16:00–16:30		*Coffee*
Chairman:	L. von Wolfersdorf	*Session 7* Room: CP-Anf.A Algebraic Techniques in Operator Theory
16:30–16:55	William Longstaff	*Single elements of operator algebras*
17:00–17:25	Bernhard Gramsch	*Analytic Fredholm families and stochastic partial differential equations*
17:30–17:55	Eizaburo Kamei	*Chaotic order and Furuta inequality*
18:00–18:25	Luís Castro	*Well-posedness of diffraction problems involving n coplanar strips*
Chairman:	B. Silbermann	*Session 8* Room: CP-Anf.B Topics in Operator Theory
16:30–16:55	Harry Dym	*On the role of the Muckenhoupt condition in the theory of J-inner matrix functions*
17:00–17:25	Vladimir Peller	*Hankel and Toeplitz Schur multipliers*
17:30–17:55	Art Frazho	*The band method, unitary dilations and positive real interpolation*
18:00–18:25	Heinz Langer	*Variational principles for selfadjoint operator pencils*
18:30–20:30		*Reception*

Wednesday, 13 September 2000

Morning Sessions

Chairman:	H. Widom	*Session 9* Room: Anf. Azul
09:00–09:45	Marinus Kaashoek	*Inverse problems for canonical systems with pseudo-exponential potentials and applications to integrable systems*

09:50–10:35	Myroslav Gorbachuk	*Boundary values of semigroups of operators on a Banach space*	
10:40–11:00		*Coffee*	
Chairman:	H. Dym	*Session 10*	Room: CP-Anf.A
11:00–11:45	Leiba Rodman	*Almost periodic factorizations of multivariable matrix functions*	
11:50–12:35	Andre C.M. Ran	*Symmetric factorizations of selfadjoint rational matrix functions*	
Chairman:	I. Kac	*Session 11*	Room: CP-Anf.B
11:00–11:45	Aad Dijksma	*The Schur algorithm for generalized Schur functions*	
11:50–12:35	Harald Upmeier	*Non-commutative Hardy spaces and Toeplitz-Berezin quantization*	
12:40–14:00		*Lunch*	

Afternoon Sessions

Chairman:	H. Upmeier	*Session 12* Topics in Operator Theory	Room: CP-Anf.A
14:00–14:25	Nikolai Vasilevski	*Bergman-Toeplitz operators: "Inside-the-domain" effects*	
14:30–14:55	Paula Cerejeiras	*About reproducing kernels on hyperbolic spaces*	
15:00–15:25	Annemarie Luger	*About generalized zeros of non-positive type of singular generalized matrix Nevanlinna functions*	
15:30–15:55	Gregory Derfel	*Spatially-chaotic configuration and functional equations with rescaling*	
Chairman:	G. Litvinchuk	*Session 13* Applications to Mathematical Physics	Room: CP-Anf.B
14:00–14:25	Alexey Karapetyants	*On the inversion of potential type operators with kernels having singularities on a sphere*	
14:30–14:55	Alexander Kozhevnikov	*Multi-weighted parameter-ellipticity and parabolicity for multi-order boundary value problems*	
15:00–15:25	Cornelis van der Mee	*Inverse scattering for a Schrödinger equation with energy dependent potential*	
15:30–15:55	Petr Kulish	*Riemann-Hilbert problem and generators of infinite-dimensional algebras*	
16:00–16:30		*Coffee*	
Chairman:	R. Duduchava	*Session 14* Factorization Theory	Room: CP-Anf.A
16:30–16:55	Maria Amélia Bastos	*The invertibility of convolution type operators on a union of intervals*	
17:00–17:25	Vladimir Kisil	*Toeplitz operators as convolutions*	

17:30–17:55	Ana Moura Santos	*Image/domain normalization of Wiener-Hopf operators*	
Chairman:	B. Gramsch	*Session 15* Topics in Operator Theory	Room: CP-Anf.B
16:30–16:55	Gerhard Jank	*Riccati differential equations and dynamical games*	
17:00–17:25	Said Mesloub	*On a singular problem with a non-local constraint*	
17:30–17:55	Carsten Trunk	*Sturm-Liouville equations whose coefficients depend rationally on the eigenvalue parameter*	

Thursday, 14 September 2000

Morning Sessions

Chairman:	A. Böttcher	*Session 16*	Room: Anf. Azul
09:00–09:45	Vladimir Maz'ya	*The Schrödinger operator on the energy space: boundedness and compactness criteria*	
09:50–10:35	Ilya Spitkovsky	*Left versus right factorization*	
10:40–11:00		*Coffee*	
Chairman:	H. Bart	*Session 17*	Room: CP-Anf.A
11:00–11:45	Roland Duduchava	*Boundary integral equations on curves with cusps*	
11:50–12:35	Nikolai Karapetyants	*Multi-dimensional integral operators with homogeneous kernels with shift*	
Chairman:	A. Ran	*Session 18*	Room: CP-Anf.B
11:00–11:45	Bernd Silbermann	*Algebraic methods in numerical analysis: theory, examples, problems*	
11:50–12:35	Vladimir Rabinovich	*Limit operators method for band-dominated and pseudodifferential operators*	
12:40–14:00		*Lunch*	

Afternoon Sessions

Chairman:	V. Rabinovich	*Session 19* Algebraic Techniques in Operator Theory	Room: CP-Anf.A
14:00–14:25	Valentyna Gorbachuk	*Operator approach to approximation problems*	
14:30–14:55	Karla Rost	*Centro-symmetric and centro-skewsymmetric Toeplitz matrices and Bezoutians*	
15:00–15:25	Leonid Lerer	*Inertia of operator polynomials and stability of differential equations*	
Chairman:	N. Karapetyants	*Session 20* Topics in Operator Theory	Room: CP-Anf.B
14:00–14:25	Stefan Samko	*Fractional powers of operators via hypersingular integrals*	

14:30–14:55	Anatoly Kilbas	*Integral operators with Bessel type kernels in spaces of summable functions*
15:00–15:25	Alexei Karlovich	*One-sided invertibility of functional operators in rearrangement-invariant spaces*
Chairman:	I. Feldman	*Session 21* Room: CP-2.21 Factorization Theory
14:00–14:25	M.C. Câmara	*$N \times N$ Daniele-Khrapkov class: explicit Wiener-Hopf factorization, structure of the factors and applications to integrable systems*
14:30–14:55	Ana Conceição	*Factorization of matrix functions and the resolvent of certain operators*
15:00–15:25	Pedro A. Santos	*An axiomatic approach to stability of approximating sequences*
16:00–16:30		*Coffee*
16:30–17:00		*Bus trip from the university to Vilamoura*
17:00–20:00		*Boat Excursion along the cost of Algarve*

Friday, 15 September 2000

Morning Sessions

Chairman:	H. Langer	*Session 22* Room: Anf. Azul
09:00–09:45	Yuri Karlovich	*New trends in the study of algebras of integral operators*
09:50–10:35	Israel Feldman	*On the continuity of the spectrum in some classes of linear operators*
Chairman:	R. Mennicken	*Session 23* Room: CP-Anf.B
09:50–10:35	George Hsiao	*On the two-dimensional inverse scattering problems in electromagnetics*
10:40–11:00		*Coffee*
Chairman:	E. Meister	*Session 24* Room: CP-Anf.A
11:00–11:45	Ernst Lüneburg	*Skew incidence half-plane diffraction problems: Factorization and restricted range*
11:50–12:35	Hamit Serbest	*Diffraction by thin isotropic chiral slabs*
Chairman:	L. Rodman	*Session 25* Room: CP-Anf.B
11:00–11:45	Georg Heinig	*Recent advances in theory and algorithms for structured matrices*
11:50–12:35	Steffen Roch	*Basic concepts in the analysis of algebras of approximation sequences*
12:40–14:00		*Lunch*

Afternoon Sessions

Chairman:	A. Dijksma	*Session 26* Topics in Operator Theory	Room: CP-Anf.A
14:00–14:25		—	
14:30–14:55	Vladimir Strauss	*On the bicommutant of a commutative symmetric algebra in Krein space*	
15:00–15:25	Ramón Bruzual	*Extensions of locally defined positive definite functions on the Heisenberg group*	
15:30–15:55	Marisela Domínguez	*Commuting unitary extensions of a pair of isometries and applications to moment problems on the discrete plane*	
Chairman:	N. Vasilevski	*Session 27* Applications to Mathematical Physics	Room: CP-Anf.B
14:00–14:25	Tatyana Shaposhnikova	*Sobolev multipliers in the non-smooth L_p-theory of boundary integral equations of elasticity*	
14:30–14:55	Natasha Samko	*Criterion for the Fredholmness of singular operators with piecewise continuous coefficients in generalized Hölder spaces with weight*	
15:00–15:25	Anna-Margarete Sändig	*The Shapiro-Lopatinskij condition for boundary and transmission problems for the Lamé system*	
15:30–15:55	Uwe Kähler	*Elliptic boundary value problems over unbounded domains*	
16:00–16:30		*Coffee*	
Chairman:	F. dos Santos	*Session 28*	Room: Anf. Azul
16:30–17:15	Erhard Meister	*Aperiodic wave diffraction by semi-infinite obstacles*	
17:20–17:40	Harold Widom	*Closing Session*	Room: Anf. Azul
18:00–19:00		*Bus trip from the university to the typical village of Alte – 50 km away from Faro*	
19:30– ...		*Conference dinner at Alte. Return late night.*	

List of Participants

1 G. Heinig
2 P. Resende
3 R. Marreiros
4 K. Rost
5 V. Strauss
6 R. Bruzual
7 P. Paúl
8 S. Roch
9 C. Mancera
10 E. Kamei
11 C. Fernandes
12 T. Ehrhardt
13 P. Santos
14 H. Bart
15 A. Dijksma
16 A. Ran
17 L. Lerer
18 H. Langer
19 L. Pessoa
20 G. Derfel
21 P. Kulish
22 Y. Karlovich
23 M.C. Câmara
24 A. Conceição
25 V. V. Kravchenko
26 M. Domínguez
27 A. Hommel
28 H. Woracek
29 M. Langer
30 M.A. Bastos
31 B. Gramsch
32 N. Karapetiants
33 I. Spitkovsky
34 L. Rodman
35 H. Dym
36 R. Mennicken
37 I. Feldman
38 M. Gorbachuk
39 J. Rodríguez
40 C. Trunk
41 A. Luger
42 A. Karlovich
43 A. Bravo
44 W. Longstaff
45 H. Upmeier
46 V. Kisil
47 U. Kähler
48 H. Widom
49 R. Duduchava
50 B. Silbermann
51 M.T. Malheiro
52 A. M. Santos
53 C. van der Mee
54 R. Picken
55 A. F. dos Santos
56 N. Vasilevski
57 N. Manojlovic
58 A. Kozhevnikov
59 P. Cerejeiras
60 A. Karapetyants
61 C. Tretter
62 A.-M. Sändig
63 G. Jank
64 E. Wegert
65 L. Oliveira
66 V. Gorbachuk
67 P. Lucena
68 A. Böttcher
69 L. von Wolfersdorf
70 A. Lebre
71 M. Kaashoek
72 I. Kac
73 H. Oviedo
74 S. Mesloub
75 G. Hsiao
76 V. Maz'ya
77 E. Lüneburg
78 G. Litvinchuk

L. Castro
A. Frazho
A. Kilbas
V. G. Kravchenko
H. Mascarenhas
E. Meister
H. Oliveira
V. Peller
V. Rabinovich
M.T. Rodríguez
N. Samko
S. Samko
H. Serbest
T. Shaposhnikova
F.-O. Speck
F. Teixeira

Operator Theory:
Advances and Applications, Vol. 142, 1–27

Covariant Symbolic Calculi on Real Symmetric Domains

Jonathan Arazy and Harald Upmeier

Abstract. We introduce the concept of "covariant symbolic calculus" on real and complex symmetric domains, prove a general product formula for the link transform (generalized Berezin transform) between two such calculi, and describe a basic example (Toeplitz calculus) in more detail.

1. Introduction

The complex Hermitian spaces of non-compact type, realized as bounded symmetric domains $D \subset \mathbb{C}^n$, are a fundamental class of non-compact Kähler manifolds whose quantization, e.g. by the well-known Berezin-Toeplitz operators, has been studied intensively [BLU], [UU]. Writing $D = G/K$ for a semi-simple Lie group G of Hermitian type and its maximal compact subgroup K, the quantization map

$$\begin{array}{rcl} \mathcal{A} : \mathcal{C}^\infty(D) & \to & \mathcal{L}(H) \\ f & \mapsto & \mathcal{A}_f \end{array} \tag{1.1}$$

realized by (possibly unbounded) operators on a complex Hilbert space H should satisfy the covariance condition

$$\mathcal{A}_{f \circ g^{-1}} = U(g)\, \mathcal{A}_f\, U(g^{-1})$$

for all $g \in G$, where U denotes an irreducible (projective) representation of G acting on H. In [AU1] a general theory concerning such "covariant quantizations" on complex symmetric domains (including the flat case $D = \mathbb{C}^n$) is developed for the weighted Bergman spaces $H = H^2_\nu(D)$ of holomorphic functions on D, belonging to the scalar holomorphic discrete series of G. The Toeplitz calculus and also the Weyl calculus are natural examples of covariant quantizations, and the main result of [AU1] gives a "product formula" for the link transform $f \mapsto \mathcal{A}^*(\mathcal{B}_f)$ on $\mathcal{C}^\infty(D)$ in terms of the spherical Fourier transform of certain K-invariant "characteristic" functions associated with the covariant quantizations $\mathcal{A}$ and $\mathcal{B}$.

While the setting of complex symmetric domains is still an active research area (e.g., concerning vector-valued representations or the Weyl calculus for domains of higher dimension), a promising new direction is to apply ideas and methods from quantization theory to *real* symmetric domains $G_\mathbb{R}/K_\mathbb{R}$, for a semi-simple or reductive Lie group $G_\mathbb{R}$ not necessarily of Hermitian type. Besides dealing with a much wider geometric framework compared to the complex case, a major application of this more general approach is to harmonic analysis of $G_{\mathbb{R}}$, yielding an explicit decomposition of certain (reducible) $G_{\mathbb{R}}$-representations into irreducible components. The complex case described above corresponds to the special case of a tensor product representation.

2. Real symmetric domains

Let
$$D = G/K$$
be a Hermitian symmetric space of non-compact type, realized as the (spectral) unit ball
$$D = \{z \in Z : \ \|z\| < 1\}$$
of a complex JB^*-triple Z of finite dimension. Then
$$G = \mathrm{Aut}(D)^\circ$$
is the identity component of the holomorphic automorphism group of D, and $K = \mathrm{Aut}(Z)^\circ$. Let
$$z \mapsto \overline{z}$$
be a conjugation of Z preserving the triple product, and define
$$\begin{aligned} Z_{\mathbb{R}} &:= \{z \in Z : \ \overline{z} = z\}, \\ D_{\mathbb{R}} &:= D \cap Z_{\mathbb{R}} = \{z \in Z_{\mathbb{R}} : \ \|z\| < 1\}, \\ G_{\mathbb{R}} &:= \{g \in G : \ \overline{g(z)} = g(\overline{z}) \ \forall \ z \in Z\} = \{g \in G : \ g(D_{\mathbb{R}}) = D_{\mathbb{R}}\}, \\ K_{\mathbb{R}} &:= K \cap G_{\mathbb{R}}. \end{aligned}$$
Then $Z_{\mathbb{R}}$ is a real JB^*-triple and
$$D_{\mathbb{R}} = G_{\mathbb{R}}/K_{\mathbb{R}}$$
is a real Riemannian symmetric space of non-compact type, called a *real bounded symmetric domain.* Up to a few low-dimensional exceptions, all Riemannian symmetric spaces of non-compact type can be realized in this way. In the following we assume that $D_{\mathbb{R}}$ is *irreducible.* The domain D is called the *complexification* of $D_{\mathbb{R}}$. It is not necessarily irreducible.

Example 2.1. Let $Z_{\mathbb{R}}$ be a complex (irreducible) JB^*-triple, considered as a real JB^*-triple. Then its unit ball $D_{\mathbb{R}}$ is a complex bounded symmetric domain, considered as a real bounded symmetric domain. Put
$$Z := Z_{\mathbb{R}} \times \overline{Z}_{\mathbb{R}}, \quad D := D_{\mathbb{R}} \times \overline{D}_{\mathbb{R}}$$
endowed with the flip conjugation
$$\overline{(z_1, \overline{z}_2)} := (z_2, \overline{z}_1) \tag{2.1}$$

for all $z_1, z_2 \in Z_{\mathbb{R}}$. The corresponding real form is the diagonal

$$\begin{aligned} \{(z_1, \overline{z}_2) \in Z : \ z_1 = z_2\} &= \{(z, \overline{z}) : \ z \in Z_{\mathbb{R}}\} \equiv Z_{\mathbb{R}}, \\ \{(z_1, \overline{z}_2) \in D : \ z_1 = z_2\} &= \{(z, \overline{z}) : \ z \in D_{\mathbb{R}}\} \equiv D_{\mathbb{R}}. \end{aligned}$$

Thus the complex case gives rise to a product domain in the complexification. Putting

$$(g_1, \overline{g}_2)(z_1, \overline{z}_2) := (g_1(z_1), \overline{g_2(z_2)})$$

for $z_1, z_2 \in D_{\mathbb{R}}$ and $g_1, g_2 \in G_{\mathbb{R}} := \mathrm{Aut}(D_{\mathbb{R}})^\circ$, we have

$$G := \mathrm{Aut}(D)^\circ = \{(g_1, \overline{g}_2) : \ g_1, g_2 \in G_{\mathbb{R}}\} \approx G_{\mathbb{R}} \times G_{\mathbb{R}}.$$

Furthermore, an automorphism $(g_1, \overline{g}_2) \in G$ commutes with (2.1) if and only if $g_1 = g_2$, since we have for $z_1, z_2 \in D_{\mathbb{R}}$

$$(g_1, \overline{g}_2)\overline{(z_1, \overline{z}_2)} = (g_1, \overline{g}_2)(z_2, \overline{z}_1) = (g_1(z_2), \overline{g_2(z_1)})$$

and

$$\overline{(g_1, \overline{g}_2)(z_1, \overline{z}_2)} = \overline{(g_1(z_1), \overline{g_2(z_2)})} = (g_2(z_2), g_1(z_1)).$$

Example 2.2. The complex matrix ball

$$D = \{z \in \mathbb{C}^{p \times q} : \ z^* z < I\}$$

in $Z := \mathbb{C}^{p \times q}$, endowed with the usual conjugation $z \mapsto \overline{z}$, gives rise to the real matrix ball

$$D_{\mathbb{R}} = \{z \in \mathbb{R}^{p \times q} : \ z^* z < I\}$$

in $Z_{\mathbb{R}} = \mathbb{R}^{p \times q}$. In particular, the interval $D_{\mathbb{R}} = (-1, 1) \subset \mathbb{R}$ has the unit disk $D = \{z \in \mathbb{C} : \ |z| < 1\}$ as complexification.

Example 2.3. The complex matrix ball

$$D = \{z \in \mathbb{C}^{2 \times 2} : \ z^* z < I\}$$

in $Z = \mathbb{C}^{2 \times 2}$, endowed with the conjugation

$$\begin{pmatrix} a & b \\ c & d \end{pmatrix}^{-} := \begin{pmatrix} \overline{d} & -\overline{c} \\ -\overline{b} & \overline{a} \end{pmatrix} = \begin{pmatrix} 0 & 1 \\ -1 & 0 \end{pmatrix} \begin{pmatrix} \overline{a} & \overline{b} \\ \overline{c} & \overline{d} \end{pmatrix} \begin{pmatrix} 0 & -1 \\ 1 & 0 \end{pmatrix},$$

has the real form

$$D_{\mathbb{R}} = \{z \in \mathbb{H} : \ z z^* < I\}$$

in $Z_{\mathbb{R}} = \mathbb{H}$. Here $\mathbb{H}$ denotes the real division algebra of quaternions.

Example 2.4. Every (irreducible) symmetric cone Ω in a Euclidean Jordan algebra X is a real symmetric domain (realized as the unit ball of X via a Cayley transform) whose complexification can be realized as the *tube domain*

$$T(\Omega) := \{x + iy : \ x \in \Omega, \ y \in X\}$$

in $Z := X^{\mathbb{C}}$.

Generalizing Example 2.4 every irreducible real bounded symmetric domain can be realized as a *real Siegel domain* as follows:

Let Z be an irreducible complex JB^*-triple with conjugation $z \mapsto \overline{z}$, and consider the Peirce decomposition

$$Z = U \oplus V, \;\; U = Z_1(e), \;\; V = Z_{1/2}(e)$$

with respect to a maximal tripotent $e = \overline{e} \in Z$. Then the respective real forms satisfy

$$Z_{\mathbb{R}} = U_{\mathbb{R}} \oplus V_{\mathbb{R}} \tag{2.2}$$

and $U_{\mathbb{R}}$ is a semi-simple (not necessarily Euclidean) real Jordan algebra with a decomposition

$$U_{\mathbb{R}} = X \oplus Y, \tag{2.3}$$

where

$$X := \{x \in U_{\mathbb{R}} : \; x^* = x\}$$

is an irreducible Euclidean Jordan algebra, and we put

$$Y := \{y \in U_{\mathbb{R}} : \; y^* = -y\}.$$

Combining (2.2) and (2.3) we obtain

$$Z_{\mathbb{R}} = X \oplus Y \oplus V_{\mathbb{R}}. \tag{2.4}$$

The self-adjoint part $X \oplus iY$ of U is also a Euclidean Jordan algebra. According to [L2, 10.1] or [U, Section 21], the unit ball of Z is equivalent (via a Cayley transformation) to a complex Siegel domain

$$D = \left\{ u + v \in Z : \; \frac{u + u^*}{2} - \{e\, v^*\, v\} \in \Omega_{X \oplus iY} \right\}$$

where $\Omega_{X \oplus iY}$ is the positive cone of $X \oplus iY$. It follows that

$$D_{\mathbb{R}} = \{x + y + v \in Z_{\mathbb{R}} : \; x - \{e\, v^*\, v\} \in \Omega\}, \tag{2.5}$$

where $\Omega = X \cap \Omega_{X \oplus iY}$ is the positive cone of X. This is the unbounded realization of $D_{\mathbb{R}}$ as a real Siegel domain.

Summing up, real symmetric domains $D_{\mathbb{R}}$ are of the following types.

Type 1 Z is irreducible and $D_{\mathbb{R}}$ has an unbounded realization

$$D_{\mathbb{R}} \approx \{x + y + v \in Z_{\mathbb{R}} : \; x - \{e\, v^*\, v\} \in \Omega\}$$

as described above.

Type 2 $Z_{\mathbb{R}}$ is a complex irreducible JB^*-triple, with unit ball $D_{\mathbb{R}}$ being a complex Hermitian symmetric domain, and the complexification is given by

$$Z = Z_{\mathbb{R}} \times \overline{Z}_{\mathbb{R}} \quad \text{and} \quad D = D_{\mathbb{R}} \times \overline{D}_{\mathbb{R}}.$$

We also include the *flat* case:

Type 3 $D = Z = \mathbb{C}^n$, endowed with the usual conjugation, so that $D_{\mathbb{R}} = Z_{\mathbb{R}} = \mathbb{R}^n$.

3. Quantization Hilbert spaces

For a real symmetric domain $D_{\mathbb{R}} = G_{\mathbb{R}}/K_{\mathbb{R}}$, with complexification $D = G/K$, let

$$H^2_\nu(D) \tag{3.1}$$

be the ν-th weighted Bergman (resp. Bargmann) space of holomorphic functions on D, endowed with the canonical irreducible (projective) representation

$$U_\nu : G \to U(H^2_\nu(D)). \tag{3.2}$$

More precisely, according to the three types of domains (cf. Section 1), (3.1) and (3.2) are defined as follows.

Case 1 Let Z be an irreducible JB^*-triple, with unit ball D of rank r, dimension n and genus p. Let $\Delta(z,w)$ denote the Jordan triple determinant and denote by $(z|w)$ the unique scalar product invariant under $K := \mathrm{Aut}(Z)^\circ$ normalized by the condition $(c|c) = 1$ for all minimal tripotents $c \in Z$. Let $dm(z)$ be the associated Lebesgue measure. Fix $\nu > p-1$ and consider the probability measure

$$d\mu_\nu(z) = \frac{\Gamma_\Omega(\nu)}{\pi^n\, \Gamma_\Omega(\nu - \frac{n}{r})}\, \Delta(z,z)^{\nu-p}\, dm(z) \tag{3.3}$$

where Γ_Ω denotes the so-called Koecher-Gindikin Γ-function of the positive cone Ω associated with Z [FK]. The *weighted Bergman space*

$$\begin{aligned} H^2_\nu(D) &:= \{h \in L^2(D,\mu_\nu) : h \text{ holomorphic}\} \\ &= \{h : D \to \mathbb{C} \text{ holomorphic } : \|h\|^2_\nu := \int_D d\mu_\nu(z)\, |h(z)|^2 < +\infty\} \end{aligned}$$

has the reproducing kernel

$$K(z,w) = \Delta(z,w)^{-\nu} \tag{3.4}$$

for all $z, w \in D$. The irreducible unitary (projective) representation U_ν of G on $H^2_\nu(D)$ is defined by

$$(U_\nu(g^{-1})h)(z) := j(g,z)\, h(g(z)) \tag{3.5}$$

for all $g \in G$, $h \in H^2_\nu(D)$ and $z \in D$, where

$$j(g,z) := [\mathrm{Det}_Z\, g'(z)]^{\nu/p}.$$

Case 2 If $D_{\mathbb{R}} \subset Z_{\mathbb{R}}$ is an irreducible *complex* symmetric domain, we consider the Hilbert space tensor product

$$H^2_\nu(D) := H^2_\nu(D_{\mathbb{R}}) \otimes \overline{H^2_\nu(D_{\mathbb{R}})} \tag{3.6}$$

for the weighted Bergman space $H^2_\nu(D_{\mathbb{R}})$ over $D_{\mathbb{R}}$, as described in Case 1. Here $\nu > p_{\mathbb{R}} - 1$ where $p_{\mathbb{R}} = $ genus $(D_{\mathbb{R}})$. $H^2_\nu(D)$ consists of sesqui-holomorphic functions $h(z_1, \overline{z}_2)$ on $D := D_{\mathbb{R}} \times \overline{D}_{\mathbb{R}} \subset Z := Z_{\mathbb{R}} \times \overline{Z}_{\mathbb{R}}$ which are square-integrable under the product measure

$$d\mu_\nu(z_1,z_2) := d\mu^{\mathbb{R}}_\nu(z_1)\, d\mu^{\mathbb{R}}_\nu(z_2). \tag{3.7}$$

An equivalent realization of (3.6) is via Hilbert-Schmidt integral operators

$$(h\phi)(z_1) = \int_{D_{\mathbb{R}}} d\mu_\nu^{\mathbb{R}}(z_2)\, h(z_1, \overline{z}_2)\, \phi(z_2) \tag{3.8}$$

on $H_\nu^2(D_{\mathbb{R}})$. The reproducing kernel of $H_\nu^2(D)$ is the product

$$K(z_1, \overline{z}_2,\ w_1, \overline{w}_2) := K^{\mathbb{R}}(z_1, w_1)\ K^{\mathbb{R}}(w_2, z_2), \tag{3.9}$$

where $K^{\mathbb{R}}(z_1, w_1)$ is the kernel function for $H_\nu^2(D_{\mathbb{R}})$. $H_\nu^2(D)$ carries an irreducible unitary (projective) representation U_ν of $G = G_{\mathbb{R}} \times G_{\mathbb{R}}$ defined by

$$U_\nu(g_1, \overline{g}_2)\, T = U_\nu^{\mathbb{R}}(g_1)\, T\, U_\nu^{\mathbb{R}}(g_2)^* \tag{3.10}$$

for all $g_1, g_2 \in G_{\mathbb{R}}$ and $T \in H_\nu^2(D)$, regarded as a Hilbert-Schmidt operator on $H_\nu^2(D_{\mathbb{R}})$. For the diagonal $(g, \overline{g}) \in G$, with $g \in G_{\mathbb{R}}$, we obtain the adjoint action

$$U_\nu(g, \overline{g})\, T = U_\nu^{\mathbb{R}}(g)\, T\, U_\nu^{\mathbb{R}}(g)^{-1}.$$

We put

$$j(g_1, \overline{g}_2;\ z_1, \overline{z}_2) := j^{\mathbb{R}}(g_1, z_1)\ \overline{j^{\mathbb{R}}(g_2, z_2)}$$

for all $g_1, g_2 \in G_{\mathbb{R}}$ and $z_1, z_2 \in D_{\mathbb{R}}$.

Case 3 For the *flat* case $D = \mathbb{C}^n$, let $(z|w)$ denote the inner product and let $dm(z)$ be the associated Lebesgue measure. Fix $\nu > 0$ and consider the probability measure

$$d\mu_\nu(z) := \left(\frac{\nu}{\pi}\right)^n e^{-\nu(z|z)} dm(z). \tag{3.11}$$

The *weighted Bargmann space*

$$\begin{aligned} H_\nu^2(\mathbb{C}^n) &:= \{h \in L^2(\mathbb{C}^n; d\mu_\nu) :\ h \text{ holomorphic}\} \\ &= \{h : \mathbb{C}^n \to \mathbb{C} \text{ holomorphic} :\ \|h\|_\nu^2 := \int_{\mathbb{C}^n} d\mu_\nu(z)\, |h(z)|^2 < \infty\} \end{aligned}$$

has the reproducing kernel

$$K(z, w) = e^{\nu(z|w)} \tag{3.12}$$

for all $z, w \in \mathbb{C}^n$. $H_\nu^2(\mathbb{C}^n)$ carries an irreducible unitary (projective) representation of the semi-direct product $G := U(n) \triangleleft \mathbb{C}^n$ via

$$(U_\nu(g)\, h)(z) := j(g^{-1}, z)\, h(g^{-1}(z)) \tag{3.13}$$

for all $g \in G,\ \ h \in H_\nu^2(\mathbb{C}^n)$ and $z \in \mathbb{C}^n$, where

$$j(g^{-1}, z) = \frac{K(z, g(0))}{K(g(0), g(0))} = \exp\left(\nu(z|g(0)) - \frac{\nu}{2}\, (g(0)|g(0))\right).$$

This completes the definition of $H_\nu^2(D)$. In all three cases we have

$$j(g, z)\ K(g(z), g(w))\ \overline{j(g, w)} = K(z, w) \tag{3.14}$$

for all $g \in G$ and $z, w \in D$. Put

$$K_w(z) := K(z, w). \tag{3.15}$$

Then $K_w \in H^2_\nu(D)$ for all $w \in D$, and (3.14) shows

$$U(g)\,K_w = \overline{j(g,w)}\;K_{g(w)}$$

for all $g \in G$ and $w \in D$ since

$$\begin{aligned}\overline{j(g,w)}\;K_{g(w)}(g(z)) &= \overline{j(g,w)}\;K(g(z),g(w))\\ &= j(g,z)^{-1}\,K(z,w) = j(g^{-1},g(z))\;K_w(z)\\ &= (U(g)\,K_w)(g(z))\end{aligned}$$

for all $z \in D$. Let

$$(h|k)_\nu := \int\limits_D d\mu_\nu(z)\,\overline{h(z)}\,k(z)$$

denote the scalar product in $H^2_\nu(D)$. Then

$$h(z) = (K_z|h)_\nu \tag{3.16}$$

for all $h \in H^2_\nu(D)$ by the reproducing kernel property.

Let $z \mapsto \overline{z}$ denote the conjugation of D with real form $D_{\mathbb{R}}$. Since D is simply connected

$$I(z) := K(z,\overline{z})^{1/2} \tag{3.17}$$

defines a holomorphic function on D (not belonging to $H^2_\nu(D)$).

Example 3.1. In the product case (Case 2) we have

$$\begin{aligned}I(z_1,\overline{z}_2) &= K(z_1,\overline{z}_2,\overline{(z_1,\overline{z}_2)})^{1/2} = K(z_1,\overline{z}_2,z_2,\overline{z}_1)^{1/2}\\ &= [K^{\mathbb{R}}(z_1,z_2)\,K^{\mathbb{R}}(z_1,z_2)]^{1/2} = K^{\mathbb{R}}(z_1,z_2)\end{aligned} \tag{3.18}$$

as a holomorphic function on $D = D_{\mathbb{R}} \times \overline{D}_{\mathbb{R}}$. Since the reproducing kernel property implies

$$\phi(z_1) = \int\limits_{D_{\mathbb{R}}} d\mu^{\mathbb{R}}_\nu(z_2)\,K^{\mathbb{R}}(z_1,z_2)\,\phi(z_2)$$

for all $\phi \in H^2_\nu(D_{\mathbb{R}})$, it follows that I corresponds to the identity operator on $H^2_\nu(D_{\mathbb{R}})$. On the other hand, the kernel

$$\begin{aligned}K_{w_1,\overline{w}_2}(z_1,\overline{z}_2) &= K(z_1,\overline{z}_2,w_1,\overline{w}_2)\\ &= K^{\mathbb{R}}(z_1,w_1)\,K^{\mathbb{R}}(w_2,z_2) = K^{\mathbb{R}}_{w_1}(z_1)\;\overline{K^{\mathbb{R}}_{w_2}(z_2)}\end{aligned} \tag{3.19}$$

for $(w_1, \overline{w}_2) \in D$ corresponds to the rank 1 operator $K^{\mathbb{R}}_{w_1}(K^{\mathbb{R}}_{w_2})^*$ acting on $H^2_\nu(D_{\mathbb{R}})$ since for all $\phi \in H^2_\nu(D_{\mathbb{R}})$

$$\begin{aligned}
&\int_{D_{\mathbb{R}}} d\mu^{\mathbb{R}}_\nu(z_2)\; K_{w_1,w_2}(z_1, \overline{z}_2)\, \phi(z_2)\\
=\;& \int_{D_{\mathbb{R}}} d\mu^{\mathbb{R}}_\nu(z_2)\; K^{\mathbb{R}}_{w_1}(z_1)\; K^{\mathbb{R}}(w_2, z_2)\, \phi(z_2)\\
=\;& K^{\mathbb{R}}_{w_1}(z_1) \int_{D_{\mathbb{R}}} d\mu^{\mathbb{R}}_\nu(z_2)\; K^{\mathbb{R}}(w_2, z_2)\, \phi(z_2)\\
=\;& K^{\mathbb{R}}_{w_1}(z_1)\, \phi(z_2) = K^{\mathbb{R}}_{w_1}(K^{\mathbb{R}}_{w_2}|\phi)_\nu .
\end{aligned}$$

Proposition 3.2. *I is the unique (up to a multiplicative constant) holomorphic function on D which is U_ν-invariant under $G_{\mathbb{R}} \subset G$.*

Proof. Let $g \in G_{\mathbb{R}}$. In order to show

$$I(z) = (U_\nu(g^{-1})\, I)(z) = j(g, z)\, I(g(z)) \tag{3.20}$$

for all $z \in D$, we may assume that $z = \zeta \in D_{\mathbb{R}}$ since both sides of (3.20) are holomorphic. Since $I(\zeta)$, $j(g, \zeta)$ and $I(g(\zeta))$ are positive, the assertion follows by taking squares

$$[j(g,\zeta)\, I(g(\zeta))]^2 = j(g,\zeta)\, K(g(\zeta),\, g(\zeta))\; j(g,\zeta) = K(\zeta,\zeta) = I(\zeta)^2.$$

□

As a consequence of (3.20) we have

$$I(\zeta) = I(g_\zeta(o)) = j(g_\zeta, o)^{-1} \tag{3.21}$$

for all $\zeta \in D_{\mathbb{R}}$, with $g_\zeta \in G_{\mathbb{R}}$ satisfying $g_\zeta(o) = \zeta$. Here o is the origin of $D_{\mathbb{R}}$.

Lemma 3.3. *For all $z \in D$ we have*

$$\int_{D_{\mathbb{R}}} d\mu_0(\zeta)\; K(z,\zeta)\; I(\zeta)^{-1} = I(z) \int_{D_{\mathbb{R}}} d\mu_0(\zeta)\; K(o,\zeta)\; I(\zeta)^{-1}. \tag{3.22}$$

Proof. Since both sides of (3.22) are holomorphic on D we may assume $z \in D_{\mathbb{R}}$. Write $z = g(o)$ for some $g \in G_{\mathbb{R}}$. Then $G_{\mathbb{R}}$-invariance of μ_0 implies

$$\begin{aligned}
&\int_{D_{\mathbb{R}}} d\mu_0(\zeta)\; K(z,\zeta)\; K(\zeta,\zeta)^{-1/2}\\
=\;& \int_{D_{\mathbb{R}}} d\mu_0(\zeta)\; K(z, g(\zeta))\; K(g(\zeta),\, g(\zeta))^{-1/2}
\end{aligned}$$

$$= \int_{D_{\mathbb{R}}} d\mu_0(\zeta)\ j(g,o)^{-1}\ K(o,\zeta)\ j(g,\zeta)^{-1}\ j(g,\zeta)\ K(\zeta,\zeta)^{-1/2}$$

$$= j(g,o)^{-1} \int_{D_{\mathbb{R}}} d\mu_0(\zeta)\ K(o,\zeta)\ K(\zeta,\zeta)^{-1/2}.$$

Since $K(z,\overline{z}) = K(g(o),g(o)) = j(g,o)^{-2}\,K(o,o) = j(g,o)^{-2}$ the assertion follows. □

NOTATION. In view of Lemma 3.3 it is natural to normalize the $G_{\mathbb{R}}$-invariant measure μ_0 on $D_{\mathbb{R}}$ by the condition

$$\int_{D_{\mathbb{R}}} d\mu_0(\zeta)\ K(o,\zeta)\ I(\zeta)^{-1} = 1. \tag{3.23}$$

This normalization (depending on ν) will be chosen in the sequel. Then

$$I(z) = \int_{D_{\mathbb{R}}} d\mu_0(\zeta)\ K(z,\zeta)\ I(\zeta)^{-1} \tag{3.24}$$

for all $z \in D$. In the case of complex bounded symmetric domains $D_{\mathbb{R}}$ the normalization (3.23) amounts to

$$d\mu_0(\zeta) = d\mu_\nu^{\mathbb{R}}(\zeta)\ K^{\mathbb{R}}(\zeta,\zeta) \tag{3.25}$$

since $K^{\mathbb{R}}(o,\zeta) = 1$.

Proposition 3.4. *For holomorphic $h \in H_\nu^2(D)$ we have*

$$\int_D d\mu_\nu(w)\ \overline{h(w)}\ I(w) = \int_{D_{\mathbb{R}}} d\mu_0(\zeta)\ \overline{h(\zeta)}\ I(\zeta)^{-1}. \tag{3.26}$$

Proof. We may assume $h = K_z$ for some $z \in D$. Since $I(z) := K(z,\overline{z})^{1/2}$ is holomorphic, Lemma 3.3 implies

$$\int_{D_{\mathbb{R}}} d\mu_0(\zeta)\ \overline{K_z(\zeta)}\ I(\zeta)^{-1} = \int_{D_{\mathbb{R}}} d\mu_0(\zeta)\ K(z,\zeta)\ I(\zeta)^{-1}$$

$$= I(z) = (K_z|I)_\nu = \int_D d\mu_\nu(w)\ \overline{K_z(w)}\ I(w).$$

□

4. Covariant symbolic calculi on real symmetric domains

We will now introduce the new concept of covariant calculus for real symmetric domains, which arises as a natural generalization of the complex case [AU1] replacing operators by Hilbert space vectors.

Definition 4.1. A *covariant symbolic calculus* is given by a linear "symbol map"

$$\begin{aligned} \sigma : H^2_\nu(D) &\rightarrow \{\text{functions on } D_{\mathbb{R}}\} \\ h &\mapsto \sigma h \end{aligned} \tag{4.1}$$

satisfying the covariance condition

$$\sigma(U_\nu(g)\, h) = (\sigma h) \circ g^{-1} \tag{4.2}$$

for all $g \in G_{\mathbb{R}}$. More precisely, σ should have a $G_{\mathbb{R}}$-invariant domain $\mathrm{Dom}(\sigma)$ of holomorphic functions on D containing all reproducing kernel vectors K_w, $w \in D$. Then the covariance condition becomes

$$(\sigma K_w) \circ g^{-1} = \sigma(U_\nu(g)\, K_w) = \overline{j(g,w)}\; \sigma(K_{gw}) \tag{4.3}$$

for all $w \in D$ and $g \in G_{\mathbb{R}}$. In addition we require that the holomorphic function $I(z) := K(z, \bar{z})^{1/2}$ on D belongs to $\mathrm{Dom}(\sigma)$. Then σI is a constant function by (4.2) and Lemma 3.3, and we normalize σ by assuming

$$\sigma I = 1. \tag{4.4}$$

The function σh is also called the *passive* (or weak) *symbol* of h. The main examples of symbolic calculi are "real" in the sense that σK_o is a real-valued function on $D_{\mathbb{R}}$. This will be assumed in the sequel.

Proposition 4.2. *For every symbolic calculus σ we have*

$$(\sigma h)(\zeta) = (\overline{(\sigma K_\sqcup)(\zeta)} | h)_\nu \tag{4.5}$$

for all $h \in \mathrm{Dom}(\sigma)$ and $\zeta \in D_{\mathbb{R}}$, where for fixed ζ the function

$$z \mapsto (\sigma K_\sqcup)(\zeta)\,(z) := (\sigma K_z)(\zeta) \tag{4.6}$$

is anti-holomorphic on D.

Proof. We may assume $h = K_w$ for $w \in D$. Then

$$\overline{(\sigma K_w)(\zeta)} = \overline{(\sigma K_\sqcup)(\zeta)}\,(w) = (K_w | \overline{(\sigma K_\sqcup)(\zeta)})_\nu$$

and hence

$$(\sigma K_w)(\zeta) = \overline{(K_w | \overline{(\sigma K_\sqcup)(\zeta)})_\nu} = (\overline{(\sigma K_\sqcup)(\zeta)} | K_w)_\nu.$$

□

Definition 4.3. The *adjoint*

$$\sigma^* : \{\text{functions on } D_{\mathbb{R}}\} \rightarrow H^2_\nu(D) \tag{4.7}$$

of a covariant symbolic calculus σ is defined by assigning to a function f on $D_{\mathbb{R}}$ (belonging to $\mathrm{Dom}(\sigma^*)$) the holomorphic function

$$(\sigma^* f)(z) := \int_{D_{\mathbb{R}}} d\mu_0(\zeta)\; f(\zeta)\; \overline{(\sigma K_z)(\zeta)} \tag{4.8}$$

on D. Here the invariant measure μ_0 is normalized by the condition (3.23), depending on ν but not on σ. We call f the *active* (or strong) *symbol* of $\sigma^* f$. The adjoint σ^*, corresponding to the map $\mathcal{A}$ of (1.1), can of course also be taken as the starting point of the theory. The "dual" view point emphasizing the symbol map σ is closer to Berezin's original approach.

Proposition 4.4. *A covariant symbolic calculus σ and its adjoint σ^* are related by duality*

$$(\sigma^* f|h)_\nu = (f|\sigma h) := \int_{D_{\mathbb{R}}} d\mu_0(\zeta)\ \overline{f(\zeta)}\ (\sigma h)(\zeta) \tag{4.9}$$

for all $f \in \mathrm{Dom}(\sigma^*)$ and $h \in \mathrm{Dom}(\sigma)$.

Proof. We may assume $h = K_z$ for some $z \in D$. Then

$$(\sigma^* f|K_z)_\nu = \overline{(K_z|\sigma^* f)_\nu} = \overline{(\sigma^* f)(z)}$$

and (4.9) follows from (4.8) by taking conjugates. □

Since (4.9) involves the $G_{\mathbb{R}}$-invariant measure μ_0 and σ is $G_{\mathbb{R}}$-covariant, it follows that σ^* satisfies the covariance condition

$$\sigma^*\left(f \circ g^{-1}\right) = U_\nu(g)(\sigma^* f) \tag{4.10}$$

for all $g \in G_{\mathbb{R}}$ and $f \in \mathrm{Dom}(\sigma^*)$.

Proposition 4.5. *For any covariant symbolic calculus σ the adjoint σ^* satisfies*

$$\sigma^* 1 = I. \tag{4.11}$$

Proof. For $\zeta \in D_{\mathbb{R}}$ and $g_\zeta \in G_{\mathbb{R}}$ satisfying $\zeta = g_\zeta(o)$ the covariance of σ implies

$$(\sigma K_o)(g_\zeta^{-1}(o)) = \sigma(U_\nu(g_\zeta)K_o)(o) = j(g_\zeta, o)(\sigma K_\zeta)(o) = I(\zeta)^{-1}(\sigma K_\zeta)(o).$$

Since $G_{\mathbb{R}}$ is unimodular and $(\sigma K_\xi)(\eta)$ is real for $\xi, \eta \in D_{\mathbb{R}}$, Proposition 3.4 and Proposition 4.2 imply

$$\begin{aligned}(\sigma^* 1)(o) &= \int_{D_{\mathbb{R}}} d\mu_0(\zeta)(\sigma K_o)(\zeta) = \int_{D_{\mathbb{R}}} d\mu_0(\zeta)(\sigma K_o)(g_\zeta(o)) \\ &= \int_{D_{\mathbb{R}}} d\mu_0(\zeta)(\sigma K_o)(g_\zeta^{-1}(o)) = \int_{D_{\mathbb{R}}} d\mu_0(\zeta)\ I(\zeta)^{-1}(\sigma K_\zeta)(o) \\ &= \int_D d\mu_\nu(z)\ I(z)(\sigma K_z)(o) = (\overline{(\sigma K_{\sqcup})(o)}\,|\,I)_\nu = (\sigma I)(o) = 1.\end{aligned}$$

Since $\sigma^* 1 = (\sigma^* 1)(o)\, I$ by covariance, the assertion follows. □

Definition 4.6. Let σ be a covariant symbolic calculus with adjoint σ^*. The composite map

$$f \mapsto (\sigma\sigma^*)\, f = \sigma(\sigma^* f)$$

acting on functions on $D_{\mathbb{R}}$ is called the *link transform* associated with σ. Note that $\sigma\sigma^*$ maps the active symbol to the passive symbol. By our assumptions, $\sigma\sigma^*$ has a $G_{\mathbb{R}}$-invariant domain and commutes with the $G_{\mathbb{R}}$-action

$$(\sigma\sigma^*)(f \circ g) = (\sigma\sigma^* f) \circ g \tag{4.12}$$

for all functions $f \in \mathrm{Dom}(\sigma\sigma^*)$ and $g \in G_{\mathbb{R}}$. More generally, one may consider transforms

$$f \mapsto (\sigma_1\sigma_2^*)\, f = \sigma_1(\sigma_2^* f)$$

linking two covariant symbolic calculi σ_1, σ_2 on $H^2_\nu(D)$, and the invariance property (4.12) still holds. In view of (4.4) and (4.11) these transforms are "stochastic" operators: $(\sigma_1\sigma_2^*)\,1 = \sigma_1(\sigma_2^* 1) = \sigma_1 I = 1$.

Proposition 4.7. *Let σ_1, σ_2 be covariant symbolic calculi. Then the link transform $\sigma_1\sigma_2^*$ has the integral kernel*

$$(\overline{(\sigma_1 K_\sqcup)(\xi)} \,|\, \overline{(\sigma_2 K_\sqcup)(\eta)})_\nu = \int_D d\mu_\nu(z) (\sigma_1 K_z)(\xi)\, \overline{(\sigma_2 K_z)(\eta)} \tag{4.13}$$

with respect to μ_0 (normalized by (3.23)*).*

Proof. Using (4.5) and (4.8) we obtain

$$\begin{aligned}
(\sigma_1\sigma_2^* f)(\xi) &= (\overline{(\sigma_1 K_\sqcup)(\xi)} \,|\, \sigma_2^* f)_\nu \\
&= \int_D d\mu_\nu(z) (\sigma_1 K_z)(\xi) (\sigma_2^* f)(z) \\
&= \int_D d\mu_\nu(z) (\sigma_1 K_z)(\xi) \int_{D_{\mathbb{R}}} d\mu_0(\eta)\, f(\eta)\, \overline{(\sigma_2 K_z)(\eta)} \\
&= \int_{D_{\mathbb{R}}} d\mu_0(\eta)\, f(\eta) \int_D d\mu_\nu(z) (\sigma_1 K_z)(\xi)\, \overline{(\sigma_2 K_z)(\eta)}.
\end{aligned}$$

□

It is well known [H] that the L^2-space $L^2(D_{\mathbb{R}}, \mu_0)$ has a multiplicity-free decomposition

$$L^2(D_{\mathbb{R}}, \mu_0) \equiv \int_{\mathfrak{a}^\#} |c(\lambda)|^{-2} d\lambda\, \langle G_{\mathbb{R}} \rangle_\lambda \tag{4.14}$$

under the group $G_{\mathbb{R}}$, where $\langle G_{\mathbb{R}} \rangle_\lambda$ denotes the principal series representation with spectral parameter $\lambda \in \mathfrak{a}^\#$, $c(\lambda)$ is Harish-Chandra's c-function and we choose an Iwasawa decomposition

$$G_{\mathbb{R}} = K_{\mathbb{R}}\, A\, N \tag{4.15}$$

of $G_{\mathbb{R}}$, with $\mathfrak{a} = \mathrm{Lie}(A)$. By $G_{\mathbb{R}}$-invariance it follows that the link transform $\sigma_1\sigma_2^*$ is diagonalized under the $G_{\mathbb{R}}$-action. More precisely, let $\phi_\lambda \in \mathcal{C}^\infty(D_{\mathbb{R}})$ be the normalized spherical function associated with λ. Then

$$\sigma_1\sigma_2^*\phi_\lambda = \widetilde{\sigma_1\sigma_2^*}(\lambda)\, \phi_\lambda \tag{4.16}$$

for all $\phi_\lambda \in \mathrm{Dom}(\sigma_1\sigma_2^*)$, and the eigenvalues $\widetilde{\sigma_1\sigma_2^*}(\lambda)$ completely characterize the link transform. For every $\lambda \in \mathfrak{a}^\#$ the Iwasawa decomposition (4.15) defines "exponential functions" [H]

$$e_\lambda(\zeta) = \exp\langle A(g) | \lambda + \rho \rangle, \quad g \in G_{\mathbb{R}}, \ g(o) = \zeta$$

on $D_{\mathbb{R}}$ satisfying $e_\lambda(o) = 1$ and

$$e_\lambda(g(\zeta)) = e_\lambda(g(o))\, e_\lambda(\zeta)$$

for all $g \in NA$, such that

$$\phi_\lambda(\zeta) = \int\limits_{K_{\mathbb{R}}} dk\ e_\lambda(k\zeta)$$

is the spherical function of type λ. In the flat case e_λ is expressed in terms of the exponential function [AU1]; in the curved setting an explicit description of e_λ can be given by realizing $D_{\mathbb{R}}$ as a real Siegel domain (cf. Section 4).

Proposition 4.8. *For every covariant symbolic calculus σ we have*

$$(\sigma^* e_\lambda)(z) = (\sigma^* e_\lambda)(o)\, e_\lambda(z)\ I(z) \tag{4.17}$$

where $e_\lambda(z)$ denotes the unique extension of e_λ to a holomorphic function on D.

Proof. Since both sides of (4.17) are holomorphic it suffices to let $z = \zeta \in D_{\mathbb{R}}$. Let $g_\zeta \in NA$ satisfy $g_\zeta(o) = \zeta$. Then

$$\begin{aligned}(\sigma^* e_\lambda)(\zeta) &= (\sigma^* e_\lambda)(g_\zeta(o)) = j(g_\zeta, o)^{-1}(U_\nu(g_\zeta^{-1})(\sigma^* e_\lambda))(o)\\ &= j(g_\zeta, o)^{-1}(\sigma^*(e_\lambda \circ g_\zeta))(o) = j(g_\zeta, o)^{-1} e_\lambda(\zeta)\,(\sigma^* e_\lambda)(o)\\ &= I(\zeta)\ e_\lambda(\zeta)\,(\sigma^* e_\lambda)(o).\end{aligned}$$

□

Our main result is a "product formula" for the link transform for any pair of covariant symbolic calculi.

Theorem 4.9. *Let σ_1, σ_2 be covariant symbolic calculi on an irreducible real symmetric domain $D_{\mathbb{R}}$. Then the $G_{\mathbb{R}}$-invariant link transform σ_1, σ_2^* has the eigenvalues*

$$\widetilde{\sigma_1\sigma_2^*}(\lambda) = \frac{\overline{(\sigma_1^* e_\lambda)(o)}\,(\sigma_2^* e_\lambda)(o)}{c_\lambda} \tag{4.18}$$

for all $\lambda \in \mathrm{Dom}(\widetilde{\sigma_1\sigma_2^*}) \subset \mathfrak{a}^\#$, where c_λ is a positive constant independent of σ_1, σ_2 which will be computed explicitly in Section 5. In particular,

$$\widetilde{\sigma\sigma^*}(\lambda) = \frac{1}{c_\lambda}\ |(\sigma^* e_\lambda)(o)|^2. \tag{4.19}$$

Proof. By Proposition 4.7, $\sigma_1\sigma_2^*$ is the adjoint of $\sigma_2\sigma_1^*$ with respect to μ_0. Using Proposition 4.8 and Proposition 4.2 it follows that

$$\begin{aligned}&(\sigma_2^* e_\lambda)(o)\,\overline{(e_\lambda I|\overline{(\sigma_1 K_\sqcup)(o)})_\nu} \\ &= (\sigma_2^* e_\lambda)(o)\,(\overline{(\sigma_1 K_\sqcup)(o)}\,|\,e_\lambda I)_\nu = (\overline{(\sigma_1 K_\sqcup)(o)}\,|\,\sigma_2^* e_\lambda)_\nu \\ &= \sigma_1(\sigma_2^* e_\lambda)(o) = \widetilde{\sigma_1\sigma_2^*}(\lambda) = \overline{\widetilde{\sigma_2\sigma_1^*}(\lambda)} \\ &= \overline{(\sigma_1^* e_\lambda)(o)}\,(e_\lambda I|\overline{(\sigma_2 K_\sqcup)(o)})_\nu.\end{aligned} \tag{4.20}$$

Therefore

$$c_\lambda := \frac{(\sigma_2^* e_\lambda)(o)}{(e_\lambda I|\overline{(\sigma_2 K_\sqcup)(o)})_\nu} = \overline{\left[\frac{(\sigma_1^* e_\lambda)(o)}{(e_\lambda I|\overline{(\sigma_1 K_\sqcup)(o)})_\nu}\right]}.$$

Taking $\sigma_1 = \sigma_2$ it follows that c_λ is real. Therefore c_λ is independent of σ_1, σ_2 and (4.18) follows from (4.20). □

Remark 4.10. In terms of harmonic analysis

$$(\sigma^* e_\lambda)(o) = \int_{D_\mathbb{R}} d\mu_0(\zeta)\, e_\lambda(\zeta)\ (\sigma K_o)(\zeta)$$

$$= \int_{D_\mathbb{R}} d\mu_0(\zeta)\, \phi_\lambda(\zeta)\ (\sigma K_o)(\zeta) = (\sigma K_o)^\sim(\lambda)$$

can be identified with the *spherical Fourier transform* [H] of the real $K_\mathbb{R}$-invariant function σK_o on $D_\mathbb{R}$. Viewed as a function of λ, we call

$$a_o(\lambda) := (\sigma^* e_\lambda)(o)$$

the *fundamental function* of σ.

5. The Toeplitz-Berezin calculus

Up to now the discussion of covariant symbolic calculi was quite general. In this section we describe a basic example, the Toeplitz-Berezin calculus, and show that it determines the value of the constant c_λ in an explicit way. Let $D_\mathbb{R} = G_\mathbb{R}/K_\mathbb{R}$ be an irreducible real symmetric domain, with complexification D and quantization Hilbert space $H_\nu^2(D)$, as introduced in Section 2. Let $o \in D_\mathbb{R} \subset D$ be the origin.

Definition 5.1. The *Toeplitz-Berezin symbol*

$$\tau : H_\nu^2(D) \to \mathcal{C}^\infty(D_\mathbb{R})$$

is defined by

$$(\tau h)(\zeta) := K(\zeta,\zeta)^{-1/2} h(\zeta) = I(\zeta)^{-1}\, h(\zeta) \tag{5.1}$$

for all $h \in H_\nu^2(D)$ and $\zeta \in D_\mathbb{R}$. In particular

$$(\tau K_z)(\zeta) = I(\zeta)^{-1}\, K(\zeta, z) \tag{5.2}$$

for all $z \in D$ and $\zeta \in D_\mathbb{R}$.

Note that τ is well defined (and injective) for *all* holomorphic functions h on D since $D_\mathbb{R} \subset D$ is a set of uniqueness. The normalization $\tau I = 1$ made in (4.4) is trivially satisfied.

By Definition 4.3 the adjoint τ^* of τ (called the *Toeplitz-Berezin calculus*) is defined by

$$(\tau^* f)(z) = \int_{D_\mathbb{R}} d\mu_0(\zeta)\, f(\zeta)\, I(\zeta)^{-1}\, K(z,\zeta) \tag{5.3}$$

for all functions $f \in L^\infty(D_{\mathbb{R}})$, since $d\mu_0(\zeta)\, I(\zeta)^{-1}$ is a finite measure for ν large enough. Here μ_0 is normalized by (3.23) so that (3.24) implies $\tau^*1 = I$. According to Proposition 4.7, the link transform $\tau\tau^*$ (called the *Berezin transform* in the complex case) has the integral kernel

$$\int_D d\mu_\nu(z)(\tau K_z)(\xi)\, \overline{(\tau K_z)(\eta)} \tag{5.4}$$

$$= \int_D d\mu_\nu(z)\, I(\xi)^{-1}\, K(\xi, z)\, I(\eta)^{-1}\, \overline{K(\eta, z)}$$

$$= I(\xi)^{-1}\, I(\eta)^{-1}\, K(\xi, \eta) = \frac{K(\xi,\eta)}{K(\xi,\xi)^{1/2}\, K(\eta,\eta)^{1/2}}$$

for all $\xi, \eta \in D_{\mathbb{R}}$.

Example 5.2. In case $D_{\mathbb{R}}$ is complex (Case 2) $H^2_\nu(D) = H^2_\nu(D_{\mathbb{R}}) \otimes \overline{H^2_\nu(D_{\mathbb{R}})}$ is the space of Hilbert-Schmidt operators via the identification (3.8), where h is a sesqui-holomorphic function on $D_{\mathbb{R}} \times D_{\mathbb{R}}$. For fixed $z \in D_{\mathbb{R}}$ we have

$$(K^{\mathbb{R}}_z \,|\, hK^{\mathbb{R}}_z)_\nu = (hK^{\mathbb{R}}_z)(z) = \int_{D_{\mathbb{R}}} d\mu^{\mathbb{R}}_\nu(w)\, h(z, \overline{w})\, K^{\mathbb{R}}_z(w) = h(z, \overline{z})$$

since $w \mapsto \overline{h(z,\overline{w})}$ is holomorphic. Therefore (5.1) amounts to

$$(\tau h)(z, \overline{z}) = I(z,\overline{z})^{-1}\, h(z,\overline{z}) = K^{\mathbb{R}}(z,z)^{-1}\, h(z,\overline{z})$$

$$= \frac{(K^{\mathbb{R}}_z | hK^{\mathbb{R}}_z)_\nu}{(K^{\mathbb{R}}_z | K^{\mathbb{R}}_z)_\nu}$$

for all $z \in D_{\mathbb{R}}$. This is the classical Berezin symbol of the operator h.

On the other hand, the sesqui-holomorphic function

$$(\tau^* f)(z, \overline{w}) = \int_{D_{\mathbb{R}}} d\mu_0(\zeta)\; f(\zeta)\; I(\zeta,\overline{\zeta})^{-1}\, K(z, \overline{w}, \zeta, \overline{\zeta})$$

$$= \int_{D_{\mathbb{R}}} d\mu_0(\zeta)\; f(\zeta)\; K^{\mathbb{R}}(\zeta,\zeta)^{-1}\; K^{\mathbb{R}}(z,\zeta)\; K^{\mathbb{R}}(\zeta, w)$$

$$= \int_{D_{\mathbb{R}}} d\mu^{\mathbb{R}}_\nu(\zeta)\; f(\zeta)\; K^{\mathbb{R}}(z,\zeta)\; K^{\mathbb{R}}(\zeta, w)$$

on $D_{\mathbb{R}} \times D_{\mathbb{R}}$, defined via (5.3), gives rise to the integral operator

$$((\tau^* f)\phi)(z) = \int_{D_{\mathbb{R}}} d\mu^{\mathbb{R}}_\nu(w)\, (\tau^* f)(z, \overline{w})\, \phi(w)$$

$$= \int_{D_{\mathbb{R}}} d\mu^{\mathbb{R}}_\nu(w) \int_{D_{\mathbb{R}}} d\mu^{\mathbb{R}}_\nu(\zeta)\; f(\zeta)\; K^{\mathbb{R}}(z,\zeta)\; K^{\mathbb{R}}(\zeta, w)\, \phi(w)$$

$$= \int_{D_{\mathbb{R}}} d\mu_\nu^{\mathbb{R}}(\zeta)\, f(\zeta)\, K^{\mathbb{R}}(z,\zeta) \int_{D_{\mathbb{R}}} d\mu_\nu^{\mathbb{R}}(w)\, K^{\mathbb{R}}(\zeta,w)\, \phi(w)$$

$$= \int_{D_{\mathbb{R}}} d\mu_\nu^{\mathbb{R}}(\zeta)\, f(\zeta)\, K^{\mathbb{R}}(z,\zeta)\, \phi(\zeta) = E(f\phi)(z)$$

where E denotes the orthogonal projection onto $H_\nu^2(D_{\mathbb{R}})$. Thus

$$(\tau^* f)\phi = E(f\phi) =: T_f \phi$$

gives the *Toeplitz operator* with symbol f acting on $H_\nu^2(D_{\mathbb{R}})$. By (5.4) the Berezin transform in the complex case has the kernel

$$\frac{K(\xi,\overline{\xi},\eta,\overline{\eta})}{K(\xi,\overline{\xi},\xi,\overline{\xi})^{1/2}\, K(\eta,\overline{\eta},\eta,\overline{\eta})^{1/2}} = \frac{K^{\mathbb{R}}(\xi,\eta)\, K^{\mathbb{R}}(\eta,\xi)}{K^{\mathbb{R}}(\xi,\xi)\, K^{\mathbb{R}}(\eta,\eta)} = \frac{|K^{\mathbb{R}}(\xi,\eta)|^2}{K^{\mathbb{R}}(\xi,\xi)\, K^{\mathbb{R}}(\eta,\eta)}$$

for all $\xi,\eta \in D_{\mathbb{R}}$.

Proposition 5.3. *For all $\lambda \in \mathfrak{a}^{\#}$ we have*

$$\widetilde{\tau\tau^*}(\lambda) = (\tau e_\lambda)(o) = c_\lambda. \tag{5.5}$$

Proof. Since

$$\overline{(\tau K_z)(o)} = \overline{K(o,o)^{-1/2}\, K_z(o)} = K_o(z)$$

for all $z \in D$, Proposition 4.2 implies

$$\widetilde{\tau\tau^*}(\lambda) = \tau(\tau^* e_\lambda)(o) = (\overline{(\tau K_\sqcup)(o)} \,|\, \tau^* e_\lambda)_\nu = (K_o|\tau^* e_\lambda)_\nu = (\tau^* e_\lambda)(o).$$

This shows that $(\tau^* e_\lambda)(o)$ is real, and the product formula (4.19) implies

$$(\tau^* e_\lambda)(o) = \widetilde{\tau\tau^*}(\lambda) = \frac{|(\tau^* e_\lambda)(o)|^2}{c_\lambda} = \frac{(\tau^* e_\lambda)(o)^2}{c_\lambda}.$$

□

Combining Proposition 5.3 with Theorem 4.9 and Remark 4.10 we obtain

Corollary 5.4. *For covariant symbolic calculi σ_1,σ_2 on $H_\nu^2(D)$ the link transform $\sigma_1\sigma_2^*$ has the eigenvalues*

$$\widetilde{\sigma_1\sigma_2^*}(\lambda) = \frac{(\sigma_2^* e_\lambda)(o)\ \overline{(\sigma_1^* e_\lambda)(o)}}{(\tau^* e_\lambda)(o)} = \frac{(\sigma_2 K_o)^\sim(\lambda)\ \overline{(\sigma_1 K_o)^\sim(\lambda)}}{(\tau K_o)^\sim(\lambda)}$$

where τ is the Toeplitz-Berezin calculus and $\sim$ denotes spherical Fourier transforms.

In view of Corollary 5.4 it is important to compute the integral

$$\begin{aligned}(\tau^* e_\lambda)(o) &= \int_{D_{\mathbb{R}}} d\mu_0(\zeta)\ e_\lambda(\zeta)\ (\tau K_o)(\zeta) \\ &= \int_{D_{\mathbb{R}}} d\mu_0(\zeta)\ e_\lambda(\zeta)\ I(\zeta)^{-1}\ K(o,\zeta) \\ &= \int_{D_{\mathbb{R}}} d\mu_0(\zeta)\ e_\lambda(\zeta)\ K(\zeta,\zeta)^{-1/2}\ K(o,\zeta)\end{aligned}$$

explicitly. In the complex case this has been carried out in [UU], verifying a long-standing conjecture of Berezin. For details on the complex case and also the flat case we refer to [AU1]. In the sequel we determine $(\tau^* e_\lambda)(o)$ for real symmetric domains. The Toeplitz-Berezin calculus has also been studied in [DP], [N], [Z]; we avoid a case-by-case separation according to the various root systems of $D_{\mathbb{R}}$, thus giving a more uniform treatment.

As in Section 2 consider the unbounded realization

$$D_{\mathbb{R}} = \{x + y + v \in Z_{\mathbb{R}} :\ x - \{ev^*v\} \in \Omega\}$$

as a real Siegel domain. For the Jordan theoretic concepts used in the sequel we refer to [FK], [L2], [U]. Let $e_1, \ldots, e_r \in X$ be a frame with $e = e_1 + \cdots + e_r$. Then we have Peirce decompositions

$$X = \sum_{1 \le i \le j \le r} X_{ij}, \quad Y = \sum_{1 \le i \le j \le r} Y_{ij}, \quad V_{\mathbb{R}} = \sum_{1 \le j \le r} V_j^{\mathbb{R}}$$

and we put

$$\begin{aligned} 1 &= \dim X_{jj} \quad (1 \le j \le r) \\ a &:= \dim X_{ij} \quad (1 \le i < j \le r). \end{aligned}$$

In case $Z_{\mathbb{R}} \ne X$ we also put

$$\begin{aligned} b &:= \dim V_j^{\mathbb{R}} \quad (1 \le j \le r) \\ c &:= \dim Y_{jj} \quad (1 \le j \le r). \end{aligned}$$

It is known that

$$\dim Y_{ij} = a \quad (1 \le i < j \le r)$$

up to one exception (root system D_2) which we omit in the sequel. The dimensions of the respective subspaces are then given by

$$\begin{aligned} n_X &:= \dim_{\mathbb{R}} X &&= r + r(r-1)\,a/2 \\ n_Y &:= \dim_{\mathbb{R}} Y &&= cr + r(r-1)\,a/2 \\ n_V &:= \dim_{\mathbb{C}} V &&= \dim_{\mathbb{R}} V_{\mathbb{R}} = rb. \end{aligned}$$

Put $n := \dim_{\mathbb{R}} Z_{\mathbb{R}} = n_X + n_Y + n_V$. Let P denote the quadratic representation of X and let $\Delta(x)$ be the Jordan determinant of X.

Lemma 5.5. *For $x \in \Omega$ we have*

$$\mathrm{Det}_X\, P_x^{1/2} = \Delta(x)^{n_X/r}.$$

Proof. Without loss of generality (due to the spectral decomposition and transitivity of $K_{\mathbb{R}}$ on frames) we may assume $x = \sum_j x_j e_j$ diagonal. Putting $t' = P_x^{1/2} t$ for $t \in X$, the respective Peirce components satisfy $t'_{ij} = x_i^{1/2}\, t_{ij}\, x_j^{1/2}$. It follows that

$$\begin{aligned}\mathrm{Det}_X\, P_x^{1/2} &= \prod_{1 \le j \le r} x_j \prod_{1 \le i < j \le r} (x_i x_j)^{a/2} \\ &= (x_1 \cdots x_r)^{1+(r-1)a/2} = \Delta(x)^{1+(r-1)a/2} = \Delta(x)^{n_X/r}.\end{aligned}$$

□

In the sequel we use the *conical functions*

$$\Delta^{\underline{\alpha}}(x) = \Delta_1(x)^{\alpha_1 - \alpha_2} \Delta_2(x)^{\alpha_2 - \alpha_3} \cdots \Delta_r(x)^{\alpha_r} \tag{5.6}$$

on Ω, associated with $\underline{\alpha} = (\alpha_1, \ldots, \alpha_r)$, where $\Delta_1, \ldots, \Delta_r$ are the Jordan theoretic minors. Let $\Delta_*^{\underline{\alpha}}$ denote the corresponding function using the minors in reverse order $e_r, \ldots, e_1$. For the following result, cf. [FK, Section VII.1].

Lemma 5.6. *Let $x \in \Omega$ and $\mathrm{Re}(\alpha_j) > -1 - (r-j)\,\frac{a}{2}$. Then*

$$\int_\Omega dt\ \Delta^{\underline{\alpha}}(t)\, e^{-(x|t)} = \Gamma_\Omega\left(\underline{\alpha} + \frac{n_X}{r}\right)\ \Delta_*^{-\underline{\alpha}^* - n_X/r}(x)$$

where

$$\Gamma_\Omega\left(\underline{\alpha} + \frac{n_X}{r}\right) = \int_\Omega dt\ \Delta^{\underline{\alpha}}(t)\, e^{-(e|t)}$$

is the (multi-variable) Koecher-Gindikin Γ-function and $\underline{\alpha}^* := (\alpha_r, \ldots, \alpha_1)$.

Lemma 5.7. *Let $\mathrm{Re}(\beta) > 1 + \frac{a}{2}(2r - j - 1) + \mathrm{Re}(\alpha_j) > \frac{a}{2}(r-1)$. Then*

$$\Gamma_\Omega(\beta) \int_\Omega dt\ \Delta^{\underline{\alpha}}(t)\, \Delta(e+t)^{-\beta} = \Gamma_\Omega\left(\underline{\alpha} + \frac{n_X}{r}\right)\ \Gamma_\Omega\left(\beta - \underline{\alpha} + \underline{\delta} - \frac{2n_X}{r}\right)$$

where $\underline{\delta} = (\delta_1, \ldots, \delta_r)$ is defined by

$$\delta_j := 1 + (j-1)a. \tag{5.7}$$

Proof. Applying Lemma 5.6 twice we obtain

$$\Gamma_\Omega(\beta) \int_\Omega dt\ \Delta^{\underline{\alpha}}(t)\ \Delta(e+t)^{-\beta}$$

$$= \int_\Omega dt\ \Delta^{\underline{\alpha}}(t) \int_\Omega ds\ \Delta(s)^{\beta-n_X/r}\, e^{-(s|e+t)}$$

$$= \int_\Omega ds\ \Delta(s)^{\beta-n_X/r}\, e^{-(s|e)} \int_\Omega dt\ \Delta^{\underline{\alpha}}(t)\, e^{-(s|t)}$$

$$= \Gamma_\Omega\left(\underline{\alpha}+\frac{n_X}{r}\right) \int_\Omega ds\ \Delta_*^{\beta-\underline{\alpha}^*-2n_X/r}(s)\, e^{-(s|e)}$$

$$= \Gamma_\Omega\left(\underline{\alpha}+\frac{n_X}{r}\right) \Gamma_\Omega\left(\beta-\underline{\alpha}^*-\frac{n_X}{r}\right).$$

Now the assertion follows from the well-known identity

$$\Gamma_\Omega(\underline{\alpha}^*) = \Gamma_\Omega\left(\underline{\alpha}+\underline{\delta}-\frac{n_X}{r}\right). \tag{5.8}$$

□

Lemma 5.8. [FK, p. 142, Exercise 5] *Let $2\gamma > 1+(r-1)\,a$. Then*

$$\Gamma_\Omega(\gamma) \int_X dx\ \Delta(e+x^2)^{-\gamma} = \pi^{n_X}\, 4^{n_X-r\gamma}\, \frac{\Gamma_\Omega(2\gamma-n_X/r)}{\Gamma_\Omega(\gamma)}.$$

Proof. Let $\Delta^{\mathbb{C}}$ be the holomorphic extension of Δ to the complexified Jordan algebra $X^{\mathbb{C}}$. By analytic continuation, Lemma 5.6 implies for every $x \in X$

$$\Gamma_\Omega(\gamma)\, \Delta^{\mathbb{C}}(e+ix)^{-\gamma} = \int_\Omega ds\ \Delta(s)^{\gamma-n_X/r}\, e^{-(e+ix|s)}$$

$$= \int_\Omega ds\ \Delta(s)^{\gamma-n_X/r}\, e^{-(e|s)}\, e^{-i(x|s)}.$$

It follows that $\Gamma_\Omega(\gamma)\, \Delta^{\mathbb{C}}(e+ix)^{-\gamma}$ is the Fourier transform of the function

$$f(s) := \chi_\Omega(s)\ \Delta(s)^{\gamma-n_X/r}\, e^{-(e|s)}$$

on X, where χ_Ω is the characteristic function of $\Omega \subset X$. Therefore Parseval's formula implies

$$\Gamma_\Omega(\gamma)^2 \int_X dx\ |\Delta^{\mathbb{C}}(e+ix)|^{-2\gamma} = (2\pi)^{n_X} \int_X ds\ |f(s)|^2$$

$$= (2\pi)^{n_X} \int ds\ \Delta(s)^{2\gamma-2n_X/r}\, e^{-2(e|s)}$$

$$= (2\pi)^{n_X} \Gamma_\Omega\left(2\gamma - \frac{n_X}{r}\right) \Delta(2e)^{-2\gamma + n_X/r}$$
$$= \pi^{n_X} 4^{n_X - r\gamma} \Gamma_\Omega\left(2\gamma - \frac{n_X}{r}\right).$$

Now the assertion follows from

$$|\Delta^{\mathbb{C}}(e+ix)|^2 = \Delta^{\mathbb{C}}(e+ix)\,\Delta^{\mathbb{C}}(e-ix) = \Delta^{\mathbb{C}}((e+ix)(e-ix)) = \Delta(e+x^2).$$

□

From now on we assume $Y \neq \{0\}$. For $t \in X$ the Jordan multiplication operator M_t leaves X and Y invariant.

Lemma 5.9. *For $t \in \Omega$ we have*

$$\Delta(t)^{n_X/r}\,\mathrm{Det}_Y\, M_t = \Delta(t)^{n_Y/r}\,\mathrm{Det}_X\, M_t.$$

Proof. Assuming $t = \sum_j t_j\, e_j$ diagonal and putting $x' = M_t\, x,\ \ y' = M_t\, y$ for $x \in X$ and $y \in Y$, the respective Peirce components satisfy $x'_{ij} = x_{ij}(t_i + t_j)/2,\ \ y'_{ij} = y_{ij}(t_i + t_j)/2$ for all $1 \le i \le j \le r$. It follows that

$$\mathrm{Det}_X\, M_t = \prod_{1\le j\le r} t_j \cdot \prod_{1\le i<j\le r} \left(\frac{t_i + t_j}{2}\right)^a,$$
$$\mathrm{Det}_Y\, M_t = \prod_{1\le j\le r} t_j^c \cdot \prod_{1\le i<j\le r} \left(\frac{t_i + t_j}{2}\right)^a.$$

Since $\Delta(t) = t_1 \cdots t_r$ and $c - 1 = \frac{n_Y - n_X}{r}$, the assertion follows. □

Let r_U and Δ_U be the rank and Jordan determinant of U. We define

$$\nu_{\mathbb{R}} := \frac{\nu r_U}{2r}, \quad p_{\mathbb{R}} := \frac{p r_U}{2r} = \frac{n_X + n_Y + n_V/2}{r}. \tag{5.9}$$

Lemma 5.10. *Let $2\nu_{\mathbb{R}} > (r-1)\,a + c$. Then*

$$2^{\nu r_U} \int_Y dy\ \Delta_U(e-y)^{-\nu} = \pi^{(n_X+n_Y)/2}\, 2^{n_X+n_Y}\ \frac{\Gamma_\Omega\left(2\nu_{\mathbb{R}} - \frac{n_Y}{r}\right)}{\Gamma_\Omega(\nu_{\mathbb{R}})\ \Gamma_\Omega\left(\nu_{\mathbb{R}} + \frac{n_X - n_Y}{2r}\right)}.$$

Proof. For $t \in \Omega,\ \ x \in X$ and $y \in Y$ we have

$$(x^2|t) = (\{xx^*e\}|t) = (x|\{xe^*t\}) = (x|M_t x),$$
$$-(y^2|t) = (\{yy^*e\}|t) = (y|\{ye^*t\}) = (y|M_t y).$$

In view of Lemma 5.9 this implies

$$\begin{aligned}
\pi^{-n_Y/2} \int\limits_Y dy\ e^{(y^2|t)} &= \pi^{-n_Y/2} \int\limits_Y dy\ e^{-(y|M_t\, y)} \\
&= \mathrm{Det}_Y^{-1/2}\ M_t = \Delta(t)^{(n_X-n_Y)/2r}\ \mathrm{Det}_X^{-1/2}\ M_t \\
&= \Delta(t)^{(n_X-n_Y)/2r}\ \pi^{-n_X/2} \int\limits_X dx\ e^{-(x|M_t\, x)} \\
&= \Delta(t)^{(n_X-n_Y)/2r}\ \pi^{-n_X/2} \int\limits_X dx\ e^{-(x^2|t)}.
\end{aligned}$$

Since $\Delta_U(e-y)$ is real for all $y \in Y$, we have

$$\Delta_U(e-y) = \overline{\Delta_U(e-y)} = \Delta_U((e-y)^*) = \Delta_U(e+y)$$

and therefore

$$\Delta_U(e-y)^2 = \Delta_U(e-y)\Delta_U(e+y) = \Delta_U((e-y)(e+y)) = \Delta_U(e-y^2) = \Delta(e-y^2)^{r_U/r}.$$

Applying Lemma 5.6 and Lemma 5.8 it follows that

$$\begin{aligned}
&\pi^{-n_Y/2}\ \Gamma_\Omega(\nu_\mathbb{R}) \int\limits_Y dy\ \Delta_U(e-y)^{-\nu} \\
= \ &\pi^{-n_Y/2}\ \Gamma_\Omega(\nu_\mathbb{R}) \int\limits_Y dy\ \Delta(e-y^2)^{-\nu_\mathbb{R}} \\
= \ &\pi^{-n_Y/2} \int\limits_Y dy \int\limits_\Omega dt\ \Delta(t)^{\nu_\mathbb{R}-n_X/r}\ e^{-(e-y^2|t)} \\
= \ &\pi^{-n_Y/2} \int\limits_\Omega dt\ \Delta(t)^{\nu_\mathbb{R}-n_X/r}\ e^{-(e|t)} \int\limits_Y dy\ e^{(y^2|t)} \\
= \ &\pi^{-n_X/2} \int\limits_\Omega dt\ \Delta(t)^{\nu_\mathbb{R}-(n_X+n_Y)/2r}\ e^{-(e|t)} \int\limits_X dx\ e^{-(x^2|t)} \\
= \ &\pi^{-n_X/2} \int\limits_X dx \int\limits_\Omega dt\ \Delta(t)^{\nu_\mathbb{R}-(n_X+n_Y)/2r}\ e^{-(e+x^2|t)} \\
= \ &\pi^{-n_X/2}\ \Gamma_\Omega\left(\nu_\mathbb{R} + \frac{n_X-n_Y}{2r}\right) \int\limits_X dx\ \Delta(e+x^2)^{(n_Y-n_X)/2r-\nu_\mathbb{R}} \\
= \ &\pi^{-n_X/2}\ \pi^{n_X}\ 4^{n_X+(n_Y-n_X-\nu r_U)/2}\ \frac{\Gamma_\Omega\left(2\nu_\mathbb{R} - \frac{n_Y}{r}\right)}{\Gamma_\Omega\left(\nu_\mathbb{R} + \frac{n_X-n_Y}{2r}\right)}.
\end{aligned}$$

□

For $x \in X$ the quadratic representation P_x acts also on Y.

Lemma 5.11. *For $x \in \Omega$ we have*

$$\mathrm{Det}_Y\ P_x^{1/2} = \Delta(x)^{n_Y/r}.$$

Proof. Assuming $x = \sum_j x_j\, e_j$ diagonal and putting $y' = P_x^{1/2}\, y$ for $y \in Y$, the respective Peirce components satisfy $y'_{ij} = x_i^{1/2}\, y_{ij}\, x_j^{1/2}$. It follows that

$$\begin{aligned}\mathrm{Det}_Y\ P_x^{1/2} &= \prod_{1\le j\le r} x_j^c \prod_{1\le i<j\le r} (x_i\, x_j)^{a/2}\\ &= (x_1\cdots x_r)^{c+(r-1)\,a/2} = \Delta(x)^{c+(r-1)\,a/2} = \Delta(x)^{n_Y/r}.\end{aligned}$$

□

Lemma 5.12. *Let $x \in \Omega$ and $2\nu_{\mathbb{R}} > (r-1)\,a + c$. Then*

$$\begin{aligned}&2^{\nu r_U} \int\limits_Y dy\ \Delta_U(x-y)^{-\nu}\\ = \quad &\Delta(x)^{-2\nu_{\mathbb{R}}+n_Y/r}\ \pi^{(n_X+n_Y)/2}\ 2^{n_X+n_Y}\ \frac{\Gamma_\Omega\left(2\nu_{\mathbb{R}} - \frac{n_Y}{r}\right)}{\Gamma_\Omega\left(\nu_{\mathbb{R}}\right)\,\Gamma_\Omega\left(\nu_{\mathbb{R}} + \frac{n_X-n_Y}{2r}\right)}.\end{aligned}$$

Proof. Putting $y' = P_x^{-1/2}\, y$ for $y \in Y$, we have

$$dy = \Delta(x)^{n_Y/r}\, dy'$$

by Lemma 5.8. Since

$$\Delta_U(x-y) = \Delta_U(P_x^{1/2}(e-y')) = \Delta_U(x)\ \Delta_U(e-y') = \Delta(x)^{r_U/r}\ \Delta_U(e-y')$$

it follows that

$$\int\limits_Y dy\ \Delta_U(x-y)^{-\nu} = \Delta(x)^{-2\nu_{\mathbb{R}}+n_Y/r} \int\limits_Y dy'\ \Delta_U(e-y')^{-\nu}.$$

Now the assertion follows from Lemma 5.10. □

Let R denote the canonical Jordan representation of X on $V_{\mathbb{R}}$ [FK], [L1].

Lemma 5.13. *For $x \in \Omega$ we have*

$$\mathrm{Det}_{V_{\mathbb{R}}}\ R_x^{1/2} = \Delta(x)^{n_V/2r}.$$

Proof. Assuming $x = \sum_j x_j\, e_j$ diagonal and putting $v' = R_x^{1/2}\, v$ for $v \in V_{\mathbb{R}}$, the respective Peirce components satisfy $v'_j = x_j^{1/2}\, v_j$. It follows that

$$\mathrm{Det}_{V_{\mathbb{R}}}\ R_x^{1/2} = \Big(\prod_{1\le j\le r} x_j^{1/2}\Big)^b = (x_1\cdots x_r)^{b/2} = \Delta(x)^{b/2} = \Delta(x)^{n_V/2r}.$$

□

For $v \in V_{\mathbb{R}}$ we have $(e|\{ev^*v\}) = (\{ee^*v\}|v) = (v|v)/2$ and therefore

$$\int_{V_{\mathbb{R}}} dv\ e^{-(e|\{ev^*v\})} = \int_{V_{\mathbb{R}}} dv\ e^{-(v|v)/2} = (2\pi)^{n_V/2}. \tag{5.10}$$

Lemma 5.14. *Let $x \in \Omega$. Then*

$$\int_{V_{\mathbb{R}}} dv\ e^{-(x|\{ev^*v\})} = (2\pi)^{n_V/2}\ \Delta(x)^{-n_V/2r}.$$

Proof. Putting $v' = R_x^{1/2}\, v$ for $v \in V_{\mathbb{R}}$, we have

$$dv = \Delta(x)^{-n_V/2r}\, dv'$$

by Lemma 5.13. Moreover

$$P_x^{1/2}\{ev^*v\} = \{e(R_x^{1/2}\, v)^*\,(R_x^{1/2}\, v)\} = \{e\overset{*}{v}'v'\}$$

and hence

$$(x|\{ev^*v\}) = (P_x^{1/2}\, e|\{ev^*v\}) = (e|P_x^{1/2}\{ev^*v\}) = (e|\{e\overset{*}{v}'v'\}).$$

Now the assertion follows from (5.10). □

Lemma 5.15. *Let $\beta > (r-1)\frac{a}{2} + \frac{b}{2}$. Then*

$$\Gamma_\Omega(\beta) \int_{V_{\mathbb{R}}} dv\ \Delta(e + \{ev^*v\})^{-\beta} = (2\pi)^{n_V/2}\ \Gamma_\Omega\left(\beta - \frac{n_V}{2r}\right).$$

Proof. Applying Lemma 5.6 and Lemma 5.14 we obtain

$$\begin{aligned}
&\Gamma_\Omega(\beta) \int_{V_{\mathbb{R}}} dv\ \Delta(e + \{ev^*v\})^{-\beta} \\
= &\int_{V_{\mathbb{R}}} dv \int_{\Omega} dx\ \Delta(x)^{\beta - n_X/r}\, e^{-(x|e+\{ev^*v\})} = \\
&\int_{\Omega} dx\ \Delta(x)^{\beta - n_X/r}\, e^{-(x|e)} \int_{V_{\mathbb{R}}} dv\ e^{-(x|\{ev^*v\})} \\
= &(2\pi)^{n_V/2} \int_{\Omega} dx\ \Delta(x)^{\beta - n_X/r - n_V/2r}\ e^{-(x|e)} \\
= &(2\pi)^{n_V/2}\ \Gamma_\Omega\left(\beta - \frac{n_V}{2r}\right).
\end{aligned}$$

□

Lemma 5.16. *Let $x \in \Omega$ and $\beta > (r-1)\frac{a}{2} + \frac{b}{2}$. Then*

$$\Gamma_\Omega(\beta) \int_{V_{\mathbb{R}}} dv\ \Delta(x + \{ev^*v\})^{-\beta} = (2\pi)^{n_V/2}\ \Gamma_\Omega\left(\beta - \frac{n_V}{2r}\right)\ \Delta(x)^{n_V/2r - \beta}.$$

Proof. Putting $v' = R_x^{-1/2}\, v$ for $v \in V_{\mathbb{R}}$, we have

$$dv = \Delta(x)^{n_V/2r}\, dv'$$

by Lemma 5.13. Moreover

$$P_x^{-1/2}\{ev^*v\} = \{e(R_x^{-1/2}\, v)^*\, (R_x^{-1/2}\, v)\} = \{e\overset{*}{v'}v'\}$$

and hence

$$\Delta(x + \{ev^*v\}) = \Delta(P_x^{1/2}\, (e + P_x^{-1/2}\{ev^*v\})) = \Delta(x)\ \Delta(e + \{e\overset{*}{v'}v'\}).$$

Therefore

$$\int\limits_{V_{\mathbb{R}}} dv\ \Delta(x + \{ev^*v\})^{-\beta} = \Delta(x)^{n_V/2r-\beta} \int\limits_{V_{\mathbb{R}}} dv'\ \Delta(e + \{e\overset{*}{v'}v'\})^{-\beta}$$

and Lemma 5.15 implies the assertion. □

Theorem 5.17. *The link transform $\tau\tau^*$ of the Toeplitz calculus with "Wallach" parameter ν has the eigenvalues*

$$\widetilde{\tau\tau^*}(\lambda) = \frac{\Gamma_\Omega\left(\underline{\lambda} + \underline{\rho} + \nu_{\mathbb{R}} - \frac{n_Y + n_V/2}{r}\right)\ \Gamma_\Omega\left(-\underline{\lambda} + \underline{\rho} + \nu_{\mathbb{R}} - \frac{n_Y + n_V/2}{r}\right)}{\Gamma_\Omega\left(\nu_{\mathbb{R}} - \frac{n_Y + n_V/2}{r}\right)\ \Gamma_\Omega(\nu_{\mathbb{R}})}.$$

Here $\underline{\lambda} = (\lambda_1, \ldots, \lambda_r)$ is the spectral parameter in $\mathfrak{a}^\#$ and $\underline{\rho} = (\rho_1, \ldots, \rho_r)$ is the half-sum of positive restricted roots [L2], [Z] *given by*

$$2\rho_j = 1 + (j-1)a + \frac{n_Y - n_X + n_V/2}{r}. \tag{5.11}$$

Proof. Writing $\zeta = x + y + v \in X \oplus Y \oplus V_{\mathbb{R}}$ according to (2.4), the domain $D_{\mathbb{R}}$ is defined by the condition

$$t := x - \{ev^*v\} \in \Omega. \tag{5.12}$$

Using (3.4) and applying Lemma 5.12, Lemma 5.16 and Lemma 5.7 we obtain in case $Z_{\mathbb{R}} \neq X$

$$\begin{aligned} I_\alpha &:= \int dx \int dv \int dy\ \Delta_U(x - \{ev^*v\})^{-p/2}\, \Delta^{\underline{\alpha}}(x - \{ev^*v\})\cdot \\ &\quad \cdot \Delta_U\left(\frac{e + x - y}{2}\right)^{-\nu}\ \Delta_U(x - \{ev^*v\})^{\nu/2} \\ &= 2^{\nu r_U} \int dx \int dv \int dy\ \Delta^{\underline{\alpha} + \nu_{\mathbb{R}} - p_{\mathbb{R}}}(x - \{ev^*v\}) \cdot \Delta_U(e + x - y)^{-\nu} \\ &= 2^{\nu r_U} \int\limits_\Omega dt\ \Delta^{\underline{\alpha} + \nu_{\mathbb{R}} - p_{\mathbb{R}}}(t) \int\limits_{V_{\mathbb{R}}} dv \int\limits_Y dy \cdot \Delta_U(e + t + \{ev^*v\} - y)^{-\nu} \end{aligned}$$

$$= 2^{n_X+n_Y}\,\pi^{(n_X+n_Y)/2}\,\frac{\Gamma_\Omega\left(2\nu_{\mathbb{R}}-\frac{n_Y}{r}\right)}{\Gamma_\Omega(\nu_{\mathbb{R}})\,\Gamma_\Omega\left(\nu_{\mathbb{R}}+\frac{n_X-n_Y}{2r}\right)}\cdot$$

$$\cdot\int_\Omega dt\;\Delta^{\underline{\alpha}+\nu_{\mathbb{R}}-p_{\mathbb{R}}}(t)\int_{V_{\mathbb{R}}} dv\;\Delta(e+t+\{ev^*v\})^{-2\nu_{\mathbb{R}}+n_Y/r}$$

$$= 2^{n_X+n_Y}\,\pi^{(n_X+n_Y)/2}\,\frac{(2\pi)^{n_V/2}\,\Gamma_\Omega\left(2\nu_{\mathbb{R}}-\frac{n_Y+n_V/2}{r}\right)}{\Gamma_\Omega(\nu_{\mathbb{R}})\,\Gamma_\Omega\left(\nu_{\mathbb{R}}+\frac{n_X-n_Y}{2r}\right)}\cdot$$

$$\cdot\int_\Omega dt\;\Delta^{\underline{\alpha}+\nu_{\mathbb{R}}-p_{\mathbb{R}}}(t)\,\Delta(e+t)^{-2\nu_{\mathbb{R}}+n_Y/r+n_V/2r} = 2^{n_X+n_Y+n_V/2}\,\pi^{n/2}.$$

$$\cdot\frac{\Gamma_\Omega\left(\underline{\alpha}+\nu_{\mathbb{R}}-p_{\mathbb{R}}+\frac{n_X}{r}\right)\,\Gamma_\Omega\left(2\nu_{\mathbb{R}}-\frac{n_Y+n_V/2}{r}-\underline{\alpha}-\nu_{\mathbb{R}}+p_{\mathbb{R}}+\underline{\delta}-\frac{2n_X}{r}\right)}{\Gamma_\Omega(\nu_{\mathbb{R}})\,\Gamma_\Omega\left(\nu_{\mathbb{R}}+\frac{n_X-n_Y}{2r}\right)}$$

$$= 2^{n_X+n_Y+n_V/2}\,\pi^{n/2}\,\frac{\Gamma_\Omega\left(\underline{\alpha}+\nu_{\mathbb{R}}-\frac{n_Y+n_V/2}{r}\right)\,\Gamma_\Omega\left(\nu_{\mathbb{R}}-\underline{\alpha}+\underline{\delta}-\frac{n_X}{r}\right)}{\Gamma_\Omega(\nu_{\mathbb{R}})\,\Gamma_\Omega\left(\nu_{\mathbb{R}}+\frac{n_X-n_Y}{2r}\right)}$$

since $rp_{\mathbb{R}} = n_X+n_Y+n_V/2$. In the remaining case $Z_{\mathbb{R}} = X$, Lemma 5.7 yields

$$\begin{aligned} I_\alpha &:= \int_\Omega dx\;\Delta(x)^{-n_X/r}\,\Delta^{\underline{\alpha}}(x)\,\Delta\left(\frac{e+x}{2}\right)^{-\nu}\;\Delta(x)^{\nu/2} \\ &= 2^{r\nu}\int_\Omega dx\;\Delta^{\underline{\alpha}-n_X/r+\nu/2}(x)\;\Delta(e+x)^{-\nu} \\ &= 2^{r\nu}\,\frac{\Gamma_\Omega\left(\frac{\nu}{2}+\underline{\alpha}\right)\,\Gamma_\Omega\left(\frac{\nu}{2}-\underline{\alpha}+\underline{\delta}-\frac{n_X}{r}\right)}{\Gamma_\Omega(\nu)}. \end{aligned}$$

In general [Z, Lemma 2.3] we have

$$e_\lambda(x+y+v) = \Delta^{\underline{\lambda}+\underline{\rho}}(x-\{ev^*v\}).$$

Since $\widetilde{\tau\tau^*}(-\rho) = (\tau\tau^*1)(e) = 1$, it follows that

$$\widetilde{\tau\tau^*}(\lambda) = I_{\lambda+\rho}/I_0 = \frac{\Gamma_\Omega\left(\underline{\lambda}+\underline{\rho}+\nu_{\mathbb{R}}-\frac{n_Y+n_V/2}{r}\right)\,\Gamma_\Omega\left(\nu_{\mathbb{R}}-\underline{\lambda}-\underline{\rho}+\underline{\delta}-\frac{n_X}{r}\right)}{\Gamma_\Omega\left(\nu_{\mathbb{R}}-\frac{n_Y+n_V/2}{r}\right)\,\Gamma_\Omega\left(\nu_{\mathbb{R}}+\underline{\delta}-\frac{n_X}{r}\right)}$$

in both cases (since $\nu_{\mathbb{R}} = \nu/2$ if $Z_{\mathbb{R}} = X$). Since

$$2\underline{\rho} = \underline{\delta} + \frac{n_Y-n_X+n_V/2}{r}$$

by (5.7) and (5.11) and

$$\Gamma_\Omega\left(\nu_{\mathbb{R}}+\underline{\delta}-\frac{n_X}{r}\right) = \Gamma_\Omega(\nu_{\mathbb{R}})$$

by (5.8), the assertion follows. □

Remark 5.18. In case $D_{\mathbb{R}}$ is complex, we have $\nu_{\mathbb{R}} = \nu$ and $d := \dim_{\mathbb{C}} D_{\mathbb{R}} = n_Y + n_V/2$. Therefore (5.10) simplifies to

$$\widetilde{\tau\tau^*}(\lambda) = \frac{\Gamma_\Omega(\underline{\rho} + \nu - \frac{d}{r} + \underline{\lambda})\, \Gamma_\Omega(\underline{\rho} + \nu - \frac{d}{r} - \underline{\lambda})}{\Gamma_\Omega(\nu - \frac{d}{r})\, \Gamma_\Omega(\nu)}$$

(cf. [AU1]).

Remark 5.19. While the Toeplitz-Berezin calculus is certainly fundamental, it is important to study other covariant symboli calculi such as the *Wick calculus* and the *Weyl calculus* and the relationship between them. In [AU1] a detailed investigation is carried out in the complex case, in particular for the Bargmann spaces over $\mathbb{C}^n$. In the curved setting the Weyl calculus, which involves the *symmetries* in a crucial way, poses many open problems, but in [AU2] the eigenvalues for the link transform of the Weyl calculus are determined for all (real and complex) symmetric domains of rank 1. The surprising new feature is the deep role played by hypergeometric functions in this context.

References

[AU1] J. Arazy, H. Upmeier, *Invariant symbolic calculi and eigenvalues of invariant operators on symmetric domains*, Proc. Lund 2000.

[AU2] J. Arazy, H. Upmeier, *Weyl calculus on rank 1 symmetric domains*, Preprint (2000).

[BLU] D. Borthwick, A. Lesniewski, H. Upmeier, *Non-perturbative deformation quantization of Cartan domains*, J. Func. Anal 113 (1993), 153–176.

[DP] G. van Dijk, M. Pevzner, *Berezin kernels and tube domains*, Preprint (1999).

[FK] J. Faraut, A. Korányi, *Analysis on Symmetric Cones*, Clarendon Press Oxford (1994).

[H] S. Helgason, *Groups and Geometric Analysis*, Academic Press (1984).

[L1] O. Loos, *Jordan Pairs*, Springer Lect. Notes 460 (1975).

[L2] O. Loos, *Bounded Symmetric Domains and Jordan Pairs*, Univ. of California, Irvine (1977).

[N] Y. Neretin, *Matrix analogs of Beta-integral and Plancherel formula for Berezin kernel representations*, Preprint (1999).

[U] H. Upmeier, *Symmetric Banach Manifolds and Jordan C^*-Algebras*, North Holland (1985).

[UU] A. Unterberger, H. Upmeier, *The Berezin transform and invariant differential operators*, Comm. Math. Phys. 164 (1994), 563–597.

[Z] G. Zhang, *Berezin transform on real bounded symmetric domains*, Preprint (1999).

Jonathan Arazy
Department of Mathematics
Haifa University
Mount Carmel
Haifa, Israel
e-mail: jarazy@mathc2.haifa.ac.il

Harald Upmeier
Fachbereich Mathematik
University of Marburg
D-35032 Marburg, Germany
e-mail: upmeier@mathematik.uni-marburg.de

2000 *Mathematics Subject Classification.*
Primary 46L65, 47A60; Secondary 53C35, 17C20.

Received: 19 June 2001.

Operator Theory:
Advances and Applications, Vol. 142, 29–57

Singular Integral Operators with Piecewise Almost Periodic Coefficients and Carleman Shift

M. Amélia Bastos, António Bravo and Yuri Karlovich

Abstract. A Fredholm criterion and an index formula for singular integral operators with fixed singularities and piecewise almost periodic matrix coefficients on the space $L_p^n(\mathbb{R})$ are obtained. The proof is based on the fact that the operators

$$I+\sum_j B_j H_j \quad \text{and} \quad I+\sum_j \mathbf{M}(B_j) H_j,$$

where B_j are almost periodic matrix functions, $\mathbf{M}(B_j)$ are their Bohr mean values and H_j are integral operators with fixed singularities at infinity, are Fredholm on $L_p^n(\mathbb{R})$ only simultaneously, and that in this case their indices coincide. Applying this result, more general singular integral operators with fixed singularities at ∞ and semi-almost periodic coefficients are studied. The Fredholm study of singular integral operators with fixed singularities and piecewise almost periodic coefficients is reduced to the study of singular integral operators with piecewise continuous matrix coefficients and singular integral operators with fixed singularities at ∞ and semi-almost periodic matrix coefficients. Based on a reduction of operators with a shift to singular integral operators without shift, a Fredholm theory for singular integral operators with a Carleman backward shift and oscillating matrix coefficients on Lebesgue spaces over the real line is developed.

1. Introduction

Equations associated with singular integral operators with shifts and convolution type equations with oscillating coefficients appear naturally when mathematical models of physical systems with delays and reflections are considered (see, e.g., [5], [8], [10], [16], [17], and the references therein). The influence of delays and reflections is essential in various fields of natural sciences.

In this paper we develop a Fredholm theory for singular integral operators with a Carleman backward shift and oscillating matrix coefficients on Lebesgue spaces over the real line. The coefficients belong to the C^*-algebra PAP of piecewise almost periodic matrix functions (see [3]), which is generated by uniform almost periodic (AP) matrix functions (see [15]) and piecewise continuous (PC) matrix functions. The study is based on the theory of singular integral operators with semi-almost periodic matrix coefficients, results on almost periodic factorization of matrix functions (see [2], [9], [10], [19]), and a reduction of the operators with a shift under consideration to some singular integral operators without shift (cf. [7]). A cornerstone new problem consists in the necessity to study products of multiplication operators by semi-almost periodic functions and integral operators with fixed singularities at infinity.

The paper is organized as follows. Section 2 contains preliminary results. In Section 3 we recall the Fredholm theory for the Banach algebra $\mathfrak{S}$ of singular integral operators with piecewise continuous coefficients on the space $L_p^n(\mathbb{R})$ following [6] and [1] (see also [3], [18], [21]). It is also shown there that a singular integral operator V with fixed singularities at 0 and ∞ belongs to a closed two-sided ideal of $\mathfrak{S}$, and its symbol is identified.

In Section 4 we develop the Fredholm theory of singular integral operators with fixed singularities and semi-almost periodic coefficients. Our results are based on the following fundamental fact. Let $\mathcal{H}_\infty$ be the closed two-sided ideal of the Banach algebra $\mathfrak{S}$ that is generated by the operator $uS - SuI$, where $u(x) = \tanh(\pi x)$ and S is the Cauchy singular integral operator over $\mathbb{R}$. If the matrix functions B_j are in AP, $\mathbf{M}(B_j)$ are their Bohr mean values, and the operators H_j belong to $\mathcal{H}_\infty$, then the operators

$$I + \sum_j B_j H_j \quad \text{and} \quad I + \sum_j \mathbf{M}(B_j) H_j$$

are Fredholm on $L_p^n(\mathbb{R})$ only simultaneously. In that case, their indices coincide.

Let $PAPW$ denote the functions in PAP with almost periodic representatives of the Wiener type. In Section 5 we establish the Fredholmness on $L_p^n(\mathbb{R})$ for singular integral operators of the form

$$aP + bQ + H_{\mathbb{R}} + cH_\infty \tag{1.1}$$

where a, b, c are $n \times n$ matrix functions in $PAPW$,

$$P = (I + S)/2, \quad Q = (I - S)/2 \tag{1.2}$$

are the usual projections, $H_\infty \in \mathcal{H}_\infty$, $H_{\mathbb{R}} \in \mathcal{H}_{\mathbb{R}}$, and $\mathcal{H}_{\mathbb{R}}$ is the closed two-sided ideal of the algebra $\mathfrak{S}$ that is generated by the commutators $cS - ScI$, $c \in PC$, $c(+\infty) = c(-\infty)$. We generalize the notion of the winding number to functions in PAP. Using this generalization, we get an index formula for Fredholm operators of the form (1.1).

In Section 6, using the previous results, we construct a Fredholm theory for singular integral operators with a Carleman shift and piecewise almost periodic coefficients. We get necessary and sufficient conditions of Fredholmness on $L_p^n(\mathbb{R})$ for the operator

$$T = (aI + bJ)P + (cI + dJ)Q$$

where the $n \times n$ matrix coefficients a, b, c, d are in $PAPW$ and J is the Carleman backward shift operator,

$$(Jf)(x) := f(-x), \quad x \in \mathbb{R}. \tag{1.3}$$

2. Preliminaries

For $1 \le p \le \infty$, let $L_p(\mathbb{R})$ be the usual Lebesgue spaces on the real line $\mathbb{R}$. As is well known (see, e.g., [4], [22]), the Cauchy singular integral operator $S = S_{\mathbb{R}}$,

$$(S_{\mathbb{R}}\varphi)(x) := \frac{1}{\pi i} v.p. \int_{\mathbb{R}} \frac{\varphi(\tau)}{\tau - x}\, d\tau, \quad x \in \mathbb{R},$$

is bounded on $L_p(\mathbb{R})$ if and only if $1 < p < \infty$.

Let $PC := PC(\dot{\mathbb{R}})$ be the set of all functions $f : \mathbb{R} \to \mathbb{C}$ which possess finite left-hand and right-hand limits at every point $x \in \dot{\mathbb{R}} := \mathbb{R} \cup \{\infty\}$. The set of discontinuities of each function $g \in PC$ is at most countable.

Let AP be the set of Bohr almost periodic functions, defined as the Banach subalgebra of $L_\infty(\mathbb{R})$ generated by the monomials $e_\lambda(x) = e^{i\lambda x}, \lambda \in \mathbb{R}$. Let APW denote the set of absolutely convergent series

$$g = \sum_j a_j e_{\lambda_j}, \quad \lambda_j \in \mathbb{R}, \quad a_j \in \mathbb{C},$$

which is a dense (non-closed) subalgebra of AP. If $g \in AP$, the following limits exist, are finite and equal:

$$\lim_{T\to+\infty} \frac{1}{2T} \int_{-T}^{T} g(x)\, dx = \lim_{T\to\pm\infty} \frac{1}{T} \int_{0}^{T} g(x)\, dx.$$

Their common value is denoted by $\mathbf{M}(g)$ and is called the Bohr mean value of g. The numbers $\widehat{g}(\lambda) = \mathbf{M}(ge_{-\lambda})$ are referred to as the Fourier coefficients of g. For a given $g \in AP$, the set

$$\Omega(g) := \left\{\lambda \in \mathbb{R} : \ \widehat{g}(\lambda) \neq 0\right\},$$

which is called the Fourier spectrum of g, is at most countable. Let

$$AP^{\pm} := \left\{g \in AP : \ \Omega(g) \subset \mathbb{R}_{\pm}\right\}, \quad APW^{\pm} := AP^{\pm} \cap APW,$$

$$APW_0^{\pm} := \left\{g \in APW^{\pm} : \ \widehat{g}(0) = 0\right\}.$$

Here, as usual, $\mathbb{R}_{\pm} = \{t \in \mathbb{R} : \ \pm t \geq 0\}$.

A function $g \in AP$ is said to be not degenerate if $\operatorname{ess\,inf} |g(x)| > 0$. For every non-degenerate function $g \in AP$, the following limits exist, are finite, equal and independent of the choice of a continuous branch of argument of g:

$$v(g) := \lim_{T\to+\infty} \frac{1}{2T} \left\{ \arg g(x) \right\}_{-T}^{T} = \lim_{T\to\pm\infty} \frac{1}{T} \left\{ \arg g(x) \right\}_{0}^{T}.$$

Their common value is denoted by $v(g)$ and is called the mean motion (the almost-periodic index) of g.

For an algebra $\mathfrak{A}$, let $\mathcal{G}\mathfrak{A}$ stand for the group of elements $a \in \mathfrak{A}$ invertible in $\mathfrak{A}$. Bohr's theorem says that if $g \in \mathcal{G}AP$, then there is an $h \in AP$ such that

$$g(x) = \exp\Big(i(v(g)x + h(x))\Big), \quad x \in \mathbb{R}.$$

In particular, $\arg g(x) - v(g)x$ is a bounded function of x.

Let $\overline{\mathbb{R}} = \mathbb{R} \cup \{\pm\infty\}$. We denote by SAP the set of semi almost-periodic functions, which is the Banach subalgebra of $L_\infty(\mathbb{R})$ generated by AP and $C(\overline{\mathbb{R}})$. Any function $g \in SAP$ can be represented in the form (see [20])

$$g = ug_+ + (1-u)g_- + g_0 \tag{2.1}$$

with

$$g_\pm \in AP, \quad g_0 \in C_0(\mathbb{R}) := \Big\{f \in C(\overline{\mathbb{R}}) : f(+\infty) = f(-\infty) = 0\Big\},$$

and $u \in C\left(\overline{\mathbb{R}}\right)$ such that

$$u(\mathbb{R}) = [0,1], \quad u(+\infty) = 1, \quad u(-\infty) = 0. \tag{2.2}$$

The set of functions (2.1) with $g_\pm \in APW$ is denoted by $SAPW$. Let $PAP = \mathrm{alg}(PC, AP)$ be the Banach subalgebra of $L_\infty(\mathbb{R})$ generated by the sets PC and AP. Every function $f \in PAP$ can be decomposed in the following form (see [3]):

$$f = uf_+ + (1-u)f_- + f_0 \tag{2.3}$$

with

$$f_\pm \in AP, \quad f_0 \in PC_0(\mathbb{R}) := \Big\{f \in PC : f(+\infty) = f(-\infty) = 0\Big\}$$

and $u \in C\left(\overline{\mathbb{R}}\right)$ defined by (2.2). The set of functions $f \in PAP$ of the form (2.3) with $f_\pm \in APW$ is denoted by $PAPW$.

The functions $g_\pm$, $f_\pm \in AP$, uniquely defined by $g \in SAP$ and $f \in PAP$, are referred to as the *almost periodic representatives* of g and f at $\pm\infty$, respectively.

Let H_∞^+ (respectively, H_∞^-) stand for the set of all functions $f \in L_\infty(\mathbb{R})$ which can be holomorphically continued to the upper (respectively, lower) complex half-plane, let $C := C(\dot{\mathbb{R}})$ and

$$H_\infty^\pm + C := \Big\{f + g : f \in H_\infty^\pm,\ g \in C\Big\}.$$

The following result was obtained by D. Sarason [19].

Lemma 2.1. *If $h \in H_\infty^+$ and $\lim\limits_{y\to+\infty} h(iy) = 0$, then for every $u \in C\left(\overline{\mathbb{R}}\right)$ the function uh is in $H_\infty^+ + C$.*

Let $\mathfrak{B}(X)$ stand for the Banach algebra of all bounded linear operators acting on a Banach space X. An operator $A \in \mathfrak{B}(X)$ is said to be Fredholm if its image is closed in X and the defect numbers $n(A) := \dim \mathrm{Ker}\, A$, $d(A) := \dim \mathrm{Coker}\, A$ are finite. In this case

$$\mathrm{ind}\, A := n(A) - d(A)$$

is referred to as the index of the operator A.

Below we will use the same notation for the sets of scalar and matrix functions. Following [10], we say that an $n \times n$ matrix functions $G \in AP$ admits a canonical right AP factorization if

$$G = G^- G^+ \tag{2.4}$$

with matrix functions $G^\pm \in \mathcal{G}AP^\pm$. If $G^\pm \in \mathcal{G}APW^\pm$, then (2.4) is said to be a canonical right APW factorization of $G(\in APW)$. If a matrix function $G \in AP$ has a canonical right AP factorization then the matrix

$$\mathbf{d}(G) := \mathbf{M}(G^-)\mathbf{M}(G^+),$$

with $\mathbf{M}(G^\pm)$ given entry-wise, does not depend on the particular choice of such a factorization and is called the geometric mean of G.

From [9, Theorem 7] we immediately get the following crucial criterion.

Theorem 2.2. *Let $1 < p < \infty$, $n \geq 1$, and let G be an $n \times n$ matrix function in APW. Then the operator $T = GP + Q$ is Fredholm (equivalently, invertible) on the space $L_p^n(\mathbb{R})$ if and only if G admits a canonical right APW factorization.*

3. The algebra of singular integral operators with piecewise continuous coefficients

Let $\mathfrak{S} = \operatorname{alg}(S, PC)$ denote the Banach algebra of singular integral operators on $L_p^n(\mathbb{R})$, which is generated by the operator $S = S_\mathbb{R}$ and all the operators aI of multiplication by $n \times n$ matrix functions $a \in PC$. As is well known [6], the ideal $\mathcal{K}$ of all compact operators on $L_p^n(\mathbb{R})$ is contained in $\mathfrak{S}$.

For $1 < p < \infty$, we introduce the set

$$\mathcal{L}_p := \left\{ \frac{\exp[\pi(x + i/p)]}{2\sinh[\pi(x + i/p)]} \; : \; x \in \mathbb{R} \right\} \cup \{0, 1\},$$

which is a circular arc with endpoints 0 and 1 if $p \neq 2$, and the line segment joining these points if $p = 2$. Following [6] (see also [21], [3], [1]), it is possible to construct a symbol calculus for the algebra $\mathfrak{S}$ and to give a Fredholm criterion and an index formula for the operators $A \in \mathfrak{S}$ in terms of their symbols.

In what follows we use the notation $\nu(\mu)$ for the branch of the function $\sqrt{\mu(1-\mu)}$ in $\mathbb{C} \setminus \big((-\infty, 0) \cup (1, +\infty)\big)$ satisfying the condition $\nu(1/2) = -1/2$. Then

$$\nu(\mu) = -i\Big(2\sinh\frac{\pi i}{p}\Big)^{-1} \quad \text{for} \quad \mu = e^{\pi i/p}\Big(2\sinh\frac{\pi i}{p}\Big)^{-1}. \tag{3.1}$$

Let I_m (0_m) be the identity (zero) $m \times m$ matrix. For $a \in PC$ and $t = \infty$, let $a(t \pm 0) := a(\mp\infty)$.

Theorem 3.1. *Let $1 < p < \infty$, $1/p + 1/q = 1$, and*

$$\mathcal{M} := \big(\mathbb{R} \times \mathcal{L}_p\big) \cup \big(\{\infty\} \times \mathcal{L}_q\big).$$

(a) *For each* $(t,\mu)\in\mathcal{M}$, *the map* $\{S\}\cup\{aI : a\in PC\}\to\mathbb{C}^{2n\times 2n}$ *given by*

$$(aI)(t,\mu)=\begin{bmatrix} a(t+0) & 0_n \\ 0_n & a(t-0)\end{bmatrix}, S(t,\mu)=\begin{bmatrix} (2\mu-1)I_n & 2\nu(\mu)I_n \\ 2\nu(\mu)I_n & (1-2\mu)I_n\end{bmatrix} \tag{3.2}$$

extends to a Banach algebra homomorphism $\mathfrak{S}\to\mathbb{C}^{2n\times 2n}$, $A\mapsto A(t,\mu)$.

(b) *An operator* $A\in\mathfrak{S}$ *is Fredholm on the space* $L_p^n(\mathbb{R})$ *if and only if*

$$\det A(t,\mu)\neq 0 \quad \textit{for all} \quad (t,\mu)\in\mathcal{M}.$$

(c) *If an operator* $A\in\mathfrak{S}$ *is Fredholm on* $L_p^n(\mathbb{R})$, *then all its regularizers belong to the algebra* $\mathfrak{S}$.

Write the symbol $A(t,\mu)$ of a Fredholm operator $A\in\mathfrak{S}$ in the form

$$A(t,\mu)=\begin{bmatrix} A_{11}(t,\mu) & A_{12}(t,\mu) \\ A_{21}(t,\mu) & A_{22}(t,\mu)\end{bmatrix},$$

and consider the functions

$$\mathcal{A}(t,\mu):=\frac{\det A(t,\mu)}{\det A_{11}(t,0)\det A_{22}(t,1)}, \quad (t,\mu)\in\mathcal{M},$$

and $\mathcal{A}_0(t):=\mathcal{A}(t,0)$, $t\in\dot{\mathbb{R}}$. Denote by $J[\mathcal{A}]$ the set of those points $t\in\dot{\mathbb{R}}$ at which the functions $\mathcal{A}(t,\cdot)$ are not constant. For each point $t\in J[\mathcal{A}]$, join the one-sided limits $\mathcal{A}_0(t-0)=\mathcal{A}(t,0)$ and $\mathcal{A}_0(t+0)=\mathcal{A}(t,1)$ of the function $\mathcal{A}_0$ by the continuous curve $\left\{\mathcal{A}(t,\mu):\mu\in\mathcal{L}_p\right\}$ if $t\in\mathbb{R}$, and $\left\{\mathcal{A}(\infty,\mu):\mu\in\mathcal{L}_q\right\}$ if $t=\infty$. Orient these curves from $\mathcal{A}_0(t-0)$ to $\mathcal{A}_0(t+0)$ for every $t\in J[\mathcal{A}]$. This induces an orientation on the continuous closed curve

$$A^{\#}:=\mathcal{A}_0\Big(\dot{\mathbb{R}}\setminus J[\mathcal{A}]\Big)\cup\Big\{\mathcal{A}(\infty,\mu):\mu\in\mathcal{L}_q\Big\}\cup\bigcup_{t\in J[\mathcal{A}]\setminus\{\infty\}}\Big\{\mathcal{A}(t,\mu):\mu\in\mathcal{L}_p\Big\}, \tag{3.3}$$

which does not pass through the point $z=0$.

Theorem 3.2. *If* A *is a Fredholm operator in the algebra* $\mathfrak{S}$, *then*

$$\operatorname{ind} A=-\operatorname{wind} A^{\#}$$

where $\operatorname{wind} A^{\#}$ *is the winding number of the continuous closed oriented curve* $A^{\#}$ *about the origin.*

From Theorem 3.1 it follows that every operator $A\in\mathfrak{S}$ can be represented in the form

$$A=aP+bQ+H \tag{3.4}$$

where a, $b\in PC$ and H is an operator in the closed two-sided ideal $\mathcal{H}\subset\mathfrak{S}$ generated by all the commutators $cS-ScI$ $(c\in PC)$.

Indeed, every operator $A\in\mathfrak{S}$ of the form

$$A=\sum_i A_{i,1}A_{i,2}\cdots A_{i,n_i},$$

where $A_{i,j}$ are generators of the algebra $\mathfrak{S}$ and i, j run through finite sets, can be represented in the form (3.4) with

$$a(t) = A_{11}(t,1), \quad b(t) = A_{11}(t,0).$$

Moreover, one can prove that

$$\max\Big\{\|aI\|, \|bI\|\Big\} \le |A| := \inf_{K\in\mathcal{K}} \|A+K\|,$$

which implies that (3.4) is valid for every operator $A \in \mathfrak{S}$.

For each $t \in \dot{\mathbb{R}}$, let $\mathcal{H}_t$ be the closed two-sided ideal of the algebra $\mathfrak{S}$ which is generated by the commutators $u_t S - S u_t I$ where $u_t \in PC \cap C(\dot{\mathbb{R}} \setminus \{t\})$ and $u_t(t \pm 0) = \pm 1$.

Let $C_p(\overline{\mathbb{R}})$ denote the closure of the set of all continuous functions $d : \overline{\mathbb{R}} \to \mathbb{C}$ with finite total variation by the norm $\|\cdot\|_{M_p}$ of Fourier multipliers on $L_p(\mathbb{R})$. In what follows we will use the operator

$$V := \chi_+ N \chi_+ I + \chi_- N \chi_- I, \tag{3.5}$$

where $\chi_\pm$ are the characteristic functions of $\mathbb{R}_\pm$ and the operator $N \in \mathfrak{B}(L_p^n(\mathbb{R}))$ is given by

$$(N\varphi)(x) := \frac{1}{\pi i}\int_{\mathbb{R}} \frac{\varphi(\tau)}{\tau + x}\, d\tau, \quad x \in \mathbb{R}.$$

Lemma 3.3. *Let* $1 < p < \infty$, $1/p + 1/q = 1$, *and*

$$\mathcal{M}_0 := \{0\} \times \mathcal{L}_p, \quad \mathcal{M}_\infty := \{\infty\} \times \mathcal{L}_q.$$

The operator (3.5) *belongs to the ideal* $\mathcal{H}_0 + \mathcal{H}_\infty$, *and*

$$V(t,\mu) = \begin{cases} \begin{bmatrix} 2i\nu(\mu) I_n & 0_n \\ 0_n & -2i\nu(\mu) I_n \end{bmatrix} & \text{if} \quad (t,\mu) \in \mathcal{M}_0 \cup \mathcal{M}_\infty, \\ 0_{2n} & \text{if} \quad (t,\mu) \in \mathcal{M} \setminus (\mathcal{M}_0 \cup \mathcal{M}_\infty). \end{cases} \tag{3.6}$$

Proof. Consider the operators $S_{\mathbb{R}_+}, N_{\mathbb{R}_+} \in \mathfrak{B}(L_p^n(\mathbb{R}_+))$ given by

$$(S_{\mathbb{R}_+}\varphi)(x) := \frac{1}{\pi i}\int_{\mathbb{R}_+} \frac{\varphi(\tau)}{\tau - x}\, d\tau, \quad (N_{\mathbb{R}_+}\varphi)(x) := \frac{1}{\pi i}\int_{\mathbb{R}_+} \frac{\varphi(\tau)}{\tau + x}\, d\tau, \ x \in \mathbb{R}_+. \tag{3.7}$$

As is known (see, e.g., [13, Theorem 2.3], [21, Chapter 1] and [18, Propositions 2.3–2.4]), the operator $N_{\mathbb{R}_+}$ belongs to $\mathrm{alg}(S_{\mathbb{R}_+}, I)$,

$$N_{\mathbb{R}_+}^2 = S_{\mathbb{R}_+}^2 - I, \tag{3.8}$$

and the Gelfand transforms of the operators $S_{\mathbb{R}_+}, N_{\mathbb{R}_+}$ are given by the following functions in $C_p(\overline{\mathbb{R}})$:

$$S_{\mathbb{R}_+}(x) = \coth[\pi(x + i/p)], \quad N_{\mathbb{R}_+}(x) = 1/(\sinh[\pi(x+i/p)]), \quad x \in \overline{\mathbb{R}}. \tag{3.9}$$

By [21, Lemmas 1.2–1.3], $C_p(\overline{\mathbb{R}})$ coincides with the closure in $C_p(\overline{\mathbb{R}})$ of the algebra generated by 1 and $z = \coth[\pi(x + i/p)]$, and every function $f \in C_p(\overline{\mathbb{R}})$ with $f(\pm\infty) = 0$ is of the form

$$f = \lim_{m\to\infty} P_m(z)(z^2 - 1)$$

where P_m is a polynomial in z and the limit is taken in the norm $\|\cdot\|_{M_p}$. According to (3.9), $N_{\mathbb{R}_+} \in C_p(\overline{\mathbb{R}})$ and $N_{\mathbb{R}_+}(\pm\infty) = 0$. Hence

$$N_{\mathbb{R}_+}(x) = \lim_{m\to\infty} \left(\sum_{k=1}^{m} a_k^{(m)} \Big(\coth[\pi(x+i/p)] \Big)^k \Big(\coth^2[\pi(x+i/p)] - 1 \Big) \right) \tag{3.10}$$

where $a_k^{(m)} \in \mathbb{C}$ and thus, by (3.8),

$$N_{\mathbb{R}_+} = \lim_{m\to\infty} \left(\sum_{k=1}^{m} a_k^{(m)} S_{\mathbb{R}_+}^k (S_{\mathbb{R}_+}^2 - I) \right). \tag{3.11}$$

From (3.11) and the equality

$$\chi_+(S_{\mathbb{R}_+}^2 - I)\chi_+ I = \chi_+ S\chi_+ S\chi_+ I - \chi_+ S^2\chi_+ I = \chi_+ S\chi_- S\chi_+ I,$$

it follows that

$$\chi_+ N\chi_+ I = \lim_{m\to\infty} \left(\sum_{k=1}^{m} a_k^{(m)} \Big(\chi_+ S\chi_+ I \Big)^k (\chi_+ S\chi_- S\chi_+ I) \right) \in \mathfrak{S}. \tag{3.12}$$

Let $J : L_p^n(\mathbb{R}) \to L_p^n(\mathbb{R})$ be the isometric isomorphism given by (1.3). Then (3.12) implies that

$$\begin{aligned} \chi_- N\chi_- I &= -J\chi_+ N\chi_+ J \qquad (3.13)\\ &= -\lim_{m\to\infty} \left(\sum_{k=1}^{m} a_k^{(m)} \Big(J\chi_+ S\chi_+ J \Big)^k (J\chi_+ S\chi_- S\chi_+ J) \right)\\ &= -\lim_{m\to\infty} \left(\sum_{k=1}^{m} a_k^{(m)} (-1)^k \Big(\chi_- S\chi_- I \Big)^k (\chi_- S\chi_+ S\chi_- I) \right) \in \mathfrak{S}. \end{aligned}$$

Consequently,

$$V = \chi_+ N\chi_+ I + \chi_- N\chi_- I \in \mathfrak{S}. \tag{3.14}$$

Furthermore, setting $\mu := \Big(1 + \coth[\pi(x+i/p)]\Big)/2$, we conclude from (3.1), (3.8) and (3.9) that $1/\Big(\sinh[\pi(x+i/p)]\Big) = 2i\nu(\mu)$, whence by (3.9)–(3.10),

$$\lim_{m\to\infty} \left(\sum_{k=1}^{m} a_k^{(m)} (2\mu - 1)^k 4\mu(\mu - 1) \right) = 2i\nu(\mu). \tag{3.15}$$

From (3.12)–(3.15) and (3.2) we now immediately get (3.6).

Finally, by (3.4),

$$V = aP + bQ + H \tag{3.16}$$

where $H \in \mathcal{H}$. Every operator $H \in \mathcal{H}$ can be represented in the form

$$H = \lim_{m\to\infty} \sum_{t\in\Lambda_m} H_t^{(m)},$$

where Λ_m are finite subsets of $\dot{\mathbb{R}}$ and $H_t^{(m)} \in \mathcal{H}_t$. Since $H(t,0) = H(t,1) = 0_{2n}$ for every $t \in \dot{\mathbb{R}}$ and since $H_t^{(m)}(\tau,\mu) = 0_{2n}$ for all $\tau \in \dot{\mathbb{R}} \setminus \{t\}$, we infer from (3.16) and (3.6) that $a(t) = b(t) = 0_n$ for all $t \in \dot{\mathbb{R}}$ and that therefore $V = H = H_0 + H_\infty$, where $H_0 \in \mathcal{H}_0$ and $H_\infty \in \mathcal{H}_\infty$. □

4. The Fredholm study of operators with semi-almost periodic coefficients

In what follows, for bounded linear operators C and D, let $C \simeq D$ mean that $C - D$ is a compact operator.

Recall that $\mathcal{H}_\infty$ denotes the closed two-sided ideal of the algebra $\mathfrak{S}$ of singular integral operators on $L_p^n(\mathbb{R})$ with matrix coefficients of the class PC which is generated by the operator $uS - SuI$, where $u(x) = \tanh(\pi x)$ and $S = S_{\mathbb{R}}$. We remark that since every function $v \in C(\overline{\mathbb{R}})$ is represented in the form $v = \lim\limits_{m\to\infty} P_m(u)$, where $P_m(u)$ are polynomials in u, and since

$$u^n S - Su^n I = \sum_{k=0}^{n-1} u^k(uS - Su)u^{n-k-1}I \quad (n \ge 1),$$

the closed two-sided ideal of $\mathfrak{S}$ generated by all the commutators $vS - SvI$ with $v \in C(\overline{\mathbb{R}})$ coincides with the ideal $\mathcal{H}_\infty$ generated by the only commutator $uS - SuI$, where $u(x) = \tanh(\pi x)$. Every operator $H \in \mathcal{H}_\infty$ can be represented in the form

$$H \simeq \lim_{k\to\infty} (uS - Su)A_k \simeq \lim_{k\to\infty} A_k'(uS - Su)I, \tag{4.1}$$

where $A_k,\ A_k' \in \mathfrak{S}$. Recall also that for a matrix function $B \in AP$, the Bohr mean value is defined entry-wise by

$$\mathbf{M}(B) = \lim_{T\to\infty} \frac{1}{2T} \int_{-T}^{T} B(x)dx.$$

Lemma 4.1. *Suppose we are given matrix functions $B_j \in AP$ and operators $H_j \in \mathcal{H}_\infty$ $(j = 1, 2, \ldots, m)$. Then the operators*

$$K = I + \sum_j B_j H_j \quad \text{and} \quad K_0 = I + \sum_j \mathbf{M}(B_j) H_j$$

are Fredholm on $L_p^n(\mathbb{R})$ only simultaneously, and in that case their indices coincide.

Proof. If K_0 is a Fredholm operator, then $K \simeq (I + R)K_0$, where

$$R = \sum_j (B_j - \mathbf{M}(B_j))H_j K_0^{(-1)}$$

and $K_0^{(-1)}$ is a regularizer of K_0. We claim that the operator

$$K^{(-1)} = K_0^{(-1)}(I - R)$$

is a regularizer of the operator K. To show this it is sufficient to prove that

$$R^2 = \Big(\sum_j (B_j - \mathbf{M}(B_j))H_j K_0^{(-1)}\Big)^2 \simeq 0, \tag{4.2}$$

where we may assume that the matrix functions B_j are in APW. Put

$$\mathcal{P}_+(e^{i\lambda t}) = \chi(\lambda)e^{i\lambda t}, \quad \mathcal{P}_- = I - \mathcal{P}_+, \tag{4.3}$$

where $\chi(\lambda) = 1$ for $\lambda \geq 0$ and $\chi(\lambda) = 0$ for $\lambda < 0$. Defining

$$B_j^\pm := \mathcal{P}_\pm(B_j - \mathbf{M}(B_j)) \in APW_0^\pm,$$

we infer from Lemma 2.1 that for $u = \tanh(\pi x) \in C(\overline{\mathbb{R}})$,

$$B_j^\pm u \in H_\infty^\pm + C. \tag{4.4}$$

Then taking into account (4.1) for the operators $H = H_k K_0^{(-1)} \in \mathcal{H}_\infty \quad (k = i, j)$, and also (4.4), we get, for every i, j,

$$\begin{aligned} &H_i K_0^{(-1)}(B_j - \mathbf{M}(B_j))H_j K_0^{(-1)} \\ &\simeq \lim_{k\to\infty} \Big[A'_{ki}(-2)(uQ - Qu)B_j^+ 2(uP - Pu)A_{kj}\Big] \\ &\quad + \lim_{k\to\infty} \Big[A'_{ki} 2(uP - Pu)B_j^-(-2)(uQ - Qu)A_{kj}\Big] \simeq 0, \end{aligned}$$

which proves (4.2).

If K_0 is a Fredholm operator, then the family $T(\tau) = I + (1-\tau)R$, $\tau \in [0,1]$, is a homotopy which connects the operator $T(0) = I + R$ with $T(1) = I$ in the class of Fredholm operators. Hence $\operatorname{ind} T(0) = 0$ and $\operatorname{ind} K = \operatorname{ind} K_0$.

If K is a Fredholm operator, but K_0 is not a Fredholm operator, then, since the matrix symbol $K_0(\infty, \mu)$ of the operator $K_0 \in \mathfrak{S}$ has zero determinant for some $\mu \in \mathcal{L}_q$, in every given neighborhood of the singular integral operator K_0 it is possible to choose two Fredholm singular integral operators of the same form with different indices, which contradicts the stability of the index of the operator K. □

The Cauchy index of a function $f \in \mathcal{G}C(\overline{\mathbb{R}})$ is defined as the real number

$$\operatorname{Ind} f := \frac{1}{2\pi}\Big(\psi(+\infty) - \psi(-\infty)\Big),$$

where ψ is any continuous argument of f. There is a reasonable extension of the notion of Cauchy index to the class

$$\mathcal{G}SAP_0 := \Big\{f \in \mathcal{G}SAP : v(f_+) = v(f_-) = 0\Big\}.$$

Namely, for $f \in \mathcal{G}SAP_0$,

$$\operatorname{Ind} f := \frac{1}{2\pi} \lim_{T\to\infty} \frac{1}{T}\int_0^T \Big(\psi(x) - \psi(-x)\Big)dx \tag{4.5}$$

where ψ is again any continuous argument of f (see [10]).

Lemma 4.2. *If an $n \times n$ matrix function $G \in SAPW$ is invertible, its almost periodic representatives $G_\pm (\in APW)$ admit canonical right APW factorizations*

$$G_\pm = G_\pm^- G_\pm^+,$$

and $H \in \mathcal{H}_\infty$, then there exists an invertible matrix function $\widetilde{G} \in C(\overline{\mathbb{R}})$ such that $\widetilde{G}(\pm\infty) = \mathbf{d}(G_\pm)$,

$$\operatorname{Ind} \det \widetilde{G} = \operatorname{Ind} \det G, \tag{4.6}$$

and the operators

$$Y = GP + Q + (I_n - G)H \quad \text{and} \quad \widetilde{Y} = \widetilde{G}P + Q + (I_n - \widetilde{G})H$$

are Fredholm on $L_p^n(\mathbb{R})$ only simultaneously. In that case

$$\operatorname{ind} Y = \operatorname{ind} \widetilde{Y}.$$

Proof. Let $G_\pm = G_\pm^- G_\pm^+$ be canonical right APW factorizations of the matrix functions $G_\pm \in APW$. Let $N^\pm$ be invertible matrix functions in $(H_\infty^\pm + C) \cap SAP$ which are defined for $\pm x > 0$ by

$$N^+(x) := [\mathbf{M}(G_\pm^+)]^{-1} G_\pm^+(x \pm ix^{-1}), \quad N^-(x) := G_\pm^-(x \mp ix^{-1})[\mathbf{M}(G_\pm^-)]^{-1}$$

and admit a homotopy to I_n in the class of invertible elements. Then

$$\widetilde{G} := (N^-)^{-1} G (N^+)^{-1} \in \mathcal{G}C(\overline{\mathbb{R}}), \quad \widetilde{G}(\pm\infty) = \mathbf{d}(G_\pm),$$

and (4.6) holds.

Obviously, the operators

$$X_1 = I - (N^- - I_n)P(N^-)^{-1}I \quad \text{and} \quad X_2 = I + [(N^+)^{-1} - I_n]P \tag{4.7}$$

are regularizers for the operators $N^-P + Q$ and $N^+P + Q$, respectively, and moreover, $\operatorname{ind} X_k = 0$ $(k = 1, 2)$. Since

$$\lim_{y \to +\infty} \Big([\mathbf{M}(G_\pm^+)]^{-1} G_\pm^+(iy) - I_n \Big) = 0_n,$$

we obtain

$$(uQ - Qu)[(N^+)^{-1} - I_n]P \simeq 0.$$

Hence, in view of (4.1),

$$\begin{aligned} HX_2 &\simeq -\lim_{k \to \infty} 2A_k'(uQ - Qu)(I + [(N^+)^{-1} - I_n]P) \\ &\simeq -\lim_{k \to \infty} 2A_k'(uQ - QuI) \simeq H. \end{aligned} \tag{4.8}$$

Analogously,

$$P(N^-)^{-1}H \simeq PH, \quad P\widetilde{G}(N^+ - I_n)H \simeq \widetilde{G}(N^+ - I_n)H,$$

and thus

$$\begin{aligned} &X_1(I_n - G)H \qquad (4.9)\\ &= H - GH - (N^- - I_n)\Big[P(N^-)^{-1}H - P\widetilde{G}H - P\widetilde{G}(N^+ - I_n)H\Big]\\ &\simeq H - GH - (N^- - I_n)(PH - P\widetilde{G}H) + (N^- - I_n)\widetilde{G}(N^+ - I_n)H\\ &= (I_n - \widetilde{G})H - \Big[(N^- - I_n)(PH - P\widetilde{G}H + \widetilde{G}H) + \widetilde{G}(N^+ - I_n)H\Big]. \end{aligned}$$

Since $(GP + Q)X_2 \simeq N^-\widetilde{G}P + Q$, we get

$$\begin{aligned} X_1(GP + Q)X_2 &\simeq \Big[I - (N^- - I_n)P(N^-)^{-1}I\Big](N^-\widetilde{G}P + Q) \qquad (4.10)\\ &\simeq N^-\widetilde{G}P + Q - (N^- - I_n)P\widetilde{G}P\\ &\simeq \widetilde{G}P + Q + (N^- - I_n)Q\widetilde{G}P. \end{aligned}$$

It follows from (4.8)–(4.10) that

$$\begin{aligned} &X_1YX_2 \qquad (4.11)\\ &\simeq \widetilde{Y} - \Big[(N^- - I_n)(PH - P\widetilde{G}H + \widetilde{G}H - Q\widetilde{G}P) + \widetilde{G}(N^+ - I_n)H\Big] =: Y_1. \end{aligned}$$

Hence the operators Y and Y_1 are Fredholm only simultaneously, and in that case $\operatorname{ind} Y = \operatorname{ind} Y_1$. As, due to (4.11), $Y_1 = \widetilde{Y} + \sum G_k H_k$, where $H_k \in \mathcal{H}_\infty$, $G_k \in AP$ and $\mathbf{M}(G_k) = 0$, we get by Lemma 4.1 that the operators Y_1 and $\widetilde{Y}$ are Fredholm only simultaneously and that in this case $\operatorname{ind} Y_1 = \operatorname{ind} \widetilde{Y}$. Indeed, let, for example, $\widetilde{Y}$ be Fredholm. Then

$$Y_1 \simeq \Big(I + \sum G_k H_k \widetilde{Y}^{(-1)}\Big)\widetilde{Y} \qquad (4.12)$$

where $\widetilde{Y}^{(-1)}$ is a regularizer of $\widetilde{Y}$, $H_k\widetilde{Y}^{(-1)} \in \mathcal{H}_\infty$, and the operator

$$K := I + \sum G_k H_k \widetilde{Y}^{(-1)}$$

satisfies the conditions of Lemma 4.1. By that lemma, K is Fredholm on $L_p^n(\mathbb{R})$ simultaneously with $K_0 = I$, and $\operatorname{ind} K = \operatorname{ind} I = 0$. It remains to use (4.12). If Y_1 is Fredholm and $Y_1^{(-1)}$ is a regularizer of Y_1, we apply Lemma 4.1 to the operator

$$K_1 = I - \sum G_k H_k Y_1^{(-1)},$$

which again gives the assertion. □

Lemma 4.3. *Let a and b be $n \times n$ matrix functions in SAP, $a_\pm$ and $b_\pm$ be their almost periodic representatives, and $H \in \mathcal{H}$. If the operator $aP + bQ + H$ is Fredholm on $L_p^n(\mathbb{R})$, then the operators $a_\pm P + b_\pm Q$ are invertible on $L_p^n(\mathbb{R})$.*

Proof. For $a \in SAP$, set $a_h(t) := a(t + h)$, $t \in \mathbb{R}$. As is known (see, e.g., [15, p. 20]), if a function d is in AP, then the set $\{d_h\}_{h\in\mathbb{R}}$ is relatively compact, that is, for every sequence $\{g_k\} \subset \mathbb{R}$ there exists a subsequence $\{h_n\} \subset \{g_k\}$ such that

$$\lim_{n\to\infty} \|d - d_{h_n}\|_{L_\infty(\mathbb{R})} = 0.$$

Consequently, there are sequences $\{h_n^\pm\} \subset \mathbb{R}$ such that $h_n^\pm \to \pm\infty$ as $n \to \infty$ and

$$\lim_{n\to\infty} \left\| \left(a_+ - (a_+)_{h_n^+}\right) I \right\|_{\mathfrak{B}(L_p^n(\mathbb{R}))} = \lim_{n\to\infty} \left\| \left(b_+ - (b_+)_{h_n^+}\right) I \right\|_{\mathfrak{B}(L_p^n(\mathbb{R}))} = 0, \quad (4.13)$$

$$\lim_{n\to\infty} \left\| \left(a_- - (a_-)_{h_n^-}\right) I \right\|_{\mathfrak{B}(L_p^n(\mathbb{R}))} = \lim_{n\to\infty} \left\| \left(b_- - (b_-)_{h_n^-}\right) I \right\|_{\mathfrak{B}(L_p^n(\mathbb{R}))} = 0. \quad (4.14)$$

Consider the invertible shift operators $W_h \in \mathfrak{B}(L_p^n(\mathbb{R}))$, $h \in \mathbb{R}$, given by

$$(W_h f)(t) = f(t+h), \quad t \in \mathbb{R}.$$

Since for an $n \times n$ matrix function $c \in PC$,

$$s\text{-}\lim_{n\to\infty} W_{h_n^\pm} c W_{h_n^\pm}^{-1} = c(\pm\infty) I, \quad (4.15)$$

we deduce from (4.13)–(4.15) that for $a \in SAP$,

$$s\text{-}\lim_{n\to\infty} W_{h_n^\pm} a W_{h_n^\pm}^{-1} = a_\pm I, \quad s\text{-}\lim_{n\to\infty} W_{h_n^\pm} b W_{h_n^\pm}^{-1} = b_\pm I. \quad (4.16)$$

Furthermore,

$$W_{h_n^\pm} S W_{h_n^\pm}^{-1} = S. \quad (4.17)$$

As the ideal $\mathcal{H}$ is generated by all the commutators $cS - ScI$ with $c \in PC$, we infer from (4.15) and (4.17) that for every $H \in \mathcal{H}$,

$$s\text{-}\lim_{n\to\infty} W_{h_n^\pm} H W_{h_n^\pm}^{-1} = 0. \quad (4.18)$$

Finally, from (4.16)–(4.18) it follows that

$$s\text{-}\lim_{n\to\infty} W_{h_n^\pm} (aP + bQ + H) W_{h_n^\pm}^{-1} = a_\pm P + b_\pm Q. \quad (4.19)$$

Now analogously to [11, Theorem 5.4] (see also [9, Theorem 5]), the Fredholmness of the operator $aP + bQ + H$ implies the invertibility of the operators $a_\pm P + b_\pm Q$. □

By virtue of (4.19), we call $a_\pm P + b_\pm Q$ the limit operators for the operator $aP + bQ + H$ at $\pm\infty$ (cf., e.g., [14]).

Theorem 4.4. *Let G be an $n \times n$ matrix function in $SAPW$ and $H \in \mathcal{H}_\infty$. The operator*

$$Y = GP + Q + (I_n - G)H$$

is Fredholm on the space $L_p^n(\mathbb{R})$ if and only if

(i) *the matrix function G is invertible in $SAPW$;*
(ii) *its almost periodic representatives $G_\pm$ admit canonical right APW factorizations;*
(iii) $\det \widetilde{Y}(\infty, \mu) \neq 0$ *for all $\mu \in \mathcal{L}_q$,*

where

$$\widetilde{Y}(\infty,\mu) = (\widetilde{G}P+Q)(\infty,\mu) + ((I_n-\widetilde{G})H)(\infty,\mu),$$

$$(\widetilde{G}P+Q)(\infty,\mu) = \begin{bmatrix} \mathbf{d}(G_-)\mu + I_n(1-\mu) & (\mathbf{d}(G_-)-I_n)\nu(\mu) \\ (\mathbf{d}(G_+)-I_n)\nu(\mu) & \mathbf{d}(G_+)(1-\mu)+I_n\mu \end{bmatrix}, \tag{4.20}$$

$$((I_n-\widetilde{G})H)(\infty,\mu) = \begin{bmatrix} I_n-\mathbf{d}(G_-) & 0_n \\ 0_n & I_n-\mathbf{d}(G_+) \end{bmatrix} H(\infty,\mu),$$

and $H(t,\mu)$ *is the symbol of the operator* $H \in \mathcal{H}_\infty$. *If the operator* Y *is Fredholm, then*

$$\operatorname{ind} Y = -\operatorname{Ind}\,\det G - \frac{1}{2\pi}\Big\{\arg\det \widetilde{Y}(\infty,\mu)\Big\}_{\mu\in\mathcal{L}_q},$$

where Ind det G *is defined by* (4.5).

Proof. Let the operator Y be Fredholm on the space $L_p^n(\mathbb{R})$. Since

$$Y(t,\mu) = \operatorname{diag}\{G(t), I_n\} \quad \text{for} \quad (t,\mu)\in\mathbb{R}\times\mathcal{L}_p,$$

we conclude that $\det G(t)\neq 0$ for all $t\in\mathbb{R}$. Further, by Lemma 4.3, the limit operators $Y_\pm = G_\pm P + Q$ of the operator Y at $\pm\infty$ are invertible on $L_p^n(\mathbb{R})$. Hence, by Theorem 2.2, the matrix functions $G_\pm \in APW$ admit canonical right APW factorizations. Now Lemma 4.2 shows that the operator $\widetilde{Y}\in\mathfrak{S}$ is Fredholm on $L_p^n(\mathbb{R})$, which implies that its symbol $\widetilde{Y}(\cdot,\cdot)$ is invertible on $\mathcal{M}$. In particular, $\det\widetilde{Y}(\infty,\mu)\neq 0$ for all $\mu\in\mathcal{L}_q$. Thus (i)–(iii) hold.

Conversely, if (iii) holds, then the operator $\widetilde{Y}\in\mathfrak{S}$ with the matrix coefficient $\widetilde{G}\in\mathcal{G}C(\overline{\mathbb{R}})$ is Fredholm on $L_p^n(\mathbb{R})$. Since (i)–(ii) also are satisfied, the operator Y is Fredholm on $L_p^n(\mathbb{R})$ by Lemma 4.2 if $\widetilde{G}(\pm\infty) = \mathbf{d}(G_\pm)$.

If Y is Fredholm on $L_p^n(\mathbb{R})$ and if (4.6) holds, then by Lemma 4.2,

$$\begin{aligned}\operatorname{ind} Y &= \operatorname{ind}\widetilde{Y} = -\operatorname{Ind}\,\det\widetilde{G} - \frac{1}{2\pi}\Big\{\arg\det\widetilde{Y}(\infty,\mu)\Big\}_{\mu\in\mathcal{L}_q} \\ &= -\operatorname{Ind}\,\det G - \frac{1}{2\pi}\Big\{\arg\det\widetilde{Y}(\infty,\mu)\Big\}_{\mu\in\mathcal{L}_q},\end{aligned}$$

which completes the proof. □

Now we generalize Lemma 4.2 and Theorem 4.4.

Lemma 4.5. *Let* G *and* F *be* $n\times n$ *matrix functions in* SAP *with almost periodic representatives* $G_\pm, F_\pm$ *at* $\pm\infty$, *let* $G_\pm$ *admit canonical right AP factorizations*

$$G_\pm = G_\pm^- G_\pm^+,$$

and let $H\in\mathcal{H}_\infty$. *Then there exist matrix functions* $\widetilde{G},\widetilde{F}\in C(\overline{\mathbb{R}})$ *such that*

$$\widetilde{G}(\pm\infty) = \mathbf{d}(G_\pm), \quad \widetilde{F}(\pm\infty) = \mathbf{M}(G_\pm^-)\mathbf{M}\Big((G_\pm^-)^{-1}F_\pm\Big),$$

$\widetilde{G}$ *is invertible in* $C(\overline{\mathbb{R}})$, (4.6) *holds, and the operators*

$$Y_0 = GP+Q+FH, \quad \widetilde{Y}_0 = \widetilde{G}P+Q+\widetilde{F}H$$

are Fredholm on $L_p^n(\mathbb{R})$ only simultaneously. In that case

$$\operatorname{ind} Y_0 = \operatorname{ind} \widetilde{Y}_0.$$

Proof. We want to prove that $X_1 Y_0 X_2 \simeq \widetilde{Y}_0$, where X_1, X_2 are defined by (4.7). From (4.10) we have

$$X_1(GP+Q)X_2 \simeq \widetilde{G}P + Q + (N^- - I_n)Q\widetilde{G}P.$$

Thus, taking into account (4.8), we only need to study the operator

$$\begin{aligned} X_1 FH &= \Big[I - (N^- - I_n)P(N^-)^{-1}I\Big]FH \\ &= \Big[I + (N^- - I_n)Q\Big](N^-)^{-1}FH. \end{aligned} \tag{4.21}$$

Let $\widetilde{F} \in C(\overline{\mathbb{R}})$ be chosen such that $\widetilde{F}(\pm\infty) = \mathbf{M}(G_\pm^-)\mathbf{M}\Big((G_\pm^-)^{-1}F_\pm\Big)$, where $F_\pm$ are the almost periodic representatives of F. The matrix function

$$D := (N^-)^{-1}F - \widetilde{F} \in SAP \tag{4.22}$$

is represented in the form

$$D = uD_+ + (1-u)D_- + D_0$$

where

$$D_\pm := \Big((N^-)^{-1}F\Big)_\pm - \mathbf{M}(G_\pm^-)\mathbf{M}\Big((G_\pm^-)^{-1}F_+\Big) \in AP \tag{4.23}$$

and $D_0 \in C_0(\mathbb{R})$. From (4.8), (4.21) and (4.22) we get

$$X_1FHX_2 \simeq X_1FH = DH + \widetilde{F}H + (N^- - I_n)QDH + (N^- - I_n)Q\widetilde{F}H. \tag{4.24}$$

Using (4.10) and (4.24) we obtain

$$X_1Y_0X_2 \simeq \widetilde{Y}_0 + DH + (N^- - I_n)(Q\widetilde{G}P + Q\widetilde{F}H + QDH) =: \widetilde{Y}_1. \tag{4.25}$$

Hence, Y_0 and $\widetilde{Y}_1$ are equivalent in the sense of Fredholmness, and in that case $\operatorname{ind} Y_0 = \operatorname{ind}\widetilde{Y}_1$. On the other hand, if one of the operators $\widetilde{Y}_0$ or $\widetilde{Y}_1$, say $\widetilde{Y}_0$, is Fredholm, and $\widetilde{Y}_0^{(-1)}$ is a regularizer of $\widetilde{Y}_0$, then

$$\widetilde{Y}_1 \simeq K\widetilde{Y}_0 \tag{4.26}$$

where

$$K = I + DH\widetilde{Y}_0^{(-1)} + (N^- - I_n)(Q\widetilde{G}P + Q\widetilde{F}H + QDH)\widetilde{Y}_0^{(-1)}. \tag{4.27}$$

It remains to prove that the operator K is Fredholm and $\operatorname{ind} K = 0$. Then it will follow from (4.26) that $\widetilde{Y}_1$ and $\widetilde{Y}_0$ are Fredholm only simultaneously and that in this case $\operatorname{ind}\widetilde{Y}_1 = \operatorname{ind}\widetilde{Y}_0$, which directly implies the assertion of lemma.

Assume first that $G_\pm$, $F_\pm \in APW$. Then using (4.3), (4.23) and setting

$$B_\pm^+ := \mathcal{P}_+D_\pm \in APW_0^+, \quad B_\pm^- := \mathcal{P}_-D_\pm \in APW_0^-, \tag{4.28}$$

by analogy with the proof of Lemma 4.1 we obtain

$$QD_+uH = Q(B_+^+ + B_+^-)uH \simeq B_+^- uH, \tag{4.29}$$
$$QD_-(1-u)H = Q(B_-^+ + B_-^-)(1-u)H \simeq B_-^-(1-u)H. \tag{4.30}$$

Since $D_0 H \simeq 0$, from (4.27)–(4.30) it follows that

$$\begin{aligned} K \;\simeq\;& I + \Big[(B_+^+ + B_+^-)uH + (B_-^+ + B_-^-)(1-u)H\Big]\widetilde{Y}_0^{(-1)} \\ &+ (N^- - I_n)(Q\widetilde{G}P + Q\widetilde{F}H + B_+^- uH + B_-^-(1-u)H)\widetilde{Y}_0^{(-1)}. \end{aligned} \tag{4.31}$$

By (4.31), the operator (4.27) has the form

$$K = I + \sum G_k H_k,$$

where the operators H_k are in $\mathcal{H}_\infty$, the matrix functions G_k belong to APW and $\mathbf{M}(G_k) = 0$. Thus, by Lemma 4.1, K and I are equivalent in the sense of Fredholmness, and $\operatorname{ind} K = \operatorname{ind} I = 0$. It remains to observe from (4.25) that the latter properties and (4.26) will be valid for general $G_\pm$, $F_\pm \in AP$. □

By analogy with Theorem 4.4, application of Lemma 4.5 immediately yields the following.

Theorem 4.6. *Let G and F be $n \times n$ matrix functions in SAP with almost periodic representatives $G_\pm \in APW$ and $F_\pm \in AP$, and let $H \in \mathcal{H}_\infty$. The operator*

$$Y_0 = GP + Q + FH$$

is Fredholm on the space $L_p^n(\mathbb{R})$ if and only if

(i) *the matrix function G is invertible in $SAPW$;*
(ii) *the matrix functions $G_\pm$ admit canonical right APW factorizations;*
(iii) $\det \widetilde{Y}_0(\infty,\mu) \neq 0$ *for all $\mu \in \mathcal{L}_q$, where*

$$\widetilde{Y}_0(\infty,\mu) = (\widetilde{G}P + Q)(\infty,\mu) + (\widetilde{F}H)(\infty,\mu),$$

$(\widetilde{G}P + Q)(\infty,\mu)$ is given by (4.20), and

$$(\widetilde{F}H)(\infty,\mu) = \operatorname{diag}\Big\{\mathbf{M}(G_-^-)\mathbf{M}\Big((G_-^-)^{-1}F_-\Big), \mathbf{M}(G_+^-)\mathbf{M}\Big((G_+^-)^{-1}F_+\Big)\Big\}H(\infty,\mu).$$

If the operator Y_0 is Fredholm, then

$$\operatorname{ind} Y_0 = -\operatorname{Ind}\det G - \frac{1}{2\pi}\Big\{\arg\det \widetilde{Y}_0(\infty,\mu)\Big\}_{\mu\in\mathcal{L}_q}.$$

5. Singular integral operators with piecewise almost periodic coefficients

Let $\mathcal{H}$ be the closed two-sided ideal of the Banach algebra $\mathfrak{S} \subset \mathfrak{B}(L_p^n(\mathbb{R}))$ that is generated by all the commutators $cS - ScI$ with $c \in PC$. Clearly, each operator $H \in \mathcal{H}$ can be represented in the form

$$H = H_{\mathbb{R}} + H_\infty$$

where $H_{\mathbb{R}} \in \mathcal{H}_{\mathbb{R}}$ and $H_\infty \in \mathcal{H}_\infty$.

In this section we construct the Fredholm theory for singular integral operators of the form

$$B = aP + bQ + H_{\mathbb{R}} + cH_\infty \tag{5.1}$$

where a, b, c are $n \times n$ matrix functions in $PAPW$, $H_{\mathbb{R}} \in \mathcal{H}_{\mathbb{R}}$ and $H_\infty \in \mathcal{H}_\infty$.

With the operator (5.1) we associate the operator

$$\widehat{B} = \widehat{a}P + \widehat{b}Q + \widehat{c}H_\infty \tag{5.2}$$

where the $n \times n$ matrix functions $\widehat{a}, \widehat{b}, \widehat{c}$ belong to $SAPW$ and have the same APW representatives $a_\pm, b_\pm, c_\pm$ at $\pm\infty$ as the matrix functions $a, b, c \in PAPW$, respectively, and, in addition, the matrix functions $\widehat{a}, \widehat{b}$ are invertible at every point $t \in \mathbb{R}$. If the operator (5.1) is Fredholm on the space $L_p^n(\mathbb{R})$, then by the Allan-Douglas local principle (see, e.g., [3]), the operator (5.2) also is Fredholm on $L_p^n(\mathbb{R})$. Indeed, the operators B and $\widehat{B}$ are locally Fredholm at ∞ due to the Fredholmness of B, and $\widehat{B}$ is locally Fredholm at every point $t \in \mathbb{R}$ simultaneously with the invertible operator $\widehat{a}(t)P + \widehat{b}(t)Q$ having invertible constant coefficients.

The operators (5.1) and (5.2) have the same limit operators at $\pm\infty$, namely,

$$B_\pm = a_\pm P + b_\pm Q \tag{5.3}$$

with $a_\pm, b_\pm \in APW$. If the operator (5.1) is Fredholm, then, by Lemma 4.3, its limit operators (5.3) are invertible on $L_p^n(\mathbb{R})$ (cf. [9]). In particular (see, e.g., [21]), $a_\pm, b_\pm$ are invertible in $L_\infty^{n\times n}(\mathbb{R})$, whence

$$a_\pm, b_\pm \in \mathcal{G}APW \tag{5.4}$$

and thus $\widehat{a}, \widehat{b} \in \mathcal{G}SAPW$.

If the operator (5.1) and, hence, the operator (5.2) are Fredholm on $L_p^n(\mathbb{R})$, then there exists an operator $A \in \mathfrak{S}$ such that A is locally equivalent to I at ∞ and

$$B \simeq A\widehat{B}. \tag{5.5}$$

Let $\widehat{B}^{(-1)}$ be a regularizer of the operator $\widehat{B}$. From (5.5) it follows that the operator $A \simeq B\widehat{B}^{(-1)}$ is also Fredholm on $L_p^n(\mathbb{R})$. Thus we have proved the following assertion.

Lemma 5.1. *The operator* (5.1) *is Fredholm on the space* $L_p^n(\mathbb{R})$ *if and only if the operators* $\widehat{B}$ *and* $A \simeq B\widehat{B}^{(-1)}$ *are Fredholm on* $L_p^n(\mathbb{R})$ *as well. In that case*

$$\operatorname{ind} B = \operatorname{ind} A + \operatorname{ind} \widehat{B}. \tag{5.6}$$

It remains to apply the results of Sections 3 and 4 to the operators A and $\widehat{B}$, respectively.

For every $t \in \mathbb{R}$, the operator B given by (5.1) with matrix coefficients $a, b, c \in PAPW$ is locally equivalent at the point t to an operator $B_t \in \mathfrak{S}$. Hence we can define the symbol $B(\cdot,\cdot)$ of B on the basis of Theorem 3.1 and the relation $B(t,\mu) = B_t(t,\mu)$. Clearly, for every fixed $\mu \in \mathcal{L}_p$, the matrix function $B(\cdot,\mu)$ is a member of $PAPW$.

Theorem 5.2. *The operator* (5.1) *is Fredholm on the space* $L_p^n(\mathbb{R})$ *if and only if*

(i) $\det B(t,\mu) \neq 0$ *for every* $(t,\mu) \in \mathbb{R} \times \mathcal{L}_p$;

(ii) *the matrix functions* $a_\pm$ *and* $b_\pm$ *are invertible in* APW;

(iii) *the matrix functions* $G_\pm = b_\pm^{-1} a_\pm$ *admit canonical* APW *factorizations*

$$G_\pm = G_\pm^- G_\pm^+;$$

(iv) $\det \widetilde{Y}(\infty,\mu) \neq 0$ *for every* $\mu \in \mathcal{L}_q$, *where*

$$\widetilde{Y} = \widetilde{G}P + Q + \widetilde{F}H_\infty \in \mathfrak{S}, \tag{5.7}$$

and $\widetilde{G}$, $\widetilde{F}$ *are matrix functions in* $C(\overline{\mathbb{R}})$ *such that*

$$\widetilde{G}(\pm\infty) = \mathbf{d}(G_\pm), \quad \widetilde{F}(\pm\infty) = \mathbf{M}(G_\pm^-)\mathbf{M}\Big((G_\pm^-)^{-1} b_\pm^{-1} c_\pm\Big). \tag{5.8}$$

Proof. Sufficiency. Fix an arbitrary $t \in \mathbb{R}$. Then there exist matrix functions $a_t, b_t, c_t \in PC$ and $\widehat{a}_t, \widehat{b}_t, \widehat{c}_t \in C(\overline{\mathbb{R}})$ such that

$$a_t(\pm\infty) = \widehat{a}_t(\pm\infty), \quad b_t(\pm\infty) = \widehat{b}_t(\pm\infty), \quad c_t(\pm\infty) = \widehat{c}_t(\pm\infty), \tag{5.9}$$

$$a_t(t\pm 0) = a(t\pm 0), \quad b_t(t\pm 0) = b(t\pm 0), \quad c_t(t\pm 0) = c(t\pm 0), \tag{5.10}$$

$$\widehat{a}_t(t) = \widehat{a}(t), \quad \widehat{b}_t(t) = \widehat{b}(t), \quad \widehat{c}_t(t) = \widehat{c}(t). \tag{5.11}$$

Then the operators

$$B_t = a_t P + b_t Q + H_{\mathbb{R}} + c_t H_\infty, \quad \widehat{B}_t = \widehat{a}_t P + \widehat{b}_t Q + \widehat{c}_t H_\infty,$$

which are locally equivalent at the point t to the operators B and $\widehat{B}$, respectively, belong to the algebra $\mathfrak{S}$. Hence, taking into account (5.9)–(5.11), we can put

$$B(t,\mu) := B_t(t,\mu), \quad \widehat{B}(t,\mu) := \widehat{B}_t(t,\mu), \quad \mu \in \mathcal{L}_p, \tag{5.12}$$

whence, due to (5.5),

$$B(t,\mu) = A(t,\mu)\widehat{B}(t,\mu), \quad \mu \in \mathcal{L}_p. \tag{5.13}$$

But $\det\Big(\widehat{a}(t)\widehat{b}(t)\Big) \neq 0$, from which, by the continuity of $\widehat{a}, \widehat{b}$ at t, we get

$$\det \widehat{B}(t,\mu) = \det\Big(\widehat{a}(t)\widehat{b}(t)\Big) \neq 0. \tag{5.14}$$

Now from (i) and (5.13) it follows that

$$\det A(t,\mu) \neq 0 \quad \text{for all} \quad (t,\mu) \in \mathbb{R} \times \mathcal{L}_p. \tag{5.15}$$

Since $A(\infty,\mu) = I_{2n}$, we infer from (5.15) and Theorem 3.1(b) that the operator $A \in \mathfrak{S}$ is Fredholm on the space $L_p^n(\mathbb{R})$.

By (5.14), the operator $\widehat{B}$ is locally Fredholm at every point $t \in \mathbb{R}$. In view of (ii) the operators $\widehat{B}$ and

$$Y = \widehat{b}^{-1}\widehat{B} = GP + Q + FH_\infty, \tag{5.16}$$

where $G := \widehat{b}^{-1}\widehat{a}$ and $F := \widehat{b}^{-1}\widehat{c}$, are locally Fredholm at ∞ only simultaneously. By virtue of (iii)–(iv) and Lemma 4.5, the operators $\widetilde{Y}, Y$ and, hence, $\widehat{B}$ are locally Fredholm at ∞. Thus $\widehat{B}$ is Fredholm on the space $L_p^n(\mathbb{R})$ and, by Lemma 5.1, the operator B is Fredholm on $L_p^n(\mathbb{R})$, as well.

Necessity. If the operator B is Fredholm on $L_p^n(\mathbb{R})$, then the local principle and Theorem 3.1 yield that (i) holds, and due to (5.4), condition (ii) is also satisfied. Thus the matrix functions $\widehat{a}, \widehat{b}$ belong to $\mathcal{G}SAPW$, and therefore the operator $\widehat{B}$ is Fredholm. Then its limit operators (5.3) are invertible on $L_p^n(\mathbb{R})$, which implies the invertibility on $L_p^n(\mathbb{R})$ of the operators $G_\pm P + Q$ with the matrix coefficients $G_\pm = b_\pm^{-1}a_\pm \in APW$. By Theorem 2.2, the matrix functions $G_\pm$ satisfy (iii).

The Fredholmness of $\widehat{B}$ implies the Fredholmness of the operator (5.16). Then from Lemma 4.5 and Theorem 3.1 we get (iv). □

If an operator B of the form (5.1) is Fredholm on the space $L_p^n(\mathbb{R})$, it follows from Theorem 5.2(i) that

$$\det B(t,\mu) \neq 0 \quad \text{for all} \quad (t,\mu) \in \mathbb{R} \times \mathcal{L}_p$$

and hence that the matrix functions

$$B_{11}(t,0) = b(t+0), \quad B_{22}(t,1) = b(t-0)$$

are invertible for every $t \in \mathbb{R}$ and that the function

$$\mathcal{B}(t,\mu) = \frac{\det B(t,\mu)}{\det B_{11}(t,0)\det B_{22}(t,1)}$$

does not have zeros and infinite values on $\mathbb{R} \times \mathcal{L}_p$.

For operators B of the form (5.1), let $\Lambda(B)$ be the set of those points $t \in \mathbb{R}$ at which the functions $\mathcal{B}(t,\cdot)$ are not constant. Then similarly to (3.3) we get the oriented continuous open curve

$$B_{\mathbb{R}}^{\#} := \mathcal{B}_0\Big(\mathbb{R} \setminus \Lambda(B)\Big) \cup \bigcup_{t \in \Lambda(B)} \Big\{\mathcal{B}(t,\mu) : \mu \in \mathcal{L}_p\Big\}$$

where $\mathcal{B}_0(t) = \mathcal{B}(t,0)$, $t \in \mathbb{R}$, and a contribution of the set $\{\infty\} \times \mathcal{L}_q$ is excluded. Thus we can choose a continuous branch of $\arg\mathcal{B}(t,\mu)$ when $\mathcal{B}(t,\mu)$ runs through the oriented continuous open curve $B^{\#}_{\mathbb{R}}$. It is clear that the function $\mathcal{B}(\cdot,\mu)$ belongs to $PAPW$ for every fixed $\mu \in \mathcal{L}_p$. Moreover, setting $G(t) = b^{-1}(t)a(t)$, we get

$$\mathcal{B}(t,0) = \frac{\det a(t-0)}{\det b(t-0)} = \det G(t-0), \quad \mathcal{B}(t,1) = \frac{\det a(t+0)}{\det b(t+0)} = \det G(t+0).$$

Thus, by the conditions (ii)–(iii) of Theorem 5.2, for every fixed $\mu \in \mathcal{L}_p$ the function $\mathcal{B}(\cdot,\mu)$ belongs to $\mathcal{G}PAPW_0$, that is, $\mathcal{B}(\cdot,\mu)$ is invertible in $PAPW$ and its almost periodic representatives $\det G_\pm = \det(b_\pm^{-1}a_\pm)$ at $\pm\infty$ admit canonical right APW factorizations. Consequently, the function

$$\psi_\mu(t) := \arg \mathcal{B}(t,\mu)$$

belongs to $PAPW$ for every fixed $\mu \in \mathcal{L}_p$, and thus the value

$$\mathrm{Ind}_{\mathbb{R}} B^{\#} := \frac{1}{2\pi} \lim_{T\to+\infty} \frac{1}{T} \int_0^T \Big(\psi_\mu(x) - \psi_\mu(-x)\Big)\,dx \tag{5.17}$$

exists and is finite. Clearly, (5.17) does not depend on the choice of $\mu \in \mathcal{L}_p$.

Let $\mathfrak{D}$ be the set of all operators B of the form (5.1) with $n \times n$ matrix coefficients $a, b, c \in PAP$ which are locally Fredholm at every point $t \in \mathbb{R}$ and which have invertible limit operators $B_\pm = a_\pm P + b_\pm Q$ with matrix coefficients $a_\pm, b_\pm \in AP$. Then $\det B(t,\mu) \neq 0$ for all $(t,\mu) \in \mathbb{R} \times \mathcal{L}_p$, and the matrix functions $a_\pm, b_\pm$ are in $\mathcal{G}AP$. Approximating the operators $B_\pm$ by invertible operators $B_\pm^{(m)} = a_\pm^{(m)}P + b_\pm^{(m)}Q$ with matrix coefficients $a_\pm^{(m)}, b_\pm^{(m)} \in \mathcal{G}APW$, we get in view of Theorem 2.2 that the matrix functions $G_\pm^{(m)} = (b_\pm^{(m)})^{-1}a_\pm^{(m)}$ admit canonical right APW factorizations, whence $v(\det G_\pm^{(m)}) = 0$. Since

$$G_\pm := b_\pm^{-1}a_\pm = \lim_{m\to\infty} G_\pm^{(m)} \in \mathcal{G}AP,$$

we conclude that $v(\det G_\pm) = 0$ as well.

Thus, for every $\mu \in \mathcal{L}_p$ the functions $\mathcal{B}(\cdot,\mu)$ are in $\mathcal{G}PAP_0$, that is, $\mathcal{B}(\cdot,\mu)$ is invertible in PAP and its almost periodic representatives $\det G_\pm$ have zero mean motions $v(\det G_\pm)$. Hence we can again define $\mathrm{Ind}_{\mathbb{R}} B^{\#}$ for the operator $B \in \mathfrak{D}$ by formula (5.17).

Lemma 5.3. *If $B \in \mathfrak{D}$, then $\mathrm{Ind}_{\mathbb{R}} B^{\#}$ defined by (5.17) possesses the following properties:*

(i) *$\mathrm{Ind}_{\mathbb{R}} B^{\#}$ is well defined and is a finite real number;*

(ii) *if $B = GP + Q$ and the matrix function G belongs to $\mathcal{G}C(\overline{\mathbb{R}})$, then the right-hand side of (5.17) is the Cauchy index of $\det G$;*

(iii) *if $B = GP + Q$, the matrix function G belongs to $\mathcal{G}SAP$ and the limit operators $B_\pm = G_\pm P + Q$ are invertible, then the right-hand side of (5.17) coincides with (4.5) for $\psi(x) = \arg\det G(x)$;*

(iv) *if $B_1, B_2 \in \mathfrak{D}$ and $B = B_1B_2 \in \mathfrak{D}$, then*

$$\mathrm{Ind}_{\mathbb{R}} B^{\#} = \mathrm{Ind}_{\mathbb{R}} B_1^{\#} + \mathrm{Ind}_{\mathbb{R}} B_2^{\#}; \tag{5.18}$$

(v) *the mapping $\mathfrak{D} \to \mathbb{R}$, $B \mapsto \mathrm{Ind}_{\mathbb{R}} B^{\#}$ is continuous.*

Proof. (i) Fix $B \in \mathfrak{D}$. As $\mathcal{B}(\cdot,\mu) \in \mathcal{G}PAP_0$, it is clear that $\operatorname{Ind}_{\mathbb{R}} B^{\#}$ defined by (5.17) is a finite real number. Indeed, according to (2.3), for every fixed $\mu \in \mathcal{L}_p$,

$$\mathcal{B}(t,\mu) = u(t)\det G_+(t) + (1-u(t))\det G_-(t) + G_0(t,\mu), \quad t \in \mathbb{R},$$

where $\det G_\pm \in \mathcal{G}AP$ and $G_0(\cdot,\mu) \in PC_0(\mathbb{R})$. Moreover, we can choose a continuous branch of $\arg\mathcal{B}(t,\mu)$ for $\mathcal{B}(t,\mu)$ running through the oriented continuous open curve $B^{\#}_{\mathbb{R}} \not\ni 0$. Then following the proof of Lemma 3.5 in [10], we can choose branches $\psi_\pm$ of $\arg\det G_\pm$ that are continuous on $\mathbb{R}$ and satisfy the condition

$$\lim_{x\to\pm\infty}\Big(\psi_\mu(x) - \psi_\pm(x)\Big) = 0. \tag{5.19}$$

This is possible because $\psi_\pm$ are determined independently of each other and of ψ_μ to within an integer multiple of 2π, and

$$\lim_{t\to\pm\infty}\frac{\mathcal{B}(t,\mu)}{\det G_\pm(t)} = 1 \quad \text{for every} \quad \mu \in \mathcal{L}_p.$$

From (5.19) it follows that

$$\lim_{T\to\pm\infty}\frac{1}{T}\int_0^T\Big(\psi_\mu(x) - \psi_\pm(x)\Big)dx = 0. \tag{5.20}$$

By the Bohr theorem (see, e.g., [15, p. 129]), the functions $\psi_\pm$ belong to the class AP together with $\det G_\pm$, because $v(\det G_\pm) = 0$. This implies that the limits

$$\lim_{T\to\pm\infty}\frac{1}{T}\int_0^T \psi_\pm(x)\,dx$$

exist and coincide with $\mathbf{M}(\psi_\pm)$, respectively. But then the limit on the right-hand side of (5.17) also exists and is equal to

$$\frac{1}{2\pi}\Big(\mathbf{M}(\psi_+) - \mathbf{M}(\psi_-)\Big).$$

Moreover, since the functions $\psi_\pm$ do not depend on $\mu \in \mathcal{L}_p$, we conclude that $\operatorname{Ind}_{\mathbb{R}} B^{\#}$ does also not depend on $\mu \in \mathcal{L}_p$.

(ii) If $G \in \mathcal{G}C(\overline{\mathbb{R}})$, then for every $\mu \in \mathcal{L}_p$, $\psi_\mu(t) = \arg\det G(t)$, $t \in \mathbb{R}$. Thus the functions ψ_μ do not depend on μ and belong to $C(\overline{\mathbb{R}})$. Consequently, $\psi_\pm$ are the constants $\arg\det G(\pm\infty)$ for our choice of the branch, and hence

$$\frac{1}{2\pi}\Big(\mathbf{M}(\psi_+) - \mathbf{M}(\psi_-)\Big) = \frac{1}{2\pi}\Big(\psi(+\infty) - \psi(-\infty)\Big) = \frac{1}{2\pi}\Big\{\arg\det G(x)\Big\}_{x\in\overline{\mathbb{R}}}.$$

(iii) If $G \in \mathcal{G}SAP$, then again $\mathcal{B}(t,\mu) = G(t)$, $(t,\mu) \in \mathbb{R}\times\mathcal{L}_p$, and the assertion follows from [10, Lemma 3.5].

(iv) If $B_1, B_2 \in \mathfrak{D}$ and also $B = B_1B_2 \in \mathfrak{D}$, then

$$B(t,\mu) = B_1(t,\mu)B_2(t,\mu) \neq 0 \quad \text{for} \quad (t,\mu) \in \mathbb{R}\times\mathcal{L}_p,$$

whence

$$\mathcal{B}(t,\mu) = \mathcal{B}_1(t,\mu)\mathcal{B}_2(t,\mu), \quad (t,\mu) \in \mathbb{R}\times\mathcal{L}_p. \tag{5.21}$$

Since $\mathcal{B}_1(\cdot,\mu)$, $\mathcal{B}_2(\cdot,\mu)$, $\mathcal{B}(\cdot,\mu) \in \mathcal{G}PAP_0$ for every $\mu \in \mathcal{L}_p$ (note that $B_1, B_2, B \in \mathfrak{D}$) and since we can choose branches of the arguments of the functions in (5.21) such that

$$\arg \mathcal{B}(t,\mu) = \arg \mathcal{B}_1(t,\mu) + \arg \mathcal{B}_2(t,\mu),$$

we immediately get (5.18) from (5.17).

(v) Let $B, B_m \in \mathfrak{D}$ and $\|B - B_m\| < \varepsilon$. Then there exists a number $C > 0$ such that

$$\sup_{(t,\mu)\in\mathbb{R}\times\mathcal{L}_p} \|B(t,\mu) - B_m(t,\mu)\|_{\mathbb{C}^{2n\times 2n}} < C\|B - B_m\| < C\varepsilon. \tag{5.22}$$

Furthermore, for the coefficients b and b_m of the operators B and B_m we derive from (5.22) that

$$\|b - b_m\|_{L_\infty(\mathbb{R})} \leq \sup_{(t,\mu)\in\mathbb{R}\times\mathcal{L}_p} \|B(t,\mu) - B_m(t,\mu)\|_{\mathbb{C}^{2n\times 2n}} < C\varepsilon. \tag{5.23}$$

Since $b, b_m \in \mathcal{G}L_\infty(\mathbb{R})$, inequalities (5.22) and (5.23) imply that there exists a constant $C_0 > 0$ such that

$$\sup_{(t,\mu)\in\mathbb{R}\times\mathcal{L}_p} |\mathcal{B}(t,\mu) - \mathcal{B}_m(t,\mu)| < C_0\varepsilon. \tag{5.24}$$

Taking into account that

$$\inf_{(t,\mu)\in\mathbb{R}\times\mathcal{L}_p} |\mathcal{B}(t,\mu)| > 0,$$

because $B \in \mathfrak{D}$, we infer from (5.24) that

$$\left|\sin\arg\frac{\mathcal{B}_m(t,\mu)}{\mathcal{B}(t,\mu)}\right| \leq \left|\frac{\mathcal{B}_m(t,\mu)}{\mathcal{B}(t,\mu)} - 1\right| < k(\varepsilon) := C_0\varepsilon\Big(\inf_{(t,\mu)\in\mathbb{R}\times\mathcal{L}_p} |\mathcal{B}(t,\mu)|\Big)^{-1}.$$

Now we can choose continuous branches of $\arg \mathcal{B}_m(t,\mu)$ and $\arg \mathcal{B}(t,\mu)$ such that

$$|\arg \mathcal{B}_m(t,\mu) - \arg \mathcal{B}(t,\mu)| < \arcsin k(\varepsilon). \tag{5.25}$$

Then from (5.25) and (5.17) it follows that

$$|\mathrm{Ind}_{\mathbb{R}} B_m^\# - \mathrm{Ind}_{\mathbb{R}} B^\#| < \arcsin k(\varepsilon),$$

whence $\mathrm{Ind}_{\mathbb{R}} B_m^\# \to \mathrm{Ind}_{\mathbb{R}} B^\#$ as $m \to 0$. □

Theorem 5.4. *If an operator B of the form* (5.1) *is Fredholm on the space $L_p^n(\mathbb{R})$, then*

$$\mathrm{ind}\, B = -\mathrm{Ind}_{\mathbb{R}} B^\# - \frac{1}{2\pi}\Big\{\arg\det \widetilde{Y}(\infty,\mu)\Big\}_{\mu\in\mathcal{L}_q}, \tag{5.26}$$

where $\mathrm{ind}_{\mathbb{R}} B^\#$ *is given by* (5.17) *and the operator* $\widetilde{Y} \in \mathfrak{S}$ *is given by* (5.7)–(5.8).

Proof. By (5.5), $B \simeq A\widehat{B}$, where $B, \widehat{B}, A \in \mathfrak{D}$. Hence

$$\mathcal{B}(t,\mu) = \mathcal{A}(t,\mu)\widehat{\mathcal{B}}(t,\mu) \quad \text{for} \quad (t,\mu) \in \mathbb{R}\times\mathcal{L}_p,$$

and we deduce from Lemma 5.3(iv) that

$$\mathrm{Ind}_{\mathbb{R}} B^\# = \mathrm{Ind}_{\mathbb{R}} A^\# + \mathrm{Ind}_{\mathbb{R}} \widehat{B}^\#. \tag{5.27}$$

Since $A \in \mathfrak{S}$ and since A is locally equivalent to I at ∞, we get

$$\operatorname{Ind}_{\mathbb{R}} A^{\#} = \operatorname{wind} A^{\#},$$

whence, by Theorem 3.2,

$$\operatorname{ind} A = -\operatorname{Ind}_{\mathbb{R}} A^{\#}. \tag{5.28}$$

On the other hand, by Theorem 5.2(ii), $\widehat{b} \in \mathcal{G}APW$, which implies that the operator

$$Y := GP + Q + FH_{\infty},$$

with $G = \widehat{b}^{-1}\widehat{a}$ and $F = \widehat{b}^{-1}\widehat{c}$ in SAP, is Fredholm together with $\widehat{B}$. Thus, by Theorem 4.6, the operators Y and $\widetilde{Y}$ are Fredholm, and

$$\operatorname{ind} \widehat{B} = \operatorname{ind} Y = -\operatorname{Ind} \det G - \frac{1}{2\pi}\left\{\arg \det \widetilde{Y}(\infty, \mu)\right\}_{\mu \in \mathcal{L}_q}. \tag{5.29}$$

By Lemma 5.3(iii),

$$\operatorname{Ind} \det G = \operatorname{Ind}_{\mathbb{R}} \widehat{B}^{\#}. \tag{5.30}$$

Since by (5.6),

$$\operatorname{ind} B = \operatorname{ind} A + \operatorname{ind} \widehat{B},$$

we derive from (5.28)–(5.30) that

$$\operatorname{ind} B = -\operatorname{Ind}_{\mathbb{R}} A^{\#} - \operatorname{Ind}_{\mathbb{R}} \widehat{B}^{\#} - \frac{1}{2\pi}\left\{\arg \det \widetilde{Y}(\infty, \mu)\right\}_{\mu \in \mathcal{L}_q},$$

which gives (5.26) due to (5.27). □

Note that Lemma 5.1 and Theorems 5.2, 5.4 remain valid if the coefficient c in (5.1) belongs to PAP only.

6. Fredholmness of singular integral operators with Carleman shift and piecewise almost periodic coefficients

Let the operator $T : L_p^n(\mathbb{R}) \to L_p^n(\mathbb{R})$ be defined by

$$T = (aI + bJ)P + (cI + dJ)Q \tag{6.1}$$

where the $n \times n$ matrix functions a, b, c, d are taken from $PAPW$ and J is the Carleman backward shift operator defined by (1.3). We begin this section with a reduction of the operator T to a singular integral operator on $L_p^{2n}(\mathbb{R})$. Let the operators $S_{\mathbb{R}_+}$ and $N_{\mathbb{R}_+}$ be given by (3.7), and put

$$P_{\mathbb{R}_+} = (I + S_{\mathbb{R}_+})/2, \quad Q_{\mathbb{R}_+} = I - P_{\mathbb{R}_+}.$$

Let $\Psi : L_p^n(\mathbb{R}) \to L_p^{2n}(\mathbb{R}_+)$ be the isomorphism defined by

$$(\Psi f)(t) = \begin{bmatrix} f(t) \\ f(-t) \end{bmatrix}, \quad t \in \mathbb{R}_+, \tag{6.2}$$

and consider the operator $A = aI \in \mathfrak{B}(L_p^n(\mathbb{R}))$, with an $n \times n$ matrix function $a \in PAPW$. We relate the operators A, J, $S \in \mathfrak{B}(L_p^n(\mathbb{R}))$ to operators in $\mathfrak{B}(L_p^{2n}(\mathbb{R}_+))$:

$$\Psi A \Psi^{-1} = \begin{bmatrix} a(t)I & 0 \\ 0 & a(-t)I \end{bmatrix}, \quad \Psi J \Psi^{-1} = \begin{bmatrix} 0 & I \\ I & 0 \end{bmatrix}, \tag{6.3}$$

$$\Psi S \Psi^{-1} = \begin{bmatrix} S_{\mathbb{R}_+} & -N_{\mathbb{R}_+} \\ N_{\mathbb{R}_+} & -S_{\mathbb{R}_+} \end{bmatrix}. \tag{6.4}$$

Then, using relations (6.3)–(6.4), for the operator T defined by (6.1) we obtain

$$T_1 := \Psi T \Psi^{-1} = \widehat{a}\, P_{\mathbb{R}_+} + \widehat{b}\, Q_{\mathbb{R}_+} + \widehat{c}\, N_{\mathbb{R}_+} \tag{6.5}$$

where

$$\widehat{a}(x) = \begin{bmatrix} a(x) & d(x) \\ b(-x) & c(-x) \end{bmatrix}, \quad \widehat{b}(x) = \begin{bmatrix} c(x) & b(x) \\ d(-x) & a(-x) \end{bmatrix}, \tag{6.6}$$

$$\widehat{c}(x) = \frac{1}{2}\begin{bmatrix} b(x) - d(x) & c(x) - a(x) \\ a(-x) - c(-x) & d(-x) - b(-x) \end{bmatrix} = \frac{1}{2}(\widehat{b}(x) - \widehat{a}(x)) \begin{bmatrix} 0_n & I_n \\ I_n & 0_n \end{bmatrix}. \tag{6.7}$$

Equality (6.5) immediately implies the following.

Lemma 6.1. *Let $T \in \mathfrak{B}(L_p^n(\mathbb{R}))$ be defined by (6.1) and let $T_1 \in \mathfrak{B}(L_p^{2n}(\mathbb{R}_+))$ be defined by (6.5)–(6.7). Then T is a Fredholm operator on $L_p^n(\mathbb{R})$ if and only if T_1 is a Fredholm operator on $L_p^{2n}(\mathbb{R}_+)$. In that case*

$$\operatorname{ind} T = \operatorname{ind} T_1. \tag{6.8}$$

With the operator $T_1 \in \mathfrak{B}(L_p^{2n}(\mathbb{R}_+))$ we associate the operator $T_2 \in \mathfrak{B}(L_p^{2n}(\mathbb{R}))$ given by

$$T_2 = \widetilde{a}P + \widetilde{b}Q + \widetilde{c}V \tag{6.9}$$

where P and Q are defined by (1.2), V is given by (3.5),

$$\widetilde{a} = \chi_+\widehat{a} + \chi_- I_{2n}, \quad \widetilde{b} = \chi_+\widehat{b} + \chi_- I_{2n}, \quad \widetilde{c} = \chi_+\widehat{c}, \tag{6.10}$$

and $\widehat{a}$, $\widehat{b}$, $\widehat{c}$ are given by (6.6)–(6.7).

Lemma 6.2. *Let $T \in \mathfrak{B}(L_p^n(\mathbb{R}))$ be defined by (6.1) and let $T_2 \in \mathfrak{B}(L_p^{2n}(\mathbb{R}))$ be defined by (6.9)–(6.10) and (3.5). Then T is a Fredholm operator on $L_p^n(\mathbb{R})$ if and only if T_2 is a Fredholm operator on $L_p^{2n}(\mathbb{R})$. In that case*

$$\operatorname{ind} T = \operatorname{ind} T_2. \tag{6.11}$$

Proof. From (6.10) it follows that

$$\chi_+\widetilde{a} = \chi_+\widehat{a}, \quad \chi_+\widetilde{b} = \chi_+\widehat{b}, \quad \chi_+\widetilde{c} = \chi_+\widehat{c}.$$

Thus, since by (3.5),

$$\chi_+ V = \chi_+ N \chi_+ I,$$

we get

$$\chi_+\widetilde{c}V = \chi_+\widehat{c}\chi_+ V = \chi_+\widehat{c}\chi_+ N \chi_+ I = \chi_+\widehat{c}\chi_+ N_{\mathbb{R}_+}\chi_+ I.$$

Therefore the operator (6.9) can be represented in the direct sum

$$L_p^{2n}(\mathbb{R}) = L_p^{2n}(\mathbb{R}_+) \oplus L_p^{2n}(\mathbb{R}_-)$$

by the operator matrix

$$\begin{bmatrix} \chi_+(\widehat{a}P_{\mathbb{R}_+} + \widehat{b}Q_{\mathbb{R}_+} + \widehat{c}N_{\mathbb{R}_+})\chi_+I & \chi_+T_2\chi_-I \\ 0 & \chi_-I \end{bmatrix},$$

which reveals that the operators $T_1 \in \mathfrak{B}(L_p^{2n}(\mathbb{R}_+))$ and $T_2 \in \mathfrak{B}(L_p^{2n}(\mathbb{R}))$ are Fredholm only simultaneously and that in the case of Fredholmness

$$\operatorname{ind} T_2 = \operatorname{ind} T_1. \tag{6.12}$$

Applying now Lemma 6.1, we conclude that the operators $T_2 \in \mathfrak{B}(L_p^{2n}(\mathbb{R}))$ and $T \in \mathfrak{B}(L_p^n(\mathbb{R}))$ are Fredholm only simultaneously. Moreover, if T is Fredholm, then (6.8) and (6.12) imply (6.11). □

By Lemma 3.3,

$$H = V \in \mathcal{H}_0 + \mathcal{H}_\infty,$$

which shows that the operator T_2 given by (6.9) is of the form (5.1) with coefficients $\widetilde{a}, \widetilde{b}, \widetilde{c} \in PAPW$. Thus we can apply Theorems 5.2 and 5.4 to the operator T_2.

Finally, from Lemma 6.2 and Theorems 5.2 and 5.4 we get the following.

Theorem 6.3. *The operator* (6.1) *is Fredholm on the space* $L_p^n(\mathbb{R})$ *if and only if*

(i) *for all* $x \in \mathbb{R}$,

$$\det \begin{bmatrix} a(x \pm 0) & d(x \pm 0) \\ b(-x \mp 0) & c(-x \mp 0) \end{bmatrix} \neq 0;$$

(ii) *the* $2n \times 2n$ *matrix functions*

$$\begin{bmatrix} a_+ & d_+ \\ b_- & c_- \end{bmatrix}, \quad \begin{bmatrix} c_+ & b_+ \\ d_- & a_- \end{bmatrix}$$

are invertible in APW;

(iii) *the matrix function*

$$G_+ = \begin{bmatrix} c_+ & b_+ \\ d_- & a_- \end{bmatrix}^{-1} \begin{bmatrix} a_+ & d_+ \\ b_- & c_- \end{bmatrix}$$

admits a canonical right APW factorization $G_+ = G_+^- G_+^+$;

(iv) *for all* $\mu \in \mathcal{L}_q$,

$$\det\left(\mathbf{d}(G_+)\begin{bmatrix} (1-\mu)I_n & i\nu(\mu)I_n \\ i\nu(\mu)I_n & (1-\mu)I_n \end{bmatrix} + \begin{bmatrix} \mu I_n & -i\nu(\mu)I_n \\ -i\nu(\mu)I_n & \mu I_n \end{bmatrix}\right) \neq 0;$$

(v) *for all* $\mu \in \mathcal{L}_p$,

$$\det\left(G(+0)\begin{bmatrix} \mu I_n & -i\nu(\mu)I_n \\ -i\nu(\mu)I_n & \mu I_n \end{bmatrix} + \begin{bmatrix} (1-\mu)I_n & i\nu(\mu)I_n \\ i\nu(\mu)I_n & (1-\mu)I_n \end{bmatrix}\right) \neq 0;$$

(vi) *for all* $(x,\mu)\in(0,+\infty)\times\mathcal{L}_p$,

$$\det\begin{bmatrix} G(x+0)\mu+I_{2n}(1-\mu) & (G(x+0)-I_{2n})\nu(\mu) \\ (G(x-0)-I_{2n})\nu(\mu) & I_{2n}\mu+G(x-0)(1-\mu)\end{bmatrix}\neq 0,$$

where

$$G(x)=\widetilde{b}^{-1}(x)\widetilde{a}(x)=\begin{bmatrix} c(x) & b(x) \\ d(-x) & a(-x)\end{bmatrix}^{-1}\begin{bmatrix} a(x) & d(x) \\ b(-x) & c(-x)\end{bmatrix},\qquad x\in\mathbb{R}_+.$$

Proof. By Lemma 6.2, the operator T is Fredholm on $L_p^n(\mathbb{R})$ if and only if the operator T_2 defined by (6.9)–(6.10) is Fredholm on $L_p^{2n}(\mathbb{R})$. According to Lemma 3.3, the operator $H=V$ belongs to $\mathcal{H}_0+\mathcal{H}_\infty$ and has the symbol $H(t,\mu)=V(t,\mu)$ defined by (3.6). Thus from Theorem 5.2 we get Theorem 6.3 as a particular case.

Let the operator T_2 be Fredholm on $L_p^{2n}(\mathbb{R})$. By (6.10) and (6.6)–(6.7),

$$\widetilde{a}_+=\begin{bmatrix} a_+ & d_+ \\ b_- & c_-\end{bmatrix},\ \widetilde{b}_+=\begin{bmatrix} c_+ & b_+ \\ d_- & a_-\end{bmatrix},\ \widetilde{c}_+=\frac{1}{2}(\widetilde{b}_+-\widetilde{a}_+)\begin{bmatrix} 0_n & I_n \\ I_n & 0_n\end{bmatrix},\tag{6.13}$$

$$\widetilde{a}_-=\widetilde{b}_-=I_{2n},\quad \widetilde{c}_-=0_{2n}.\tag{6.14}$$

So from conditions (i)–(iii) of Theorem 5.2 we directly get conditions (i)–(iii) and (vi) of Theorem 6.3, where the matrix function $G=\widetilde{b}^{-1}\widetilde{a}\in PAPW$ has almost periodic representatives $G_\pm=\widetilde{b}_\pm^{-1}\widetilde{a}_\pm\in APW$ of the form

$$G_+=\begin{bmatrix} c_+ & b_+ \\ d_- & a_-\end{bmatrix}^{-1}\begin{bmatrix} a_+ & d_+ \\ b_- & c_-\end{bmatrix},\qquad G_-=I_{2n}.\tag{6.15}$$

Furthermore, by (3.2) and (3.6), for $\mu\in\mathcal{L}_p$ we obtain

$$T_2(0,\mu)=\begin{bmatrix} \widehat{a}(+0)\mu+\widehat{b}(+0)(1-\mu)+2\widehat{c}(+0)i\nu(\mu) & 2(\widehat{a}(+0)-\widehat{b}(+0))\nu(\mu) \\ 0_{2n} & I_{2n}\end{bmatrix}.$$

Hence, by condition (i) of Theorem 5.2,

$$\det T_2(0,\mu)=\det\Big(\widehat{a}(+0)\mu+\widehat{b}(+0)(1-\mu)+2\widehat{c}(+0)i\nu(\mu)\Big)\neq 0 \text{ for } \mu\in\mathcal{L}_p.\tag{6.16}$$

Taking into account that $G(+0)=[\widehat{b}(+0)]^{-1}\widehat{a}(+0)$ and that, due to (6.7),

$$\widehat{b}(+0)^{-1}\widehat{c}(+0)=\frac{1}{2}\Big(I_{2n}-G(+0)\Big)\begin{bmatrix} 0_n & I_n \\ I_n & 0_n\end{bmatrix}$$

we derive condition (v) of Theorem 6.3 from (6.16).

According to Theorem 5.2(iv) we obtain the operator

$$\widetilde{Y}=\widetilde{G}P+Q+\widetilde{F}H_\infty\ \in\mathfrak{S},$$

where

$$H_\infty=\chi_+V\in\mathcal{H}_\infty,\quad \widetilde{G}(+\infty)=\mathbf{d}(G_+),\quad \widetilde{F}(+\infty)=\mathbf{M}(G_+^-)\mathbf{M}((G_+^-)^{-1}\widetilde{b}_+^{-1}\widetilde{c}_+),$$

and by (6.14), $\widetilde{G}(-\infty) = I_{2n}$, $\widetilde{F}(-\infty) = 0_{2n}$. Now from (3.2) and (3.6) it follows that for $\mu \in \mathcal{L}_q$,

$$\widetilde{Y}(\infty,\mu) \tag{6.17}$$
$$= \begin{bmatrix} I_{2n} & 0_{2n} \\ (\widetilde{G}(+\infty) - I_{2n})\nu(\mu) & \widetilde{G}(+\infty)(1-\mu) + I_{2n}\mu - 2i\widetilde{F}(+\infty)\nu(\mu) \end{bmatrix}.$$

Hence, by Theorem 5.2(iv),

$$\det \widetilde{Y}(\infty,\mu) = \det\Big(\widetilde{G}(+\infty)(1-\mu) + I_{2n}\mu - 2i\widetilde{F}(+\infty)\nu(\mu)\Big) \neq 0 \text{ for } \mu \in \mathcal{L}_q. \tag{6.18}$$

From (6.13) and (6.15) it results that

$$\widetilde{b}_+^{-1}\widetilde{c}_+ = \frac{1}{2}(I_{2n} - G_+)\begin{bmatrix} 0_n & I_n \\ I_n & 0_n \end{bmatrix}.$$

Thus

$$\begin{aligned} \widetilde{F}(+\infty) &= \mathbf{M}(G_+^-)\mathbf{M}\left((G_+^-)^{-1}\frac{1}{2}(I_{2n} - G_+)\begin{bmatrix} 0_n & I_n \\ I_n & 0_n \end{bmatrix}\right) \\ &= \frac{1}{2}\Big(I_{2n} - \mathbf{d}(G_+)\Big)\begin{bmatrix} 0_n & I_n \\ I_n & 0_n \end{bmatrix}. \end{aligned} \tag{6.19}$$

Since $\widetilde{G}(+\infty) = \mathbf{d}(G_+)$, from (6.18) and (6.19) we immediately obtain condition (iv) of Theorem 6.3.

Conversely, if all the conditions (i)–(vi) of Theorem 6.3 hold, we conclude from Theorem 5.2 that the operator T_2 is Fredholm on the space $L_p^{2n}(\mathbb{R})$, which implies the Fredholmness of T on the space $L_p^n(\mathbb{R})$. □

Let an operator T of the form (6.1) be Fredholm on the space $L_p^n(\mathbb{R})$. Then all the conditions of Theorem 6.3 are fulfilled, and from (6.11) and Theorem 5.4 we get

$$\operatorname{ind} T = \operatorname{ind} T_2 = -\operatorname{Ind}_{\mathbb{R}} T_2^{\#} - \frac{1}{2\pi}\Big\{\arg\det\widetilde{Y}(\infty,\mu)\Big\}_{\mu\in\mathcal{L}_q}.$$

Since $T_2(t,\mu) = I_{4n}$ for $(t,\mu) \in (-\infty,0)\times\mathcal{L}_p$, and hence $\mathcal{T}_2(t,\mu) = 1$ for those points, we infer from (5.17) that

$$\operatorname{Ind}_{\mathbb{R}} T_2^{\#} = \frac{1}{2\pi}\lim_{T\to+\infty}\frac{1}{T}\int_0^T \psi_\mu(x)\,dx, \tag{6.20}$$

where $\psi_\mu(t) = \arg\mathcal{T}_2(t,\mu)$ and for $t < 0$, $\psi_\mu(t) = 1$. On the other hand, as $\widetilde{G}(+\infty) = \mathbf{d}(G_+)$, from (6.18) and (6.19) it follows that

$$\det\widetilde{Y}(\infty,\mu) \tag{6.21}$$
$$= \det\left(\mathbf{d}(G_+)\begin{bmatrix} (1-\mu)I_n & i\nu(\mu)I_n \\ i\nu(\mu)I_n & (1-\mu)I_n \end{bmatrix} + \begin{bmatrix} \mu I_n & -i\nu(\mu)I_n \\ -i\nu(\mu)I_n & \mu I_n \end{bmatrix}\right).$$

Thus we have proved the following.

Theorem 6.4. *If an operator T of the form* (6.1) *is Fredholm on the space $L_p^n(\mathbb{R})$, then its index is calculated by*

$$\operatorname{ind} T = -\operatorname{Ind}_{\mathbb{R}} T_2^{\#} - \frac{1}{2\pi}\left\{\arg \det \widetilde{Y}(\infty,\mu)\right\}_{\mu\in\mathcal{L}_q},$$

where $\operatorname{Ind}_{\mathbb{R}} T_2^{\#}$ *and* $\det \widetilde{Y}(\infty,\mu)$ *are given by* (6.20) *and* (6.21), *respectively.*

Acknowledgements

The authors were partially supported by CONACYT grant, Cátedra Patrimonial, Nivel II, No. 990017–R2000 and by CONACYT project 32726–E, México, and by FCT (Portugal) under project POCTI/34222/MAT/2000-FEDER.

References

[1] Böttcher, A., Karlovich, Yu.I., *Carleson Curves, Muckenhoupt Weights, and Toeplitz Operators*, Birkhäuser Verlag, Basel, Boston, Berlin 1997.

[2] Böttcher, A., Karlovich, Yu.I., Spitkovsky, I.M., Toeplitz operators with semi-almost periodic matrix symbols on Hardy spaces, *Acta Applicandae Mathematica* **65** (2001), 115–136.

[3] Böttcher, A., Silbermann, B., *Analysis of Toeplitz Operators*, Springer-Verlag, Berlin, Heidelberg, New York 1990.

[4] Garnett, J.B., *Bounded Analytic Functions*, Academic Press, New York 1981.

[5] Gohberg, I., Feldman, I.A., *Convolution Equations and Projection Methods for Their Solution*, Math. Monogr. 41 AMS, Providence 1974.

[6] Gohberg, I., Krupnik, N., Singular integral operators with piecewise continuous coefficients and their symbols, *Math. USSR Izv.* **5** (1971), 955–979.

[7] ———, On algebras of singular integral operators with a shift, *Mat. Issled. (Kishinev)* **8** (1973), no. 2, 170–175 [Russian].

[8] Karapetiants, N.K., Samko, S.G., *Equations with Involutive Operators*, Birkhäuser, Boston 2001.

[9] Karlovich, Yu.I., On the Haseman problem, *Demonstratio Math.* **26** (1993), 581–595.

[10] Karlovich, Yu.I., Spitkovsky, I.M., Factorization of almost periodic matrix-valued functions and the Noether theory for a certain classes of equations of convolution type, *Math. USSR Izv.* **34** (1990), 281–316.

[11] ———, (Semi)-Fredholmness of convolution operators on the spaces of Bessel potentials, *Operator Theory: Advances and Applications* **71** (1994), 122–152.

[12] Kravchenko, V.G., Litvinchuk, G.S., *Introduction to the Theory of Singular Integral Operators with Shift*, Kluwer Academic Publishers, Dordrecht, Boston, London 1994.

[13] Krupnik, N.Ya., Nyaga, V.I., Singular integral operators with a shift along a piecewise Lyapunov contour, *Soviet Math. (Iz. VUZ)* **19** (1975), no. 6, 49–59.

[14] Lange, B.V., Rabinovich, V.S., Pseudodifferential operators on R^n and limit operators, *Math. USSR-Sb.* **57** (1987), 183–194.

[15] Levitan, B.M., *Almost Periodic Functions*, GITTL, Moscow 1956 [Russian].

[16] Litvinchuk, G.S., *Boundary Value Problems and Singular Integral Equations with Shift*, Nauka, Moscow 1977 [Russian].

[17] ———, *Solvability Theory of Boundary Value Problems and Singular Integral Equations with Shift*, Kluwer Academic Publishers, Dordrecht, Boston, London 2000.

[18] Roch, S., Silbermann, B., *Algebras of Convolution Operators and Their Image in the Calkin Algebra*, Report R-Math-05/90, Karl-Weierstrass-Inst. f. Math., Berlin 1990.

[19] Sarason, D., Approximation of piecewise continuous functions by quotients of bounded analytic functions, *Canadian J. Math.* **24** (1972), 642–657.

[20] ———, Toeplitz operators with semi almost-periodic symbols, *Duke Math. J.* **44** (1977), 357–364.

[21] Simonenko, I.B., Chin Ngok Min, *The Local Method in the Theory of One-Dimensional Singular Integral Equations with Piecewise Continuous Coefficients. Noethericity*, Rostov University Press, Rostov-on-Don 1986 [Russian].

[22] Stein, E.M., *Singular Integrals and Differentiability Properties of Functions*, Princeton University Press, Princeton 1970.

M.A. Bastos, A. Bravo
Departamento de Matemática
Instituto Superior Técnico
Av. Rovisco Pais
1049 – 001 Lisboa, Portugal
e-mail: abastos@math.ist.utl.pt
e-mail: abravo@math.ist.utl.pt

Yu. I. Karlovich
Departamento de Matemáticas
CINVESTAV del I.P.N.
Apartado Postal 14–740
07000 México, D.F., México
e-mail: karlovich@servm.uaem.mx

2000 *Mathematics Subject Classification.*
Primary 45F15, 47A53; Secondary 47G10, 47A68.

Received: 8 June 2001.

Operator Theory:
Advances and Applications, Vol. 142, 59–77

On the Essential Spectrum of Toeplitz Operators with Semi-Almost Periodic Symbols

Albrecht Böttcher, Sergei Grudsky and Ilya Spitkovsky

Abstract. The essential spectra of Toeplitz operators with piecewise continuous and almost periodic symbols admit pretty nice geometric descriptions. In contrast to this, little is known about the essential spectra of Toeplitz operators with semi-almost periodic symbols, that is, with symbols belonging to the algebra generated by the continuous functions on the real line with a jump at infinity and by the continuous almost periodic functions. A classical result by Sarason enables us to decide whether the Toeplitz operator $T(a-\lambda)$ is Fredholm for a given point λ in the plane, but the problem of characterizing the set of *all* λ for which $T(a-\lambda)$ is not Fredholm is nevertheless intricate. This question is studied in the present paper.

1. Introduction and main results

We denote by $H^2(\mathbf{R})$ the usual Hardy space of all functions on the real line $\mathbf{R}$ which are of the form

$$f(x) = \int_0^\infty g(t)e^{itx}dt \quad (x \in \mathbf{R})$$

with $g \in L^2(0,\infty)$. For $a \in L^\infty(\mathbf{R})$, the Toeplitz operator $T(a)$ is defined on $H^2(\mathbf{R})$ by $f \mapsto P(af)$, where P is the orthogonal projection of $L^2(\mathbf{R})$ onto $H^2(\mathbf{R})$. The function a is in this context usually referred to as the symbol of the operator $T(a)$. We remark that $T(a)$ is unitarily equivalent to the operator generated on $l^2(\mathbf{Z}_+)$ by the Toeplitz matrix $(a_{j-k})_{i,j=0}^\infty$ in which $\{a_n\}_{n=-\infty}^\infty$ is the sequence of the Fourier coefficients of the function

$$\mathbf{T} \to \mathbf{C}, \quad \frac{x-i}{x+i} \mapsto a(x).$$

We let $\mathrm{sp}_{\mathrm{ess}}T(a)$ denote the essential spectrum of $T(a)$:

$$\mathrm{sp}_{\mathrm{ess}}T(a) = \{\lambda \in \mathbf{C} : T(a) - \lambda I \text{ is not Fredholm}\}.$$

Recall that an operator A is said to be Fredholm if it is invertible modulo compact operators. In that case the kernel and cokernel dimensions of A are finite, and their difference is called the index of A and denoted by $\operatorname{Ind} A$. Also notice that $T(a) - \lambda I = T(a-\lambda)$.

The essential spectra of Toeplitz operators are known for many symbol classes; see, for example, the books [2], [4], [8]. In particular, one has nice geometric descriptions of the essential spectra for piecewise continuous or almost periodic symbols. In [1], when answering a question raised in [5] and [7], we observed that in the case of semi-almost periodic symbols there are no satisfactory results on the geometric nature of the essential spectra of the corresponding Toeplitz operators in the literature, although Sarason himself already understood the local nature of these spectra [9]. These spectra can have unexpectedly beautiful shapes and they are the subject of this paper.

Let AP be the closure in $L^\infty(\mathbf{R})$ of the set of all almost periodic polynomials, that is, let AP be the smallest closed subalgebra of $L^\infty(\mathbf{R})$ which contains the set $\cup_{\lambda\in\mathbf{R}}\{e_\lambda\}$ with $e_\lambda(x) := e^{i\lambda x}$. Let further $C(\overline{\mathbf{R}})$ denote the collection of all functions $\varphi \in C(\mathbf{R})$ with finite limits $\varphi(\pm\infty)$ at $\pm\infty$. The smallest closed subalgebra of $L^\infty(\mathbf{R})$ which contains $AP \cup C(\overline{\mathbf{R}})$ is referred to as the algebra of all semi-almost periodic functions and is denoted by SAP. This algebra was introduced by Sarason [9].

Every function $a \in SAP$ has two so-called almost periodic representatives $a_l \in AP$ and $a_r \in AP$ at $-\infty$ and $+\infty$, respectively. One can show that

$$\mathrm{sp}_{AP}\, T(a) := a(\mathbf{R}) \cup \mathrm{sp}_{\mathrm{ess}} T(a_l) \cup \mathrm{sp}_{\mathrm{ess}} T(a_r)$$

is always contained in $\mathrm{sp}_{\mathrm{ess}} T(a)$. The interesting part of $\mathrm{sp}_{\mathrm{ess}} T(a)$ is the set

$$\mathrm{sp}_{PC}\, T(a) := \mathrm{sp}_{\mathrm{ess}} T(a) \setminus \mathrm{sp}_{AP}\, T(a).$$

We call a set $X \subset \mathbf{C}$ an *analytic star* if there are a point $\lambda_0 \in \mathbf{C}$, an open neighborhood $U \subset \mathbf{C}$ of λ_0, and a nonconstant analytic function $f : U \to \mathbf{C}$ such that $f(\lambda_0) = 0$ and

$$X = \{\lambda \in U : f(\lambda) \in \mathbf{R}\}.$$

If $f'(\lambda_0) \neq 0$, then $X \cap V$ is an analytic arc through λ_0 whenever $V \subset \mathbf{C}$ is a sufficiently small open neighborhood of λ_0. If

$$f'(\lambda_0) = \cdots = f^{(k-1)}(\lambda_0) = 0, \quad f^{(k)}(\lambda_0) \neq 0,$$

then, for every sufficiently small open neighborhood $V \subset \mathbf{C}$ of the point λ_0, the set $(X \cap V) \setminus \{\lambda_0\}$ is composed of $2k$ pairwise disjoint analytic arcs whose closures all contain the point λ_0. We call λ_0 the *center* of the star and refer to the number of arcs in $(X \cap V) \setminus \{\lambda_0\}$ as the *valency* of the star.

The following result exhibits some characteristic features of the subset $\mathrm{sp}_{PC}\, T(a)$ of the essential spectrum of $T(a)$.

Theorem A (Sarason). *Let a be a function in SAP and let Ω be a connected component of the (open) set $\mathbf{C} \setminus \mathrm{sp}_{AP} T(a)$. We then have exactly one of the following three possibilities:*

(i) $\Omega \cap \mathrm{sp}_{PC} T(a) = \emptyset$;

(ii) $\Omega \subset \mathrm{sp}_{PC} T(a)$;

(iii) $\Omega \cap \mathrm{sp}_{PC} T(a) \neq \emptyset$ *and each point* $\lambda_0 \in \Omega \cap \mathrm{sp}_{PC} T(a)$ *has an open neighborhood* $U \subset \Omega$ *such that* $U \cap \mathrm{sp}_{PC} T(a)$ *is an analytic star.*

We remark that $\mathrm{sp}_{PC} T(a)$ may be empty; this happens, for instance, if a is in $C(\dot{\mathbf{R}})$, that is, if a is continuous on $\mathbf{R}$ and the limits $a(\pm\infty)$ exist and coincide. Here is an immediate consequence of Theorem A.

Corollary B (Sarason). *Let* $a \in SAP$, *suppose* $\mathrm{sp}_{PC} T(a) \neq \emptyset$, *and let* λ_0 *be a point in* $\mathrm{sp}_{PC} T(a)$. *Then there exists an open neighborhood* $U \subset \mathbf{C}$ *of* λ_0 *such that* $\mathrm{sp}_{PC} T(a) \cap U$ *is either all of* U *or an analytic star.*

The next two theorems show that Corollary B is sharp and that this corollary provides us with a complete description of the local geometry of the essential spectra of Toeplitz operators with SAP symbols. Recall that a *closed Jordan curve* is a subset of the plane that is homeomorphic to the unit circle.

Theorem 1.1. *Given any bounded open subset* G *of the complex plane whose boundary* ∂G *is a closed Jordan curve, there exist* $a \in SAP$ *such that*

$$\mathrm{sp}_{PC} T(a) = G.$$

Theorem 1.2. *Given any analytic star* X, *there exist* $a \in SAP$ *such that*

$$\mathrm{sp}_{PC} T(a) \cap U = X \cap U$$

for some open set $U \subset \mathbf{C}$ *containing the center of* X.

The following results give an idea of the global geometry of the essential spectra of Toeplitz operators with symbols in SAP.

Theorem 1.3. *Given any distinct points* $\lambda_1, \dots, \lambda_n$ *in the complex plane and any even natural numbers* $2k_1, \dots, 2k_n$, *there exist* $a \in SAP$ *such that* $\lambda_1, \dots, \lambda_n$ *are all located in a single component* Ω *of* $\mathbf{C} \setminus \mathrm{sp}_{AP} T(a)$ *and such that, for each* j, *the set* $\mathrm{sp}_{PC} T(a)$ *is an analytic star with the center* λ_j *and the valency* $2k_j$ *in some neighborhood of* λ_j.

We put $\mathbf{T} = \{\lambda \in \mathbf{C} : |\lambda| = 1\}$ and $\mathbf{D} = \{\lambda \in \mathbf{C} : |\lambda| < 1\}$.

Theorem 1.4. *Let* $G \subset \mathbf{C}$ *be a bounded open set whose boundary is a closed Jordan curve and let* f *be a conformal map of* $\mathbf{D}$ *onto* G. *Then there exist* $a \in SAP$ *such that*

$$\mathrm{sp}_{PC} T(a) = f\big((0,1)\big).$$

A subset $\mathcal{S}$ of the plane is called a *simple curve* if there exists an injective and continuous map $\varphi : [0,1] \to \mathbf{C}$ such that $\mathcal{S} = \varphi((0,1))$. The point $\varphi(0)$ and $\varphi(1)$ are referred to as the endpoints of $\mathcal{S}$. The next result is a consequence of Theorem 1.4, and it shows that $\mathrm{sp}_{PC} T(a)$ may mimic any prescribed simple curve in the plane.

Corollary 1.5. *Given any simple curve $\mathcal{S} \subset \mathbf{C}$ with the endpoints A and B and any $\varepsilon > 0$, there exist $a \in SAP$ such that $\mathrm{sp}_{PC}\, T(a)$ is a simple curve that lies entirely in the ε-neighborhood of $\mathcal{S}$ and that has points in the ε-neighborhood of A and in the ε-neighborhood of B.*

For $a \in L^\infty(\mathbf{R})$, we denote by $\mathcal{R}(a)$ the essential range of a. In other terms, $\mathcal{R}(a)$ is the spectrum of a as an element of $L^\infty(\mathbf{R})$.

Theorem 1.6. *Let $a \in SAP$ have the almost periodic representatives a_l and a_r. Then every straight line in the plane which separates $\mathcal{R}(a_l)$ and $\mathcal{R}(a_r)$ contains at most one point of $\mathrm{sp}_{PC}\, T(a)$.*

This theorem can be restated in slightly sharper form. Let $\mathrm{conv}\, E$ denote the convex hull of a set $E \in \mathbf{C}$. We call a simple curve $\mathcal{S} \subset \mathbf{C}$ a *simple analytic curve* if every point $\lambda_0 \in \mathcal{S}$ has an open neighborhood $U \subset \mathbf{C}$ such that $\mathcal{S} \cap U$ is an analytic arc (= analytic star of the valency 2).

Theorem 1.6′. *Let $a \in SAP$, let a_l and a_r be the almost periodic representatives of a, and suppose*

$$\mathrm{conv}\, \mathcal{R}(a_l) \cap \mathrm{conv}\, \mathcal{R}(a_r) = \emptyset.$$

Put

$$\Sigma = \mathbf{C} \setminus \Big(\mathrm{conv}\, \mathcal{R}(a_l) \cup \mathrm{conv}\, \mathcal{R}(a_r)\Big).$$

Then the intersection of $\mathrm{sp}_{\mathrm{ess}} T(a)$ and Σ is the union of $a(\mathbf{R}) \cap \Sigma$ and a simple analytic curve $\mathcal{S}$ with one endpoint on $\mathrm{conv}\, \mathcal{R}(a_l)$ and the other endpoint on $\mathrm{conv}\, \mathcal{R}(a_r)$. Moreover, every straight line which separates $\mathcal{R}(a_l)$ and $\mathcal{R}(a_r)$ contains exactly one point of $\mathcal{S}$.

We remark that Theorems 1.6 and 1.6′ are in particular applicable if $\mathcal{R}(a_l)$ and $\mathcal{R}(a_r)$ are circles whose convex hulls have no common points. Here is another nice special case.

Corollary 1.7. *Let $a \in SAP$ and suppose $\mathcal{R}(a_l)$ and $\mathcal{R}(a_r)$ are line segments that form two opposite sides AB and CD of a convex quadrangle $ABCD$. Then $\mathrm{sp}_{\mathrm{ess}} T(a)$ is the union of $a(\mathbf{R})$, $\mathcal{R}(a_l)$, $\mathcal{R}(a_r)$, and a simple analytic curve $\mathcal{S}$ with one endpoint on $\mathcal{R}(a_l)$ and the other endpoint on $\mathcal{R}(a_r)$. The curve $\mathcal{S}$ is contained in the quadrangle $ABCD$, and each line segment having one endpoint on BC and one endpoint of DA intersects $\mathcal{S}$ exactly once.*

In Section 2 we cite Sarason's Fredholm criterion for Toeplitz operators with symbols in SAP, and in Section 3 we give a proof of Theorem A. These two sections do not contain new results and are merely included for the reader's convenience. The remaining sections are devoted to the proofs of the other results listed above.

2. Sarason's criterion

The mean value $M(a)$ of a function $a \in AP$ is defined by

$$M(a) = \lim_{T\to\infty} \frac{1}{2T} \int_{-T}^{T} a(x)dx.$$

Let GAP denote the set of all functions in AP which are invertible in $L^\infty(\mathbf{R})$ (and thus also in AP). A famous theorem by H. Bohr says that every function $a \in GAP$ is of the form

$$a(x) = e^{i\kappa(a)x+b(x)}$$

where $\kappa(a) \in \mathbf{R}$ and $b \in AP$. The number $\kappa(a)$ is referred to as the mean motion of a and can be obtained from the formula

$$\kappa(a) = \lim_{T\to\infty} \frac{1}{2T}(\arg a(T) - \arg a(-T)),$$

where $\arg a \in C(\mathbf{R})$ is any continuous argument of a.

Theorem C (Gohberg/Feldman/Coburn/Douglas). *If $a \in AP$ then the essential spectrum $\mathrm{sp}_{\mathrm{ess}}T(a)$ equals*

$$\{\lambda \in \mathbf{C} : a - \lambda \notin GAP\} \cup \{\lambda \in \mathbf{C} : a - \lambda \in GAP \text{ and } \kappa(a-\lambda) \neq 0\}.$$

This theorem was established in [3] and [6]. A proof is also in [2, Theorem 9.20].

Sarason showed that every function $a \in SAP$ can be written in the form

$$a(x) = (1 - u(x))a_l(x) + u(x)a_r(x) + a_0(x)$$

where $u \in C(\overline{\mathbf{R}})$ is any fixed function such that $u(-\infty) = 0$ and $u(+\infty) = 1$, a_l and a_r are functions in AP, and $a_0 \in C(\dot{\mathbf{R}})$ is a function vanishing at infinity. One has

$$\|a_l\|_\infty \leq \|a\|_\infty, \quad \|a_r\|_\infty \leq \|a\|_\infty,$$

which implies that a_l and a_r are uniquely determined by a. The functions a_l and a_r are called the almost periodic representatives of a at $-\infty$ and $+\infty$, respectively.

If $a \in GAP$ and $\kappa(a) = 0$, then $a = e^b$ with some $b \in AP$. This function b is unique up to an additive constant in $2\pi i\mathbf{Z}$ and is usually denoted by $\log a$. Clearly,

$$d(a) := e^{M(\log a)}$$

is independent of the particular choice of $\log a \in AP$. The number $d(a)$ is called the geometric mean of a.

Theorem D (Sarason). *Let $a \in SAP$. The operator $T(a)$ is Fredholm if and only if*

$$0 \notin a(\mathbf{R}), \quad a_l \in GAP, \quad a_r \in GAP, \quad \kappa(a_l) = 0, \quad \kappa(a_r) = 0, \quad 0 \notin [d(a_l), d(a_r)];$$

the last condition means that the origin does not belong to the line segment between $d(a_l)$ and $d(a_r)$.

A proof of Theorem D is in [9].

We finally remark that for every $a \in L^\infty(\mathbf{R})$ the essential spectrum $\mathrm{sp}_{\mathrm{ess}}T(a)$ is a connected subset of the convex hull of $\mathcal{R}(a)$ (see, e.g., [4]).

3. The local shape of the essential spectrum

Let $a \in SAP$. To obtain $\mathrm{sp_{ess}}T(a)$, we must apply Theorem D to $T(a-\lambda)$. We divide $\mathrm{sp_{ess}}T(a)$ into two disjoint parts,

$$\mathrm{sp_{ess}}T(a) = \mathrm{sp}_{AP}\,T(a) \cup \mathrm{sp}_{PC}\,T(a),$$

where $\mathrm{sp}_{AP}\,T(a)$ and $\mathrm{sp}_{PC}\,T(a)$ are as in Section 1. By Theorem C,

$$\begin{aligned} \mathrm{sp}_{AP}\,T(a) &= a(\mathbf{R}) \cup \{\lambda \in \mathbf{C} \setminus a(\mathbf{R}) : a_l - \lambda \notin GAP \text{ or } \kappa(a_l - \lambda) \neq 0\} \\ &\quad \cup \{\lambda \in \mathbf{C} \setminus a(\mathbf{R}) : a_r - \lambda \notin GAP \text{ or } \kappa(a_r - \lambda) \neq 0\}. \end{aligned}$$

From Theorem D we therefore deduce that

$$\mathrm{sp}_{PC}\,T(a) = \{\lambda \in \mathbf{C} \setminus \mathrm{sp}_{AP}\,T(a) : 0 \in (d(a_l - \lambda), d(a_r - \lambda))\}.$$

Obviously, the last equality can also be written in the form

$$\mathrm{sp}_{PC}\,T(a) = \left\{\lambda \in \mathbf{C} \setminus \mathrm{sp}_{AP}\,T(a) : \frac{d(a_l - \lambda)}{d(a_r - \lambda)} \in (-\infty, 0)\right\}. \tag{3.1}$$

Proof of Theorem A. Fix a connected component Ω of $\mathbf{C} \setminus \mathrm{sp}_{AP}\,T(a)$. Let $\lambda_0 \in \Omega$. If $\lambda \in \Omega$ and $|\lambda - \lambda_0|$ is sufficiently small, then

$$\begin{aligned} d(a_r - \lambda) &= d(a_r - \lambda_0 - (\lambda - \lambda_0)) \\ &= d(a_r - \lambda_0) \exp\left(M(\log(1 - (\lambda - \lambda_0)(a_r - \lambda_0)^{-1}))\right) \\ &= d(a_r - \lambda_0) \exp\left(-\sum_{k=1}^{\infty} \frac{(\lambda - \lambda_0)^k}{k} M\left((a_r - \lambda_0)^{-k}\right)\right) \end{aligned} \tag{3.2}$$

and, analogously,

$$d(a_l - \lambda) = d(a_l - \lambda_0) \exp\left(-\sum_{k=1}^{\infty} \frac{(\lambda - \lambda_0)^k}{k} M\left((a_l - \lambda_0)^{-k}\right)\right) \tag{3.3}$$

This shows that

$$f(\lambda) := \frac{d(a_r - \lambda)}{d(a_l - \lambda)}$$

is analytic in Ω and that

$$f(\lambda) = \frac{d(a_r - \lambda_0)}{d(a_l - \lambda_0)} g(\lambda) \tag{3.4}$$

with

$$g(\lambda) := \exp\left(-\sum_{k=1}^{\infty} \frac{(\lambda - \lambda_0)^k}{k} \left(M\left((a_r - \lambda_0)^{-k}\right) - M\left((a_l - \lambda_0)^{-k}\right)\right)\right). \tag{3.5}$$

Suppose (i) is not satisfied. Then there is a point $\lambda_0 \in \Omega \cap \mathrm{sp}_{PC}\,T(a)$. Pick $\lambda \in \Omega$ in a sufficiently small open neighborhood $U \subset \Omega$ of λ_0. By (3.1), $\lambda \in \mathrm{sp}_{PC}\,T(a)$ if and only if

$$f(\lambda) \in (-\infty, 0). \tag{3.6}$$

If

$$M\left((a_l-\lambda_0)^{-k}\right)=M\left((a_r-\lambda_0)^{-k}\right) \text{ for all } k\in\mathbf{N}, \tag{3.7}$$

then (3.4) and (3.5) imply that f is constant in U. As f is analytic in Ω, it results that $f(\lambda)=f(\lambda_0)\in(-\infty,0)$ for all $\lambda\in\Omega$. Thus, we have (ii). So assume (3.7) does not hold. Let $n\in\mathbf{N}$ be the smallest number for which

$$\delta_n:=M\left((a_r-\lambda_0)^{-n}\right)-M\left((a_l-\lambda_0)^{-n}\right)\neq 0. \tag{3.8}$$

Then, by (3.4) and (3.5),

$$f(\lambda)=f(\lambda_0)-\frac{f(\lambda_0)\delta_n}{n}(\lambda-\lambda_0)^n+O\left((\lambda-\lambda_0)^{n+1}\right). \tag{3.9}$$

Because $f(\lambda_0)\in(-\infty,0)$, the set of all λ given by (3.6) and (3.9) is an analytic star of the valency $2n$ and hence this set is as described by (iii). □

4. Essential spectra of prescribed local shapes

Throughout this section we assume that a is a unimodular function in SAP, that is, $|a(x)|=1$ for all $x\in\mathbf{R}$. It is easily seen that then $|a_l(x)|=|a_r(x)|=1$ for all $x\in\mathbf{R}$. Furthermore, we assume that $\kappa(a_l)=\kappa(a_r)=0$. This implies that $\kappa(a_l-\lambda)=\kappa(a_r-\lambda)=0$ for all $\lambda\in\mathbf{C}\setminus\mathbf{T}$. For $\lambda\in\mathbf{C}\setminus\mathbf{T}$, we can write

$$a_l(x)-\lambda=e^{ib_{l,\lambda}(x)},\quad a_r(x)-\lambda=e^{ib_{r,\lambda}(x)},$$

where $b_{l,\lambda}, b_{r,\lambda}$ are functions in SAP. It follows that

$$d(a_l-\lambda)=e^{iM(b_{l,\lambda})},\quad d(a_r-\lambda)=e^{iM(b_{r,\lambda})}.$$

From Theorem C and the remark at the end of Section 2 we therefore deduce that

$$\mathrm{sp}_{AP}\,T(a)\subset\mathbf{T},\quad \mathrm{sp}_{PC}\,T(a)\subset\mathbf{D}.$$

In Section 3 we showed that

$$\frac{d(a_r-\lambda)}{d(a_l-\lambda)}=-e^{h(\lambda)},$$

where $h(\lambda)$ is analytic for $|\lambda|<1$. Because

$$\frac{d(a_r)}{d(a_l)}=e^{i(M(b_{r,0})-M(b_{l,0}))}$$

and the functions $b_{l,0}$ and $b_{r,0}$ can be chosen to be real-valued, we see that $h(0)$ is in $i\mathbf{R}$. Consequently, we arrive at the following result.

Proposition 4.1. *If $a\in SAP$ is unimodular and $\kappa(a_l)=\kappa(a_r)=0$, then*

$$\mathrm{sp}_{PC}\,T(a)=\{\lambda\in\mathbf{D}:h(\lambda)\in\mathbf{R}+2\pi i\mathbf{Z}\}$$

with some function h which is analytic in $\mathbf{D}$ and satisfies $h(0)\in i\mathbf{R}$. □

The construction of symbols whose spectra have prescribed local shapes is based on the following lemma.

Lemma 4.2. *Let $R > 1$. If $h(\lambda)$ is analytic for $|\lambda| < R$, $h(0) \in i\mathbf{R}$, and*

$$\max_{|t|=1} |h'(t)| < \frac{1}{2},$$

then there is a unimodular function $a \in SAP$ such that $\kappa(a_l - \lambda) = \kappa(a_r - \lambda) = 0$ and

$$\frac{d(a_r - \lambda)}{d(a_l - \lambda)} = -e^{h(\lambda)}$$

for $|\lambda| < 1$.

Proof. Let $h(\lambda) = i\mu + h_1\lambda + h_2\lambda^2 + \cdots$ ($|\lambda| < R$). For $t \in \mathbf{T}$, put

$$p(t) = 1 - \sum_{m=1}^{\infty} \left(m h_m t^m + m\overline{h_m} t^{-m} \right);$$

note that the series converges absolutely, because $h(\lambda)$ was supposed to be analytic in $|\lambda| < R$ for some $R > 1$. Since

$$\left| \sum_{m=1}^{\infty} m h_m t^m \right| = |th'(t)| < 1/2 \text{ for } t \in \mathbf{T},$$

we see that $p(t) > 0$ for $t \in \mathbf{T}$. Choose a real number δ and define a function c by

$$c(y) = -\pi + \frac{1}{2} \int_{-\delta}^{y} p(e^{iy}) dy.$$

Obviously,

$$c(-\delta) = -\pi, \quad c(2\pi - \delta) = -\pi + \frac{1}{2} \int_{-\delta}^{2\pi-\delta} p(e^{iy}) dy = 0,$$

and

$$c'(y) = \frac{1}{2} p(e^{iy}) > 0 \text{ for } y \in [-\delta, 2\pi - \delta].$$

Thus, $c : [-\delta, 2\pi - \delta] \to [-\pi, 0]$ is strictly monotonous. Let $b_r : [-\pi, 0] \to [-\delta, 2\pi - \delta]$ be the inverse function and extend b_r to a 2π-periodic and even function on all of $\mathbf{R}$. For $|\lambda| < 1$, we have

$$\begin{aligned} &\frac{1}{2\pi} \int_{-\pi}^{\pi} \log \left(e^{ib_r(x)} - \lambda \right) dx \qquad (4.1) \\ &= \frac{i}{2\pi} \int_{-\pi}^{\pi} b_r(x) dx + \frac{1}{2\pi} \int_{-\pi}^{\pi} \log \left(1 - \lambda e^{-ib_r(x)} \right) dx \end{aligned}$$

and

$$\frac{1}{2\pi}\int_{-\pi}^{\pi}\log\left(1-\lambda e^{-ib_r(x)}\right)dx=-\frac{1}{2\pi}\sum_{m=1}^{\infty}\frac{\lambda^m}{m}\int_{-\pi}^{\pi}e^{-imb_r(x)}dx$$

$$=-\frac{1}{\pi}\sum_{m=1}^{\infty}\frac{\lambda^m}{m}\int_{-\pi}^{0}e^{-imb_r(x)}dx=-\frac{1}{\pi}\sum_{m=1}^{\infty}\frac{\lambda^m}{m}\int_{-\delta}^{2\pi-\delta}e^{-imy}c'(y)dy$$

$$=-\frac{1}{2\pi}\sum_{m=1}^{\infty}\frac{\lambda^m}{m}\int_{-\delta}^{2\pi-\delta}e^{-imy}p(e^{iy})dy=-\frac{1}{2\pi}\sum_{m=1}^{\infty}\frac{\lambda^m}{m}\int_{0}^{2\pi}e^{-imy}p(e^{iy})dy$$

$$=-\frac{1}{2\pi}\sum_{m=1}^{\infty}\frac{\lambda^m}{m}(-2\pi m h_m)=\sum_{m=1}^{\infty}h_m\lambda^m.$$

Hence, (4.1) equals

$$\frac{i}{2\pi}\int_{-\pi}^{\pi}b_r(x)dx+\sum_{m=1}^{\infty}h_m\lambda^m. \tag{4.2}$$

On appropriately choosing $\delta\in\mathbf{R}$, we can achieve that $(2\pi)^{-1}\int_{-\pi}^{\pi}b_r(x)dx$ assumes any prescribed value $\mu\in\mathbf{R}$. From (4.1) and (4.2) we therefore obtain that

$$\frac{1}{2\pi}\int_{-\pi}^{\pi}\log\left(e^{ib_r(x)}-\lambda\right)dx=i\mu+\sum_{m=1}^{\infty}h_m\lambda^m. \tag{4.3}$$

Repeating the above construction in the special case where $h_1=h_2=\cdots=0$, we get a function b_l such that

$$\frac{1}{2\pi}\int_{-\pi}^{\pi}\log\left(e^{ib_l(x)}-\lambda\right)dx=\frac{i}{2\pi}\int_{-\pi}^{\pi}b_l(x)dx=-i\pi, \tag{4.4}$$

Now define $a\in SAP$ by

$$a(x)=\exp\Big((1-u(x)ib_l(x)+u(x)ib_r(x)\Big),$$

where $u\in C(\overline{\mathbf{R}})$ is any function for which $u(-\infty)=0$ and $u(+\infty)=1$. Clearly, the almost periodic representatives of a are

$$a_l(x)=e^{ib_l(x)},\quad a_r(x)=e^{ib_r(x)},$$

and $\kappa(a_l)=\kappa(a_r)=0$. The function a_l is 2π-periodic. This implies that

$$d(a_l-\lambda)=\exp\lim_{T\to\infty}\frac{1}{2T}\int_{-T}^{T}\log(a_l(x)-\lambda)dx=\exp\frac{1}{2\pi}\int_{-\pi}^{\pi}\log(a_l(x)-\lambda)dx.$$

An analogous formula is true for $d(a_r-\lambda)$. Hence, from (4.3) and (4.4) we infer that

$$\frac{d(a_r-\lambda)}{d(a_l-\lambda)}=e^{i(\mu+\pi)+h_1\lambda+h_2\lambda^2+\cdots}=-e^{h(\lambda)}.$$

□

The following result in conjunction with Proposition 4.1 yields a fairly good idea of the variety of sets that can be assumed by $\mathrm{sp}_{PC}\, T(a)$.

Proposition 4.3. *Let $R > 1$ and suppose that $h(\lambda)$ is analytic for $|\lambda| < R$. Then there is a function $a \in SAP$ such that*

$$\mathrm{sp}_{PC}\, T(a) = \{\lambda \in \mathbf{D} : h(\lambda) \in \mathbf{R}\}.$$

Proof. Let $h(0) = \nu + i\mu$ with $\nu, \mu \in \mathbf{R}$ and put $g(\lambda) = h(\lambda) - \nu$. Then $g(0) = i\mu \in i\mathbf{R}$. If $M > 0$ is sufficiently large, then

$$\max_{|t|=1} \frac{|g'(t)|}{M} < \frac{1}{2}, \quad \max_{|\lambda|\le 1} \frac{|g(\lambda)|}{M} < 2\pi.$$

Thus, Lemma 4.2 ensures the existence of a function $a \in SAP$ such that

$$\frac{d(a_r - \lambda)}{d(a_l - \lambda)} = -e^{g(\lambda)/M} \text{ for } |\lambda| < 1.$$

It results that

$$\mathrm{sp}_{PC}\, T(a) = \{\lambda \in \mathbf{D} : g(\lambda)/M \in \mathbf{R} + 2\pi i \mathbf{Z}\}.$$

Since $|g(\lambda)|/M < 2\pi$ for $|\lambda| < 1$, we see that actually

$$\mathrm{sp}_{PC}\, T(a) = \{\lambda \in \mathbf{D} : g(\lambda)/M \in \mathbf{R}\} = \{\lambda \in \mathbf{D} : h(\lambda) \in \mathbf{R}\}.$$

□

Proof of Theorem 1.2. Let X have its center at the origin and suppose $X = \{\lambda \in V : f(\lambda) \in \mathbf{R}\}$, where f is analytic in V and $f(0) = 0$. Choose $N > 0$ so large that $f(\lambda/N)$ is analytic for $|\lambda| < R$ with some $R > 1$. From Proposition 4.3 we deduce that there is an a in SAP such that $\mathrm{sp}_{PC}\, T(a)$ is the set $\{\lambda \in \mathbf{D} : f(\lambda/N) \in \mathbf{R}\}$. Hence, if $U \subset \mathbf{C}$ is a sufficiently small neighborhood of the origin, then $\mathrm{sp}_{PC}\, T(a) \cap U = NX \cap U$. Since

$$\mathrm{sp}_{PC}\, T(\alpha + \beta a) = \mathrm{sp}_{PC}\,(\alpha I + \beta T(a)) = \alpha + \beta\, \mathrm{sp}_{PC}\, T(a)$$

for all $\alpha, \beta \in \mathbf{C}$, we obtain that

$$\mathrm{sp}_{PC}\, T(\lambda_0 + a/N) \cap (\lambda_0 + U) = (\lambda_0 + X) \cap (\lambda_0 + U).$$

□

Proof of Theorem 1.3. If $N > 0$ is large enough, then $\lambda_j/N \in \mathbf{D}$ for all j. Put

$$h(\lambda) = (\lambda - \lambda_1/N)^{k_1} \cdots (\lambda - \lambda_n/N)^{k_n}.$$

Proposition 4.3 tells us that there exist $a \in SAP$ such that

$$\mathrm{sp}_{PC}\, T(a) = \{\lambda \in \mathbf{D} : h(\lambda) \in \mathbf{R}\}. \tag{4.5}$$

Clearly, $\mathrm{sp}_{PC}\, T(a)$ is an analytic star with the center λ_j/N and the valency $2k_j$ in a neighborhood of λ_j/N. This implies that $\mathrm{sp}_{PC}\, T(Na) = N\mathrm{sp}_{PC}\, T(a)$ is an analytic star centered at λ_j and of the valency $2k_j$ in some neighborhood of λ_j. □

Figures 1 and 2 show the set (4.5) for two particular choices of $h(\lambda)$.

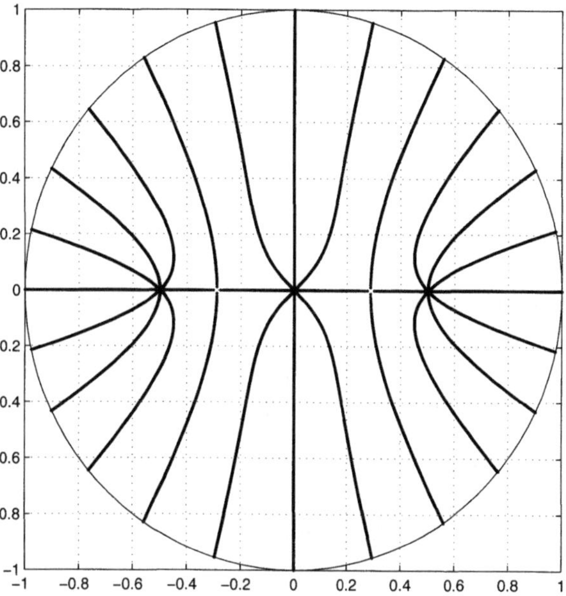

FIGURE 1. $h(\lambda) = (\lambda - 1/2)^4 \lambda^4 (\lambda + 1/2)^4$

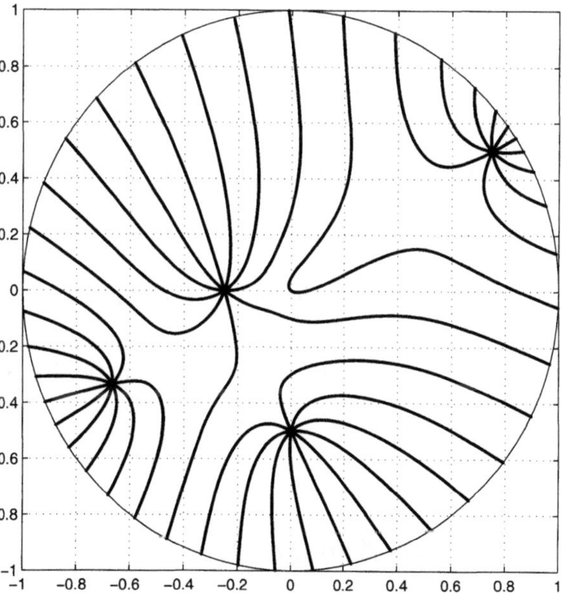

FIGURE 2. $h(\lambda) = \lambda(\lambda + 1/4)^5(\lambda + i/2)^5(\lambda - 3/4 - i/2)^5(\lambda + 2/3 + i/3)^5$

5. Essential spectra of prescribed global shapes

In this section we prove Theorems 1.1 and 1.4 and Corollary 1.5.

Let Γ be a closed Jordan curve in the complex plane and let D_Γ^+ and D_Γ^- denote the bounded and the unbounded component of $\mathbf{C} \setminus \Gamma$, respectively. We suppose that $0 \in D_\Gamma^+$. Let $f_\pm$ be any conformal maps of $\mathbf{D}$ onto $D_\Gamma^\pm$ such that

$$f_+(0) = 0, \quad f_-(z) = \frac{c_{-1}}{z} + c_0 + c_1 z + \cdots \quad (|z| < 1),$$

and denote by $F_\pm : D_\Gamma^\pm \to \mathbf{D}$ the inverse maps of $f_\pm$. Our assumptions that Γ be a closed Jordan curve guarantees that $f_\pm$ extend to homeomorphisms of $\mathbf{T}$ onto Γ. Define $\varphi_\pm : \mathbf{T} \to \Gamma$ by

$$\varphi_\pm(e^{ix}) = \begin{cases} f_\pm(e^{-2ix}) & \text{for } x \in (-\pi, 0], \\ f_\pm(e^{2ix}) & \text{for } x \in [0, \pi]. \end{cases} \tag{5.1}$$

Cut the complex plane along any curves $\gamma_\pm \subset D_\Gamma^-$ joining $f_\pm(1)$ and the point at infinity and fix any continuous branches of $\arg(f_\pm(1) - \lambda)$ for $\lambda \in \mathbf{C} \setminus \gamma_\pm$. Then, for each $\lambda \in \mathbf{C} \setminus (\Gamma \cup \gamma_\pm)$, we can choose $\arg(f_\pm(e^{iy}) - \lambda)$ and $\arg(\varphi_\pm(e^{ix}) - \lambda)$ so that the functions

$$(0, 2\pi] \to \mathbf{R}, \quad y \mapsto \arg(f_\pm(e^{iy}) - \lambda)$$
$$(-\pi, \pi] \to \mathbf{R}, \quad x \mapsto \arg(\varphi_\pm(e^{ix}) - \lambda)$$

are continuous. Finally, for $\lambda \in \mathbf{C} \setminus (\Gamma \cup \gamma_\pm)$, put

$$\log(f_\pm(e^{iy}) - \lambda) = \log|f_\pm(e^{iy}) - \lambda| + i \arg(f_\pm(e^{iy}) - \lambda),$$
$$\log(\varphi_\pm(e^{ix}) - \lambda) = \log|\varphi_\pm(e^{ix}) - \lambda| + i \arg(\varphi_\pm(e^{ix}) - \lambda),$$

and consider

$$\delta_\pm(\lambda) := \frac{1}{2\pi} \int_{-\pi}^{\pi} \log(\varphi_\pm(e^{ix}) - \lambda) dx.$$

Obviously,

$$\begin{aligned} \delta_\pm(\lambda) &= \frac{1}{2\pi} \int_{-\pi}^{0} \log(f_\pm(e^{-2ix}) - \lambda) dx + \frac{1}{2\pi} \int_{0}^{\pi} \log(f_\pm(e^{2ix}) - \lambda) dx \\ &= \frac{1}{\pi} \int_{0}^{\pi} \log(f_\pm(e^{2ix}) - \lambda) dx = \frac{1}{2\pi} \int_{0}^{2\pi} \log(f_\pm(e^{iy}) - \lambda) dy \\ &= \frac{1}{2\pi i} \int_{\mathbf{T}} \log(f_\pm(t) - \lambda) \frac{dt}{t}. \end{aligned} \tag{5.2}$$

Lemma 5.1. *We have*

$$e^{\delta_+(0)} = -f_+'(0), \quad e^{\delta_-(0)} = -c_{-1}.$$

Proof. From (5.2) we see that

$$\delta_+(0) = \frac{1}{2\pi i} \int_{\mathbf{T}} \log f_+(t) \frac{dt}{t}. \tag{5.3}$$

Let $\log z$ be the branch of the logarithm which is analytic in $\mathbf{C}\setminus[0,\infty)$ and assumes the value $i\pi$ at the point $z=-1$. By the construction of $\log f_+(t)$, the function

$$\log\frac{f_+(t)}{t} := \log f_+(t) - \log t$$

is continuous on $\mathbf{T}\setminus\{1\}$ and has equal one-sided limits at $t=1$. Hence, this function can be continued to a function $\log f_+(z)/z$ which is analytic for all $z\in\mathbf{D}$. From (5.3) we therefore get

$$\begin{aligned}\delta_+(0) &= \frac{1}{2\pi i}\int_{\mathbf{T}}\log\frac{f_+(t)}{t}\frac{dt}{t} + \frac{1}{2\pi i}\int_{\mathbf{T}}\log t\frac{dt}{t}\\ &= \log\left(\frac{f_+(z)}{z}\right)\bigg|_{z=0} + \frac{1}{2\pi}\int_0^{2\pi} ix\,dx\\ &= \log f_+'(0) + i\pi,\end{aligned}$$

whence $e^{\delta_+(0)} = -f_+'(0)$. Analogously one can show that

$$\begin{aligned}\delta_-(0) &= \frac{1}{2\pi i}\int_{\mathbf{T}}\log f_-(t)\frac{dt}{t}\\ &= \frac{1}{2\pi i}\int_{\mathbf{T}}\log\Big(f_-(t)t\Big)\frac{dt}{t} - \frac{1}{2\pi i}\int_{\mathbf{T}}\log t\frac{dt}{t}\\ &= \log\Big(f_-(z)z\Big)\Big|_{z=0} - \frac{1}{2\pi}\int_0^{2\pi} ix\,dx\\ &= \log c_{-1} - i\pi,\end{aligned}$$

which gives $e^{\delta_-(0)} = -c_{-1}$. □

Lemma 5.2. *We have*

$$e^{\delta_+(\lambda)} = \begin{cases} -\frac{\lambda}{F_+(\lambda)} & \text{for } \lambda\in D_\Gamma^+\setminus\{0\},\\ -\lambda & \text{for } \lambda\in D_\Gamma^- \end{cases}$$

and

$$e^{\delta_-(\lambda)} = \begin{cases} -c_{-1} & \text{for } \lambda\in D_\Gamma^+,\\ -\frac{c_{-1}}{F_-(\lambda)} & \text{for } \lambda\in D_\Gamma^-. \end{cases}$$

Proof. From (5.2) we infer that

$$\delta_+'(\lambda) = -\frac{1}{2\pi i}\int_{\mathbf{T}}\frac{dt}{(f_+(t)-\lambda)t}. \tag{5.4}$$

If $\lambda\in D_\Gamma^+\setminus\{0\}$, then the residue theorem shows that the right-hand side of (5.4) is

$$-\left(\frac{1}{f_+'(F_+(\lambda))\,F_+(\lambda)} - \frac{1}{\lambda}\right) = \frac{1}{\lambda} - \frac{F_+'(\lambda)}{F_+(\lambda)}.$$

It follows that

$$\left(e^{\delta_+(\lambda)}\frac{F_+(\lambda)}{\lambda}\right)' = e^{\delta_+(\lambda)}\left(\delta_+'(\lambda)\frac{F_+(\lambda)}{\lambda} + \frac{F_+'(\lambda)\lambda - F_+(\lambda)}{\lambda^2}\right)$$
$$= e^{\delta_+(\lambda)}\left(\left(\frac{1}{\lambda} - \frac{F_+'(\lambda)}{F_+(\lambda)}\right)\frac{F_+(\lambda)}{\lambda} + \frac{F_+'(\lambda)\lambda - F_+(\lambda)}{\lambda^2}\right) = 0.$$

Thus,

$$e^{\delta_+(\lambda)} = \mu\frac{\lambda}{F_+(\lambda)} \text{ for } \lambda \in D_\Gamma^+ \setminus \{0\}$$

with some constant μ. As

$$\lim_{\lambda\to 0} e^{\delta_+(\lambda)} = e^{\delta_+(0)}, \quad \lim_{\lambda\to 0}\frac{\lambda}{F_+(\lambda)} = \frac{1}{F_+'(0)} = f_+'(0),$$

we deduce from Lemma 5.1 that $\mu = -1$.

Now let $\lambda \in D_\Gamma^-$. There is a cut γ_+ such that $\lambda \in D_\Gamma^- \setminus \gamma_+$. From (5.4) and the residue theorem we obtain that $\delta_+'(\lambda) = 1/\lambda$, whence

$$\left(\frac{e^{\delta_+(\lambda)}}{\lambda}\right)' = e^{\delta_+(\lambda)}\left(\delta_+'(\lambda)\frac{1}{\lambda} - \frac{1}{\lambda^2}\right) = 0$$

and thus

$$e^{\delta_+(\lambda)} = \mu\lambda \text{ for } \lambda \in D_\Gamma^- \setminus \gamma_+$$

with some constant μ. By (5.2),

$$\begin{aligned}\delta_+(\lambda) &= \frac{1}{2\pi i}\int_{\mathbf{T}}\left(\log(-\lambda) + \log\left(1 - \frac{f_+(t)}{\lambda}\right)\right)\frac{dt}{t}\\ &= \log(-\lambda) + O(1/\lambda) \text{ as } \lambda \to \infty.\end{aligned}$$

Consequently, $e^{\delta_+(\lambda)} = -\lambda + O(1)$ as $\lambda \to \infty$. This gives $\mu = -1$.

Again by (5.2),

$$\delta_-'(\lambda) = -\frac{1}{2\pi i}\int_{\mathbf{T}}\frac{dt}{(f_-(t) - \lambda)t}. \tag{5.5}$$

If $\lambda \in D_\Gamma^+$, then $f_-(z) - \lambda \neq 0$ for $|z| \le 1$, and hence (5.5) shows that $\delta_-'(\lambda) = 0$. Thus, $e^{\delta_-(\lambda)}$ is some constant μ for $\lambda \in D_\Gamma^+$. By Lemma 5.1, $\mu = -c_{-1}$. Now suppose $\lambda \in D_\Gamma^-$. We can find a cut γ_- such that $\lambda \in D_\Gamma^- \setminus \gamma_-$. Then (5.5) in conjunction with the residue theorem gives

$$\delta_-'(\lambda) = -\frac{1}{f_-'(F_-(\lambda))\,F_-(\lambda)} = -\frac{F_-'(\lambda)}{F_-(\lambda)}.$$

Hence

$$\left(e^{\delta_-(\lambda)}F_-(\lambda)\right)' = e^{\delta_-(\lambda)}\left(\delta_-'(\lambda)F_-(\lambda) + F_-'(\lambda)\right) = 0,$$

and it results that

$$e^{\delta_-(\lambda)} = \frac{\mu}{F_-(\lambda)} \text{ for } \lambda \in D_\Gamma^- \setminus \gamma_-$$

with some constant μ. From (5.2) we obtain as above that

$$e^{\delta_-(\lambda)} = -\lambda + O(1) \text{ as } \lambda \to \infty,$$

and since $F_-(\lambda) = c_{-1}/\lambda + O(1)$ as $\lambda \to \infty$, we conclude that $\mu = -c_{-1}$. □

Proof of Theorem 1.1. Without loss of generality assume that $0 \in G$. Let Γ be the boundary of G and let $D_\Gamma^\pm$, f_-, F_- be as above. Clearly, $D_\Gamma^+ = G$. Define $a \in SAP$ by

$$a(x) = (1 - u(x))a_l(x) + u(x)a_r(x) + a_0(x),$$

where $u \in C(\overline{\mathbf{R}})$, $u(-\infty) = 0$, $u(+\infty) = 1$, $a_0 \in C(\dot{\mathbf{R}})$, $a_0(\infty) = 0$,

$$a_l(x) = \begin{cases} f_-(e^{-2ix}) & \text{for } x \in (-\pi, 0] + 2\pi\mathbf{Z}, \\ f_-(e^{2ix}) & \text{for } x \in [0, \pi] + 2\pi\mathbf{Z}, \end{cases}$$

$$a_r(x) = \begin{cases} f_-(-e^{-2ix}) & \text{for } x \in (-\pi, 0] + 2\pi\mathbf{Z}, \\ f_-(-e^{2ix}) & \text{for } x \in [0, \pi] + 2\pi\mathbf{Z}. \end{cases}$$

The function a_0 is chosen so that $a(x) \in \Gamma$ for all $x \in \mathbf{R}$. This in conjunction with Theorem C implies that $\mathrm{sp}_{AP}\, T(a) = \Gamma$. Given $\lambda \in \mathbf{C} \setminus \Gamma$, we have

$$d(a_l - \lambda) = e^{\delta_-(\lambda)},$$

and Lemma 5.2 shows that

$$d(a_l - \lambda) = \begin{cases} -c_{-1} & \text{for } \lambda \in G, \\ -\frac{c_{-1}}{F_-(\lambda)} & \text{for } \lambda \in \mathbf{C} \setminus (G \cup \Gamma). \end{cases}$$

To obtain $d(a_r - \lambda)$, we apply Lemma 5.2 to the situation where $f_-(z)$ and $F_-(\lambda)$ are replaced by $f_-(-z)$ and $-F_-(\lambda)$, respectively. This gives

$$d(a_r - \lambda) = e^{\delta_-(\lambda)} = \begin{cases} -(-c_{-1}) & \text{for } \lambda \in G, \\ -\frac{-c_{-1}}{-F_-(\lambda)} & \text{for } \lambda \in \mathbf{C} \setminus (G \cup \Gamma). \end{cases}$$

In summary,

$$\frac{d(a_r - \lambda)}{d(a_l - \lambda)} = \begin{cases} -1 & \text{for } \lambda \in G, \\ 1 & \text{for } \lambda \in \mathbf{C} \setminus (G \cup \Gamma), \end{cases}$$

and consequently, by (3.1), $\mathrm{sp}_{PC}\, T(a) = G$. □

Proof of Theorem 1.4. We can without loss of generality assume that $0 \in G$ and that $f(0) = 0$. Let Γ be the boundary of G and let $D_\Gamma^+ = G$, $D_\Gamma^- = \mathbf{C} \setminus (\Gamma \cup G)$ be as above. The map $f_+(z) := f(-z)$ is a conformal map of $\mathbf{D}$ onto G and $f_+((-1, 0)) = f((0, 1))$. Define $a \in SAP$ by

$$a(x) = u(x)\varphi_+(x) + a_0(x),$$

where $u \in C(\overline{\mathbf{R}})$, $u(-\infty) = 0$, $u(+\infty) = 1$, $a_0 \in C(\dot{\mathbf{R}})$, $a_0(\infty) = 0$, and φ_+ is given by (5.1). Obviously,

$$a_l(x) = 0, \quad a_r(x) = \varphi_+(x).$$

We can choose a_0 so that $a(\mathbf{R}) = \Gamma \cup \mathcal{L}$ where $\mathcal{L} \subset G$ is some curve starting at the origin and terminating on Γ such that $\mathcal{L} \cap f((0, 1)) = \emptyset$. In this way we get

$\mathrm{sp}_{AP}\, T(a) = \Gamma \cup \mathcal{L}$ due to Theorem C. We have $d(a_l - \lambda) = -\lambda$ for all $\lambda \in \mathbf{C} \setminus \Gamma$. For $\lambda \in \mathbf{C} \setminus (\Gamma \cup \mathcal{L})$, we obtain from Lemma 5.2 that

$$d(a_r - \lambda) = e^{\delta_+(\lambda)} = \begin{cases} -\frac{\lambda}{F_+(\lambda)} \text{ for } \lambda \in G \setminus \mathcal{L}, \\ -\lambda \text{ for } \lambda \in \mathbf{C} \setminus (G \cup \Gamma), \end{cases}$$

and hence

$$\frac{d(a_r - \lambda)}{d(a_l - \lambda)} = \begin{cases} \frac{1}{F_+(\lambda)} \text{ for } \lambda \in G \setminus \mathcal{L}, \\ 1 \text{ for } \lambda \in \mathbf{C} \setminus (G \cup \Gamma). \end{cases}$$

Thus, by (3.1),

$$\begin{aligned} \mathrm{sp}_{PC}\, T(a) &= \{\lambda \in G \setminus \mathcal{L} : 1/F_+(\lambda) \in (-\infty, 0)\} \\ &= \{\lambda \in G \setminus \mathcal{L} : F_+(\lambda) \in (-1, 0)\} = f_+((-1,0)). \end{aligned}$$

□

Proof of Corollary 1.5. Let $\overline{\mathcal{S}} = \mathcal{S} \cup \{A, B\}$ and $V_\varepsilon = \{\lambda \in \mathbf{C} : \mathrm{dist}\,(\lambda, \mathcal{S}) < \varepsilon\}$. For each point $\lambda \in \overline{\mathcal{S}}$, we have $U_\varepsilon(\lambda) := \{z \in \mathbf{C} : |z - \lambda| < \varepsilon\} \subset V_\varepsilon$. As $\overline{\mathcal{S}}$ is compact, there exist finitely many $\lambda_1, \ldots, \lambda_n \in \overline{\mathcal{S}}$ such that $\overline{\mathcal{S}} \subset G \subset V_\varepsilon$ with $G := \cup_{j=1}^n U_\varepsilon(\lambda_j)$. Obviously, the boundary of G is a closed Jordan curve. Let $\varphi : \mathbf{D} \to G$ be a conformal map for which $\varphi(0) = A$. There is a point $t_0 \in \mathbf{T}$ such that $\varphi(t_0) \in U_\varepsilon(B)$. Letting $f(z) := \varphi(t_0^{-1} z)$, we get a conformal map $f : \mathbf{D} \to G$ satisfying $f(0) = A$ and $f(1) \in U_\varepsilon(B)$. By Theorem 1.4, there are $a \in SAP$ such that $\mathrm{sp}_{PC}\, T(a) = f((0,1))$. Clearly, these a have the required properties. □

6. Separated almost periodic representatives

This section is devoted to the proofs of Theorems 1.6 and 1.6′. Notice that Corollary 1.7 is an immediate consequence of Theorem 1.6′ and the remark at the end of Section 2.

Proof of Theorem 1.6. Let $\Pi = \{\alpha + t\beta : t \in (-\infty, +\infty)\}$ be a straight line separating $\mathcal{R}(a_l)$ and $\mathcal{R}(a_r)$. We give Π the orientation induced by the movement of t from $-\infty$ to $+\infty$. Assume that $\mathcal{R}(a_l)$ lies in the open half-plane on the left of Π and that $\mathcal{R}(a_r)$ is contained in the open half-plane on the right of Π.

Clearly, we can choose arguments so that the two functions

$$\begin{aligned} \mathbf{R} \times \mathbf{R} \to \mathbf{R}, &\quad (x,t) \mapsto \arg(a_l(x) - \alpha - t\beta), \\ \mathbf{R} \times \mathbf{R} \to \mathbf{R}, &\quad (x,t) \mapsto \arg(a_r(x) - \alpha - t\beta), \end{aligned}$$

are continuous. It is easily seen that, for each x,

$$t_1 < t_2 \implies \arg(a_l(x) - \alpha - t_1\beta) < \arg(a_l(x) - \alpha - t_2\beta).$$

Moreover, $\arg(a_l(x) - \alpha - t_2\beta) - \arg(a_l(x) - \alpha - t_1\beta)$ is the angle at which the segment $\{\alpha + t\beta : t_1 \le t \le t_2\}$ is seen at the point $a_l(x)$, and because $\mathcal{R}(a_l)$ is compact, there is an $\varepsilon(t_1, t_2) > 0$ such that

$$\arg(a_l(x) - \alpha - t_2\beta) - \arg(a_l(x) - \alpha - t_1\beta) \ge \varepsilon(t_1, t_2)$$

for all $x \in \mathbf{R}$. Hence

$$\begin{aligned} &M\Big(\arg(a_l(x)-\alpha-t_2\beta)\Big)-M\Big(\arg(a_l(x)-\alpha-t_1\beta)\Big) \\ &\qquad=\lim_{T\to\infty}\frac{1}{2T}\int_{-T}^{T}\Big(\arg(a_l(x)-\alpha-t_2\beta)-\arg(a_l(x)-\alpha-t_1\beta)\Big)dx \\ &\qquad\geq\lim_{T\to\infty}\frac{1}{2T}\int_{-T}^{T}\varepsilon(t_1,t_2)dx=\varepsilon(t_1,t_2)>0. \end{aligned}$$

This shows that

$$L(t):=M\Big(\arg(a_l-\alpha-t\beta)\Big)$$

is a strictly monotonously increasing continuous function. Analogously one can prove that

$$R(t):=M\Big(\arg(a_r-\alpha-t\beta)\Big)$$

is a strictly monotonously decreasing continuous function.

Now let $\lambda=\alpha+t\beta\in\Pi$. We have

$$\frac{d(a_r-\lambda)}{d(a_l-\lambda)}\in(-\infty,0) \tag{6.1}$$

if and only if

$$\arg d(a_r-\lambda)-\arg d(a_l-\lambda)\in\pi+2\pi\mathbf{Z},$$

and because

$$\begin{aligned} \arg d(a_r-\lambda)&=M(\arg(a_r-\lambda))=R(t),\\ \arg d(a_l-\lambda)&=M(\arg(a_l-\lambda))=L(t), \end{aligned}$$

condition (6.1) is satisfied if and only if

$$R(t)-L(t)\in\pi+2\pi\mathbf{Z}. \tag{6.2}$$

A moment's thought reveals that there is a $k\in\mathbf{Z}$ such that

$$R(-\infty)-L(-\infty)=2k\pi,\quad R(+\infty)-L(+\infty)=2k\pi-2\pi.$$

As $R(t)-L(t)$ is a strictly monotonously decreasing continuous function, it follows that (6.2) has exactly one solution $t=t_0\in(-\infty,+\infty)$. Thus, there is exactly one point $\lambda=\lambda_0\in\Pi$ which satisfies (6.1). As

$$\mathrm{sp}_{PC}\,T(a)\cap\Pi=\left\{\lambda\in\Pi\setminus a(\mathbf{R}):\frac{d(a_r-\lambda)}{d(a_l-\lambda)}\in(-\infty,0)\right\},$$

we see that $\mathrm{sp}_{PC}\,T(a)\cap\Pi$ is the empty set if $\lambda_0\in a(\mathbf{R})$ and that $\mathrm{sp}_{PC}\,T(a)\cap\Pi=\{\lambda_0\}$ if $\lambda_0\notin a(\mathbf{R})$. □

Proof of Theorem 1.6′. The set $\mathrm{sp}_{\mathrm{ess}}T(a)\cap\Sigma$ is the union of $a(\mathbf{R})\cap\Sigma$ and the set

$$\mathcal{S}:=\left\{\lambda\in\Sigma:\frac{d(a_r-\lambda)}{d(a_l-\lambda)}\in(-\infty,0)\right\}. \tag{6.3}$$

The argument of the proof of Theorem A shows that there are exactly three possibilities:

(i) $\mathcal{S}$ is empty;

(ii) $\mathcal{S}$ is all of Σ;

(iii) $\mathcal{S}$ is an analytic star in some neighborhood of each of its points.

When proving Theorem 1.6, we observed that $\mathcal{S} \cap \Pi$ is a singleton whenever Π is a straight line which separates $\mathcal{R}(a_l)$ and $\mathcal{R}(a_r)$. This implies that we actually have case (iii) and that $\mathcal{S}$ is a simple analytic curve whose endpoints are located on $\operatorname{conv} \mathcal{R}(a_l)$ and $\operatorname{conv} \mathcal{R}(a_r)$. □

Acknowledgements

The work of Böttcher and Grudsky was supported by DFG-Kooperationsprojekt 436 RUS 113/426 for German and Russian scientists within the "Memorandum of Understanding" between DFG and RFFI. Grudsky also acknowledges support by RFFI Grant 98-01-01023. Spitkovsky was partially supported by NSF Grant DMS 9988579.

References

[1] A. Böttcher, S. Grudsky, and I. Spitkovsky: The spectrum is discontinuous on the manifold of Toeplitz operators. *Arch. Math.* **75** (2000), 46–52.

[2] A. Böttcher and B. Silbermann: *Analysis of Toeplitz Operators.* Akademie-Verlag, Berlin 1989 and Springer-Verlag, Berlin, Heidelberg, New York 1990.

[3] L.A. Coburn and R.G. Douglas: Translation operators on a half-line. *Proc. Nat. Acad. Sci. USA* **62** (1969), 1010–1013.

[4] R.G. Douglas: *Banach Algebra Techniques in Operator Theory.* Academic Press, New York 1972; 2nd edition: Springer-Verlag, New York 1998.

[5] D.R. Farenick and W.Y. Lee: Hyponormality and spectra of Toeplitz operators. *Trans. Amer. Math. Soc.* **348** (1996), 4153–4174.

[6] I. Gohberg and I.A. Feldman: On Wiener-Hopf integro-difference equations. *Soviet Math. Dokl.* **9** (1968), 1312–1316.

[7] S. Hwang and W.Y. Lee: On the continuity of spectra of Toeplitz operators. *Arch. Math.* **70** (1998), 66–73.

[8] N.K. Nikolski: *Treatise on the shift operator.* Springer-Verlag, Berlin, Heidelberg, New York 1986.

[9] D. Sarason: Toeplitz operators with semi-almost periodic symbols. *Duke Math. J.* **44** (1977), 357–364.

A. Böttcher
Fakultät für Mathematik
TU Chemnitz
D-09107 Chemnitz, Germany
e-mail: aboettch@mathematik.tu-chemnitz.de

S.M. Grudsky
Faculty of Mechanics and Mathematics
Rostov-on-Don State University
Bolshaya Sadovaya 105
344711 Rostov-on-Don, Russian Federation
e-mail: grudsky@aaanet.ru

I.M. Spitkovsky
Department of Mathematics
The College of William and Mary
Williamsburg, VA 23187-8795, USA
e-mail: ilya@math.wm.edu

2000 *Mathematics Subject Classification.*
Primary: 47B35; Secondary 42A75, 47A10.

Received: 29 January 2001.

Operator Theory:
Advances and Applications, Vol. 142, 79–90

Well-Posedness of Diffraction Problems Involving n Coplanar Strips

Luís Pinheiro de Castro and Frank-Olme Speck

Abstract. A class of problems of wave diffraction by a union of n infinite strips with first and second kind boundary conditions is studied in an operator theoretical way. They are considered as boundary-transmission problems in a Sobolev space setting. Conditions are presented in order to obtain existence and uniqueness of the solution and continuous dependence on the data.

1. Introduction

In recent years, a very efficient operator theoretical approach was developed to investigate certain canonical boundary value and transmission problems [13, 14, 15, 16, 22, 23] and, in particular, their well-posedness [2, 4, 5, 7, 15, 22]. In certain cases this leads also to explicit solutions in analytical form [13, 14, 15, 16, 23]. The strategy is very transparent (although based on highly complicated formulas): (1) An operator acting in suitable Sobolev spaces is associated to the problem, (2) a certain relation to a (matrix) Wiener-Hopf operator is proved, which is an operator matrix identity between appropriate classes of operators, (3) the invertibility of the related Wiener-Hopf operator is proved, e.g., by estimation, and finally (4) the relation is shown to be strong enough to transfer this property to the original operator.

In the present paper we give an application to a problem from diffraction theory which would be rather complicated to be treated by classical methods.

2. Formulation of the problem

We will consider problems of wave diffraction by a union of n infinite strips, here represented by $\Omega =]\gamma_1, \gamma_2[\cup \cdots \cup]\gamma_{2n-1}, \gamma_{2n}[$, with $0 = \gamma_1 < \cdots < \gamma_{2n}$ and $n \in \mathbb{N}$. They will be posed as boundary-transmission problems for the two-dimensional Helmholtz equation where the dependence on one dimension is dropped already due to perpendicular wave incidence (so the strips appear as intervals).

We will use the Bessel potential spaces $H^\sigma(\mathbb{R})$, with $\sigma \in \mathbb{R}$, formed by the tempered distributions φ such that $\|\varphi\|_{H^\sigma(\mathbb{R})} = \|\mathcal{F}^{-1}(1+\xi^2)^{\sigma/2} \cdot \mathcal{F}\varphi\|_{L^2(\mathbb{R})}$ is finite (where $\mathcal{F}$ represents the Fourier transformation). Moreover, we denote by $\widetilde{H}^\sigma(\Omega)$ [24, §2.10.3] the closed subspace of $H^\sigma(\mathbb{R})$ defined by the distributions with support contained in $\overline{\Omega}$, and $H^\sigma(\Omega)$ will denote the space of generalized functions on Ω which have extensions onto $\mathbb{R}$ that belong to $H^\sigma(\mathbb{R})$. The space $\widetilde{H}^\sigma(\Omega)$, sometimes denoted by $H^\sigma_{\overline{\Omega}}$, is endowed with the subspace topology, and we put the norm of the quotient space $H^\sigma(\mathbb{R})/\widetilde{H}^\sigma\left(\mathbb{R}\backslash\overline{\Omega}\right)$ on $H^\sigma(\Omega)$. In particular, we will denote the spaces $H^0(\mathbb{R}_+)$ and $\widetilde{H}^0(\mathbb{R}_+)$ by $L^2(\mathbb{R}_+)$ and $L^2_+(\mathbb{R})$, respectively. All those definitions can be extended to the multi-index case $\sigma = (\sigma_1, \sigma_2, \ldots, \sigma_m) \in \mathbb{R}^m$ by taking the product topology.

The problem is inspired by a class of Sommerfeld type problems considered in [15, 21, 22] for the half-line case instead of the present Ω.

More concretely, we are interested in studying well-posedness of the problem to find $u \in L^2(\mathbb{R}^2)$, with $u_{|\mathbb{R}^2_\pm} \in H^1(\mathbb{R}^2_\pm)$, so that

$$\left(\Delta + k^2\right) u = 0 \quad \text{in} \quad \mathbb{R}^2_\pm, \tag{2.1}$$

$$\begin{cases} a_0 u_0^+ + b_0 u_0^- = h_0 \\ a_1 u_1^+ + b_1 u_1^- = h_1 \end{cases} \quad \text{on} \quad \Omega, \tag{2.2}$$

$$\begin{cases} a_0' u_0^+ + b_0' u_0^- = 0 \\ a_1' u_1^+ + b_1' u_1^- = 0 \end{cases} \quad \text{on} \quad \mathbb{R}\setminus\overline{\Omega}, \tag{2.3}$$

where $\mathbb{R}^2_\pm$ represents the upper/lower half-plane, $k \in \mathbb{C}$ (with $\Im m k > 0$), $u_0^\pm = u_{|y=\pm 0}$, $u_1^\pm = (\partial u/\partial y)_{|y=\pm 0}$, $a_0, b_0, a_1, b_1, a_0', b_0', a_1', b_1'$ are given complex numbers such that $a_0 b_1 + a_1 b_0, a_0' b_1' + a_1' b_0' \neq 0$, and (for $l = 0,1$) the elements $h_l \in H^{-l+1/2}(\Omega)$ are arbitrarily given.

This problem can be described by the use of a linear operator

$$\mathcal{P} \;:\; D(\mathcal{P}) \to H^s(\Omega), \qquad s = (1/2, -1/2),$$

if we define $D(\mathcal{P})$ as the subspace of $H^1(\mathbb{R}^2_+) \times H^1(\mathbb{R}^2_-)$ whose functions fulfill the Helmholtz equation (2.1) and the homogeneous transmission conditions (2.3) whereas the action $\mathcal{P}u = (h_0, h_1)^T$ results from the non-homogeneous conditions (2.2).

In this sense, we will say that the operator $\mathcal{P}$ is associated to the problem and our aim is to prove that $\mathcal{P}$ is bounded and invertible if the coefficients are suitable. For this purpose, in the first instance, it is necessary to understand the structure of the operator $\mathcal{P}$ (see the next theorem).

Let

$$t(\xi) = \left(\xi^2 - k^2\right)^{1/2}, \qquad \xi \in \mathbb{R},$$

denote the branch of the square root that tends to $+\infty$ as $\xi \to +\infty$ with branch cuts along $\pm k \pm i\eta$, $\eta \geq 0$.

Theorem 2.1. *The operator $\mathcal{P}$ is (toplinearly* [11, Chapter IV - §1]*) equivalent to the convolution type operator*

$$W_{\Phi,\Omega} = r_{\mathbb{R}\to\Omega}\mathcal{F}^{-1}\Phi \cdot \mathcal{F} : \widetilde{H}^s(\Omega) \to H^s(\Omega), \tag{2.4}$$

$$\Phi = \frac{1}{a_0' b_1' + b_0' a_1'} \begin{bmatrix} a_0 b_1' + b_0 a_1' & (-a_0 b_0' + b_0 a_0') t^{-1} \\ (-a_1 b_1' + b_1 a_1') t & a_1 b_0' + b_1 a_0' \end{bmatrix},$$

where $r_{\mathbb{R}\to\Omega} : H^s(\mathbb{R}) \to H^s(\Omega)$ *is the restriction operator and* $s = (1/2, -1/2)$, *i.e., the two operators coincide up to homeomorphic linear transformations, cf.* (2.6).

Proof. Similarly to what happens in the half-line case [15], a function $u \in L^2(\mathbb{R}^2)$ with $u_{|\mathbb{R}^2_\pm} \in H^1(\mathbb{R}^2_\pm)$ satisfies the Helmholtz equation (2.1) if and only if it is representable by

$$u(x,y) = \mathcal{F}^{-1}_{\xi\mapsto x} e^{-t(\xi)y} \mathcal{F}_{x\mapsto\xi} u_0^+(x)\chi_+(y) + \mathcal{F}^{-1}_{\xi\mapsto x} e^{t(\xi)y} \mathcal{F}_{x\mapsto\xi} u_0^-(x)\chi_-(y) \quad (2.5)$$

for $(x,y) \in \mathbb{R}^2$, where $\mathcal{F}_{x\mapsto\xi} u(x,y) = \int_{\mathbb{R}} u(x,y) e^{i\xi x} dx$ and χ_+, χ_- denote the characteristic functions of the positive and negative half-lines, respectively.

Let

$$Z = \Big\{ (\phi,\psi) \in \left[H^{1/2}(\mathbb{R})\right]^2 : a_0'\phi + b_0'\psi \in \widetilde{H}^{1/2}(\Omega),$$
$$\mathcal{F}^{-1} t \cdot \mathcal{F}(-a_1'\phi + b_1'\psi) \in \widetilde{H}^{-1/2}(\Omega) \Big\}.$$

Taking into account the representation formula (2.5), we see that the trace operator,

$$\begin{array}{rccl} T_0 & : & D(\mathcal{P}) & \to & Z \\ & & u & \mapsto & u_0 = \begin{bmatrix} u_0^+ \\ u_0^- \end{bmatrix}, \end{array}$$

is continuously invertible by the Poisson operator $K : u_0 \mapsto u$ defined by (2.5).

The latter operators help us to arrive at

$$\mathcal{P} = W_{\Phi,\Omega}\, \mathcal{C}\, T_0, \quad (2.6)$$

where $\mathcal{C}$ is the convolution operator (on the full line)

$$\mathcal{F}^{-1} \begin{bmatrix} a_0' & b_0' \\ -a_1' t & b_1' t \end{bmatrix} \cdot \mathcal{F} \quad : \quad Z \to \widetilde{H}^s(\Omega).$$

Thus, (2.6) shows (toplinear) operator equivalence between $\mathcal{P}$ and the convolution type operator $W_{\Phi,\Omega}$ defined by (2.4), because $\mathcal{C}T_0$ is continuously invertible by $K\mathcal{C}^{-1}$. □

3. Relations with Wiener-Hopf operators

We are now interested in relating the operators of the last section with operators that act between spaces of distributions supported on a half-line by certain operator matrix identities. In addition, if we obtain invertibility conditions for the new operators, the relations must be strong enough so that we shall be able to transfer these conditions to the original operators.

Theorem 3.1. *The convolution type operator $W_{\Phi,\Omega}$ introduced in (2.4) is algebraically equivalent after (left-) extension to the Wiener-Hopf operator*

$$W_{\Psi,\mathbb{R}_+} = r_{\mathbb{R}\to\mathbb{R}_+}\mathcal{F}^{-1}\Psi\cdot\mathcal{F} : \widetilde{H}^\sigma(\mathbb{R}_+) \to H^\sigma(\mathbb{R}_+), \tag{3.1}$$

$$\Psi = \begin{bmatrix} \tau_{-\gamma_2} & & & & & & & \\ & \tau_{-(\gamma_4-\gamma_3)} & & & & & & \\ & & \ddots & & & & & \\ & & & \tau_{-(\gamma_{2n}-\gamma_{2n-1})} & & & & \\ & & & & \tau_{-(\gamma_3-\gamma_2)} & & & \\ & & & & & \ddots & & \\ & & & & & & \tau_{-(\gamma_{2n-1}-\gamma_{2n-2})} & \\ \Phi & \Phi\tau_{\gamma_3} & \cdots & \Phi\tau_{\gamma_{2n-1}} & \tau_{\gamma_2} & \cdots & \tau_{\gamma_{2n-2}} & \tau_{\gamma_{2n}} \end{bmatrix},$$

where the empty entries (i.e., those outside of the main diagonal and the last row) are zero, $\tau_a(\xi) = \mathrm{diag}[e^{i\xi a}, e^{i\xi a}]$, $\xi \in \mathbb{R}$, and $\sigma = (s, \ldots, s)$.

This means that there is an additional Banach space X and invertible (but not necessarily bounded) operators E and F such that

$$\begin{bmatrix} W_{\Phi,\Omega} & 0 \\ 0 & I_X \end{bmatrix} = E\, W_{\Psi,\mathbb{R}_+}\, F. \tag{3.2}$$

Proof. Following the method introduced by A.B. KUIJPER [10] the above result was proved in [3] (even for a wider class of Sobolev spaces). We note that in [3] the algebraic equivalence after extension appears as a relation between $W_{\Phi,\Omega}$ and a corresponding operator $W_{\Psi_{-1},\mathbb{R}_+} : \widetilde{H}^\sigma(\mathbb{R}_+) \to H^\sigma(\mathbb{R}_+)$ with symbol equal to (3.1) up to minus signs in the last n places of the matrix; this operator is obviously (toplinearly) equivalent to our $W_{\Psi,\mathbb{R}_+}$. In this context we use

$$I_X = \mathrm{diag}\left[I_{[H^s(\mathbb{R}_+)]^n}, I_{[\widetilde{H}^s(\mathbb{R}_+)]^n}\right],$$

in order to obtain (3.2). Formulas for E and F can be derived from [3] as well, but they are not explicitly needed. □

In fact, we can guarantee more than what is stated in Theorem 3.1.

Theorem 3.2. *There are Banach spaces X_1, Y_1 and linear homeomorphisms E_1 and F_1 so that*

$$\begin{bmatrix} W_{\Phi,\Omega} & 0 \\ 0 & I_{X_1} \end{bmatrix} = E_1 \begin{bmatrix} W_{\Psi,\mathbb{R}_+} & 0 \\ 0 & I_{Y_1} \end{bmatrix} F_1.$$

I.e., $W_{\Phi,\Omega}$ and $W_{\Psi,\mathbb{R}_+}$ are operators which are (toplinearly) equivalent after extension.

Proof. Taking into account that bounded linear Hilbert space operators with closed ranges are generalized invertible, the result can be derived from Theorem 2 in [1] and Theorem 3.1: the first of these theorems says that generalized invertible operators in Banach spaces are (toplinearly) equivalent after extension if and only if their defect spaces are homeomorphic, and this is guaranteed by Theorem 3.1 since $W_{\Psi,\mathbb{R}_+}$ is Fredholm [3]. □

We shall use the functions

$$\lambda_\pm^\sigma(\xi) = \mathrm{diag}[(\xi \pm k)^{\sigma_1}, \dots, (\xi \pm k)^{\sigma_m}], \qquad \xi \in \mathbb{R} \qquad (k \in \mathbb{C},\ \Im m k > 0)$$

in the representation of $\widetilde{H}^\sigma(\mathbb{R}_\pm)$, as well as $\zeta^\sigma = \lambda_-^\sigma\, \lambda_+^{-\sigma}$.

Theorem 3.3. *The Wiener-Hopf operator $W_{\Psi,\mathbb{R}_+}$ is (toplinearly) equivalent to*

$$W_{\Psi_0,\mathbb{R}_+} = r_{\mathbb{R}\to\mathbb{R}_+}\mathcal{F}^{-1}\Psi_0 \cdot \mathcal{F} : [L_+^2(\mathbb{R})]^{4n} \to [L^2(\mathbb{R}_+)]^{4n}, \tag{3.3}$$

$$\Psi_0 = \begin{bmatrix} \zeta^s\tau_{-\gamma_2} & & & & & & & \\ & \zeta^s\tau_{-(\gamma_4-\gamma_3)} & & & & & & \\ & & \ddots & & & & & \\ & & & \zeta^s\tau_{-(\gamma_{2n}-\gamma_{2n-1})} & & & & \\ & & & & \zeta^s\tau_{-(\gamma_3-\gamma_2)} & & & \\ & & & & & \ddots & & \\ & & & & & & \zeta^s\tau_{-(\gamma_{2n-1}-\gamma_{2n-2})} & \\ \Phi_0 & \Phi_0\tau_{\gamma_3} & \cdots & \Phi_0\tau_{\gamma_{2n-1}} & \zeta^s\tau_{\gamma_2} & \cdots & \zeta^s\tau_{\gamma_{2n-2}} & \zeta^s\tau_{\gamma_{2n}} \end{bmatrix},$$

where

$$\Phi_0 = \frac{1}{a_0'b_1' + b_0'a_1'} \begin{bmatrix} (a_0b_1' + b_0a_1')\,\zeta^{1/2} & -a_0b_0' + b_0a_0' \\ -a_1b_1' + b_1a_1' & (a_1b_0' + b_1a_0')\,\zeta^{-1/2} \end{bmatrix}.$$

Proof. Let us consider the operators

$$\begin{aligned} E_2 &= r_{\mathbb{R}\to\mathbb{R}_+}\mathcal{F}^{-1}\lambda_-^{-\sigma} \cdot \mathcal{F} l_0 : [L^2(\mathbb{R}_+)]^{4n} \to H^\sigma(\mathbb{R}_+), \\ F_2 &= l_0 r_{\mathbb{R}\to\mathbb{R}_+}\mathcal{F}^{-1}\lambda_+^\sigma \cdot \mathcal{F} : \widetilde{H}^\sigma(\mathbb{R}_+) \to [L_+^2(\mathbb{R})]^{4n}, \end{aligned}$$

where $l_0 : [L^2(\mathbb{R}_+)]^{4n} \to [L_+^2(\mathbb{R})]^{4n}$ is the zero extension operator.

These operators are boundedly invertible (see [24, §2.10.3]). Moreover, by employing the structure of E_2 and F_2 [24], we get

$$W_{\Psi,\mathbb{R}_+} = E_2\, W_{\Psi_0,\mathbb{R}_+}\, F_2,$$

which demonstrates the operator equivalence between $W_{\Psi,\mathbb{R}_+}$ and $W_{\Psi_0,\mathbb{R}_+}$. □

Corollary 3.4. *The convolution type operator $W_{\Phi,\Omega}$ belongs to the same regularity classes [6, 20] as the Wiener-Hopf operators $W_{\Psi,\mathbb{R}_+}$ and $W_{\Psi_0,\mathbb{R}_+}$. I.e., they are invertible, one-sided invertible, Fredholm, semi-Fredholm, one-sided regularizable, generalized invertible or normally solvable only at the same time.*

Proof. From the above relations we derive that $\ker W_{\Phi,\Omega}$, $\ker W_{\Psi,\mathbb{R}_+}$, and $\ker W_{\Psi_0,\mathbb{R}_+}$ are (toplinearly) isomorphic and that the ranges of these operators are closed only at the same time. In addition, the presented relations allow us to conclude that $H^s(\Omega)/\operatorname{im} W_{\Phi,\Omega}$, $[H^\sigma(\mathbb{R}_+)]/\operatorname{im} W_{\Psi,\mathbb{R}_+}$, and $[L^2(\mathbb{R}_+)]^{4n}/\operatorname{im} W_{\Psi_0,\mathbb{R}_+}$ are also isomorphic. This implies the assertion. □

4. Sufficient conditions for the invertibility

We begin this section with the introduction of an auxiliary operator derived from $W_{\Psi_0,\mathbb{R}_+}$:

$$W_{\zeta^{-\sigma}\Psi_0,\mathbb{R}_+} = r_{\mathbb{R}\to\mathbb{R}_+}\mathcal{F}^{-1}\zeta^{-\sigma}\Psi_0\cdot\mathcal{F} : [L^2_+(\mathbb{R})]^{4n} \to [L^2(\mathbb{R}_+)]^{4n}. \tag{4.1}$$

This new operator will help us to arrive at the desired invertibility conditions. For this purpose, we will first uncover some symmetries in the structure of (4.1).

Theorem 4.1. *The Wiener-Hopf operator $W_{\zeta^{-\sigma}\Psi_0,\mathbb{R}_+}$ is (toplinearly) equivalent to*

$$W_{\Psi_1,\mathbb{R}_+}\, l_0\, W_{\Psi_2,\mathbb{R}_+}\, l_0\, W_{\Psi_3,\mathbb{R}_+} : [L^2_+(\mathbb{R})]^{4n} \to [L^2(\mathbb{R}_+)]^{4n}$$

where, for $j = 1, 2, 3$,

$$W_{\Psi_j,\mathbb{R}_+} = r_{\mathbb{R}\to\mathbb{R}_+}\mathcal{F}^{-1}\Psi_j\cdot\mathcal{F} : [L^2_+(\mathbb{R})]^{4n} \to [L^2(\mathbb{R}_+)]^{4n}$$

and

$$\Psi_1 = \left[\begin{array}{c|cccc|cccc}
 & & & & & \tau_{-\gamma_{2n}} & & & \\
 & & & & & & \tau_{-\gamma_{2n-2}} & & \\
 & & & & & & & \ddots & \\
 & & & & & & & & \tau_{-\gamma_2} \\ \hline
 & \tau_{-\gamma_{2n-1}} & & & & & & & \\
 & & \tau_{-\gamma_{2n-3}} & & & & & & \\
 & & & \ddots & & & & & \\
 & & & & \tau_{-\gamma_3} & & & & \\ \cline{1-5}
 nI_2 & I_2 & I_2 & \cdots & I_2 & & & &
\end{array}\right],$$

$$\Psi_2 = \left[\begin{array}{ccc|ccc}
I_2 & & & \zeta^{-s}\Phi_0/n & \cdots & \zeta^{-s}\Phi_0/n \\
 & \ddots & & & & \\
 & & I_2 & & & \\ \hline
 & & & I_2 & & \\
 & & & & \ddots & \\
 & & & & & I_2
\end{array}\right],$$

$$\Psi_3 = \left[\begin{array}{c|cccc|cccc}
 & & & & & \tau_{\gamma_{2n}} & & & \\
 & & & & & & \tau_{\gamma_{2n-2}} & & \\
 & & & & & & & \ddots & \\
 & & & & & & & & \tau_{\gamma_2} \\ \hline
 & \tau_{\gamma_{2n-1}} & & & & & & & \\
 & & \tau_{\gamma_{2n-3}} & & & & & & \\
 & & & \ddots & & & & & \\
 & & & & \tau_{\gamma_3} & & & & \\ \cline{1-5}
 nI_2 & I_2 & I_2 & \cdots & I_2 & & & &
\end{array}\right],$$

with $I_2 = \mathrm{diag}[1, 1]$.

Proof. From the identity

$$
\begin{aligned}
\zeta^{-\sigma}\Psi_0 &= \left[\begin{array}{ccc|ccc|c}
 & & I_2 & & & & \\
 & \iddots & & & & & \\
I_2 & & & & & & \\
\hline
 & & & & & I_2 & \\
 & & & & \iddots & & \\
 & & & I_2 & & & \\
\hline
 & & & & & & I_2
\end{array}\right] \Psi_1\,\Psi_2\,\Psi_3 \\
&\times \left[\begin{array}{c|ccc|cccc}
(1/n)I_2 & -(1/n)I_2 & \cdots & -(1/n)I_2 & & & & \\
\hline
 & & & I_2 & & & & \\
 & & \iddots & & & & & \\
 & I_2 & & & & & & \\
\hline
 & & & & & & & (1/n)I_2 \\
 & & & & & & I_2 & \\
 & & & & & \iddots & & \\
 & & & & I_2 & & &
\end{array}\right]
\end{aligned}
$$

it follows that $W_{\zeta^{-\sigma}\Psi_0,\mathbb{R}_+}$ and $W_{\Psi_1\Psi_2\Psi_3,\mathbb{R}_+}$ are (toplinearly) equivalent operators.

Therefore, in order to obtain the present result, we only have to observe that the *minus* and *plus* semi-almost periodic entries of Ψ_1 and Ψ_3, respectively, lead us to

$$
\begin{aligned}
W_{\Psi_1\Psi_2\Psi_3,\mathbb{R}_+} &= \left(r_{\mathbb{R}\to\mathbb{R}_+}\mathcal{F}^{-1}\Psi_1\cdot\mathcal{F}\right) l_0 \left(r_{\mathbb{R}\to\mathbb{R}_+}\mathcal{F}^{-1}\Psi_2\cdot\mathcal{F}\right) l_0 \left(r_{\mathbb{R}\to\mathbb{R}_+}\mathcal{F}^{-1}\Psi_3\cdot\mathcal{F}\right) \\
&= W_{\Psi_1,\mathbb{R}_+}\, l_0\, W_{\Psi_2,\mathbb{R}_+}\, l_0\, W_{\Psi_3,\mathbb{R}_+}.
\end{aligned}
$$

□

Definition 4.2. An essentially bounded $m\times m$ matrix-valued function $\psi(\xi) = [\psi_{ij}(\xi)]$ is said to be *strongly elliptic* if there exist constants $\eta\in\mathbb{C}$ and $C>0$ so that

$$
\Re e\ \eta\sum_{i,j=1}^{m}\psi_{ij}(\xi)\mu_j\overline{\mu}_i \geq C\sum_{i=1}^{m}|\mu_i|^2 \qquad \forall\mu=(\mu_1,\ldots,\mu_m)\in\mathbb{C}^m.
$$

Theorem 4.3. *If the piecewise continuous matrix-valued function Ψ_2 is strongly elliptic, then the Wiener-Hopf operator introduced in* (4.1),

$$
W_{\zeta^{-\sigma}\Psi_0,\mathbb{R}_+} : [L^2_+(\mathbb{R})]^{4n} \to [L^2(\mathbb{R}_+)]^{4n},
$$

is an invertible operator.

Proof. Firstly we note that

$$
\Psi_3 = \overline{\Psi_1}.
$$

Secondly, the non-negativity of the mean motions [19] of the semi-almost periodic entries of Ψ_3 ensure (by [19, Theorem 1]) the left invertibility of $W_{\overline{\Psi_1},\mathbb{R}_+}$.

These two observations in conjunction with the strong ellipticity of Ψ_2 imply that, for $\psi \in [L^2(\mathbb{R}_+)]^{4n}$,

$$\begin{aligned}
\Re e\langle \eta(W_{\Psi_1,\mathbb{R}_+} l_0)(W_{\Psi_2,\mathbb{R}_+} l_0)(W_{\overline{\Psi_1},\mathbb{R}_+} l_0)\psi, \psi\rangle_{[L^2(\mathbb{R}_+)]^{4n}} &= \\
&= \Re e\langle \eta(W_{\Psi_2,\mathbb{R}_+} l_0)(W_{\overline{\Psi_1},\mathbb{R}_+} l_0)\psi, (W_{\overline{\Psi_1},\mathbb{R}_+} l_0)\psi\rangle_{[L^2(\mathbb{R}_+)]^{4n}} \\
&= \Re e\langle \left(r_{\mathbb{R}\to\mathbb{R}_+} \mathcal{F}^{-1}\eta\Psi_2 \cdot \mathcal{F} l_0\right)(W_{\overline{\Psi_1},\mathbb{R}_+} l_0)\psi, (W_{\overline{\Psi_1},\mathbb{R}_+} l_0)\psi\rangle_{[L^2(\mathbb{R}_+)]^{4n}} \\
&= \Re e\langle \eta\Psi_2 \cdot \mathcal{F} l_0(W_{\overline{\Psi_1},\mathbb{R}_+} l_0)\psi, \mathcal{F} l_0(W_{\overline{\Psi_1},\mathbb{R}_+} l_0)\psi\rangle_{[L^2(\mathbb{R})]^{4n}} \\
&\geq C\,\|\mathcal{F} l_0(W_{\overline{\Psi_1},\mathbb{R}_+} l_0)\psi\|^2_{[L^2(\mathbb{R})]^{4n}} \\
&= C\,\|W_{\overline{\Psi_1},\mathbb{R}_+} l_0\psi\|^2_{[L^2(\mathbb{R}_+)]^{4n}} \\
&\geq C\,C_1\|\psi\|^2_{[L^2(\mathbb{R}_+)]^{4n}},
\end{aligned}$$

where $C_1 > 0$ due to the left invertibility of $W_{\overline{\Psi_1},\mathbb{R}_+}$. This inequality allows us to conclude that $W_{\Psi_1,\mathbb{R}_+} l_0 W_{\Psi_2,\mathbb{R}_+} l_0 W_{\overline{\Psi_1},\mathbb{R}_+} : [L^2_+(\mathbb{R})]^{4n} \to [L^2(\mathbb{R}_+)]^{4n}$ is a left invertible operator.

Noticing that the conjugate operator

$$\begin{aligned}
\left(l_0 W_{\Psi_1,\mathbb{R}_+} l_0 W_{\Psi_2,\mathbb{R}_+} l_0 W_{\overline{\Psi_1},\mathbb{R}_+}\right)^* &= \left(l_0 W_{\overline{\Psi_1},\mathbb{R}_+}\right)^* \left(l_0 W_{\Psi_2,\mathbb{R}_+}\right)^* \left(l_0 W_{\Psi_1,\mathbb{R}_+}\right)^* \\
&= l_0 W_{\Psi_1,\mathbb{R}_+} l_0 W_{\overline{\Psi_2},\mathbb{R}_+} l_0 W_{\overline{\Psi_1},\mathbb{R}_+}
\end{aligned}$$

has a structure like $l_0 W_{\Psi_1,\mathbb{R}_+} l_0 W_{\Psi_2,\mathbb{R}_+} l_0 W_{\overline{\Psi_1},\mathbb{R}_+}$, by a similar reasoning we obtain that this is also a left invertible operator. Thus $W_{\Psi_1,\mathbb{R}_+} l_0 W_{\Psi_2,\mathbb{R}_+} l_0 W_{\overline{\Psi_1},\mathbb{R}_+}$ is an invertible operator and from Theorem 4.1 we now obtain the assertion. □

Corollary 4.4. *Let Ψ_2 be strongly elliptic. The Wiener-Hopf operator $W_{\Psi_0,\mathbb{R}_+}$ defined by (3.3) is a Fredholm operator with zero Fredholm index.*

Proof. From Theorem 4.3 we know that, under the present assumptions, $W_{\zeta^{-\sigma}\Psi_0,\mathbb{R}_+}$ is an invertible operator. Thus, the result is a consequence of the fact that $W_{\Psi_0,\mathbb{R}_+}$ and $W_{\zeta^{-\sigma}\Psi_0,\mathbb{R}_+}$ are homotopic operators in the class of Fredholm operators acting from $[L^2_+(\mathbb{R})]^{4n}$ to $[L^2(\mathbb{R}_+)]^{4n}$. □

Theorem 4.5. *Let $W_{\Psi,\mathbb{R}_+}$ be the operator defined by* (3.1) *and*

$$a_0 b_1' + b_0 a_1' \neq 0, \tag{4.2}$$
$$-a_0 b_0' + b_0 a_0' \neq 0, \tag{4.3}$$
$$-a_1 b_1' + b_1 a_1' \neq 0, \tag{4.4}$$
$$a_1 b_0' + b_1 a_0' \neq 0. \tag{4.5}$$

If

$$\rho = \frac{(a_0 b_1' + b_0 a_1')(a_1 b_0' + b_1 a_0')}{(-a_0 b_0' + b_0 a_0')(-a_1 b_1' + b_1 a_1')} \notin [1,\infty), \tag{4.6}$$

then the codimension of the image of $W_{\Psi,\mathbb{R}_+}$ is zero.

Proof. For negative parameters $-\gamma$ and every $s \in \mathbb{R}^n$ (here $s = (1/2, -1/2)$), the operators

$$r_{\mathbb{R}\to\mathbb{R}_+}\mathcal{F}^{-1}\tau_{-\gamma}\cdot\mathcal{F} : \widetilde{H}^s(\mathbb{R}_+) \to H^s(\mathbb{R}_+)$$

are nothing else than left shift operators composed with the restriction operator $r_{\mathbb{R}\to\mathbb{R}_+}$. It follows that these operators are surjective. Therefore, taking into account the structure of Ψ, we obtain that $W_{\Psi,\mathbb{R}_+}$ is a surjective operator whenever

$$W_{\Phi,\mathbb{R}_+} = r_{\mathbb{R}\to\mathbb{R}_+}\mathcal{F}^{-1}\Phi\cdot\mathcal{F} : \widetilde{H}^s(\mathbb{R}_+) \to H^s(\mathbb{R}_+)$$

is a surjective operator. It remains to notice that the latter is satisfied in the present conditions, cf. Theorem 4.5 in [15]. □

Corollary 4.6. *If Ψ_2 is strongly elliptic and the conditions* (4.2)–(4.6) *are fulfilled, then the operator $\mathcal{P}$ (associated to the problem) is boundedly invertible.*

Proof. From Corollary 3.4, Corollary 4.4, and Theorem 4.5 we deduce that, under the given conditions, $W_{\Phi,\Omega}$ is invertible. Therefore, we only have to take into consideration Theorem 2.1 to arrive at the desired conclusion. □

5. Example: The Rawlins problem for n strips

In this final section we like to present a concrete example of a particularly well-known wave diffraction problem. We will consider the Rawlins problem [9, 12, 17, 18, 21, 22], but in the present geometry instead of the usual half-line case. Thus, we now study the problem of the existence and uniqueness of an element $u \in L^2(\mathbb{R}^2)$ with $u_{|\mathbb{R}^2_\pm} \in H^1(\mathbb{R}^2_\pm)$ so that

$$\begin{aligned}
(\Delta + k^2)\, u &= 0 && \text{in} \quad \mathbb{R}^2_\pm,\\
\left\{\begin{array}{l} u_0^+ = h_0 \\ u_1^- = h_1 \end{array}\right. & && \text{on} \quad \Omega,\\
\left\{\begin{array}{l} u_0^+ - u_0^- = 0 \\ u_1^+ - u_1^- = 0 \end{array}\right. & && \text{on} \quad \mathbb{R}\setminus\overline{\Omega},
\end{aligned}$$

where $k \in \mathbb{C}$ (with $\Im m k > 0$) and the elements $h_l \in H^{-l+1/2}(\Omega)$ are given, for $l = 0, 1$.

This is a special case of the general class of problems considered in the above sections. In particular, we take here $a_0 = b_1 = a_0' = a_1' = 1$, $b_0 = a_1 = 0$ and $b_0' = b_1' = -1$. Thus, in the case at hand we have

$$\Phi = -\frac{1}{2}\begin{bmatrix} -1 & t^{-1} \\ t & 1 \end{bmatrix},$$

$$\Phi_0 = -\frac{1}{2}\begin{bmatrix} -\zeta^{1/2} & 1 \\ 1 & \zeta^{-1/2} \end{bmatrix},$$

and

$$\rho = -1 \notin [1, \infty).$$

Let us verify that the corresponding Ψ_2 is strongly elliptic. For that we begin by taking into account that, for any $\mu_j \in \mathbb{C}$,

$$\begin{aligned}
\left|\frac{1}{2n}\mu_j\overline{\mu_1}\right| &\leq \frac{1}{4n}\left(|\mu_j|^2+|\mu_1|^2\right), && j=2n+1,2n+3,\ldots,4n-1,\\
\left|-\frac{1}{2n}\mu_j\overline{\mu_2}\right| &\leq \frac{1}{4n}\left(|\mu_j|^2+|\mu_2|^2\right), && j=2n+2,2n+4,\ldots,4n,\\
\left|-\frac{1}{2n}\zeta^{1/2}\mu_j\overline{\mu_2}\right| &\leq \frac{1}{4n}\left(|\mu_j|^2+|\mu_2|^2\right), && j=2n+1,2n+3,\ldots,4n-1,\\
\left|-\frac{1}{2n}\zeta^{-1/2}\mu_j\overline{\mu_1}\right| &\leq \frac{1}{4n}\left(|\mu_j|^2+|\mu_1|^2\right), && j=2n+2,2n+4,\ldots,4n.
\end{aligned}$$

From here we derive that, for $\mu_j \in \mathbb{C}$,

$$\begin{aligned}
\frac{1}{2}\sum_{j=1}^{4n}|\mu_j|^2 &\leq \sum_{j=1}^{2}|\mu_j|^2\left(1-\frac{1}{4}-\frac{1}{4}\right)+\sum_{j=3}^{2n}|\mu_j|^2+\sum_{j=n}^{2n-1}|\mu_{2j+1}|^2\left(1-\frac{1}{4n}-\frac{1}{4n}\right)\\
&\quad+\sum_{j=n}^{2n-1}|\mu_{2j+2}|^2\left(1-\frac{1}{4n}-\frac{1}{4n}\right)\\
&\leq \Re e\left[\sum_{j=1}^{4n}|\mu_j|^2+\frac{1}{2n}\left(\sum_{j=n}^{2n-1}\mu_{2j+1}\,\overline{\mu_1}-\zeta^{-1/2}\sum_{j=n}^{2n-1}\mu_{2j+2}\,\overline{\mu_1}\right.\right.\\
&\quad\left.\left.-\zeta^{1/2}\sum_{j=n}^{2n-1}\mu_{2j+1}\,\overline{\mu_2}-\sum_{j=n}^{2n-1}\mu_{2j+2}\,\overline{\mu_2}\right)\right]\\
&= \Re e\sum_{i,j=1}^{4n}\psi_{ij}\,\mu_j\overline{\mu_i},
\end{aligned}$$

where $[\psi_{ij}]_{i,j=1,\ldots,4n}$ is the corresponding Ψ_2 for the present Φ_0.

Therefore, this Ψ_2 is strongly elliptic and so, according to Corollary 4.6, the associated operator $\mathcal{P}$ to the Rawlins problem for n strips is boundedly invertible.

Remark 5.1. Various boundary-transmission problems with other conditions on Ω can be treated by analogy or as a perturbation of a well-posed problem. Another challenge is to consider non-homogeneous conditions in (2.3). This case requires the consideration of appropriate extensions to the exterior of Ω (and onto the full line), which represents an additional difficulty when considering a convolution type operator (toplinearly) equivalent to the corresponding operator $\mathcal{P}$ associated to the problem, cf. [8].

References

[1] Bart, H., Tsekanovskii, V.E., Matricial Coupling and Equivalence After Extension, *Oper. Theory Adv. Appl.* **59** (1992), 143–160.

[2] Bastos, M.A., Santos, A.F. dos, Duduchava, R., Finite Interval Convolution Operators on the Bessel Potential Spaces H_p^s, *Math. Nachr.* **173** (1995), 49–63.

[3] Castro, L.P., Wiener-Hopf Operators on Unions of Finite Intervals: Relations and Generalized Inversion, *Proceedings of the Meeting on Matrix Analysis and Applications*, F.J. Cobos (ed.) et al., Department of Applied Mathematics I, University of Sevilla, Sevilla (1997), 148–155.

[4] ———, A Relation Between Convolution Type Operators on Intervals in Sobolev Spaces, *Appl. Anal.* **74** (2000), 393–412.

[5] Castro, L.P., Speck, F.-O., A Fredholm Study for a Sommerfeld Diffraction Problem, *Pitman Res. Notes Math. Ser.* **379** (1998), 47–51.

[6] ———, Regularity Properties and Generalized Inverses of Delta-related Operators, *Z. Anal. Anwendungen* **17** (1998), 577–598.

[7] ———, An Operator Approach to a Boundary Value Problem for the Helmholtz Equation, *Operator Theoretical Methods: 17th International Conference on Operator Theory*, A. Gheondea (ed.) et al., The Theta Foundation, Bucharest (2000), 67–86.

[8] ———, Relations Between Convolution Type Operators on Intervals and on the Half-line, *Integral Equations Operator Theory* **37** (2000), 169–207.

[9] Heins, A.E., The Sommerfeld Half-plane Problem Revisited. II: The Factoring of a Matrix of Analytic Functions, *Math. Methods Appl. Sci.* **5** (1983), 14–21.

[10] Kuijper, A.B., A Note on First Kind Convolution Equations on a Finite Interval, *Integral Equations Operator Theory* **14** (1991), 146–152.

[11] Lang, S., *Real and Functional Analysis, 3rd ed.*, Grad. Texts in Math. **142**, Springer-Verlag, New York 1993.

[12] Meister, E., Some Multiple-part Wiener-Hopf Problems in Mathematical Physics, *Mathematical Methods and Models in Mechanics, Banach Center Publ.* **15** (1985), 359–407.

[13] Meister, E., Penzel, F., Speck, F.-O., Teixeira, F.S., Two Canonical Wedge Problems for the Helmholtz Equation, *Math. Methods Appl. Sci.* **17** (1994), 877–899.

[14] ———, Some Interior and Exterior Boundary-value Problems for the Helmholtz Equation in a Quadrant, *Proc. Roy. Soc. Edinburgh Sect. A* **123** (1993), 275–294.

[15] Meister, E., Speck, F.-O., Modern Wiener-Hopf Methods in Diffraction Theory, *Pitman Res. Notes Math. Ser.* **216** (1989), 130–171.

[16] Meister, E., Speck, F.-O., Teixeira, F.S., Wiener-Hopf-Hankel Operators for some Wedge Diffraction Problems with Mixed Boundary Conditions, *J. Integral Equations Appl.* **4** (1992), 229–255.

[17] Rawlins, A.D., The Solution of a Mixed Boundary Value Problem in the Theory of Diffraction by a Semi-infinite Plane, *Proc. Roy. Soc. London Ser. A* **346** (1975), 469–484.

[18] ———, The Explicit Wiener-Hopf Factorisation of a Special Matrix, *ZAMM Z. Angew. Math. Mech.* **61** (1981), 527–528.

[19] Sarason, D., Toeplitz Operators with Semi Almost Periodic Symbols, *Duke Math. J.* **44** (1977), 357–364.

[20] Speck, F.-O., *General Wiener-Hopf Factorization Methods*, Research Notes in Mathematics **119**, Pitman Advanced Publishing Program, London 1985.

[21] ———, Mixed Boundary Value Problems of the Type of Sommerfeld's Half-plane Problem, *Proc. Roy. Soc. Edinburgh Sect. A* **104** (1986), 261–277.

[22] ———, Sommerfeld Diffraction Problems with First and Second Kind Boundary Conditions, *SIAM J. Math. Anal.* **20** (1989), 396–407.

[23] Teixeira, F.S., Diffraction by a Rectangular Wedge: Wiener-Hopf-Hankel Formulation, *Integral Equations Operator Theory* **14** (1991), 436–454.

[24] Triebel, H., *Interpolation Theory, Function Spaces, Differential Operators, 2nd ed.*, Barth, Leipzig 1995.

Luís Pinheiro de Castro
Department of Mathematics
University of Aveiro
3810-193 Aveiro, Portugal
e-mail: `lcastro@mat.ua.pt`

Frank-Olme Speck
Department of Mathematics
Instituto Superior Técnico
Technical University of Lisbon
1049-001 Lisboa, Portugal
e-mail: `fspeck@math.ist.utl.pt`

2000 *Mathematics Subject Classification.*
Primary 78A45; Secondary 35J05, 47B35.

Received: 24 May 2001.

Operator Theory:
Advances and Applications, Vol. 142, 91–100

Factorization of Matrix Functions and the Resolvents of Certain Operators

Ana C. Conceição, Viktor G. Kravchenko and Francisco S. Teixeira

Abstract. The explicit factorization of matrix functions of the form

$$A_\gamma(b) = \begin{pmatrix} e & b \\ b^* & b^*b + \gamma e \end{pmatrix},$$

where b is an $n \times n$ matrix function, e represents the identity matrix, and γ is a complex constant, is studied. To this purpose some relations between a factorization of A_γ and the resolvents of the self-adjoint operators

$$N_+(b) = P_+bP_-b^*P_+ \quad \text{and} \quad N_-(b) = P_-b^*P_+bP_-$$

are analyzed. The main idea is to show that if b is a matrix function that can be represented through the decomposition $b = b_- + b_+$ where at least one of the summands is a rational matrix, then it is possible to construct an algorithm that allows us to determine an effective canonical factorization of the matrix function A_γ.

1. Introduction

The explicit factorization problem for matrix functions has applications in different areas, such as the theory of singular integral operators, contour problems, the theory of non-linear differential equations (see, for instance, [CG, LS, FT]). However, only for some restricted classes of matrix functions (such as rational matrix functions) there exists an algorithm that allows us to determine their explicit factorization.

In this work we intend to obtain an explicit factorization of matrix functions of the type

$$A_\gamma(b) = \begin{pmatrix} e & b \\ b^* & b^*b + \gamma e \end{pmatrix},$$

where γ is an arbitrary complex constant, e represents the identity matrix function, b is a matrix function which belongs to $[L_\infty(\mathbb{T})]_{n,n}$, where $\mathbb{T}$ denotes the unit circle, and b^* is the Hermitian adjoint of b.

The inherent idea of our work (see [KM]) is to disclose a relation between a factorization of $A_\gamma(b)$ and the resolvents of the self-adjoint operators

$$N_+(b) = P_+bP_-b^*P_+ \quad \text{and} \quad N_-(b) = P_-b^*P_+bP_-,$$

$$N_\pm(b) : [L_2(\mathbb{T})]_{n,n} \to [L_2(\mathbb{T})]_{n,n}.$$

In the first part, the resolvent of the operator $N_-(b)$ ($N_+(b)$) is obtained when the matrix function $A_\gamma(b)$ admits a left (right) canonical generalized factorization. In the second part, perhaps the most important, an explicit left canonical generalized factorization of $A_\gamma(b)$ is constructed using the resolvent of the operator $N_+(b)$ (the same factorization could be obtained through the operator $N_-(b)$). If a right factorization is wanted, then a similar procedure can be developed by making use of the resolvents of the operators $N_-(b^*)$ or $N_+(b^*)$.

In the case where $b \in [C(\mathbb{T})+L_\infty^+(\mathbb{T})]_{n,n}$, the self-adjoint operators N_+ and N_- are compact operators and the corresponding resolvent operators can be represented through their eigenvalues and respective eigenfunctions. Furthermore, in some particular cases, when b belongs to a decomposing subalgebra of $[L_\infty(\mathbb{T})]_{n,n}$, a canonical generalized factorization can be obtained in terms of the eigenfunctions of the operator considered. Moreover, if a decomposition $b = b_- + b_+$ is given, then the construction of a left (right) canonical factorization does not depend on the knowledge of b_- (b_+, respectively). This fact is very important, because it leads to the conclusion that there exists a quite general class of matrix functions for which the problem of the existence of an explicit left (right) canonical factorization depends only on the matrix function b_+ (b_-, respectively). In particular, if b_+ (b_-) is a rational matrix function then $N_\pm(b)$ ($N_\pm(b^*)$) are finite-dimensional operators and, consequently, it is possible to determine their eigenvalues and eigenfunctions. These can in turn be used to design an algorithm for obtaining a left (right) factorization of $A_\gamma(b)$ which is similar to the known algorithm for the factorization of rational matrix functions (see [P]).

Finally, some examples are given which clarify the results obtained.

2. Preliminaries

Let $\mathbb{T}$ denote the unit circle. We will consider the space $L_2(\mathbb{T})$. As usual, $P_\pm = (I \pm S)/2$ denote the Cauchy projectors associated with the singular integral operator S,

$$S\varphi(t) = \frac{1}{\pi i}\int_{\mathbb{T}} \frac{\varphi(\tau)}{\tau - t}\, d\tau,\ t \in \mathbb{T},$$

where the integral is understood in the sense of its principal value. We introduce the spaces

$$L_2^+(\mathbb{T}) = \mathrm{im}P_+,\ L_2^{-,0}(\mathbb{T}) = \mathrm{im}P_-,\ L_2^-(\mathbb{T}) = L_2^{-,0}(\mathbb{T}) \oplus \mathbb{C}.$$

Further, we need some standard definitions (see, for instance, [CG, GK, K, KL, LS]). Let us consider a matrix function $A \in G[L_\infty(\mathbb{T})]_{n,n}$. We say that A admits a left (right) canonical generalized factorization in $L_2(\mathbb{T})$ if it can be represented as

$$A = A_+A_- \quad (A = A_-A_+),$$

where

(i) $A_{+}^{\pm 1} \in [L_2^{+}(\mathbb{T})]_{n,n}$ and $A_{-}^{\pm 1} \in [L_2^{-}(\mathbb{T})]_{n,n}$

(ii) The operator $A_{-}^{-1}P_{-}A_{+}^{-1}I \quad (A_{+}^{-1}P_{+}A_{-}^{-1}))$ is a bounded linear operator in $[L_2(\mathbb{T})]_n$.

Moreover, if condition (i) holds and condition (ii) is dropped, we simply say that A admits a canonical factorization.

In the following sections we shall be dealing with the matrix function

$$A_\gamma(b) = \begin{pmatrix} e & b \\ b^* & b^*b + \gamma e \end{pmatrix}$$

with $b \in [L_\infty(\mathbb{T})]_{n,n}$. It is easy to see that the matrix function can be expressed in the form

$$A_\gamma(b) = \begin{pmatrix} e & 0 \\ b^* & \gamma e \end{pmatrix} \begin{pmatrix} e & b \\ 0 & e \end{pmatrix},$$

from which it follows that $\det A_\gamma(b) = \gamma^n$. Consequently, A_γ is an invertible matrix function if and only if γ is a non-zero complex constant and in that case its inverse is given by the matrix function

$$A_\gamma^{-1}(b) = \frac{1}{\gamma} \begin{pmatrix} \gamma e + bb^* & -b \\ -b^* & e \end{pmatrix}.$$

From now on, we will assume $\gamma \neq 0$.

3. Resolvents of N_- and N_+ through the factorization of A_γ

Let $b \in [L_\infty(\mathbb{T})]_{n,n}$. Considering the equation

$$\gamma\varphi + N_-(b)\varphi = f,$$

where

$$\varphi = \varphi_+ + \varphi_-, \quad \varphi_\pm \in [L_2^{\pm}(\mathbb{T})]_{n,n}, \quad f \in [L_2(\mathbb{T})]_{n,n},$$

we obtain immediately that $\varphi_+ = \gamma^{-1}P_+f$. It remains to solve the equation

$$\gamma\varphi_- + N_-(b)\varphi_- = f_-,$$

where $f_- = P_-f$. Writing $b\varphi_- = \omega_+ - \omega_-$, it follows that $\gamma\varphi_- + b^*\omega_+ - \chi_+ = f_-$, where $\omega_\pm \in [L_2^{\mp}(\mathbb{T})]_{n,n}$ and $\chi_+ \in [L_2^{+}(\mathbb{T})]_{n,n}$. It is easy to prove that

$$\begin{pmatrix} e & 0 \\ b^* & -e \end{pmatrix} \Phi_+ = \begin{pmatrix} e & b \\ 0 & -\gamma e \end{pmatrix} \Phi_- + \Psi_-,$$

where

$$\Phi_+ = \begin{pmatrix} \omega_+ \\ \chi_+ \end{pmatrix}, \quad \Phi_- = \begin{pmatrix} \omega_- \\ \varphi_- \end{pmatrix}, \quad \Psi_- = \begin{pmatrix} 0 \\ f_- \end{pmatrix}.$$

Therefore, we associate the Riemann-Hilbert problem

$$\Phi_+ = A_\gamma(b)\Phi_- - \Psi_-$$

with the matrix function $A_\gamma(b)$ (see, for instance, [KL]). Let us suppose that the matrix $A_\gamma(b)$ admits a left canonical generalized factorization of the form $A_\gamma(b) = A_\gamma^+ A_\gamma^-$. If we consider the operators

$$\hat{\pi}\begin{pmatrix} x_1 \\ x_2 \end{pmatrix} = x_2 \quad \text{and} \quad \check{\pi}x = \begin{pmatrix} 0 \\ x \end{pmatrix},$$

the consideration of the above Riemann-Hilbert problem shows that

$$\varphi_- = \hat{\pi}(A_\gamma^-)^{-1}P_-(A_\gamma^+)^{-1}\check{\pi}f_-.$$

Using a similar procedure for the operator $N_+(b)$ and denoting the resolvent of $N_\pm(b)$ by

$$R(N_\pm(b), -\gamma I) = (N_\pm(b) + \gamma I)^{-1}$$

and the resolvent set of $N_\pm$ by $\rho(N_\pm)$, we obtain the following result.

Theorem 3.1.

(i) *If $A_\gamma(b)$ admits a canonical generalized factorization of the form*

$$A_\gamma(b) = A_\gamma^+ A_\gamma^-,$$

then $-\gamma \in \rho(N_-(b))$. In that case,

$$R(N_-(b), -\gamma I) = \hat{\pi}(A_\gamma^-)^{-1}P_-(A_\gamma^+)^{-1}\check{\pi}P_- + \frac{1}{\gamma}P_+.$$

(ii) *If $A_\gamma(b^*)$ admits a canonical generalized factorization of the form*

$$A_\gamma(b^*) = A_\gamma^- A_\gamma^+,$$

then $-\gamma \in \rho(N_+(b))$. In that case,

$$R(N_+(b), -\gamma I) = \hat{\pi}(A_\gamma^+)^{-1}P_+(A_\gamma^-)^{-1}\check{\pi}P_+ + \frac{1}{\gamma}P_-.$$

4. Factorization of A_γ through the resolvent of N_- and N_+

4.1. The cases $b \in [L_\infty(\mathbb{T})]_{n,n}$ and $b \in [C(\mathbb{T}) + L_\infty^+(\mathbb{T})]_{n,n}$

Let us study the Riemann-Hilbert problem

$$\begin{cases} \Psi_+ = A_\gamma(b)(E + \Psi_-) \\ \Psi_-(\infty) = 0, \end{cases}$$

where $b \in [L_\infty(\mathbb{T})]_{n,n}$ and $E = \operatorname{diag}(e, e)$. The initial objective is to determine matrix functions $\Psi_\pm \in [L_2^\pm(\mathbb{T})]_{2n,2n}$ satisfying the above equations.

Assuming that $-\gamma$ does not belong to the spectrum of the operator $N_+(b)$, that is, $-\gamma \in \rho(N_+(b))$, and writing $\Psi_\pm$ as

$$\Psi_\pm = \begin{pmatrix} x_\pm & y_\pm \\ z_\pm & v_\pm \end{pmatrix},$$

it can be shown, after some computations, that the components of the matrices $\Psi_\pm$ can be represented through the resolvent of the operator $N_+(b)$ by

$$\begin{cases} x_+ = \gamma R(N_+(b), -\gamma I)e \\ y_+ = \gamma R(N_+(b), -\gamma I)P_+b \\ z_- = -P_-b^* R(N_+(b), -\gamma I)e \\ v_- = -P_-b^* R(N_+(b), -\gamma I)P_+b, \end{cases}$$

which yields the following result.

Lemma 4.1. *If $-\gamma \in \rho(N_+(b))$, then a solution to the Riemann-Hilbert problem can be represented by*

$$\Psi_+ = \begin{pmatrix} \gamma R(N_+(b), -\gamma I)e & \gamma R(N_+(b), -\gamma I)P_+b \\ \gamma P_+b^* R(N_+(b), -\gamma I)e & \gamma e + \gamma P_+b^* R(N_+(b), -\gamma I)P_+b \end{pmatrix},$$

and

$$\Psi_- = \begin{pmatrix} P_-bP_-b^* R(N_+(b), -\gamma I)e & -P_-b + P_-bP_-b^* R(N_+(b), -\gamma I)P_+b \\ -P_-b^* R(N_+(b), -\gamma I)e & -P_-b^* R(N_+(b), -\gamma I)P_+b \end{pmatrix}.$$

Let us remember that our goal is to determine a generalized factorization of $A_\gamma(b)$. If we introduce the notation

$$B_\gamma^+(b) = \Psi_+ \quad \text{and} \quad B_\gamma^-(b) = E + \Psi_-,$$

then the essential question is to verify if $A_\gamma(b) = B_\gamma^+(b)(B_\gamma^-(b))^{-1}$ is a canonical generalized factorization of the matrix function $A_\gamma(b)$.

It is obvious that

$$\det(B_\gamma^\pm(b))(t) \neq 0 \ \forall t \in \mathbb{T} \quad \text{and} \quad B_\gamma^\pm(b) \in [L_2^\pm(\mathbb{T})]_{2n,2n}.$$

Therefore, it remains to check that the following conditions hold:

i) $\det(B_\gamma^\pm(b))(z) \neq 0 \ \forall z \in \mathbb{T}_\pm$

ii) $(B_\gamma^\pm(b))^{-1} \in [L_2^\pm(\mathbb{T})]_{2n,2n}$

iii) $B_\gamma^-(b)P_-(B_\gamma^+(b))^{-1}I$ is a bounded operator on $[L_2(\mathbb{T})]_{2n}$

In general, we cannot give a complete answer. But in some particular cases it can be shown that

$$A_\gamma(b) = B_\gamma^+(b)(B_\gamma^-(b))^{-1}$$

is really a canonical generalized factorization of $A_\gamma(b)$.

Let us now look at the case where b belongs to the algebra of all matrices whose components can be represented as a sum of a continuous function and an essentially bounded function which admits a bounded analytic extension to the interior of $\mathbb{T}$, that is, $b \in [C(\mathbb{T}) + L_\infty^+(\mathbb{T})]_{n,n}$. In that case, N_+ is a self-adjoint and compact operator (see [LS]). Hence, the resolvent of N_+ can be represented in terms of the eigenfunctions and eigenvalues. Thus, using the spectral theorem for compact self-adjoint operators and introducing the operator

$$K(\psi) = \sum_k \frac{\lambda_k(\psi, \nu_k^+)}{\lambda_k + \gamma} \nu_k^+,$$

where $\{\nu_k^+\}$ is the orthonormal system in $[L_2(\mathbb{T})]_n$ formed by the eigenfunctions of the operator $N_+(b)$ and where λ_k are the corresponding eigenvalues, we get

$$R(N_+(b), -\gamma I)\Psi = \frac{1}{\gamma}\Psi - \frac{1}{\gamma}K(\Psi),$$

where $\Psi = [\psi_1, \ldots, \psi_n]$ and $K(\Psi) = [K(\psi_1), \ldots, K(\psi_n)]$.

We obtain the following lemma.

Lemma 4.2. *If $-\gamma \in \rho(N_+(b))$, then a solution to the Riemann-Hilbert problem*

$$\begin{cases} B_\gamma^+(b) = A_\gamma(b) B_\gamma^-(b) \\ B_\gamma^-(b)(\infty) = E \end{cases}$$

is given by

$$B_\gamma^+(b) = \begin{pmatrix} e - K(e) & P_+b - K(b) \\ P_+b^*(e - K(e)) & \gamma e + P_+b^*(P_+b - K(b)) \end{pmatrix}$$

and

$$B_\gamma^-(b) = \begin{pmatrix} e + \frac{1}{\gamma}P_-bP_-b^*(e - K(e)) & -P_-b + \frac{1}{\gamma}P_-bP_-b^*(P_+b - K(b)) \\ -\frac{1}{\gamma}P_-b^*(e - K(e)) & e - \frac{1}{\gamma}P_-b^*(P_+b - K(b)) \end{pmatrix},$$

where $e = [e_1, \ldots, e_n]$ and

$$K(e_j) = \sum_k \frac{2\pi\lambda_k \overline{\nu_{kj}^+(0)}}{\lambda_k + \gamma} \nu_k^+.$$

This lemma shows that $B_\gamma^\pm(b)$ can be obtained through the eigenvalues and eigenfunctions of $N_+(b)$. However, our question remains open, i.e., we do not yet know if $A_\gamma(b) = B_\gamma^+(b)(B_\gamma^-(b))^{-1}$ is or is not a canonical generalized factorization of $A_\gamma(b)$. This motivates the next subsection.

4.2. The case $b \in [\mathcal{A}(\mathbb{T})]_{n,n}$

Let us consider a decomposing algebra $\mathcal{A}(\mathbb{T})$ of continuous functions and let

$$\mathcal{A}^\pm(\mathbb{T}) = \mathcal{A}(\mathbb{T}) \cap \mathcal{C}^\pm(\mathbb{T}),$$

where $\mathcal{C}^\pm(\mathbb{T})$ refer to the classes of continuous functions on $\mathbb{T}$ which have a holomorphic extension to $\mathbb{T}_\pm$ and are continuous on $\mathbb{T}_\pm \cup \mathbb{T}$. Here $\mathbb{T}_+$ is the open unit disk and $\mathbb{T}_-$ is the exterior region of the unit circle (∞ included).

It is well known that the following result holds (see, for instance, [LS, GK]).

Lemma 4.3. *If* $A(z) \in [\mathcal{A}^{\pm}(\mathbb{T})]_{n,n}$ *and* $\det A(z) \neq 0$ *for all* $z \in \mathbb{T}_{\pm}$, *then* $A^{-1}(z) \in [\mathcal{A}^{\pm}(\mathbb{T})]_{n,n}$.

Let $b \in [\mathcal{A}(\mathbb{T})]_{n,n}$. In this case, if we can prove that i) is satisfied, then our problem is solved, i.e.,

$$A_{\gamma}(b) = B_{\gamma}^{+}(b)(B_{\gamma}^{-}(b))^{-1},$$

is a canonical generalized factorization of $A_{\gamma}(b)$.

Thus, we will establish necessary and sufficient conditions for the factorization $A_{\gamma}(b) = B_{\gamma}^{+}(b)(B_{\gamma}^{-}(b))^{-1}$ to be a canonical generalized factorization of $A_{\gamma}(b)$ in terms of the resolvent of $N_{+}(b)$. These conditions amount to the linear independence of the columns of the matrix functions $B_{\gamma}^{+}(b)$ and $(B_{\gamma}^{-}(b))^{-1}$ at every point of $\mathbb{T}_{+}$ and $\mathbb{T}_{-}$, respectively. After some straightforward computations, we can prove the following result.

Theorem 4.4. *Let* $-\gamma \in \rho(N_{+}(b))$, $b_{-} + b_{+} = b \in [\mathcal{A}(\mathbb{T})]_{n,n}$,

$$c_{ik} = \frac{2\pi\lambda_k \overline{\nu_{ki}^{+}(0)}}{\lambda_k + \gamma}, \quad d_{ik} = \frac{\lambda_k (P_{+}b_i, \nu_k^{+})}{\lambda_k + \gamma}, \quad f_i(z) = e_i - \sum_k c_{ik}\nu_k^{+}(z),$$

and $g_i(z) = P_{+}b_i(z) - \sum_k d_{ik}\nu_k^{+}(z)$. *Then* $A_{\gamma}(b) = B_{\gamma}^{+}(b)(B_{\gamma}^{-}(b))^{-1}$ *is a left canonical generalized factorization of the matrix function* $A_{\gamma}(b)$ *if and only if the vector systems*

$$\left\{ \begin{pmatrix} f_i(z) \\ P_{+}b_{+}^{*}(z)f_i(z) \end{pmatrix}, \begin{pmatrix} g_i(z) \\ \gamma e_i + P_{+}b_{+}^{*}(z)g_i(z) \end{pmatrix} \right\}_{i=\overline{1,n}}$$

and

$$\left\{ \begin{pmatrix} e_i + \frac{1}{\gamma}P_{-}b_{+}(z)P_{-}b_{+}^{*}(z)f_i(z) \\ -\frac{1}{\gamma}P_{-}b_{+}^{*}(z)f_i(z) \end{pmatrix}, \begin{pmatrix} \frac{1}{\gamma}P_{-}b_{+}(z)P_{-}b_{+}^{*}(z)g_i(z) \\ e_i - \frac{1}{\gamma}P_{-}b_{+}^{*}(z)g_i(z) \end{pmatrix} \right\}_{i=\overline{1,n}}$$

are linearly independent at each point of $\mathbb{T}_{\pm}$.

In general, it is hard to verify these conditions. However, we will present some concrete examples in which these conditions can be checked and thus yield an explicit left canonical generalized factorization of $A_{\gamma}(b)$.

Remark 4.5. Instead of $N_{+}(b)$, we could consider the operator $N_{-}(b)$ from the very beginning of this section, obtaining similar results. Furthermore, it is possible to prove that $\sigma(N_{+}(b)) = \sigma(N_{-}(b))$ and to show that the factorizations obtained in either way are equal for the case where $b \in [C(\mathbb{T}) + L_{\infty}^{+}(\mathbb{T})]_{n,n}$.

It should be noted that if a decomposition $b = b_{-} + b_{+}$ is considered, then the construction of an explicit left generalized factorization of the matrix function $A_{\gamma}(b)$ (where the resolvent of $N_{+}(b)$ or $N_{-}(b)$ is used) does not depend on the knowledge of b_{-}. This fact is very important, because it reveals that we have a quite general class of matrix functions where we only have to worry about the term b_{+}. In particular, if b_{+} is a rational matrix function, then $N_{+}(b)$ and $N_{-}(b)$

are finite-dimensional operators and, consequently, it is possible to determine their eigenvalues and eigenfunctions, which makes the conditions of Theorem 4.4 easier to verify. Therefore, this result can lead us to an algorithm similar to the well-known algorithm for the factorization of rational matrix functions (see [P]). Roughly speaking, if some "problems" arise at points of $\mathbb{T}_+$, it should be possible to remove them separately, one by one, until a generalized (although perhaps not canonical) factorization is obtained. In fact, let $b = b_- + b_+$ where $b_\pm \in \mathcal{A}^\pm(\mathbb{T})$. Then

$$\begin{pmatrix} e & b \\ b^* & b^*b+\gamma e \end{pmatrix} = \begin{pmatrix} e & 0 \\ b_-^* & e \end{pmatrix} \begin{pmatrix} e & b_+ \\ b_+^* & b_+^*b_+ + \gamma e \end{pmatrix} \begin{pmatrix} e & b_- \\ 0 & e \end{pmatrix}.$$

Thus, if b_+ is a rational matrix function with its poles in $\mathbb{T}_-$ and vanishing at infinity, then one can always factorize the middle factor on the right-hand side, since this involves the factorization of a rational matrix function. Then, no matter what b_- is (rational or non-rational), one can factorize $A_\gamma(b)$. In other words, a factorization of $A_\gamma(b_+)$ always leads to a factorization of $A_\gamma(b)$, independently of how complicated b_- is.

Finally, we note that the study of the Riemann-Hilbert problem

$$\begin{cases} \Psi_+ = A_\gamma^{-1}(b)(E+\Psi_-) \\ \Psi_-(\infty) = 0 \end{cases}$$

and the use of the operator $N_+(b^*)$ or $N_-(b^*)$ allows us to state a theorem analogous to Theorem 4.4, which provides us with a necessary and sufficient condition for the existence of a right canonical generalized factorization of $A_\gamma(b)$, written explicitly in the form

$$A_\gamma(b) = C_\gamma^-(b^*)(C_\gamma^+(b^*))^{-1}$$

where the factors $C_\gamma^\pm(b^*)$ are given by

$$C_\gamma^-(b^*) = \begin{pmatrix} 0 & e \\ -e & 0 \end{pmatrix} B_\gamma^-(b^*) \begin{pmatrix} 0 & -e \\ e & 0 \end{pmatrix}$$

and

$$C_\gamma^+(b^*) = \frac{1}{\gamma} \begin{pmatrix} 0 & e \\ -e & 0 \end{pmatrix} B_\gamma^+(b^*) \begin{pmatrix} 0 & -e \\ e & 0 \end{pmatrix}.$$

Here K is the operator defined by

$$\mathrm{K}(\Psi) = \Psi - \gamma R(N_+(b^*), -\gamma I)\Psi,$$

which can be expressed, as before, in terms of the eigenvalues and eigenfunctions of the operator $N_+(b^*)$. In this case, the construction of an explicit right canonical generalized factorization depends only on the knowledge of the function b_- and comments similar to those presented after Remark 4.5 still hold.

4.3. An example

We will now present an example to clarify the results obtained (for the case of left factorization). Let us consider the case where

$$b_- + \operatorname{diag}\left[\beta_i \frac{1}{t-a_i}\right]_{i=\overline{1,n}} = b \in [\mathcal{A}(\mathbb{T})]_{n,n}$$

with $a_i, \beta_i \in \mathbb{C}$ and $|a_i| > 1$ for all i.

With the matrix functions

$$A_{11}^+ = \operatorname{diag}\left[1 - \tfrac{c_{ii}}{k_i(t-a_i)}\right]_{i=\overline{1,n}},$$

$$A_{12}^+ = \operatorname{diag}\left[\tfrac{\gamma\alpha_i}{k_i(\lambda_i+\gamma)(t-a_i)}\right]_{i=\overline{1,n}},$$

$$A_{21}^+ = b_-^* A_{11}^+ + \operatorname{diag}\left[-\tfrac{\overline{\alpha_i}}{k_i\overline{a_i}} + \tfrac{\overline{\alpha_i}a_i c_{ii}}{k_i^2(|a_i|^2-1)(t-a_i)}\right]_{i=\overline{1,n}},$$

$$A_{22}^+ = b_-^* A_{12}^+ + \gamma e + \operatorname{diag}\left[-\tfrac{\gamma|\alpha_i|^2 a_i}{k_i^2(\lambda_i+\gamma)(|a_i|^2-1)(t-a_i)}\right]_{i=\overline{1,n}},$$

$$A_{11}^- = -b_- A_{21}^- + e + \operatorname{diag}\left[\left(1 + \tfrac{c_{ii}\overline{a_i}}{k_i(|a_i|^2-1)}\right)\tfrac{|\alpha_i|^2}{\gamma k_i^2\overline{a_i}(|a_i|^2-1)\left(t-\frac{1}{\overline{a_i}}\right)}\right]_{i=\overline{1,n}},$$

$$A_{12}^- = -b_- A_{22}^- + \operatorname{diag}\left[-\tfrac{|\alpha_i|^2\alpha_i}{k_i^3(|a_i|^2-1)^2(\lambda_i+\gamma)\left(t-\frac{1}{\overline{a_i}}\right)}\right]_{i=\overline{1,n}},$$

$$A_{21}^- = \operatorname{diag}\left[\left(1 + \tfrac{c_{ii}\overline{a_i}}{k_i(|a_i|^2-1)}\right)\tfrac{\overline{\alpha_i}}{\gamma k_i\overline{a_i}^2\left(t-\frac{1}{\overline{a_i}}\right)}\right]_{i=\overline{1,n}},$$

$$A_{22}^- = \operatorname{diag}\left[1 - \tfrac{|\alpha_i|^2}{k_i^2\overline{a_i}(|a_i|^2-1)(\lambda_i+\gamma)\left(t-\frac{1}{\overline{a_i}}\right)}\right]_{i=\overline{1,n}},$$

$$k_i = \sqrt{\tfrac{2\pi}{|a_i|^2-1}},\ \alpha_i = k_i\beta_i,\ \lambda_i = \tfrac{|\alpha_i|^2}{k_i^2(|a_i|^2-1)^2},\ \text{and } c_{ii} = -\tfrac{k_i\lambda_i(|a_i|^2-1)}{\overline{a_i}(\lambda_i+\gamma)},\ \forall i = \overline{1,n},$$

we have

$$A_\gamma(b) = A_\gamma^+ A_\gamma^-,$$

where

$$A_\gamma^+ = \begin{pmatrix} A_{11}^+ & A_{12}^+ \\ A_{21}^+ & A_{22}^+ \end{pmatrix} \quad \text{and} \quad A_\gamma^- = \begin{pmatrix} A_{11}^- & A_{12}^- \\ A_{21}^- & A_{22}^- \end{pmatrix}^{-1}.$$

For any $\gamma \neq -\lambda_i, \forall i = \overline{1,n}$, this represents a left canonical generalized factorization of $A_\gamma(b)$, since the conditions of Theorem 4.4 are satisfied. Besides that, it is easy to see that A_{21}^- and A_{22}^- commute and that $A_{11}^- A_{22}^- - A_{12}^- A_{21}^- = e$. Thus, if we consider, for example, the case where $\beta_i \neq 0, \forall i = \overline{1,n}$, then A_{21}^- is invertible on $\mathbb{T}_-$ and we get (see [H])

$$A_\gamma^- = \begin{pmatrix} (A_{21}^-)^{-1} A_{22}^- A_{21}^- & -(A_{21}^-)^{-1}(A_{22}^- A_{21}^- A_{11}^- (A_{21}^-)^{-1} - e) \\ -A_{21}^- & A_{21}^- A_{11}^- (A_{21}^-)^{-1} \end{pmatrix}.$$

References

[CG] Clancey, K. and Gohberg, I., *Factorization of Matrix Functions and Singular Integral Operators*, Operator Theory: Advances and Applications, Vol. 3, Birkhäuser Verlag, Basel-Boston 1981.

[FT] Faddeev, L.D. and Takhtayan, L.A., *Hamiltonian Methods in the Theory of Solitons*, Springer-Verlag, Berlin 1987.

[GK] Gohberg, I. and Krupnik, N.Ya., *One-Dimensional Linear Singular Integral Operators*, Operator Theory: Advances and Applications, Vols. 53 and 54, Birkhäuser Verlag, Basel 1992.

[H] Halmos, P.R., *A Hilbert Space Problem Book*, second edition, Springer-Verlag, New York-Berlin 1982.

[K] Krupnik, N.Ya., *Banach Algebras with Symbol and Singular Integral Operators*, Operator Theory: Advances and Applications, Vol. 26, Birkhäuser Verlag, Basel 1987.

[KL] Kravchenko, V.G. and Litvinchuk, G.S., *Introduction to the Theory of Singular Integral Operators with Shift*, Mathematics and its Applications, Vol. 289, Kluwer Academic Publishers, Dordrecht 1994.

[KM] Kravchenko, V.G. and Migdal'skii, A.I., A regularization algorithm for some boundary-value problems of linear conjugation (English), *Dokl. Math.* **52** (1995), 319–321.

[LS] Litvinchuk, G.S. and Spitkovskii, I.M., *Factorization of Measurable Matrix Functions*, Operator Theory: Advances and Applications, Vol. 25, Birkhäuser Verlag, Basel 1987.

[P] Prössdorf, S., *Some Classes of Singular Equations*, North-Holland, Amsterdam 1978.

Ana C. Conceição and Viktor G. Kravchenko
Área Dep. de Matemática
Universidade do Algarve
Campus de Gambelas
8000-810 Faro, Portugal
e-mail: aconcei@ualg.pt
e-mail: vkravch@ualg.pt

Francisco S. Teixeira
Departamento de Matemática
Instituto Superior Técnico
Av. Rovisco Pais
1049-001 Lisboa, Portugal
e-mail: fteix@math.ist.utl.pt

2000 *Mathematics Subject Classification.*
Primary 47A68; Secondary 47A10.

Received: 5 June 2001 .

Operator Theory:
Advances and Applications, Vol. 142, 101–115

Finite Difference Cauchy-Riemann Operators and Their Fundamental Solutions in the Complex Case

Klaus Gürlebeck and Angela Hommel

Abstract. Using finite differences, discrete analogues of the Cauchy-Riemann operator in the complex case can be described in form of 2×2 matrix operators. By the help of the discrete Fourier transform the fundamental solution of these difference operators is calculated. The approximation error of the fundamental solution can be estimated in the space l_p as well as in the space L_p.

1. Introduction

While the idea to approximate the Cauchy-Riemann operator by finite differences is already very old, the idea to consider classes of solutions of finite difference equations as analogues to holomorphic functions is more difficult to realize. First attempts are contained in [D2], [DD], [Fe], [Hay], [I1] and [I2]. For similar developments in the theory of discrete harmonic functions we refer to [DS], [D1], [Ho], [Rya] and [Th]. Main problems of discrete function theories are related to the requirements that discrete holomorphic functions should be discrete harmonic functions, too, that one needs a discrete version of the Cauchy integral, and the fact that discrete holomorphic functions do not form an algebra. First hints how to overcome these problems are related to the theory of discrete harmonic functions and to the theory of difference potential operators (see [Rya] and the references therein). The theory of difference potentials was extended in [Ho], where convolution operators based on fundamental solutions of finite difference operators are considered. This approach allows one to define discrete versions of single layer and double layer potentials. Some years ago similar ideas were used in [GS1], [GS2] to study a function theory of solutions of a finite difference Cauchy-Riemann operator in three dimensions. Here, already the monogenic or hyperholomorphic functions themselves (in the continuous case) do not form an algebra and we will use some of the ideas from this theory for the following consideration.
In this paper finite difference approximations of the Cauchy-Riemann operator in the complex plane will be studied, so that the factorization of the real Laplacian into two adjoint Cauchy-Riemann operators is preserved in the discrete case, too. Fundamental solutions of these finite difference operators are calculated and their properties are studied in detail. The main goal is to prepare the basic tools for the

definition of difference analogues of the complex T-operator and of the Cauchy integral, including some convergence results.

2. Cauchy-Riemann operators and fundamental solutions in the complex case

Let $\mathbb{R}^2$ be the two-dimensional Euclidean space and $x = (x_1, x_2)$ be an element of this space. In the following the Cauchy-Riemann operators

$$D^1 = (-i)\left(\frac{\partial}{\partial x_1} + i\,\frac{\partial}{\partial x_2}\right) \quad \text{and} \quad D^2 = i\left(\frac{\partial}{\partial x_1} - i\,\frac{\partial}{\partial x_2}\right)$$

are considered, which have the important property

$$D^1\, D^2 = D^2\, D^1 = \Delta \quad \text{with} \quad \Delta\, u = \sum_{i=1}^{2} \frac{\partial^2 u}{\partial x_i^2}\,.$$

Let $\mathcal{D}$ be the space of all infinitely differentiable functions with compact support. In the distributional sense the solutions $E^k(x)$ of the equations

$$D^k\, E^k(x) = \delta(x) \quad \text{with} \quad (\delta, \varphi) = \varphi(0), \;\; \varphi \in \mathcal{D} \quad \text{and} \quad k \in \{1, 2\}$$

are called fundamental solutions. These fundamental solutions have the representation

$$E^1(x) = \frac{i}{2\pi(x_1 + i\,x_2)} \quad \text{and} \quad E^2(x) = \frac{i}{2\pi(x_1 - i\,x_2)}\,.$$

Using the Fourier transform

$$(Fu(\xi))(x) = (2\pi)^{-1} \int\limits_{\xi \in \mathbb{R}^2} u(\xi) e^{-i\langle x,\, \xi\rangle}\, d\xi$$

and the inverse transform

$$(F^{-1}v(x))(\xi) = (2\pi)^{-1} \int\limits_{x \in \mathbb{R}^2} v(x) e^{i\langle x,\, \xi\rangle}\, dx$$

with the Euclidean scalar product $\langle x,\, \xi\rangle = x_1\xi_1 + x_2\xi_2$, it can be calculated that

$$(F^{-1}E^1(x))(\xi) = \frac{i}{2\pi}\left(\frac{\xi_2 + i\xi_1}{\xi_1^2 + \xi_2^2}\right)$$

and

$$(F^{-1}E^2(x))(\xi) = \frac{i}{2\pi}\left(\frac{\xi_2 - i\xi_1}{\xi_1^2 + \xi_2^2}\right).$$

Based on the group homeomorphism between the complex numbers $a+ib$ and the matrices $\begin{pmatrix} a & -b \\ b & a \end{pmatrix}$, the fundamental solutions can be written in matrix form.

We obtain the representations

$$E^1(x) = FF^{-1}E^1(x) = \frac{i}{2\pi}\begin{pmatrix} F\left(\xi_2/|\xi|^2\right) & -F\left(\xi_1/|\xi|^2\right) \\ F\left(\xi_1/|\xi|^2\right) & F\left(\xi_2/|\xi|^2\right) \end{pmatrix}$$

and

$$E^2(x) = FF^{-1}E^1(x) = \frac{i}{2\pi}\begin{pmatrix} F\left(\xi_2/|\xi|^2\right) & F\left(\xi_1/|\xi|^2\right) \\ -F\left(\xi_1/|\xi|^2\right) & F\left(\xi_2/|\xi|^2\right) \end{pmatrix},$$

where the functions $\xi_j|\xi|^{-2}$, $j = 1, 2$, are locally integrable. As regular distributions, these functions will be identified with the functional $(\xi_j|\xi|^{-2}, \varphi)$.

3. Discrete theory

3.1. Discrete Cauchy-Riemann operators and fundamental solutions

Let $b_1 = (1,0)$ and $b_2 = (0,1)$ be unit vectors in the space $\mathbb{R}^2$. An equidistant lattice with the mesh width $h > 0$ is defined by $\mathbb{R}^2_h = \{mh = (m_1h, m_2h)$ with $m_1, m_2 \in \mathbb{Z}\}$. For each complex valued function

$$f_h(mh) = \text{ Re } f_h(mh) + i \text{ Im } f_h(mh) = (f^0_h(mh), f^1_h(mh))$$

we define forward differences $D^j_h\, f^k_h = h^{-1}(f^k_h(mh+hb_j) - f^k_h(mh))$ and backward differences $D^{-j}_h\, f^k_h = h^{-1}(f^k_h(mh) - f^k_h(mh - hb_j))$ with $j \in \{1,2\}$ and $k \in \{0,1\}$. Approximations of the Cauchy-Riemann operators defined in Section 2 are given by the finite difference operators

$$\begin{pmatrix} D^{-2}_h & D^1_h \\ -D^{-1}_h & D^2_h \end{pmatrix}\begin{pmatrix} \text{Re } f_h \\ \text{Im } f_h \end{pmatrix}$$
$$= (-i)\,[\,(D^{-1}_h \text{ Re } f_h + i\, D^1_h \text{ Im } f_h) + i\,(D^{-2}_h \text{ Re } f_h + i\, D^2_h \text{ Im } f_h)\,],$$

where, for sufficiently smooth functions, the right-hand side converges to $(-i)(\frac{\partial f}{\partial x_1} + i\,\frac{\partial f}{\partial x_2})$ and

$$\begin{pmatrix} D^2_h & -D^1_h \\ D^{-1}_h & D^{-2}_h \end{pmatrix}\begin{pmatrix} \text{Re } f_h \\ \text{Im } f_h \end{pmatrix}$$
$$= i\,[\,(D^{-1}_h \text{ Re } f_h + i\, D^1_h \text{ Im } f_h) - i\,(D^2_h \text{ Re } f_h + i\, D^{-2}_h \text{ Im } f_h)\,],$$

where the right-hand side converges to $i(\frac{\partial f}{\partial x_1} - i\,\frac{\partial f}{\partial x_2})$. For other approximations of the Cauchy-Riemann operators see for instance [D2], [Fe], [I1], and [I2]. The above definition of the discrete operators is used here, because the composition of the two operators exactly describes the discrete Laplacian in the form

$$\Delta_h u_h(mh) = \sum_{k\in K} a_k\, u_h(mh - kh) \text{ with } K = \{(0,0), (-1,0), (1,0), (0,-1), (0,1)\}$$

and

$$a_k = \begin{cases} 1/h^2 & \text{for } k \in K,\, k \neq (0,0) \\ -4/h^2 & \text{for } k = (0,0). \end{cases}$$

A discrete version of the δ functional is defined by

$$\delta_h(mh) = \begin{cases} h^{-2} & \text{for } mh = 0 \\ 0 & \text{for } mh \neq 0. \end{cases}$$

Each 2×2 matrix $E_h^1(mh)$ which is a solution of the system

$$\begin{pmatrix} D_h^{-2} & D_h^1 \\ -D_h^{-1} & D_h^2 \end{pmatrix} \begin{pmatrix} E_{h11}^1(mh) & E_{h12}^1(mh) \\ E_{h21}^1(mh) & E_{h22}^1(mh) \end{pmatrix} = \begin{pmatrix} \delta_h(mh) & 0 \\ 0 & \delta_h(mh) \end{pmatrix}$$

and each 2×2-matrix $E_h^2(mh)$ which solves the system

$$\begin{pmatrix} D_h^2 & -D_h^1 \\ D_h^{-1} & D_h^{-2} \end{pmatrix} \begin{pmatrix} E_{h11}^2(mh) & E_{h12}^2(mh) \\ E_{h21}^2(mh) & E_{h22}^2(mh) \end{pmatrix} = \begin{pmatrix} \delta_h(mh) & 0 \\ 0 & \delta_h(mh) \end{pmatrix}$$

is called a *discrete fundamental solution.* Here the discrete Cauchy-Riemann operator is thought of as acting on each column of the matrices $E_h^1(mh)$ or $E_h^2(mh)$.

Using for each matrix entry the discrete Fourier transform (see e.g. [St])

$$(F_h u_h)(\xi) = \begin{cases} \frac{h^2}{2\pi} \sum\limits_{mh \in \mathbb{R}_h^2} u_h(mh)\, e^{ih\langle m,\, \xi\rangle} & \xi \in Q_h \\ 0 & \xi \in \mathbb{R}^2 \setminus Q_h \end{cases}$$

with $Q_h = \{\xi \in \mathbb{R}^2 : -\pi/h < \xi_j < \pi/h, \;\; j = 1, 2\}$ and the properties

$$\begin{aligned} F_h(D_h^j u_h) &= -\xi_{-j}^{\,h} F_h u_h, & \xi_{-j}^{\,h} &:= h^{-1}(1 - e^{-ih\xi_j}), & j &= 1, 2 \\ F_h(D_h^{-j} u_h) &= \xi_j^h F_h u_h, & \xi_j^h &:= h^{-1}(1 - e^{ih\xi_j}), & j &= 1, 2 \end{aligned}$$

it follows that

$$\begin{pmatrix} \xi_2^h & -\xi_{-1}^{\,h} \\ -\xi_1^h & -\xi_{-2}^{\,h} \end{pmatrix} \begin{pmatrix} F_h\, E_{h11}^1 & F_h\, E_{h12}^1 \\ F_h\, E_{h21}^1 & F_h\, E_{h22}^1 \end{pmatrix} = \begin{pmatrix} F_h\, \delta_h & 0 \\ 0 & F_h\, \delta_h \end{pmatrix}$$

and

$$\begin{pmatrix} -\xi_{-2}^{\,h} & \xi_{-1}^{\,h} \\ \xi_1^h & \xi_2^h \end{pmatrix} \begin{pmatrix} F_h\, E_{h11}^2 & F_h\, E_{h12}^2 \\ F_h\, E_{h21}^2 & F_h\, E_{h22}^2 \end{pmatrix} = \begin{pmatrix} F_h\, \delta_h & 0 \\ 0 & F_h\, \delta_h \end{pmatrix}$$

with $F_h\, \delta_h = (2\pi)^{-1}$. Introducing the notations

$$A_1 = \begin{pmatrix} \xi_2^h & -\xi_{-1}^{\,h} \\ -\xi_1^h & -\xi_{-2}^{\,h} \end{pmatrix} \quad \text{and} \quad A_2 = \begin{pmatrix} -\xi_{-2}^{\,h} & \xi_{-1}^{\,h} \\ \xi_1^h & \xi_2^h \end{pmatrix},$$

the Fourier transformed discrete fundamental solutions can be written in the form

$$(F_h\, E_h^1)(\xi) = (2\pi)^{-1}(A_1)^{-1} \quad \text{and} \quad (F_h\, E_h^2)(\xi) = (2\pi)^{-1}(A_2)^{-1}.$$

As a consequence of the property

$$\begin{pmatrix} D_h^{-2} & D_h^1 \\ -D_h^{-1} & D_h^2 \end{pmatrix} \begin{pmatrix} D_h^2 & -D_h^1 \\ D_h^{-1} & D_h^{-2} \end{pmatrix} = \begin{pmatrix} D_h^2 & -D_h^1 \\ D_h^{-1} & D_h^{-2} \end{pmatrix} \begin{pmatrix} D_h^{-2} & D_h^1 \\ -D_h^{-1} & D_h^2 \end{pmatrix}$$

$$= \begin{pmatrix} \Delta_h^{12} & 0 \\ 0 & \Delta_h^{12} \end{pmatrix} \qquad \text{with} \quad \Delta_h^{12} = D_h^1\, D_h^{-1} + D_h^2\, D_h^{-2}$$

it can be shown that in the Fourier transformed domain $A_1 A_2 = A_2 A_1 = -d^2 I$, where $d^2 = 4h^{-2}\left(\sin^2 \frac{h\xi_1}{2} + \sin^2 \frac{h\xi_2}{2}\right)$ and I is the identity matrix. Therefore, the Fourier transforms of $E_h^1(mh)$ and $E_h^2(mh)$ have the representations

$$(F_h E_h^1)(\xi) = -\frac{1}{2\pi}\frac{1}{d^2} A_2 \quad \text{and} \quad (F_h E_h^2)(\xi) = -\frac{1}{2\pi}\frac{1}{d^2} A_1 .$$

An integral representation for the matrix elements of the discrete fundamental solution is obtained using the inverse discrete Fourier transform $(F_h)^{-1} = R_h F$, where $R_h u$ denotes the restriction of the function u to the lattice $\mathbb{R}_h^2$ and the symbol F stands for the classical Fourier transform. In detail the representation formulas

$$E_h^1(mh) = \frac{1}{2\pi}\begin{pmatrix} R_h F\left(\xi_{-2}^h/d^2\right) & R_h F\left(-\xi_{-1}^h/d^2\right) \\ R_h F\left(-\xi_1^h/d^2\right) & R_h F\left(-\xi_2^h/d^2\right) \end{pmatrix}$$

and

$$E_h^2(mh) = \frac{1}{2\pi}\begin{pmatrix} R_h F\left(-\xi_2^h/d^2\right) & R_h F\left(\xi_{-1}^h/d^2\right) \\ R_h F\left(\xi_1^h/d^2\right) & R_h F\left(\xi_{-2}^h/d^2\right) \end{pmatrix}$$

are obtained. The fact that the discrete Fourier transform is a mapping from $\mathbb{R}_h^2$ to Q_h implies that the integration variable in the expression $R_h F F_h u_h(mh)$ is restricted to the square Q_h. In order to simplify the last two formulas, the characteristic function χ_{Q_h} was neglected.

In the next step the matrix elements of the discrete fundamental solutions will be estimated.

Lemma 3.1. *At each mesh point $mh = (m_1 h, m_2 h)$ the elements of the matrices $E_h^1(mh)$ and $E_h^2(mh)$ can be estimated in the form*

$$\left|\frac{1}{(2\pi)^2}\int\limits_{\xi\in Q_h} \frac{\xi_{\pm l}^h}{d^2}\, e^{-ih\langle m,\,\xi\rangle}\, d\xi\right| \le \frac{C_1\, h}{(|mh|+h)^2} + \frac{C_2}{|mh|+h} \qquad \forall\, l \in \{1,2\}.$$

The proof is based on the following theorem of Thomée (see e.g. [Th]).

Theorem 3.2. *Let d be the dimension of the Euclidean space and p_1 and p_2 be two positive integers with $p_2 < p_1 + d$. For $N > 0$ let κ_N be the set of functions of the form $T(\Theta) = \frac{T_1(\Theta)}{T_2(\Theta)}$, $0 \ne \Theta \in Q_\pi$, where $T_j(\Theta)$ are trigonometric polynomials*

$$T_j(\Theta) = \sum_{\mu} t_{j,\mu}\, e^{i\langle \mu,\,\Theta\rangle}, \quad j = 1,\, 2,$$

which satisfy the following conditions:

(i) *there are ordinary homogeneous polynomials $P_j(\Theta)$ of degree p_j, $j = 1, 2$, such that $T_j(\Theta) = P_j(\Theta) + o(|\Theta|^{p_j})$ when $\Theta \to 0$*

(ii) $|T_2(\Theta)| \ge N^{-1}|\Theta|^{p_2}, \quad \Theta \in Q_\pi$

(iii) $|t_{j,\mu}| \le N$

(iv) $t_{j,\mu} = 0$ *for* $|\mu| > N$.

For any $N > 0$ there is a constant C such that for all μ (with integer components) and $T \in \kappa_N$,

$$\left| \int\limits_{Q_\pi} T(\Theta)\, e^{i\langle \mu,\, \Theta\rangle}\, d\Theta \right| \leq C(|\mu|+1)^{-(d+p_1-p_2)}.$$

Proof of Lemma 3.1. Using the substitution $\Theta = (\Theta_1, \Theta_2) = (h\xi_1, h\xi_2)$ we get

$$I = \pm \frac{1}{(2\pi)^2} \int\limits_{\xi \in Q_h} \frac{\xi_{\pm l}^{h}}{d^2}\, e^{-ih\langle m,\, \xi\rangle} d\xi = \pm \frac{1}{(2\pi)^2} \frac{1}{h} \int\limits_{\Theta \in Q_\pi} \frac{1 - e^{\pm i\Theta_l}}{d_\Theta^2}\, e^{-i\langle m,\, \Theta\rangle} d\Theta$$

$$= \pm \frac{1}{(2\pi)^2} \frac{1}{h} \int\limits_{\Theta \in Q_\pi} \frac{1 - \cos\Theta_l}{d_\Theta^2}\, e^{-i\langle m,\, \Theta\rangle} d\Theta \pm \frac{1}{(2\pi)^2} \frac{1}{h} \int\limits_{\Theta \in Q_\pi} \frac{\mp i \sin\Theta_l}{d_\Theta^2}\, e^{-i\langle m,\, \Theta\rangle} d\Theta$$

with $Q_\pi = \{\Theta \in \mathbb{R}^2 : -\pi < \Theta_j < \pi,\ j = 1, 2\}$ and $d_\Theta^2 = 4\left(\sin^2 \frac{\Theta_1}{2} + \sin^2 \frac{\Theta_2}{2}\right)$. All of the conditions listed in Theorem 3.2 can be fulfilled. For the first summand in the previous formula we can take $p_1 = p_2 = 2$ while in the second summand everything works with $p_1 = 1$ and $p_2 = 2$. By this way Lemma 3.1 is proved. □

3.2. Approximation error of the discrete fundamental solution

Let $k \in \{1,2\}$ and $\hat{E}^k(mh) = \begin{cases} E^k(mh) & \text{if} \quad mh \neq (0,0) \\ 0 & \text{if} \quad mh = (0,0). \end{cases}$

Each matrix element can be written in the form

$$E_{hlj}^k(mh) - \hat{E}_{lj}^k(mh) = \frac{1}{(2\pi)^2}\Bigg(\pm \int\limits_{\xi \in Q_h} \frac{1 - \cos(h\xi_s)}{h\, d^2}\, e^{-ih\langle m,\, \xi\rangle}\, d\xi$$

$$\pm\, i \int\limits_{\xi \in Q_h} \left(\frac{\sin(h\xi_s)}{h\, d^2} - \frac{\xi_s}{|\xi|^2} \right) e^{-ih\langle m,\, \xi\rangle}\, d\xi \pm \int\limits_{\xi \in \mathbb{R}^2 \setminus Q_h} -\frac{i\xi_s}{|\xi|^2}\, e^{-ih\langle m,\, \xi\rangle} d\xi \Bigg)$$

with $l, j \in \{1,2\}$, $s = 2$ if $l = j$ and $s = 1$ if $l \neq j$.

Lemma 3.3. *At each mesh point $mh = (m_1 h, m_2 h) \neq (0,0)$ we have the estimate $|E_{hlj}^k(mh) - \hat{E}_{lj}^k(mh)| \leq C\, h\, |mh|^{-2}$ $\forall\, l, j, k \in \{1,2\}$. In the case $mh = (0,0)$ the inequality $|E_{hlj}^k(mh) - \hat{E}_{lj}^k(mh)| \leq C\, h^{-1}$ is fulfilled.*

Proof. The upper bound in the case $mh = (0,0)$ results from Lemma 3.1. In the following the case $mh \neq (0,0)$ is considered. The proof will be done for $s = 1$ and can be carried over to $s = 2$. Using the substitution $\Theta = (\Theta_1, \Theta_2) = (h\xi_1, h\xi_2)$ and the notation as well as the result of Lemma 3.1 we obtain

$$\begin{aligned} |I_1| &= \left| \frac{1}{(2\pi)^2} \int\limits_{\xi \in Q_h} \frac{1 - \cos(h\xi_1)}{h\, d^2}\, e^{-ih\langle m,\, \xi\rangle}\, d\xi \right| \\ &= \left| \frac{1}{(2\pi)^2} \frac{1}{h} \int\limits_{\Theta \in Q_\pi} \frac{1 - \cos\Theta_1}{d_\Theta^2}\, e^{-i\langle m,\, \Theta\rangle}\, d\Theta \right| \leq \frac{C\, h}{(|mh| + h)^2}. \end{aligned}$$

In the last step the integral

$$
\begin{aligned}
|I_2| &= (2\pi)^{-2}\Big| \int\limits_{\xi\in Q_h} \left(\frac{\sin(h\xi_1)}{h\,d^2} - \frac{\xi_1}{|\xi|^2}\right) e^{-ih\langle m,\,\xi\rangle} d\xi - \int\limits_{\xi\in\mathbb{R}^2\setminus Q_h} \frac{\xi_1}{|\xi|^2}\, e^{-ih\langle m,\,\xi\rangle} d\xi \Big| \\
&= \frac{(2\pi)^{-2}}{h}\Big| \int\limits_{\Theta\in Q_\pi} \left(\frac{\sin\Theta_1}{d_\Theta^2} - \frac{\Theta_1}{|\Theta|^2}\right) e^{-i\langle m,\,\Theta\rangle} d\Theta - \int\limits_{\Theta\in\mathbb{R}^2\setminus Q_\pi} \frac{\Theta_1}{|\Theta|^2}\, e^{-i\langle m,\,\Theta\rangle} d\Theta \Big| \\
&= |[Fv_1](m)|
\end{aligned}
$$

with

$$
v_1(\Theta) = \frac{(2\pi)^{-1}}{h}\left[\left(\frac{\sin\Theta_1}{d_\Theta^2} - \frac{\Theta_1}{|\Theta|^2}\right)\chi_{Q_\pi} - \frac{\Theta_1}{|\Theta|^2}\,\chi_{\mathbb{R}^2\setminus Q_\pi}\right]
$$

is to estimate. The function $v_1(\Theta)$ is locally integrable and can be identified with the functional (v_1, φ) for all $\varphi \in \mathcal{D}$. Therefore

$$
([Fv_1](x), \varphi(x)) = (v_1(\Theta), [F\varphi](\Theta)) = \left(-\frac{1}{x_1^2}\, F\left(\frac{\partial^2 v_1(\Theta)}{\partial\Theta_1^2}\right), \varphi(x)\right). \tag{3.1}
$$

Let Γ_{Q_π} be the boundary of the square Q_π and define $\overline{Q_\pi} = Q_\pi \cup \Gamma_{Q_\pi}$. Using the same notation as in [Wla], the distributional derivatives of the function $v_1(\Theta) \in C^2(\overline{Q_\pi}) \cap C^2(\mathbb{R}^2 \setminus \overline{Q_\pi})$ can be written in the form

$$
\begin{aligned}
\frac{\partial^2 v_1}{\partial\Theta_i\partial\Theta_j} &= \left\{\frac{\partial^2 v_1}{\partial\Theta_i\partial\Theta_j}\right\} + \frac{\partial}{\partial\Theta_j}\Big([v_1]_{\Gamma_{Q_\pi}} \cos(\vec{n}, \Theta_i)\chi_{\Gamma_{Q_\pi}}\Big) \\
&\quad + \left[\left\{\frac{\partial v_1}{\partial\Theta_i}\right\}\right]_{\Gamma_{Q_\pi}} \cos(\vec{n}, \Theta_j)\,\chi_{\Gamma_{Q_\pi}},
\end{aligned}
$$

where $\vec{n}$ is the outer normal unit vector at the points $\Theta \in \Gamma_{Q_\pi}$, $\left\{\frac{\partial^2 v_1}{\partial\Theta_i\partial\Theta_j}\right\}$ and $\left\{\frac{\partial v_1}{\partial\Theta_i}\right\}$ are the classical parts of the distributions $\frac{\partial^2 v_1}{\partial\Theta_i\partial\Theta_j}$ and $\frac{\partial v_1}{\partial\Theta_i}$, respectively, and $[v_1]_{\Gamma_{Q_\pi}}$ as well as $\left[\left\{\frac{\partial v_1}{\partial\Theta_i}\right\}\right]_{\Gamma_{Q_\pi}}$ denote the jump of v_1 and $\left\{\frac{\partial v_1}{\partial\Theta_i}\right\}$ when passing through Γ_{Q_π} from $\mathbb{R}^2 \setminus Q_\pi$ to Q_π along the direction of $\vec{n}$. Based on this formula it follows

$$
\begin{aligned}
\frac{\partial^2 v_1}{\partial\Theta_1^2} &= \frac{1}{2\pi}\frac{1}{h}\Bigg[\left(-\frac{\sin\Theta_1}{d_\Theta^2} - \frac{3\sin(2\Theta_1)}{d_\Theta^4} + \frac{8\sin^3\Theta_1}{d_\Theta^6} + \frac{6\Theta_1}{|\Theta|^4} - \frac{8\Theta_1^3}{|\Theta|^6}\right)\chi_{Q_\pi} \\
&\quad + \left(\frac{6\Theta_1}{|\Theta|^4} - \frac{8\Theta_1^3}{|\Theta|^6}\right)\chi_{\mathbb{R}^2\setminus Q_\pi} + \frac{\partial}{\partial\Theta_1}\left(-\frac{\sin\Theta_1}{d_\Theta^2}\right)\cos(\vec{n}, \Theta_1)\chi_{\Gamma_{Q_\pi}} \\
&\quad + \left(\frac{2\sin^2\Theta_1}{d_\Theta^4} - \frac{\cos\Theta_1}{d_\Theta^2}\right)\cos(\vec{n}, \Theta_1)\chi_{\Gamma_{Q_\pi}}\Bigg].
\end{aligned}
$$

The Fourier integral at the points $x = m = (m_1, m_2)$ can be written in the form

$$-\frac{1}{m_1^2}F\Big(\frac{\partial^2 v_1(\Theta)}{\partial\Theta_1^2}\Big) = -\frac{1}{(2\pi)^2 h m_1^2}\Big[\int\limits_{\Theta\in Q_\pi}\Big(-\frac{\sin\Theta_1}{d_\Theta^2} - \frac{3\sin(2\Theta_1)}{d_\Theta^4}$$

$$+\frac{8\sin^3\Theta_1}{d_\Theta^6} + \frac{6\Theta_1}{|\Theta|^4} - \frac{8\Theta_1^3}{|\Theta|^6}\Big)e^{-i\langle m,\Theta\rangle}d\Theta + \int\limits_{\Theta\in\mathbb{R}^2\setminus Q_\pi}\Big(\frac{6\Theta_1}{|\Theta|^4} - \frac{8\Theta_1^3}{|\Theta|^6}\Big)e^{-i\langle m,\Theta\rangle}d\Theta\Big],$$

because the Fourier transforms of the other two summands become zero. The integrand of the first integral has the upper bound $C\,|\Theta|^{-1}$ which together with representation (3.1) implies that $|I_2| = |[Fv_1](m)| \le C\,h|m_1 h|^{-2}$. In analogy, the inequality $|I_2| \le C\,h|m_2 h|^{-2}$ is fulfilled. If $k \in \{1,2\}$ is chosen as $|m_k| = \max\limits_{i=1,2}|m_i| > 0$, then $|I_2| \le C\,h|mh|^{-2}$ follows from the inequality $|m| \le \sqrt{2}\,|m_k|$. Based on the estimate

$$|E_{hlj}^k(mh) - \hat{E}_{lj}^k(mh)| \le |I_1| + |I_2| \le C\,h\,|mh|^{-2} \quad \forall\, l,j,k \in \{1,2\}$$

Lemma 3.3 is completely proved. □

3.3. Convergence in the space l_p

Let $G \subset \mathbb{R}^2$ be a bounded domain and $G_h = (G \cap \mathbb{R}_h^2) \subset \mathbb{R}_h^2$ and let $Q(G_h)$ denote the smallest square parallel to the axes which has the center $(0,0)$ and covers the domain G_h. The length of its side is $L = 2lh$. In the following the matrix norm

$$\|E_h^k(mh) - \hat{E}^k(mh)\|_M = \Big(\sum_{i,j=1}^{2}|E_{hij}^k(mh) - \hat{E}_{ij}^k(mh)|^2\Big)^{1/2}$$

will be used.

Theorem 3.4. *For each mesh width* $h \le e^{-1}$,

$$\|E_h^k(mh) - \hat{E}^k(mh)\|_{l_p(G_h)} \le \begin{cases} C_1\,h^{-1+2/p} & \text{if} \quad 1 < p < 2 \\ C_2\,h\,|\ln h| & \text{if} \quad p = 1 \end{cases} \qquad (k = 1,2).$$

Proof. Using Lemma 3.3 we estimate

$$\|E_h^k(mh) - \hat{E}^k(mh)\|_{l_p(G_h)}^p = \sum_{mh\in G_h}\|E_h^k(mh) - \hat{E}^k(mh)\|_M^p\,h^2$$

$$\le \sum_{mh\in G_h\setminus(0,0)}\Big(\sum_{i,j=1}^{2}\frac{C_3\,h^2}{|mh|^4}\Big)^{p/2}h^2 + \sum_{i,j=1}^{2}C_4\,h^{-p+2}$$

$$= C_5(p)\Big[\sum_{mh\in G_h\setminus(0,0)}\Big(\frac{h}{|mh|^2}\Big)^p h^2 + h^{-p+2}\Big]$$

$$\le C_5(p)\Big(4\sum_{m_1=1}^{l}\sum_{m_2=1}^{l}(m_1^2+m_2^2)^{-p}\,h^{-p+2} + 4\sum_{m=1}^{l}m^{-2p}\,h^{-p+2} + h^{-p+2}\Big).$$

In more detail it can be proved that, for each $h \le e^{-1}$,

$$\sum_{m_1=1}^{l} \sum_{m_2=1}^{l} (m_1^2 + m_2^2)^{-p} h^{-p+2}$$

$$\le 2^{-p} h^{-p+2} + 2 h^{-p+1} \int_{x=h}^{lh} \left(\frac{h}{x}\right)^{2p} dx + h^{-p} \int_{r=h}^{\sqrt{2}\,lh} \int_{\varphi=0}^{\pi/2} \left(\frac{h}{r}\right)^{2p} r\, d\varphi\, dr$$

$$\le \begin{cases} C_6\, h^{-p+2}, & \text{if } 1 < p < 2 \\ C_7\, h\, |\ln h|, & \text{if } p = 1 \end{cases}$$

and

$$\sum_{m=1}^{l} m^{-2p} h^{-p+2} \le h^{-p+2} + h^{-p+1} \int_{x=h}^{lh} \left(\frac{h}{x}\right)^{2p} dx \le C_8\, h^{-p+2}$$

if $1 \le p < 2$. □

Corollary 3.5. *For each* $h \le e^{-1}$ *and* $1 < p < 2$,

$$\|E_h^k(mh) - \hat{E}^k(mh)\|_{l_p(\mathbb{R}_h^2)} \le C\, h^{-1+2/p} \qquad (k = 1, 2).$$

Proof. This inequality results from Theorem 3.4, because only in the case $p = 1$ the length of the side of the square $Q(G_h)$ cannot be arbitrarily large. □

3.4. Convergence in the space L_p

The matrix elements of the discrete fundamental solutions $E_h^1(mh)$ and $E_h^2(mh)$ are defined as the restriction of a Fourier integral to the lattice $\mathbb{R}_h^2$. Neglecting this restriction, the difference $E_h^1(x) - E^1(x)$ as well as $E_h^2(x) - E^2(x)$ can be estimated in the space L_p with $1 < p < 4/3$. This extension from the lattice $\mathbb{R}_h^2$ to the space $\mathbb{R}^2$ seems to be natural. The upper bound for p is caused by the method of the proof.

Theorem 3.6. *Let* G *be a bounded domain and* $1 < p < 4/3$. *Then for each mesh width* $h < h_0$,

$$\|E_h^k(x) - E^k(x)\|_{L_p(G)} \le C(h_0)\, h^{2-2/p} \qquad (k = 1,\, 2).$$

Proof. In analogy to Theorem 3.4 let $E_{hmn}^k(x)$ and $E_{mn}^k(x)$, $m, n = 1, 2$ be the elements of the matrices $E_h^k(x)$ and $E^k(x)$ with $k = 1, 2$. The norm in the space $L_p(G)$ is defined by

$$\|E_h^k(x) - E^k(x)\|_{L_p(G)}^p = \int_{x \in G} \left(\sum_{m,n=1}^{2} |E_{hmn}^k(x) - E_{mn}^k(x)|^2 \right)^{p/2} dx.$$

For $k \in \{1,2\}$ the difference of the matrix elements can be written as a Fourier transformed distribution, that is,

$$|E^k_{hmn}(x) - E^k_{mn}(x)| = |[Ff^l(\xi)](x)| \quad \text{with} \tag{3.2}$$
$$f^l(\xi) = \frac{1}{2\pi}\left[\frac{1-e^{\pm ih\xi_l}}{hd^2} \pm \frac{i\xi_l}{|\xi|^2}\right]\chi_{Q_h} + \frac{1}{2\pi}\left[\pm \frac{i\xi_l}{|\xi|^2}\right]\chi_{\mathbb{R}^2\setminus Q_h}.$$

The parameter l is chosen as $l = 2$ if $m = n$ and $l = 1$ if $m \neq n$. The case $l = 1$ will be studied in detail. Let Γ_{Q_h} be the boundary of the square Q_h and $\overline{Q_h} = Q_h \cup \Gamma_{Q_h}$. Using again the notation of [Wla], the first distributional derivatives of the function $f^1 \in C^1(\overline{Q_h}) \cap C^1(\mathbb{R}^2 \setminus \overline{Q_h})$ can be written in the form

$$\frac{\partial f^1}{\partial \xi_i} = \left\{\frac{\partial f^1}{\partial \xi_i}\right\} + [f^1]_{\Gamma_{Q_h}} \cos(\vec{n}, \xi_i)\chi_{\Gamma_{Q_h}},$$

where $\vec{n}$ is the outer normal unit vector at the points $\xi \in \Gamma_{Q_h}$, $\left\{\frac{\partial f^1}{\partial \xi_i}\right\}$ is the classical part of the distribution $\frac{\partial f^1}{\partial \xi_i}$, and $[f^1]_{\Gamma_{Q_h}}$ denotes the jump of f^1 when passing through Γ_{Q_h} from $\mathbb{R}^2\setminus Q_h$ to Q_h along the direction of $\vec{n}$. For the Fourier integral of the distribution f^1 we have that

$$\begin{aligned} F(f^1) &= [i(x_1+ix_2)]^{-1}F\left(\frac{\partial f^1}{\partial \xi_1} + i\frac{\partial f^1}{\partial \xi_2}\right) \\ &= [i(x_1+ix_2)]^{-1}\Bigg[F\left\{\frac{\partial f^1}{\partial \xi_1} + i\frac{\partial f^1}{\partial \xi_2}\right\} \\ &\quad + \frac{1}{2\pi}F\left(-\frac{1-e^{\pm ih\xi_1}}{hd^2}\cos(\vec{n},\xi_1)\chi_{\Gamma_{Q_h}}\right) \\ &\quad + \frac{i}{2\pi}F\left(-\frac{1-e^{\pm ih\xi_1}}{hd^2}\cos(\vec{n},\xi_2)\chi_{\Gamma_{Q_h}}\right)\Bigg]. \end{aligned} \tag{3.3}$$

The term $\left\{\frac{\partial f^1}{\partial \xi_1} + i\frac{\partial f^1}{\partial \xi_2}\right\}$ has the representation

$$\left\{\frac{\partial f^1}{\partial \xi_1} + i\frac{\partial f^1}{\partial \xi_2}\right\} = w(\xi)\,\chi_{Q_h} + v(\xi)\,\chi_{\mathbb{R}^2\setminus Q_h} \quad \text{with}$$

$$\begin{aligned} w(\xi) &= \frac{1}{2\pi}\Bigg[\mp\frac{i\,e^{\pm ih\xi_1}}{d^2} - 2\frac{\sin(h\xi_1)(1-e^{\pm ih\xi_1})}{h^2d^4} \pm \frac{i}{|\xi|^2} \mp \frac{2i\xi_1^2}{|\xi|^4} \\ &\quad -2i\frac{\sin(h\xi_2)(1-e^{\pm ih\xi_1})}{h^2d^4} \pm \frac{2\xi_1\xi_2}{|\xi|^4}\Bigg] \quad \text{and} \\ v(\xi) &= \frac{1}{2\pi}\left[\pm\frac{i}{|\xi|^2} \mp \frac{2i\xi_1^2}{|\xi|^4} \pm \frac{2\xi_1\xi_2}{|\xi|^4}\right]. \end{aligned}$$

Outside of the square Q_h it can be proved that

$$\int_{\xi\in\mathbb{R}^2\setminus Q_h} |v(\xi)|^{p_1} d\xi \le C_1 \int_{r=\pi/h}^{\infty} \int_{\varphi=0}^{2\pi} r^{-2p_1+1} d\varphi\, dr \le C_2\, h^{2p_1-2} \quad \text{if} \quad p_1 > 1$$

and inside the square Q_h the inequality $|w(\xi)| \le C\, h|\xi|^{-1}$ can be used to obtain

$$\int_{\xi\in Q_h} |w(\xi)|^{p_1} d\xi \le C_3\, h^{2p_1-2} \quad \text{if} \quad p_1 < 2.$$

From both estimates it follows that

$$\left\| \left\{ \frac{\partial f^1}{\partial \xi_1} + i\frac{\partial f^1}{\partial \xi_2} \right\} \right\|_{L_{p_1}(\mathbb{R}^2)} \le C\, h^{2-2/p_1} \quad \text{if} \quad 1 < p_1 < 2.$$

Based on the theorem of Hausdorff and Young (see e.g. [Hoe]), the Fourier integral of the expression $\left\{ \frac{\partial f^1}{\partial \xi_1} + i\frac{\partial f^1}{\partial \xi_2} \right\}$ can be estimated by

$$\left\| F\left\{ \frac{\partial f^1}{\partial \xi_1} + i\frac{\partial f^1}{\partial \xi_2} \right\} \right\|_{L_{q_1}(\mathbb{R}^2)} \le C \left\| \left\{ \frac{\partial f^1}{\partial \xi_1} + i\frac{\partial f^1}{\partial \xi_2} \right\} \right\|_{L_{p_1}(\mathbb{R}^2)} \le C\, h^{2-2/p_1} \tag{3.4}$$

if $q_1 > 2$ and $\frac{1}{p_1} + \frac{1}{q_1} = 1$. Using the Hölder inequality it is possible to write for an arbitrary bounded domain $G \in \mathbb{R}^2$

$$\begin{aligned} &\left\| [i(x_1+ix_2)]^{-1} F\left\{ \frac{\partial f^1}{\partial \xi_1} + i\frac{\partial f^1}{\partial \xi_2} \right\} \right\|_{L_p(G)} \\ &\le \|\, |x_1+ix_2|^{-p} \,\|_{L_{p_2}(G)}^{1/p} \left\| \left| F\left\{ \frac{\partial f^1}{\partial \xi_1} + i\frac{\partial f^1}{\partial \xi_2} \right\} \right|^p \right\|_{L_{q_2}(G)}^{1/p} \end{aligned} \tag{3.5}$$

with $\frac{1}{p_2} + \frac{1}{q_2} = 1$. The first factor in inequality (3.5) is finite if the property $pp_2 < 2$ is fulfilled. Let $q_1 = pq_2 > 2$. Using the inequalities (3.4) and (3.5) it can be concluded that

$$\begin{aligned} \left\| [i(x_1+ix_2)]^{-1} F\left\{ \frac{\partial f^1}{\partial \xi_1} + i\frac{\partial f^1}{\partial \xi_2} \right\} \right\|_{L_p(G)} &\le C \left\| F\left\{ \frac{\partial f^1}{\partial \xi_1} + i\frac{\partial f^1}{\partial \xi_2} \right\} \right\|_{L_{q_1}(\mathbb{R}^2)} \\ &\le C\, h^{2-2/p_1} \end{aligned}$$

for $1 < p_1 < 2$ and $q_1 > 2$, and since the inequality

$$\frac{1}{p} = \frac{1}{pp_2} + \frac{1}{pq_2} > \frac{1}{2} + \frac{1}{q_1} = \frac{3}{2} - \frac{1}{p_1}$$

must be fulfilled, it is possible to choose $p = p_1$ for all p with $1 < p < 4/3$. By this way the first summand of equation (3.3) is estimated in the form

$$\left\| [i(x_1+ix_2)]^{-1} F\left\{ \frac{\partial f^1}{\partial \xi_1} + i\frac{\partial f^1}{\partial \xi_2} \right\} \right\|_{L_p(G)} \le C\, h^{2-2/p} \tag{3.6}$$

for $1 < p < 4/3$. Now the L_p-norm of the last two summands in equation (3.3) will be considered. The Fourier integral can be written in the form

$$\begin{aligned} \|I^{(h)}\|_{L_p(G)} &= \Bigg\| |x_1 + ix_2|^{-1} \Bigg| \frac{1}{2\pi} F\Big(- \frac{1 - e^{\pm ih\xi_1}}{hd^2} \cos(\vec{n}, \xi_1) \chi_{\Gamma_{Q_h}} \Big) \\ &\quad + \frac{i}{2\pi} F\Big(- \frac{1 - e^{\pm ih\xi_1}}{hd^2} \cos(\vec{n}, \xi_2) \chi_{\Gamma_{Q_h}} \Big) \Bigg\|_{L_p(G)} \\ &= \Bigg\| \frac{i\,h}{|x_1 + ix_2|} \Bigg| \frac{\sin(x_1\pi/h)}{(2\pi)^2} \int_{\xi_2=-\pi/h}^{\pi/h} \frac{e^{-ix_2\xi_2}}{1 + \sin^2(h\xi_2/2)} d\xi_2 \\ &\quad + \frac{i\,\sin(x_2\pi/h)}{2\,(2\pi)^2} \int_{\xi_1=-\pi/h}^{\pi/h} \frac{1 - e^{\pm ih\xi_1}}{\sin^2(h\xi_1/2) + 1} e^{-ix_1\xi_1}\, d\xi_1 \Bigg\|_{L_p(G)} \end{aligned} \tag{3.7}$$

Let $G_1 = \{(x_1, x_2) : |x_1| < a \text{ and } |x_2| < b\}$ be the smallest rectangle which covers the domain G. Using the relation $I^{(h/t)}(x_1, x_2) = t\, I^{(h)}(t\, x_1, t\, x_2)$ it follows for $t \geq 1$ that

$$\begin{aligned} \|I^{(h/t)}\|^p_{L_p(G_1)} &= t^{p-2}\, \|I^{(h)}\|^p_{L_p(t\,G_1)} \\ &= t^{p-2}[\, \|I^{(h)}\|^p_{L_p(G_1)} + \|I^{(h)}\|^p_{L_p(t\,G_1 \setminus G_1)}]. \end{aligned}$$

In the special case $h_0 = ht > h$ this can be written as

$$\begin{aligned} \|I^{(h)}\|^p_{L_p(G)} &\leq \|I^{(h)}\|^p_{L_p(G_1)} \\ &= \|I^{(h_0/t)}\|^p_{L_p(G_1)} \\ &\leq t^{(p-2)} \big[\, \|I^{(h_0)}\|^p_{L_p(G_1)} + \|I^{(h_0)}\|^p_{L_p(\mathbb{R}^2 \setminus G_1)} \big]\,. \end{aligned} \tag{3.8}$$

To estimate the norm $\|I^{(h_0)}\|^p_{L_p(\mathbb{R}^2 \setminus G_1)}$, the domain $\mathbb{R}^2 \setminus G_1$ is split into $B_1 = \{(x_1, x_2) : |x_1| \geq a,\, |x_2| \geq b\}$, $B_2 = \{(x_1, x_2) : |x_1| \geq a,\, |x_2| < b\}$, and $B_3 = \{(x_1, x_2) : |x_1| < a,\, |x_2| \geq b\}$. In the first step the subdomain B_1 is considered. Using partial integration with respect to ξ_2 the first integral in representation (3.7) can be estimated by

$$\begin{aligned} &\Bigg\| \frac{i\,h_0}{|x_1 + ix_2|} \Bigg| \frac{\sin(x_1\pi/h_0)}{(2\pi)^2} \int_{\xi_2=-\pi/h_0}^{\pi/h_0} \frac{e^{-ix_2\xi_2}}{1 + \sin^2(h_0\xi_2/2)} d\xi_2 \Bigg\|^p_{L_p(A_1)} \\ &\leq C_1(h_0) \int_{x_1=a}^{\infty} \int_{x_2=b}^{\infty} \left(\frac{1}{x_2\sqrt{x_1^2 + x_2^2}} \right)^p dx_2\, dx_1 \leq C_2(h_0) \quad \text{if } p > 1, \end{aligned}$$

while with partial integration with respect to ξ_1 it is possible to estimate the second integral of (3.7) in the form

$$\left\| \frac{i\,h_0}{|x_1+ix_2|} \left| \frac{i\,\sin(x_2\pi/h_0)}{2\,(2\pi)^2} \int\limits_{\xi_1=-\pi/h_0}^{\pi/h_0} \frac{(1-e^{\pm i h_0 \xi_1})\,e^{-ix_1\xi_1}}{\sin^2(h_0\xi_1/2)+1}\,d\xi_1 \right\|\right\|^p_{L_p(A_1)}$$

$$\leq C_3(h_0) \int\limits_{x_1=a}^{\infty} \int\limits_{x_2=b}^{\infty} \left(\frac{1}{x_1\sqrt{x_1^2+x_2^2}} \right)^p dx_2\,dx_1 \leq C_4(h_0) \quad \text{if} \;\; p>1.$$

In the subdomain B_2 we deduce from (3.7) that

$$\|I^{(h_0)}\|^p_{L_p(A_2)} \leq C_5 \int\limits_{x_1=a}^{\infty} \int\limits_{x_2=0}^{b} \left(\frac{1}{\sqrt{x_1^2+x_2^2}} \right)^p dx_2\,dx_1 \leq C_6 \quad \text{if} \;\; p>1$$

and in the subdomain B_3 it can be proved that

$$\|I^{(h_0)}\|^p_{L_p(A_3)} \leq C_7 \int\limits_{x_1=0}^{a} \int\limits_{x_2=b}^{\infty} \left(\frac{1}{\sqrt{x_1^2+x_2^2}} \right)^p dx_2\,dx_1 \leq C_8 \quad \text{if} \;\; p>1.$$

Therefore it results from (3.8) that

$$\|I^{(h)}\|^p_{L_p(G)} \leq h_0^{(p-2)}\,h^{-p+2}[\,\|I^{(h_0)}\|^p_{L_p(G_1)} + C_9(h_0)] \quad \text{if} \;\; p>1. \tag{3.9}$$

Let K be a disk with radius R, which covers the domain G_1, then it is possible to estimate the term $\|I^{(h_0)}\|^p_{L_p(G_1)}$ on the right-hand side. If we replace in representation (3.7) h by h_0 and G by K then we obtain

$$\|I^{(h_0)}\|^p_{L_p(G_1)} \leq C_{10} \int\limits_{r=0}^{R} \int\limits_{\varphi=0}^{2\pi} r^{-p+1} d\varphi\,dr \leq C_{11} \;\; \text{for}\; p<2. \tag{3.10}$$

From (3.3), (3.6), (3.7), (3.9), and (3.10) it follows in the case $l=1$ that

$$\|(Ff^l)(x)\|_{L_p(G)} \leq C\,h^{-1+2/p} \quad \text{if} \quad 1<p<2$$

and in the special case $1<p<4/3$ that

$$\|(Ff^l)(x)\|_{L_p(G)} \leq C\,h^{-1+2/p} \leq C\,h^{2-2/p}.$$

The proof can be carried over to the case $l=2$. In this case the upper bound will have the same order of convergence. Finally, from (3.2), the definition of the matrix norm, and the Minkowski inequality we obtain

$$\begin{aligned} \|E_h^k(x)-E^k(x)\|_{L_p(G)} &\leq \left\| \sum_{i,j=1}^{2} |E^k_{hij}(x)-E^k_{ij}(x)| \right\|_{L_p(G)} \\ &\leq C\,h^{2-2/p} \quad \text{if} \quad 1<p<4/3. \end{aligned}$$

□

References

[DL] Deeter, C.R., Lord, M.E., Further theory of operational calculus on discrete analytic functions, *J. Math. Anal. Appl.* **26** (1969), 92–113.

[DS] Deeter, C.R., Springer, G., Discrete harmonic kernels, *J. Math. Mech.* **14** (1965), 413–438.

[D1] Duffin, R.J., Discrete potential theory, *Duke Math. J.* **20** (1953), 233–251.

[D2] Duffin, R.J., Basic properties of discrete analytic functions, *Duke Math. J.* **23** (1956), 335–363.

[DD] Duffin, R.J., Duris, C.S., A convolution product for discrete function theory, *Duke Math. J.* **31** (1964), 199–220.

[Fe] Ferrand, J., Fonctions préharmonique et fonctions préholomorphes, *Bulletin des Sciences Mathématique sec. series* **68** (1944), 152–180.

[GS1] Gürlebeck, K., Sprößig, W., *Quaternionic Analysis and Elliptic Boundary Value Problems*, ISNM 89, Birkhäuser Verlag, Basel 1990.

[GS2] Gürlebeck, K., Sprößig, W., *Quaternionic and Clifford calculus for Engineers and Physicists*, John Wiley & Sons, Chichester 1997.

[Hay] Hayabara, S., Operational calculus on the discrete analytic functions, *Math. Japon.* **11** (1966), 35–65.

[Ho] Hommel, A., *Fundamentallösungen partieller Differenzenoperatoren und die Lösung diskreter Randwertprobleme mit Hilfe von Differenzenpotentialen*, Dissertation, Bauhaus-Universität Weimar 1998.

[Hoe] Hörmander, L., *The Analysis of Linear Partial Differential Operators I: Distribution Theory and Fourier Analysis*, Springer-Verlag, Berlin, Heidelberg, New York, Tokyo 1983.

[I1] Isaacs, R., Monodiffric functions, *National Bureau of Standards Applied Mathematics Series* **18** (1952), 257–266.

[I2] Isaacs, R.P., A finite difference function theory, *Universidad Nacional Tucomán, Revista* **2** (1941), 177–201.

[KS] Kravchenko, V.V., Shapiro, M.V., *Integral Representations for Spatial Models of Mathematical Physics*, Pitman Research Notes in Mathematics Series 351, Harlow Longman 1996.

[Rya] Ryabenkij, V. S., *The Method of Difference Potentials for Some Problems of Continuum Mechanics*, Moscow, Nauka 1987, Russian.

[St] Stummel, F., Elliptische Differenzenoperatoren unter Dirichletrandbedingungen, *Math. Z.* **97** (1967), 169–211.

[Th] Thomée, V., Discrete interior Schauder estimates for elliptic difference operators, *SIAM J. Numer. Anal.* **5** (1968), 626–645.

[Wla] Wladimirow, W.S., *Gleichungen der mathematischen Physik*, Deutscher Verlag der Wissenschaften, Berlin 1972.

Klaus Gürlebeck and Angela Hommel
Bauhaus-Universität Weimar
Institut für Mathematik und Physik
Coudraystraße 13b
D-99421 Weimar, Germany
e-mail: `guerlebe@fossi.uni-weimar.de`
e-mail: `angela.hommel@bauing.uni-weimar.de`

2000 *Mathematics Subject Classification.*
Primary 30G25; Secondary 39A70.

Received: 31 May 2001.

Operator Theory:
Advances and Applications, Vol. 142, 117–132

Factorization Theorem for the Transfer Function Associated with an Unbounded Non-Self-Adjoint 2 × 2 Operator Matrix

Volker Hardt, Reinhard Mennicken and Alexander K. Motovilov

Abstract. We construct operators which factorize the transfer function associated with a non-self-adjoint 2×2 operator matrix whose diagonal entries can have overlapping spectra and whose off-diagonal entries are unbounded operators.

1. Introduction

In the present work we extend some of the results of [HMM1, HMM2] obtained for a class of unbounded self-adjoint block operator matrices to the case where these matrices are unbounded non-self-adjoint operators.

Throughout the paper we consider a 2×2 block operator matrix having the form

$$H_0 = \begin{pmatrix} A & B \\ D & C \end{pmatrix} \tag{1.1}$$

and acting in the orthogonal sum $\mathcal{H} = \mathcal{H}_A \oplus \mathcal{H}_C$ of separable Hilbert spaces $\mathcal{H}_A$ and $\mathcal{H}_C$. The entry C is a possibly unbounded self-adjoint operator in $\mathcal{H}_C$ on the domain $\mathrm{dom}(C)$. This operator is assumed to be semibounded from below, that is,

$$C \geq \lambda_C \quad \text{for some} \quad \lambda_C \in \mathbb{R}. \tag{1.2}$$

In the following we assume without loss of generality that the lower bound λ_C for the entry C in assumption (1.2) is positive,

$$\lambda_C > 0$$

(otherwise one could simply shift the origin of the spectral parameter axis). The entry A is supposed to be a bounded and not necessarily self-adjoint operator in $\mathcal{H}_A$, that is, $A \in \mathcal{B}(\mathcal{H}_A)$.

Regarding the couplings B and D we assume the following hypotheses:

(i) *D is a densely defined closable operator from* $\mathrm{dom}(D) \subset \mathcal{H}_A$ *to $\mathcal{H}_C$ such that the product $C^{-1/2}D$ admits extension from* $\mathrm{dom}(D)$ *to the whole space $\mathcal{H}_A$ as a bounded operator in $\mathcal{B}(\mathcal{H}_A, \mathcal{H}_C)$,*

(ii) *B is a densely defined closable operator from* $\mathrm{dom}(B) \subset \mathcal{H}_C$ *to $\mathcal{H}_A$ such that*

$$\mathrm{dom}(B) \supset \mathrm{dom}(C^{1/2}). \tag{1.3}$$

Since $\mathrm{dom}(C^{1/2}) \supset \mathrm{dom}(C)$, these assumptions imply that the matrix H_0 is a densely defined closable operator on the domain $\mathrm{dom}(H_0) = \mathrm{dom}(D) \oplus \mathrm{dom}(C)$. We denote its closure by H, that is, $H = \overline{H_0}$.

From assumption (i) one infers that

$$\widetilde{D} := \overline{C^{-1/2}D} \in \mathcal{B}(\mathcal{H}_A, \mathcal{H}_C).$$

Also one concludes from (i) that

$$\mathrm{ran}\left(\widetilde{D}\Big|_{\mathrm{dom}(D)}\right) \subset \mathrm{dom}(C^{1/2})$$

and

$$D = C^{1/2}\widetilde{D}\Big|_{\mathrm{dom}(D)}.$$

At the same time, assumption (1.3) yields that the product $\widetilde{B} := BC^{-1/2}$ is a bounded linear operator between $\mathcal{H}_C$ and $\mathcal{H}_A$, i.e. $\widetilde{B} \in \mathcal{B}(\mathcal{H}_C, \mathcal{H}_A)$, and that $B|_{\mathrm{dom}(C^{1/2})} = \widetilde{B}C^{1/2}$.

In addition, the hypotheses (i) and (ii) imply that for z in the resolvent set $\varrho(C)$ of C the operator-valued function $A - z - B(C-z)^{-1}D$ is densely defined on $\mathrm{dom}(D)$ and admits a bounded extension onto the whole space $\mathcal{H}_A$. We denote this extension by $M(z)$,

$$M(z) := A - z - \overline{B(C-z)^{-1}D}, \tag{1.4}$$

and call it the transfer function associated with the operator H. By definition,

$$\begin{aligned} M(z)|_{\mathrm{dom}(D)} &= A - z - B(C-z)^{-1}D \\ &= A - z - \widetilde{B}C^{1/2}(C-z)^{-1}C^{1/2}\widetilde{D} \\ &= \widetilde{A} - z + V(z), \end{aligned}$$

where

$$\widetilde{A} := A - \widetilde{B}\widetilde{D} \quad \text{and} \quad V(z) := z\widetilde{B}(z-C)^{-1}\widetilde{D}. \tag{1.5}$$

Meanwhile $\widetilde{A}$ is a bounded operator on $\mathcal{H}_A$ while $V(z)$, $z \in \varrho(C)$, is an operator-valued function with values in $\mathcal{B}(\mathcal{H}_A)$. Then it follows that

$$M(z) = \widetilde{A} - z + V(z). \tag{1.6}$$

Before going into more details we mention the recent paper [LMMT] studying by means of the new concept of the quadratic numerical range, introduced in [LT], in particular the factorization properties of the transfer functions associated with bounded non-self-adjoint 2×2 block operator matrices. As a matter of fact, the operators factorizing the transfer functions for bounded non-self-adjoint operator matrices have been also constructed in [ALT] and for unbounded non-self-adjoint operator matrices with bounded entries B and D in [LT]. The results of [ALT, LMMT, LT] are obtained under the assumption that the spectra of A and C do not overlap.

We are studying the transfer function (1.6), associated with the unbounded non-self-adjoint operator matrix (1.1) satisfying assumptions (i) and (ii), in the case where the numerical range $\nu(\widetilde{A})$ of the operator $\widetilde{A}$ (and in particular its spectrum $\sigma(\widetilde{A})$) may have a non-empty intersection with the spectrum $\sigma(C)$ of C. Notice that, since the resolvent of the operator H can be expressed explicitly in terms of $\left[M(z)\right]^{-1}$ (see, e. g., [ALMSa, MS, MM]), when studying the spectral properties of the transfer function M one studies at the same time the spectral properties of the operator matrix H.

Obviously, the transfer function (1.6), considered for $z \in \varrho(C)$, represents a holomorphic operator-valued function. (We refer to the standard definition of holomorphy of an operator-valued function with respect to the operator norm topology; see, e. g., [ALMSa].) We study the transfer function $M(z)$ under the assumption that it admits analytic continuation through the absolutely continuous spectrum of the entry C at least in a neighborhood of the numerical range $\nu(\widetilde{A})$ of the operator $\widetilde{A}$. In order to present our main ideas in a more transparent form, we suppose, for the sake of simplicity, that the entry C only has the absolutely continuous spectrum, that is,

$$\sigma(C) \;=\; \sigma_{\rm ac}(C).$$

Section 2 includes a description of the conditions making the analytic continuation of $M(z)$ through the set $\sigma_{\rm ac}(C)$ to be possible. In Section 3 we introduce the nonlinear transformation equations (3.4) making a rigorous sense to the formal operator equation $M(Z) = 0$, $Z \in \mathcal{B}(\mathcal{H}_A)$. We explicitly show that eigenvalues and accompanying eigenvectors of a solution Z to equation (3.4) are eigenvalues and eigenvectors of the analytically continued transfer function M. Further, we prove the solvability of (3.4) under smallness conditions concerning the operator $\widetilde{B}$ and $\widetilde{D}$, see (3.8). In Section 4 a factorization theorem (Theorem 4.1) is proven for the analytically continued transfer function. This theorem implies in particular that there exist certain domains in $\mathbb{C}$ lying partly on the unphysical sheet(s) where the spectrum of the analytically continued transfer function is represented by the spectrum of the corresponding solutions to the transformation equations (3.4). Finally, in Section 5 we present an example.

2. Analytic continuation

Throughout this paper we assume that the spectrum $\sigma(C)$ of the entry C is absolutely continuous and that it fills the semiaxis $[\lambda_C, +\infty)$. By $\nu(\widetilde{A})$ we denote the numerical range of the (bounded) operator $\widetilde{A}$,

$$\nu(\widetilde{A}) \;=\; \overline{\left\{\langle \widetilde{A}x, x\rangle : \, x \in \mathcal{H}_A, \; \|x\| = 1\right\}}.$$

We suppose that the set $\nu(\widetilde{A})$ lies inside a neighborhood

$$\mathcal{O}_\eta\big([\alpha_1, \alpha_2]\big) \;:=\; \{z \in \mathbb{C} : \, \mathrm{dist}(z, [\alpha_1, \alpha_2]) \le \eta\}, \quad \eta > 0,$$

of a finite real interval $[\alpha_1, \alpha_2] \subset \mathbb{R}$, $\alpha_1 < \alpha_2$, and $\alpha_1 - \eta > \lambda_C$. Notice that the numerical range $\nu(\widetilde{A})$ is a convex set containing the spectrum $\sigma(\widetilde{A})$ (see, e. g., [GK], § V.6) and, moreover,

$$\|(\widetilde{A} - zI_A)^{-1}\| \leq \frac{1}{\operatorname{dist}\left(z, \nu(\widetilde{A})\right)}, \quad z \in \mathbb{C} \setminus \nu(\widetilde{A}), \tag{2.1}$$

where I_A stands for the identity operator in $\mathcal{H}_A$ (see Lemma V.6.1 in [GK]).

Let $\{E_C(\mu)\}_{\mu\in\mathbb{R}}$ be the spectral family for the entry C, $C = \int_{\sigma(C)} \mu \, dE_C(\mu)$. Then the function $V(z)$ can be written as

$$V(z) = \int_{\lambda_C}^{\infty} dK(\mu) \frac{z}{z-\mu}$$

with

$$K(\mu) := \widetilde{B} E_C(\mu) \widetilde{D}.$$

We assume that the function $K(\mu)$ is continuously differentiable in $\mu \in (\lambda_C, +\infty)$ in the operator norm topology and, moreover, that it admits analytic continuation from the interval (λ_C, β), $\beta > \alpha_2 + \eta$, to a simply connected domain $\mathcal{D} \subset \mathbb{C}$, $\mathcal{D} \supset \mathcal{O}_\eta([\alpha_1, \alpha_2])$. For the continuation we keep the same notation $K(\mu)$ and by $K'(\mu)$ denote the derivative of K. We suppose that at the points $\lambda = \lambda_C$ and $\lambda = \beta$ the function $K'(\mu)$ satisfies the condition

$$\|K'(\mu)\| \leq c\,|\mu - \lambda|^{\gamma}, \quad \mu \in \mathcal{D},$$

with some $c > 0$ and $\gamma \in (-1, 0]$. We also assume that the operators $\widetilde{B}$ and $\widetilde{D}$ are such that

$$\int_{\beta}^{\infty} |d\mu| \, (1 + |\mu|)^{-1} \|K'(\mu)\| < \infty. \tag{2.2}$$

In the following we use the notation

$$\mathcal{D}^{(+1)} := \mathcal{D} \cap \mathbb{C}^+ \quad \text{and} \quad \mathcal{D}^{(-1)} := \mathcal{D} \cap \mathbb{C}^-.$$

Lemma 2.1. *Let $\widetilde{\Gamma}_l$ ($l = \pm 1$) be a rectifiable Jordan curve in $\mathcal{D}^{(l)}$ resulting from continuous deformation of the interval (λ_C, β), the end points of this interval being fixed, and let $\Gamma_l = \widetilde{\Gamma}_l \cup [\beta, +\infty)$. Then the analytic continuation of the transfer function $M(z)$, $z \in \mathbb{C} \setminus [\lambda_C, +\infty)$, through the spectral interval (λ_C, β) into the subdomain $\mathcal{D}(\Gamma_l) \subset \mathcal{D}^{(l)}$ ($l = \pm 1$) bounded by (λ_C, β) and $\widetilde{\Gamma}_l$ is given by*

$$M_{\Gamma_l}(z) := \widetilde{A} - z + V_{\Gamma_l}(z), \tag{2.3}$$

where

$$V_{\Gamma_l}(z) := \int_{\Gamma_l} d\mu \, K'(\mu) \frac{z}{z-\mu}. \tag{2.4}$$

For $z \in \mathcal{D}(\Gamma_l)$ the function $M_{\Gamma_l}(z)$ may be written as

$$M_{\Gamma_l}(z) = M(z) + 2\pi \mathrm{i} \, l z K'(z). \tag{2.5}$$

Proof. The function (2.4) is well defined for $z \notin \Gamma_l$ since (2.2) holds and for any $z \in \mathbb{C} \setminus \Gamma_l$ there exists a number $c(z) > 0$ such that the estimate $|(z-\mu)^{-1}| < c(z)\left(1+|\mu|\right)^{-1}$ $(\mu \in \Gamma_l)$ is valid. Then the proof of this lemma is reduced to the observation that the function $M_{\Gamma_l}(z)$ is holomorphic for $z \in \mathbb{C} \setminus \Gamma_l$ and coincides with $M(z)$ for $z \in \mathbb{C} \setminus \overline{\mathcal{D}(\Gamma_l)}$. Equation (2.5) is obtained from (2.4) using the Residue Theorem. □

Remark 2.2. From formula (2.5) one concludes that the transfer function $M(z)$ has a Riemann surface larger than a single sheet of the spectral parameter plane. The sheet of the complex plane where the transfer function $M(z)$ together with the resolvent $(H-z)^{-1}$ is initially considered is said to be the physical sheet. Hence, formula (2.5) implies that the domains $\mathcal{D}^{(+1)}$ and $\mathcal{D}^{(-1)}$ should be placed on additional sheets of the complex plane different from the physical sheet. Recall that these additional sheets are usually called unphysical sheets (see, e. g., [RS]).

Remark 2.3. For $z \in \mathbb{C} \setminus \Gamma_l$, formula (2.4) defines values of the function $V_{\Gamma_l}(\cdot)$ in $\mathcal{B}(\mathcal{H}_A)$. The inverse transfer function $\left[M(z)\right]^{-1}$ coincides with the compressed resolvent $P_{\mathcal{H}_A}\,(H-z)^{-1}\big|_{\mathcal{H}_A}$ where $P_{\mathcal{H}_A}$ stands for the orthogonal projection on $\mathcal{H}_A$. Thus, $\left[M(z)\right]^{-1}$ is holomorphic in $\mathbb{C} \setminus \sigma(H)$. Since $M_{\Gamma_l}(z)$ coincides with $M(z)$ for all $z \in \mathbb{C} \setminus \overline{\mathcal{D}(\Gamma_l)}$, one concludes that $[M_{\Gamma_l}(z)]^{-1}$ exists as a bounded operator and is holomorphic in z at least for $z \in \left(\mathbb{C} \setminus \sigma(H) \cup \overline{\mathcal{D}(\Gamma_l)}\right)$.

3. The transformation equations

Let $\Gamma_l \subset \mathcal{D}^{(l)}$ $(l = \pm 1)$ be a contour as described in the hypothesis of Lemma 2.1. Assume that $Z \in \mathcal{B}(\mathcal{H}_A)$ is a bounded operator such that its spectrum $\sigma(Z)$ is separated from Γ_l,

$$\operatorname{dist}\left(\sigma(Z), \Gamma_l\right) > 0.$$

Following to [MM, M1, M2] (cf. [ALT]), for such Z we introduce the "right", $V_{\Gamma_l}^{\triangleright}(Z)$, and "left", $V_{\Gamma_l}^{\triangleleft}(Z)$, transformations respectively by

$$V_{\Gamma_l}^{\triangleright}(Z) := \int_{\Gamma_l} d\mu\, K'(\mu)\, Z(Z-\mu)^{-1} \tag{3.1}$$

and

$$V_{\Gamma_l}^{\triangleleft}(Z) := \int_{\Gamma_l} Z(Z-\mu)^{-1} K'(\mu)\, d\mu. \tag{3.2}$$

Obviously, for both "right" and "left" symbols $\varkappa = \triangleright$ and $\varkappa = \triangleleft$,

$$\begin{aligned} \left\|V_{\Gamma_l}^{\varkappa}(Z)\right\| &\le \|Z\| \sup_{\mu \in \Gamma_l} \left[(1+|\mu|)\|(Z-\mu)^{-1}\|\right] \\ &\quad \times \int_{\Gamma_l} |d\mu|\,(1+|\mu|)^{-1}\|K'(\mu)\| < \infty. \end{aligned} \tag{3.3}$$

In what follows we consider the operator transformation equations on $\mathcal{B}(\mathcal{H}_A)$ (cf. [ALT, MM, M1, M2])

$$Z = \widetilde{A} + V_{\Gamma_l}^{\varkappa}(Z), \quad \varkappa = \triangleright \quad \text{or} \quad \varkappa = \triangleleft. \tag{3.4}$$

In particular, for $\varkappa = \triangleright$ equation (3.4) possesses the following characteristic property: If an operator $Z_\triangleright$ is a solution of (3.4) and $u \in \mathcal{H}_A$ is an eigenvector of $Z_\triangleright$, i. e., $Z_\triangleright u = zu$ for some $z \in \sigma(Z_\triangleright)$, then

$$\begin{aligned} zu &= \widetilde{A}u + V_\Gamma^\triangleright(Z_\triangleright)u = \widetilde{A}u + \int_{\Gamma_l} d\mu\, K'(\mu) Z_\triangleright (Z_\triangleright - \mu)^{-1} u \\ &= \widetilde{A}u + \int_{\Gamma_l} d\mu\, K'(\mu) \frac{z}{z-\mu} u = \widetilde{A}u + V_{\Gamma_l}(z)u. \end{aligned}$$

Hence, any eigenvalue z of such an operator $Z_\triangleright$ is automatically an eigenvalue for the analytically continued transfer function $M_{\Gamma_l}(\cdot)$ and u is a corresponding eigenvector. One can easily see that a similar relation holds between the operator $Z_\triangleleft^*$, adjoint of a solution $Z_\triangleleft$ to the transformation equation (3.4) for $\varkappa = \triangleleft$, and the adjoint transfer function $[M_{\Gamma_l}(\cdot)]^*$. This means that, having found the solutions of equations (3.4) for $\varkappa = \triangleright$ and/or $\varkappa = \triangleleft$, one obtains an effective means of studying the spectral properties of the transfer function $M_{\Gamma_l}(z)$, referring to well-known facts of operator theory [GK, K].

Let Γ_l ($l = \pm 1$) be a contour as described in the hypothesis of Lemma 2.1 and let the numerical range of the operator $\widetilde{A}$ be separated from Γ_l, that is,

$$d(\Gamma_l) := \operatorname{dist}\big(\nu(\widetilde{A}), \Gamma_l\big) > 0. \tag{3.5}$$

Then it is obvious that the quantity

$$\operatorname{Var}_{\widetilde{A}}(K, \Gamma_l) := \int_{\Gamma_l} |d\mu| \frac{\|K'(\mu)\|}{\operatorname{dist}\big(\mu, \nu(\widetilde{A})\big)} \tag{3.6}$$

is finite,

$$\begin{aligned} \operatorname{Var}_{\widetilde{A}}(K, \Gamma_l) &\le \sup_{\mu \in \Gamma_l} \frac{1 + |\mu|}{\operatorname{dist}\big(\mu, \nu(\widetilde{A})\big)} \\ &\quad \times \int_{\Gamma_l} |d\mu| \, (1 + |\mu|)^{-1} \|K'(\mu)\| < \infty. \end{aligned} \tag{3.7}$$

Hypothesis 3.1. *Assume all the assumptions of Section 2 concerning the operators $\widetilde{A}$ and C, and the operator-valued function K. Assume, in addition, that for both $l = +1$ and $l = -1$ there are contours $\Gamma_l = \widetilde{\Gamma}_l \cup [\beta, +\infty)$, as described in the hypothesis of Lemma 2.1, such that the following estimates hold:*

$$\operatorname{Var}_{\widetilde{A}}(K, \Gamma_l) < 1, \quad \operatorname{Var}_{\widetilde{A}}(K, \Gamma_l) \|\widetilde{A}\| < \frac{1}{4} d(\Gamma_l) \, [1 - \operatorname{Var}_{\widetilde{A}}(K, \Gamma_l)]^2. \tag{3.8}$$

In the following such contours Γ_l ($\lambda = \pm 1$) are called admissible contours.

It is convenient to rewrite equations (3.4), $\varkappa = \triangleleft$ or $\varkappa = \triangleright$, in the equivalent form

$$X = V_{\Gamma_l}^{\varkappa}(\widetilde{A} + X), \tag{3.9}$$

where $X := Z - \widetilde{A}$. We have the following statement regarding the solvability of the transformation equation (3.9).

Theorem 3.2. *Assume Hypothesis* 3.1 *and let* Γ_l $(l = \pm 1)$ *be an admissible contour. Then for each* $\varkappa = \triangleright$ *and* $\varkappa = \triangleleft$ *equation* (3.9) *has a solution* $X_\varkappa \in \mathcal{B}(\mathcal{H}_A)$ *satisfying the estimate*

$$\|X_\varkappa\| \leq r_{\min}(\Gamma_l),$$

where

$$\begin{aligned} r_{\min}(\Gamma_l) := {} & \frac{1}{2} d(\Gamma_l)\,[1 - \operatorname{Var}_{\widetilde{A}}(K, \Gamma_l)] \\ & - \sqrt{\frac{1}{4} d^2(\Gamma_l)\,[1 - \operatorname{Var}_{\widetilde{A}}(K, \Gamma_l)]^2 - d(\Gamma_l) \operatorname{Var}_{\widetilde{A}}(K, \Gamma_l)\, \|\widetilde{A}\|}. \end{aligned} \tag{3.10}$$

Moreover, for given $\varkappa = \triangleright$ *or* $\varkappa = \triangleleft$ *this solution is a unique solution to equation* (3.9) *in the open ball*

$$\{Y \in \mathcal{B}(\mathcal{H}_A) \,:\, \|Y\| < r_{\max}(\Gamma_l)\},$$

where

$$r_{\max}(\Gamma_l) := d(\Gamma_l) - \sqrt{\operatorname{Var}_{\widetilde{A}}(K, \Gamma_l)\, d(\Gamma_l)\,[d(\Gamma_l) + \|\widetilde{A}\|]}. \tag{3.11}$$

Proof. One can prove this theorem making use of Banach's Fixed Point Theorem. The proof is very similar to the proof of Theorem 3.1 in [HMM2] (also cf. [MM]). Thus we omit it. □

Remark 3.3. Conditions (3.8) imply the following inequalities:

$$r_{\min}(\Gamma_l) < \frac{1}{2} d(\Gamma_l)\,[1 - \operatorname{Var}_{\widetilde{A}}(K, \Gamma_l)] < r_{\max}(\Gamma_l)\,.$$

Lemma 3.4. *Assume Hypothesis* 3.1. *Then for given* $l = \pm 1$ *and* $\varkappa = \triangleright$ *or* $\varkappa = \triangleleft$ *the solution* $X_\varkappa$ *of equation* (3.9) *guaranteed by Theorem* 3.2 *is the same for any admissible contour* Γ_l. *Moreover, this solution satisfies the estimate*

$$\|X\| \leq r_0^{(l)}(K)$$

where

$$r_0^{(l)}(K) := \inf\{r_{\min}(\Gamma_l) \,:\, \operatorname{Var}_{\widetilde{A}}(K, \Gamma_l) < 1\,,\, \omega(K, \Gamma_l) > 0\}$$

with $r_{\min}(\Gamma_l)$ *given by* (3.10) *and*

$$\omega(K, \Gamma_l) := d(\Gamma_l)\,[1 - \operatorname{Var}_{\widetilde{A}}(K, \Gamma_l)]^2 - 4\|\widetilde{A}\| \operatorname{Var}_{\widetilde{A}}(K, \Gamma_l).$$

Proof. This statement can be proven essentially in the same way as Theorem 3.3 in [MM]. □

Therefore, for a given holomorphy domain $\mathcal{D}^{(l)}$ $(l = \pm 1)$ and fixed $\varkappa = \triangleright$ or $\varkappa = \triangleleft$ the solution $X_\varkappa$ to equation (3.9) and the solution $Z_\varkappa = \widetilde{A} + X_\varkappa$ to equation (3.4) do not depend on the admissible contours Γ_l. But when the index l changes, $X_\varkappa$ and $Z_\varkappa$ can also change. For this reason we shall supply them in the following, when it is necessary, with the index l writing $X_\varkappa^{(l)}$ and $Z_\varkappa^{(l)}$, $Z_\varkappa^{(l)} = \widetilde{A} + X_\varkappa^{(l)}$. Surely, equations (3.4) and (3.9) are nonlinear equations and, outside the balls $\|X\| < r_{\max}(\Gamma_l)$, they may, in principle, have other solutions, different from the $X_\varkappa^{(l)}$ or $Z_\varkappa^{(l)}$ the existence of which is guaranteed by Theorem 3.2. In the following we only deal with the solutions $X_\varkappa^{(l)}$ or $Z_\varkappa^{(l)}$ guaranteed by Theorem 3.2.

4. Factorization

In this section we prove a *factorization theorem* for the transfer function $M_{\Gamma_l}(z)$. Note that this theorem resembles the corresponding statements from [MrMt, VM]. It is an extension of Theorem 4.1 in [HMM2].

Theorem 4.1. *Assume Hypothesis* 3.1 *and let* Γ_l $(l = \pm 1)$ *be an admissible contour. Let for* $\varkappa = \triangleright$ *and* $\varkappa = \triangleleft$ *the operators* $X_\varkappa^{(l)}$ *be the solutions of the transformation equations* (3.9), $\|X_\varkappa^{(l)}\| \le r_0^{(l)}(K)$, *and* $Z_\varkappa^{(l)} = \widetilde{A} + X_\varkappa^{(l)}$. *Then, for* $z \in \mathbb{C} \setminus \Gamma_l$, *the transfer function* $M_{\Gamma_l}(z)$ *admits the factorizations*

$$M_{\Gamma_l}(z) \;=\; W_{\Gamma_l}^{\triangleright}(z)\,(Z_{\triangleright}^{(l)} - z) \;=\; (Z_{\triangleleft}^{(l)} - z)\,W_{\Gamma_l}^{\triangleleft}(z), \tag{4.1}$$

where $W_{\Gamma_l}^{\varkappa}(z)$, $\varkappa = \triangleright$ *or* $\varkappa = \triangleleft$, *are bounded operators in* $\mathcal{H}_A$ *which read*

$$\begin{aligned} W_{\Gamma_l}^{\triangleright}(z) \;&=\; I_A - \int_{\Gamma_l} d\mu\, K'(\mu)\,(Z_{\triangleright}^{(l)} - \mu)^{-1} \\ &\quad + z\int_{\Gamma_l} d\mu\, K'(\mu)(z-\mu)^{-1}(Z_{\triangleright}^{(l)} - \mu)^{-1} \end{aligned} \tag{4.2}$$

and

$$\begin{aligned} W_{\Gamma_l}^{\triangleleft}(z) \;&=\; I_A - \int_{\Gamma_l} (Z_{\triangleleft}^{(l)} - \mu)^{-1}\,K'(\mu)\,d\mu \\ &\quad + z\int_{\Gamma_l} (Z_{\triangleleft}^{(l)} - \mu)^{-1}\,K'(\mu)\,d\mu \end{aligned} \tag{4.3}$$

If, in addition,

$$\operatorname{dist}\big(z, \nu(\widetilde{A})\big) \;\le\; d(\Gamma_l)[1 - \operatorname{Var}_{\widetilde{A}}(K,\Gamma_l)]/2, \tag{4.4}$$

then

$$\|W_{\Gamma_l}^{\varkappa}(z) - I_A\| \;<\; 1 \quad \varkappa = \triangleright \text{ or } \varkappa = \triangleleft \tag{4.5}$$

and the operator $W_{\Gamma_l}^{\varkappa}(z)$ *is boundedly invertible, that is,* $\|[W_{\Gamma_l}^{\varkappa}(z)]^{-1}\| < \infty$.

Proof. For both $\varkappa = \triangleright$ and $\varkappa = \triangleleft$ this statement is proven in the same way. Thus, we only present the proof for the "right" case $\varkappa = \triangleright$.

First we prove the factorization (4.1). Note that, according to (3.1) and (3.9),

$$\widetilde{A} = Z_\triangleright^{(l)} - V_{\Gamma_l}(\widetilde{A} + X^{(l)}) = Z_\triangleright^{(l)} - \int_{\Gamma_l} d\mu\, K'(\mu) Z_\triangleright^{(l)} (Z_\triangleright^{(l)} - \mu)^{-1}. \qquad (4.6)$$

Thus, in view of (2.3) and (2.4), the value of $M_{\Gamma_l}(z)$ can be written as

$$\begin{aligned} M_{\Gamma_l}(z) &= \widetilde{A} - z + \int_{\Gamma_l} d\mu\, K'(\mu) \frac{z}{z-\mu} \\ &= Z_\triangleright^{(l)} - z - \int_{\Gamma_l} d\mu\, K'(\mu)(Z_\triangleright^{(l)} - \mu)^{-1}(Z_\triangleright^{(l)} - z) \\ &\quad + z \int_{\Gamma_l} d\mu\, K'(\mu) \left[\frac{1}{z-\mu} - (Z_\triangleright^{(l)} - \mu)^{-1} \right] \\ &= (Z_\triangleright^{(l)} - z) - \int_{\Gamma_l} d\mu\, K'(\mu)(Z_\triangleright^{(l)} - \mu)^{-1}(Z_\triangleright^{(l)} - z) \\ &\quad + z \int_{\Gamma_l} d\mu\, K'(\mu)(z-\mu)^{-1}(Z_\triangleright^{(l)} - \mu)^{-1}(Z_\triangleright^{(l)} - z). \end{aligned}$$

which proves equation (4.1). The boundedness of the operator $W_{\Gamma_l}^{\triangleright}(z)$ for $z \in \mathbb{C}\backslash\Gamma_l$ is obvious.

Further, assume that condition (4.4) holds. Using the triangle inequality, it is easy to see that this condition yields

$$\sup_{\mu \in \Gamma_l} |z - \mu|^{-1} \le \frac{2}{d(\Gamma_l)[1 + \mathrm{Var}_{\widetilde{A}}(K, \Gamma_l]}.$$

For any $x \in \mathcal{H}_A$, $\|x\| = 1$, we have

$$z = \langle \widetilde{A}x, x\rangle + \langle (z - \widetilde{A})x, x\rangle$$

and, hence,

$$|z| \le \|\widetilde{A}\| + \inf_{\|x\|=1} |\langle (z - \widetilde{A})x, x\rangle| = \|\widetilde{A}\| + \inf_{\zeta \in \nu(\widetilde{A})} |z - \zeta| = \|\widetilde{A}\| + \mathrm{dist}\,\big(z, \nu(\widetilde{A})\big).$$

Then it follows from (4.4) that

$$|z| \le \|\widetilde{A}\| + \frac{1}{2} d(\Gamma_l)[1 - \mathrm{Var}_{\widetilde{A}}(K, \Gamma_l)].$$

Meanwhile the estimate (2.1) implies

$$\begin{aligned} \|(\widetilde{A} + X^{(l)} - \mu)^{-1}\| &= \|[I_A + (\widetilde{A} - \mu)^{-1} X_\triangleright^{(l)}]^{-1} (\widetilde{A} - \mu)^{-1}\| \\ &\le \frac{1}{\mathrm{dist}\,\big(\mu, \nu(\widetilde{A})\big) - \|X_\triangleright^{(l)}\|}. \end{aligned} \qquad (4.7)$$

Therefore, taking into account the definitions of $d(\Gamma_l)$ and $r_{\min}(\Gamma_l)$, and Remark 3.3 we find

$$\left\| \int_{\Gamma_l} d\mu\, K'(\mu)\,(Z_{\triangleright}^{(l)} - \mu)^{-1} \right\| \le \frac{2\,\mathrm{Var}_{\widetilde{A}}(K,\Gamma_l)}{1+\mathrm{Var}_{\widetilde{A}}(K,\Gamma_l)}. \tag{4.8}$$

Using inequality (4.7) and Remark 3.3 one also finds

$$\left\| z \int_{\Gamma_l} d\mu\, K'(\mu)\,(Z_{\triangleright}^{(l)} - \mu)^{-1}(z-\mu)^{-1} \right\| \le |z| \frac{2\,\mathrm{Var}_{\widetilde{A}}(K,\Gamma_l)}{1+\mathrm{Var}_{\widetilde{A}}(K,\Gamma_l)} \sup_{\mu\in\Gamma_l} |z-\mu|^{-1}.$$

Finally, taking into consideration the second assumption in (3.8), we obtain

$$\begin{aligned} \|W_{\Gamma_l}^{\triangleright}(z) - I_A\| &\le \frac{2\,\mathrm{Var}_{\widetilde{A}}(K,\Gamma_l)}{1+\mathrm{Var}_{\widetilde{A}}(K,\Gamma_l)} \\ &\quad + \frac{4\,\mathrm{Var}_{\widetilde{A}}(K,\Gamma_l)\left\{\|\widetilde{A}\| + \frac{1}{2}\, d(\Gamma_l)[1-\mathrm{Var}_{\widetilde{A}}(K,\Gamma_l)]\right\}}{d(\Gamma_l)\,[1+\mathrm{Var}_{\widetilde{A}}(K,\Gamma_l)]^2} \\ &= \frac{4\,\mathrm{Var}_{\widetilde{A}}(K,\Gamma_l)\,[d(\Gamma_l)+\|\widetilde{A}\|]}{d(\Gamma_l)\,[1+\mathrm{Var}_{\widetilde{A}}(K,\Gamma_l)]^2} < 1 \end{aligned}$$

and, thus, if z satisfies (4.4), then the operator $W_{\Gamma_l}^{\triangleright}(z)$ has a bounded inverse.

The proof is complete. □

Corollary 4.2. *The spectra $\sigma(Z_{\triangleright}^{(l)})$ and $\sigma(Z_{\triangleleft}^{(l)})$ of the operators $Z_{\varkappa}^{(l)} = \widetilde{A} + X_{\varkappa}^{(l)}$, $\varkappa = \triangleright$ or $\varkappa = \triangleleft$, coincide and belong to the closed $r_0^{(l)}(K)$-neighborhood*

$$\mathcal{O}_{r_0^{(l)}(K)}(\widetilde{A}) := \left\{z\in\mathbb{C} : \mathrm{dist}\,\big(z,\nu(\widetilde{A})\big) \le r_0^{(l)}(K)\right\}$$

of the numerical range $\nu(\widetilde{A})$. Moreover, for any admissible contour Γ_l these spectra coincide with a subset of the spectrum of the transfer function $M_{\Gamma_l}(\cdot)$. More precisely,

$$\sigma\big(M_{\Gamma_l}(\cdot)\big)\cap\mathcal{O}(\widetilde{A},\Gamma_l) = \sigma(Z_{\triangleright}^{(l)}) = \sigma(Z_{\triangleleft}^{(l)}), \tag{4.9}$$

where

$$\mathcal{O}(\widetilde{A},\Gamma_l) := \left\{z\in\mathbb{C} : \mathrm{dist}\,\big(z,\nu(\widetilde{A})\big) \le d(\Gamma_l)\,[1-\mathrm{Var}_{\widetilde{A}}(K,\Gamma_l)]/2\right\}.$$

Proof. This statement is an immediate consequence of the factorizations (4.1) and the bounded invertibility of the operators $W_{\Gamma_l}^{\triangleright}(z)$ and $W_{\Gamma_l}^{\triangleleft}(z)$ whenever (4.4) holds. □

Let us introduce the operator

$$\Omega^{(l)} := \int_{\Gamma_l} d\mu\,\mu\,(Z_{\triangleleft}^{(l)} - \mu)^{-1} K'(\mu)\,(Z_{\triangleright}^{(l)} - \mu)^{-1}, \quad l = \pm 1, \tag{4.10}$$

where Γ_l stands for an admissible contour and, as before, $Z_{\varkappa}^{(l)} = \widetilde{A} + X_{\varkappa}^{(l)}$ where $X_{\varkappa}^{(l)}$, $\varkappa = \triangleright$ or $\varkappa = \triangleleft$, are solutions of the transformation equations (3.9). It is obvious that $\Omega^{(l)}$ does not depend on the choice of the admissible contour Γ_l.

Theorem 4.3. *The operators $\Omega^{(l)}$ $(l = \pm 1)$ possess the following properties (cf.* [HMM1, MrMt, MM, MS, VM]):

$$\|\Omega^{(l)}\| < 1, \tag{4.11}$$

$$-\frac{1}{2\pi i}\int_\gamma dz\,[M_{\Gamma_l}(z)]^{-1} = (I_A+\Omega^{(l)})^{-1}, \tag{4.12}$$

$$\begin{aligned} -\frac{1}{2\pi i}\int_\gamma dz\,z\,[M_{\Gamma_l}(z)]^{-1} &= (I_A+\Omega^{(l)})^{-1}Z_\triangleleft^{(l)} \\ &= Z_\triangleright^{(l)}(I_A+\Omega^{(l)})^{-1}, \end{aligned}\tag{4.13}$$

where γ stands for an arbitrary rectifiable closed contour encircling the spectrum $\sigma(Z_\triangleright^{(l)}) = \sigma(Z_\triangleright^{(l)})$ inside the set $\mathcal{O}(\widetilde{A},\Gamma_l)$ in the anticlockwise direction. The integration along γ is understood in the sense of the operator norm topology.

Proof. The estimate in (4.11) can be proved by using relations (4.1) following the proof of the inequality (4.5). This estimate yields that the sum $I_A + \Omega^{(l)}$ is a boundedly invertible operator in $\mathcal{H}_A$.

To prove formula (4.12) we note that by (4.1) the following factorization holds if $z \in \mathcal{O}(\widetilde{A},\Gamma_l)\setminus\sigma(Z_\triangleright^{(l)})$:

$$\begin{aligned} \left[M_{\Gamma_l}(z)\right]^{-1} &= \left(Z_\triangleright^{(l)} - z\right)^{-1}\left[W_{\Gamma_l}^{\triangleright}(z)\right]^{-1} \\ &= \left[W_{\Gamma_l}^{\triangleleft}(z)\right]^{-1}\left(Z_\triangleleft^{(l)} - z\right)^{-1}, \end{aligned}\tag{4.14}$$

where $[W_{\Gamma_l}^{\triangleright}(z)]^{-1}$ and $[W_{\Gamma_l}^{\triangleleft}(z)]^{-1}$ are holomorphic functions with values in $\mathcal{B}(\mathcal{H}_A)$. By the resolvent equation and the definition (4.10) the product $\Omega^{(l)}(Z_\triangleright^{(l)} - z)^{-1}$ can be written as

$$\Omega^{(l)}(Z_\triangleright^{(l)} - z)^{-1} = F_1(z) + F_2(z), \tag{4.15}$$

where

$$F_1(z) := \int_{\Gamma_l} d\mu\,\mu\,(Z_\triangleleft^{(l)} - \mu)^{-1}K'(\mu)\,(Z_\triangleright^{(l)} - \mu)^{-1}(\mu - z)^{-1} \tag{4.16}$$

and

$$\begin{aligned} F_2(z) &:= \left(-\int_{\Gamma_l} d\mu \frac{\mu}{\mu - z}(Z_\triangleleft^{(l)} - \mu)^{-1}K'(\mu)\right)(Z_\triangleright^{(l)} - z)^{-1} \\ &= \left(W_{\Gamma_l}^{\triangleleft}(z) - I_A\right)(Z_\triangleright^{(l)} - z)^{-1}. \end{aligned}\tag{4.17}$$

Further, formula (4.14) yields

$$(I_A+\Omega^{(l)})\left[M_{\Gamma_l}(z)\right]^{-1} = F_1(z)\left[W_{\Gamma_l}^{\triangleright}(z)\right]^{-1} + (Z_\triangleleft^{(l)} - z)^{-1}.$$

The function $F_1(z)$ is holomorphic inside the contour $\gamma \subset \mathcal{O}(\widetilde{A},\Gamma_l)$ since the argument μ of the integrand in (4.16) belongs to Γ_l and thereby

$$|z-\mu| \ge \left[d(\Gamma_l) + \mathrm{Var}_{\widetilde{A}}(K,\Gamma_l)\right]/2 > 0.$$

Thus the term $F_1(z)[W^{\triangleright}_{\Gamma_l}(z)]^{-1}$ does not contribute to the integral

$$-\frac{1}{2\pi \mathrm{i}}\int\limits_{\gamma} dz(I_A+\Omega^{(l)})\big[M_{\Gamma_l}(z)\big]^{-1}$$

while the resolvent $(Z^{(l)}_{\triangleleft}-z)^{-1}$ gives the identity I_A, which proves (4.12).

Similarly, to prove the first equality in (4.13) we calculate

$$\begin{aligned}&-\frac{1}{2\pi \mathrm{i}}\int\limits_{\gamma} dz(I_A+\Omega^{(l)})\, z\,\big[M_{\Gamma_l}(z)\big]^{-1}\\ &= \;-\frac{1}{2\pi \mathrm{i}}\int\limits_{\gamma} dz\, z\, F_1(z)\,\big[W^{\triangleright}_{\Gamma_l}(z)\big]^{-1}-\frac{1}{2\pi \mathrm{i}}\int\limits_{\gamma} dz\, z\,(Z^{(l)}_{\triangleleft}-z)^{-1}.\end{aligned}$$

The first integral vanishes whereas the second integral equals $Z^{(l)}_{\triangleleft}$. The second equality of (4.13) can be checked in the same way. □

Corollary 4.4. *Given $l=\pm1$, formula* (4.13) *implies that the operators $Z^{(l)}_{\triangleright}$ and $Z^{(l)}_{\triangleleft}$ are similar to each other:*

$$Z^{(l)}_{\triangleleft} \;=\; (I_A+\Omega^{(l)})\,Z^{(l)}_{\triangleright}\,(I_A+\Omega^{(l)})^{-1}.$$

Remark 4.5. Formulae (4.12) and (4.13) allow us, in principle, to construct the operators $Z^{(l)}_{\triangleright}$ ($l=\pm1$, $\varkappa=\triangleright$ or $\varkappa=\triangleleft$) and, thus, to resolve equations (3.9) by a contour integration of the inverse of the transfer function $M_{\Gamma_l}(z)$.

Theorem 4.6. *Let λ be an isolated eigenvalue of $Z^{(l)}_{\triangleright}$ and, hence, of $Z^{(l)}_{\triangleleft}$ and $M_{\Gamma_l}(\cdot)$ where Γ_l is an admissible contour. Denote by $P^{(l)}_{\triangleright,\lambda}$ and $P^{(l)}_{\triangleleft\lambda}$ the eigenprojections of the operators $Z^{(l)}_{\triangleright}$ and $Z^{(l)}_{\triangleleft}$, respectively, and by $P^{(l)}_{M,\lambda}$ the residue of $M_{\Gamma_l}(z)$ at $z=\lambda$,*

$$P^{(l)}_{\varkappa,\lambda} \;:=\; -\frac{1}{2\pi \mathrm{i}}\int\limits_{\gamma} dz\;(Z^{(l)}_{\varkappa}-z)^{-1}\quad (\varkappa=\triangleright \text{ or } \varkappa=\triangleleft), \tag{4.18}$$

and

$$P^{(l)}_{\lambda} \;:=\; -\frac{1}{2\pi \mathrm{i}}\int\limits_{\gamma} dz\;[M_{\Gamma_l}(z)]^{-1}, \tag{4.19}$$

where γ stands for an arbitrary rectifiable closed contour going around λ in the positive direction in a sufficiently close neighborhood such that $\gamma\cap\Gamma_l=\emptyset$ and no points of the spectrum of $M_{\Gamma_l}(\cdot)$, except the eigenvalue λ, lie inside γ. Then the following relations hold:

$$P^{(l)}_{M,\lambda} \;=\; P^{(l)}_{\triangleright,\lambda}\,(I_A+\Omega^{(l)})^{-1}=(I_A+\Omega^{(l)})^{-1}\,P^{(l)}_{\triangleleft,\lambda}\,. \tag{4.20}$$

Proof. The proof of this statement can be done in the same way as the proof of relation (4.12), only the path of integration is changed. □

5. An example

Let $\mathcal{H}_A = \mathcal{H}_C = L_2(\mathbb{R})$ and $C = \mathsf{P}^2 + \lambda_C I_C$ where $\mathsf{P} = \mathrm{i}\dfrac{d}{dx}$, λ_C is some positive number, and I_C denotes the identity operator in $\mathcal{H}_C$. It is assumed that the domain $\mathrm{dom}(\mathsf{P})$ is the Sobolev space $W_2^1(\mathbb{R})$ and the domain $\mathrm{dom}(C)$ is the Sobolev space $W_2^2(\mathbb{R})$. The spectrum of C is absolutely continuous and fills the semi-axis $[\lambda_C, +\infty)$. By the operator A we understand the multiplication by a bounded complex-valued function a, $Af = af$, $f \in \mathcal{H}_A$.

The operators B and D are defined on $\mathrm{dom}(B) = \mathrm{dom}(D) = W_2^1(\mathbb{R})$ by

$$B \;=\; S\mathsf{P} \quad \text{and} \quad D \;=\; \mathsf{P}Q,$$

where S and Q are multiplications by bounded functions $s \in L_2(\mathbb{R})$ and $q \in W_2^1(\mathbb{R})$, that is, $Qf = qf$ and $Sg = sg$ where $f, g \in L_2(\mathbb{R})$. Both S and D are densely defined closable operators.

Notice that $\mathrm{dom}(C^{1/2}) = W_2^1(\mathbb{R})$. The proof of this statement is based on the second representation theorem for quadratic forms, see Theorem VI.2.23 in [K]. It is similar to the proof of Proposition 2.4 in [FMM].

Further, we assume that the functions s and q are exponentially decaying at infinity, so that the estimates

$$\big|s(x)\big| \;\le\; c\,\exp\big(-\alpha_0|x|\big) \quad \text{and} \quad \big|q(x)\big| \;\le\; c\,\exp\big(-\alpha_0|x|\big) \qquad (x \in \mathbb{R}) \tag{5.1}$$

hold with some $c \ge 0$ and $\alpha_0 > 0$.

For this example the operators $\widetilde{B}$ and $\widetilde{D}$ are given by

$$\begin{aligned}
\widetilde{B} \;&=\; S\,\mathsf{P}\,(\mathsf{P}^2 + \lambda_C I_C)^{-1/2} \;=\; S\int_{\mathbb{R}} \frac{\mu}{(\mu^2 + \lambda_C)^{1/2}}\, dE_{\mathsf{P}}(\mu),\\
\widetilde{D} \;&=\; \overline{(\mathsf{P}^2 + \lambda_C I_C)^{-1/2}\,\mathsf{P}\,Q} \;=\; \int_{\mathbb{R}} \frac{\mu}{(\mu^2 + \lambda_C I_C)^{1/2}}\, dE_{\mathsf{P}}(\mu)\, Q,
\end{aligned}$$

where $\{E_{\mathsf{P}}(\mu)\}_{\mu\in\mathbb{R}}$ denotes the spectral family of the selfadjoint operator P. Thus

$$\begin{aligned}
\widetilde{A} \;&=\; A - \widetilde{B}\widetilde{D}\\
&=\; A - S\int_{\mathbb{R}} \frac{\mu}{(\mu^2 + \lambda_C)^{1/2}}\, dE_{\mathsf{P}}(\mu) \int_{\mathbb{R}} \frac{\widetilde{\mu}}{(\widetilde{\mu}^2 + \lambda_C)^{1/2}}\, dE_{\mathsf{P}}(\widetilde{\mu})\, Q\\
&=\; A - S\int_{\mathbb{R}} \frac{\mu^2}{\mu^2 + \lambda_C}\, dE_{\mathsf{P}}(\mu)\, Q\\
&=\; A - SQ + \lambda_C S(\mathsf{P}^2 + \lambda_C I_C)^{-1} Q.
\end{aligned}$$

The operator $A - SQ$ is the multiplication by the function

$$\widetilde{a}(x) \;=\; a(x) - s(x)q(x)$$

while the term $S(\mathsf{P}^2 + \lambda_C I_C)^{-1}Q$ is a compact (even Hilbert-Schmidt) operator in $L_2(\mathbb{R})$. Indeed, the inverse operator $C^{-1} = (\mathsf{P}^2 + \lambda_C I_C)^{-1}$ is the integral operator whose kernel reads

$$C^{-1}(x, x') \;=\; \frac{1}{2\sqrt{\lambda_C}}\, \exp\big(-\sqrt{\lambda_C}\,|x - x'|\big).$$

Thus, the double integral $\int_{\mathbb{R}}\int_{\mathbb{R}}\left|(SC^{-1}Q)(x,x')\right|^2 dxdx'$ is convergent. Obviously,

$$\int_{\mathbb{R}}\int_{\mathbb{R}}\left|(SC^{-1}Q)(x,x')\right|^2 dx\,dx' \;\leq\; \frac{1}{4\lambda_C}\,\|s\|^2_{L_2(\mathbb{R})}\|q\|^2_{L_2(\mathbb{R})}.$$

Therefore, the essential spectrum of $\widetilde{A}$ coincides with the range of the function $\widetilde{a}$. In the following we assume that there are an interval $[\alpha_1,\alpha_2]\subset(\lambda_C,+\infty)$ with $\alpha_1<\alpha_2$ and a number $\eta>0$ such that all the numerical range $\nu(\widetilde{A})$ of $\widetilde{A}$ lies inside the domain

$$\mathcal{O}_\eta([\alpha_1,\alpha_2]) \;:=\; \{z\in\mathbb{C}:\ \operatorname{dist}(z,[\alpha_1,\alpha_2])\leq\eta\},\qquad \eta>0,$$

of a finite real interval $[\alpha_1,\alpha_2]\subset\mathbb{R}$, and, moreover, $\alpha_1-\eta>\lambda_C$.

It is easy to check that the spectral projections $E_C(\mu)$ of the operator $C=\mathsf{P}^2+\lambda_C I$ are given by the integral operator whose kernel reads

$$E_C(\mu;x,x') \;=\; \begin{cases} 0 & \text{if } \mu<\lambda_C,\\[2ex] \dfrac{1}{\sqrt{2\pi}}\displaystyle\int_{\lambda_C}^{\mu}\frac{\cos[(\mu'-\lambda_C)^{1/2}(x-x')]}{(\mu'-\lambda_C)^{1/2}}\,d\mu' & \text{if } \mu\geq\lambda_C.\end{cases}$$

Thus, the derivative $K'(\mu)$ is also an integral operator in $L_2(\mathbb{R})$. Its kernel $K'(\mu;x,x')$ is only nontrivial for $\mu>\lambda_C$ and, moreover, for these μ

$$K'(\mu;x,x') \;=\; \frac{(\mu-\lambda_C)^{1/2}}{\sqrt{2\pi}\,\mu}\,\cos\big[(\mu-\lambda_C)^{1/2}(x-x')\big]\;s(x)\,q(x').$$

Obviously, this kernel is degenerate for $\mu>\lambda_C$,

$$K'(\mu;x,x') \;=\; \frac{(\mu-\lambda_C)^{1/2}}{2\sqrt{2\pi}\,\mu}\,\big[s_+(\mu,x)\,q_-(\mu,x')+s_-(\mu,x)\,q_+(\mu,x')\big], \tag{5.2}$$

where $s_\pm(\mu,x)=\mathrm{e}^{\pm\mathrm{i}\,(\mu-\lambda_C)^{1/2}x}\,q(x)$ and $q_\pm(\mu,x)=\mathrm{e}^{\pm\mathrm{i}\,(\mu-\lambda_C)^{1/2}x}\,q(x)$. From assumptions (5.1) on s and q we conclude that in the domain $\pm\operatorname{Im}\sqrt{\mu-\lambda_C}<\alpha_0$, i. e., inside the parabola

$$\mathcal{D} \;=\; \left\{\mu\in\mathbb{C}:\ \operatorname{Re}\mu>\lambda_C-\alpha_0^2+\frac{1}{4\alpha_0^2}(\operatorname{Im}\mu)^2\right\}, \tag{5.3}$$

the functions $s_\pm(\mu,\cdot)$ and $q_\pm(\mu,\cdot)$ are elements of $L_2(\mathbb{R})$. The function $K'(\mu)$ admits an analytic continuation into this domain (cut along the interval $\lambda_C-\alpha_0^2<\mu\leq\lambda_C$) as a holomorphic function with values in $\mathcal{B}(\mathcal{H}_A)$ and equation (5.2) implies that

$$\|K'(\mu)\| \;\leq\; \frac{|\mu-\lambda_C|^{1/2}}{2\sqrt{2\pi}\,|\mu|}\,\big[\|s_-(\mu,\cdot)\|\,\|q_-(\mu,\cdot)\|+\|s_+(\mu,\cdot)\|\,\|q_+(\mu,\cdot)\|\big].$$

Obviously, for $\mu\geq\lambda_C$ we have $\|s_\pm(\mu,\cdot)\|=\|s\|$ and $\|q_\pm(\mu,\cdot)\|=\|q\|$.

Let us make our final assumption that $\mathcal{D}\supset\mathcal{O}_\eta([\alpha_1,\alpha_2])$. In this case one can always choose a contour $\Gamma=\widetilde{\Gamma}\cup[\beta,+\infty)$ where $\beta>\alpha_2+\eta$ and the rectifiable Jordan curve $\widetilde{\Gamma}\subset\mathcal{D}\setminus\mathcal{O}_\eta([\alpha_1,\alpha_2])$ results from continuous deformation of the interval (λ_C,β), the end points being fixed. Assume, in addition, that the functions

s and q are sufficiently small in the sense that conditions (3.8) hold. In such a case the contour Γ is an admissible contour (see Hypothesis 3.1) and, thus, one can apply all the statements of the Sections 3 and 4 to the corresponding transfer function $M_\Gamma(z)$.

Acknowledgements

Support of this work by the Deutsche Forschungsgemeinschaft, the Heisenberg–Landau Program, and the Russian Foundation for Basic Research is gratefully acknowledged.

References

[AL] Adamjan, V. M., and Langer, H., Spectral properties of a class of operator-valued functions, *J. Operator Theory* **33** (1995), 259–277.

[ALMSa] Adamyan, V. M., Langer, H., Mennicken, R., and Saurer, J., Spectral components of selfadjoint block operator matrices with unbounded entries, *Math. Nachr.* **178** (1996), 43–80.

[ALT] Adamyan, V., Langer, H., and Tretter, C., Existence and uniqueness of contractive solutions of some Riccati equations, *J. Funct. Anal.* **179** (2001), 448–473.

[FMM] Faierman, M., Mennicken, R., and Möller, M., The essential spectrum of a system of singular ordinary differential operators of mixed order. Part I: The general problem and an almost regular case, *Math. Nachr.* **208** (1999), 101–115.

[GK] Gohberg, I. C., and Krein, M. G., *Introduction to the theory of linear nonselfadjoint operators*, American Mathematical Society, Providence, 1988.

[HMM1] Hardt, V., Mennicken, R., and Motovilov, A. K., A factorization theorem for the transfer function of a 2×2 operator matrix having unbounded couplings, in *Spectral and evolution problems* (Natl. Taurida Univ. "V. Vernadsky", Simferopol) **10** (2000), 56–65 (*LANL e-print math.SP/9912220*).

[HMM2] Hardt, V., Mennicken, R., and Motovilov, A. K., A factorization theorem for the transfer function associated with a 2×2 operator matrix having unbounded couplings, to appear in *J. Operator Theory*.

[K] Kato, T., *Perturbation theory for linear operators*, Springer-Verlag, New York, 1966.

[LMMT] Langer, H., Markus, A. S., Matsaev, V. I., and Tretter, C., A new concept for block oprator matrices: The quadratic numerical range, preprint, to appear in *Linear Algebra Appl.*

[LT] Langer, H., and Tretter, C., Spectral decomposition of some nonselfadjoint block operator matrices, *J. Operator Theory* **39** (1998), 1–20.

[MrMt] Markus, A. S., and Matsaev, V. I., On the basis property for a certain part of the eigenvectors and associated vectors of a selfadjoint operator pencil, *Math. USSR Sb.* **61** (1988), 289–307.

[MM] Mennicken, R., and Motovilov, A. K., Operator interpretation of resonances arising in spectral problems for 2×2 operator matrices, *Math. Nachr.* **201** (1999), 117–181 (*LANL E-print funct-an/9708001*).

[MS] Mennicken, R., and Shkalikov, A. A., Spectral decomposition of symmetric operator matrices, *Math. Nachr.* **179** (1996), 259–273.

[M1] Motovilov, A. K., Potentials appearing after removal of the energy-dependence and scattering by them, In: *Proc. of the Intern. Workshop "Mathematical aspects of the scattering theory and applications"*, St. Petersburg University, St. Petersburg (1991), 101–108.

[M2] Motovilov, A. K., Removal of the resolvent-like energy dependence from interactions and invariant subspaces of a total Hamiltonian, *J. Math. Phys.* **36** (1995), 6647–6664 (*LANL E-print funct-an/9606002*).

[RS] Reed, M., and Simon, B., *Methods of modern mathematical physics, III: Scattering theory*, Academic Press, N.Y., 1979.

[VM] Virozub, A. I., and Matsaev, V. I., The spectral properties of a certain class of selfadjoint operator functions, *Funct. Anal. Appl.* **8** (1974), 1–9.

Volker Hardt and Reinhard Mennicken
Department of Mathematics
University of Regensburg
D-93040 Regensburg, Germany
e-mail: `reinhard.mennicken@mathematik.uni-regensburg.de`

Alexander K. Motovilov
Laboratory of Theoretical Physics
Joint Institute for Nuclear Research
141980 Dubna (Moscow Region), Russia
e-mail: `motovilv@thsun1.jinr.ru`

2000 *Mathematics Subject Classification.*
Primary 47A56; Secondary 47Nxx, 47N50.

Received: 14 September 2001.

Operator Theory:
Advances and Applications, Vol. 142, 133–145

Boundedness of Some Convolution and Twisted Convolution Operators

Alexey N. Karapetyants and Enrique Ramírez de Arellano

Abstract. We consider $L_p - L_q$ boundedness problems for some special integral operators with oscillations both in the symbols and in the kernels. Such kernels have singularities on the unit sphere S^{n-1} and at infinity and are smooth on $\mathbb{R}^n \setminus S^{n-1}$. The main purpose is to show how these features affect the boundedness picture. Some further problems are discussed.

1. Introduction

Consider the well-known potential type operator

$$I_\theta^\alpha f(x) = \int_{\mathbb{R}^n} \frac{\theta(y)}{|y|^{n-\alpha}} f(x-y)d\mu(y), \quad 0 < \operatorname{Re}\alpha < n,$$

with smooth bounded characteristic function θ. Here $d\mu(y)$ denotes the usual Lebesgue volume measure in $\mathbb{R}^n$. The kernels of such operators have singularities only at the origin and at infinity. For $\theta(x) = 1$ the symbol of I_1^α is $\gamma_\alpha|\xi|^{-\alpha}$, where γ_α is some normalizing constant. It is well known that this operator is bounded from L_p to L_q if and only if p, q, α satisfy the Sobolev relation, i.e., $1/q = 1/p - \operatorname{Re}\alpha/n$, where $0 < \operatorname{Re}\alpha < n$. This condition is also sufficient for the boundedness of I_θ^α.

The natural continuation of this study is the consideration of potential type operators with kernels having singularities on sets in $\mathbb{R}^n$. As a matter of fact, there has been a great deal of research into the boundedness properties of convolutions with such kernels, in particular, with kernels having singularities on the unit sphere. Such operators arise naturally in concrete problems of mathematical physics (without claiming completeness we refer to [10]–[13], [17], [21]). In particular, Miyachi [10], [11] investigated operators with symbols $a(|\xi|)|\xi|^{-\alpha}e^{i|\xi|}$, $0 \le \alpha < n$ in connection with the Cauchy problem for the wave equation. Here $a(|\xi|)$ is a smooth cut-off function, $a(|\xi|) = 1$ for $|\xi| > 2$ and $a(|\xi|) = 0$, $|\xi| \le 1$. See also the paper [21] by Strichartz on boundedness results for the convolution operator with kernel $k(x) = (1-|x|^2)^{-\alpha/2}$, $|x| \le 1$ and $k(x) = 0$, if $|x| > 1$, $0 < \alpha < (n+1)/2$. Note further that the kernels of such operators have no singularities at infinity.

We start by considering the convolution operators $A^\alpha f = \Omega_\alpha * f$, where Ω_α is the distributional (inverse) Fourier transform of the symbol $m_\alpha(\xi) = |\xi|^{-\alpha}e^{i|\xi|}$.

They are represented, at least for $(n+1)/2 < \operatorname{Re}\alpha < n$, as

$$A^\alpha f(x) = \int_{\mathbb{R}^n} \Omega_\alpha(y)\, f(x-y)\, d\mu(y), \quad f \in S. \tag{1.1}$$

Introducing oscillation in the multiplier (i.e., substituting $|\xi|^{-\alpha}$ with $|\xi|^{-\alpha}e^{i|\xi|}$) we change the boundedness properties essentially. It appears that the kernels Ω_α have singularities on the sphere S^{n-1} and at infinity and are smooth on $\mathbb{R}^n \setminus S^{n-1}$. As a consequence, the set of $(1/p, 1/q)$ for which the corresponding operator is $L_p - L_q$ bounded has positive Lebesgue plane measure ($\operatorname{Re}\alpha \neq 0$). This is what Theorem 3.1 states, and the conditions in this theorem are sufficient and necessary.

The second step is to introduce, in addition, some kind of oscillation in the kernel and to investigate how much it affects on the boundedness properties. To this end we consider a special type of operators, namely, twisted convolution operators. Thus, we consider operators $B^\alpha f = \Omega_\alpha \times f$, which, for $(n+1)/2 < \operatorname{Re}\alpha < n$, are given by

$$B^\alpha f(x) = \int_{\mathbb{R}^n} \Omega_\alpha(y)\, f(x-y)\, e^{i\,\operatorname{Im}\langle x,y\rangle}\, d\mu(y), \quad f \in S. \tag{1.2}$$

Speaking about twisted convolutions we always naturally assume the dimension n is even and identify $\mathbb{R}^n$ with $\mathbb{C}^{n/2}$. Thus, $\langle x, y\rangle$ denotes the inner product in $\mathbb{C}^{n/2}$ of vectors x, y regarded as elements of $\mathbb{C}^{n/2}$. We obtain Theorem 4.2, where we give sufficient conditions for boundedness of B^α. It naturally appears that the set of points $(1/p, 1/q)$ of "boundedness" for B^α is considerably large, containing that for A^α. Needless to say, investigation of the necessity of our conditions seems to be a very difficult problem, because in this case we can use neither the theory of multipliers nor many other methods of harmonic analysis, which work effectively in the usual convolution case. For example, twisted convolution has, in general, no longer the symmetry property, a very important feature for the usual ones (see [2], [8] for examples of such operators).

The results can be obtained for more general integral operators, like those considered in Phong and Stein [15] (see the remarks below), but we prefer the twisted convolution case due to its simplicity of explanation and its connection with classical objects. Twisted convolutions have an interest in their own right, combining properties of both singular integral operators and pseudodifferential ones. They appear to play the same role in the Weil formalism as the usual convolution for the Fourier transform. One can find more details and further features in the paper [9] (see also [19] and the bibliography given there).

Note that $L_p - L_p$ boundedness for twisted convolutions with a kernel whose Fourier transform is Miyachi's symbol $a(|\xi|)m_\alpha(\xi)$ was considered in [16]. This operator differs from an operator with compactly supported kernel by a very "nice" operator and as observed by Cowling [1], it should have the same $L_p - L_p$ boundedness properties as the usual convolution operator. Actually, the "global" part is where the twisted convolution properties enter.

The aim of the paper is to provide a survey of what was done in this topic during the last years along with the presentation of new results.

We use the following notation: $Ff(\xi) \equiv \widehat{f}(\xi) = \int_{\mathbb{R}^n} f(x)e^{ix\cdot\xi}d\mu(x)$ is the Fourier transform and $F^{-1}f(x) = (2\pi)^{-n}Ff(-x)$ is the inverse Fourier transform; throughout the paper we assume $a(|\xi|)$ to be a smooth function as described above; $[a]$ denotes the integral part of the real number a; $L_p \equiv L_p(\mathbb{R}^n)$; $S \equiv S(\mathbb{R}^n)$ is the Schwartz class of rapidly decreasing smooth functions and S' is the space of tempered distributions; C_0^∞ is the space consisting of functions from S with compact support; L_∞^0 is the subspace of L_∞ of functions vanishing at infinity.

2. Integral representations for the operators A^α, B^α and some properties of the kernel $\Omega_\alpha(x)$

For $f \in S$ and $(n+1)/2 < \operatorname{Re}\alpha < n$ we define the convolution operator by (1.1), where the kernel $\Omega_\alpha(x)$ is given by

$$\Omega_\alpha(x) = (2\pi)^{-n/2}|x|^{\alpha-n}\int_0^\infty t^{n/2-\alpha}e^{it/|x|}\mathcal{J}_{n/2-1}(t)dt. \tag{2.1}$$

Here we have applied the Bochner formula for the Fourier transform of a radial function, and $\mathcal{J}_\nu(x)$ denotes the first kind Bessel function. Due to the estimate ([21])

$$\left| t^{-(a+ib)}\mathcal{J}_{a+ib}(t) \right| \le c_a e^{c|b|}(1+t)^{-a-1/2}, \quad 0 < t < \infty,$$

the integral (2.1) converges absolutely for $(n+1)/2 < \operatorname{Re}\alpha < n$, and, moreover, using formulae 6.699.1, 6.699.2 from [3], we obtain

$$\Omega_\alpha(x) = i^{n-\alpha}\frac{\Gamma(n-\alpha)}{2^{n-1}\pi^{n/2}\Gamma(n/2)}F\left(\frac{n-\alpha}{2}, \frac{n+1-\alpha}{2}; \frac{n}{2}; |x|^2\right) \tag{2.2}$$

for $|x| < 1$, and this is understood in the usual way for $|x| > 1$ (see [3], for instance). Here $F(a, b; c; d)$ is the Gauss hypergeometric function. Actually, the right-hand side of (2.2) for $|x| < 1$ (and its analytic continuation for $|x| > 1$) represents almost everywhere the function $\Omega_\alpha(x)$ for all α, $0 \le \operatorname{Re}\alpha < n$. This can be seen by applying the analytic continuation principle.

The above representation implies that the kernel $\Omega_\alpha(x)$ is continuous on the set $\mathbb{R}^n \setminus S^{n-1}$ and has the following asymptotic:

$$\Omega_\alpha(x) \sim c\,(1-|x|)^{\operatorname{Re}\alpha-(n+1)/2}, \quad (n-1)/2 < \operatorname{Re}\alpha < (n+1)/2, \tag{2.3}$$

$$\Omega_\alpha(x) \sim c\,\ln|1-|x||, \quad \alpha = (n+1)/2, \tag{2.4}$$

as $|x| \longrightarrow 1$, and

$$\Omega_\alpha(x) \sim c\,|x|^{\operatorname{Re}\alpha-n}, \quad 0 \le \operatorname{Re}\alpha < n, \tag{2.5}$$

as $|x| \longrightarrow \infty$. For $(n+1)/2 < \operatorname{Re}\alpha < n$ or $\operatorname{Re}\alpha = (n+1)/2$, $\operatorname{Im}\alpha \ne 0$ the kernel $\Omega_\alpha(x)$ is bounded.

Thus, for $0 \le \operatorname{Re}\alpha \le (n+1)/2$ the integral (1.1) does no longer make sense. We will understand the operator A^α for non-integral α in the sense of analytic continuation. From the representation (1.1) for a $(n+1)/2 < \operatorname{Re}\alpha < n$ and by interchanging the order of integration we have

$$\begin{aligned} A^\alpha f(x) &= (2\pi)^{-n/2} \int_{\mathbb{R}^n} |y|^{\alpha-n} f(x-y)\, d\mu(y) \int_0^\infty t^{n/2-\alpha} e^{it/|y|} \mathcal{J}_{n/2-1}(t)\, dt. \\ &= (2\pi)^{-n/2} \int_{\mathbb{R}^n} |y|^{\alpha-n} e^{i/|y|}\, d\mu(y) \int_0^\infty t^{n/2} \mathcal{J}_{n/2-1}(t)\, f(x-ty)\, dt. \end{aligned}$$

Due to the formula $\frac{d}{dt}\left[t^\nu \mathcal{J}_\nu(t)\right] = t^\nu \mathcal{J}_{\nu-1}(t)$, on integrating by parts and neglecting the boundary terms we get

$$\begin{aligned} &\int_0^\infty t^{n/2} \mathcal{J}_{n/2-1}(t) f(x-yt)\, dt \\ &= (-1)^k \int_0^\infty t^{n/2+k} \mathcal{J}_{n/2+k-1}(t) \left[\left(\frac{1}{t}\frac{d}{dt}\right)^k f(x-yt)\right] dt. \end{aligned}$$

Thus, for α in the range $\max\{0,(n+1)/2-k\} \le \operatorname{Re}\alpha \le (n+1)/2-k+1$ we define, for $f \in S$,

$$A^\alpha f(x) = \int_{\mathbb{R}^n} \Omega_\alpha^k(y) \left[\left(\frac{1}{s}\frac{d}{ds}\right)^k f(x-sy)\right]_{s=1} d\mu(y),$$

where $s \in (0,\infty)$, and

$$\begin{aligned} \Omega_\alpha^k(x) &= (-1)^k (2\pi)^{-n/2} |x|^{\alpha-n} \int_0^\infty t^{n/2-\alpha-k} e^{it/|x|} \mathcal{J}_{n/2+k-1}(t) dt \\ &= i^{n+2k-\alpha} \frac{\Gamma(n-\alpha)}{2^{n-1}\pi^{n/2}\Gamma(n/2)} F\left(\frac{n-\alpha}{2}, \frac{n-\alpha+1}{2}; \frac{n}{2}+k; |x|^2\right) \end{aligned}$$

for $|x| < 1$, understanding it as above for $|x| > 1$.

We can give the integral representation for the twisted convolution operator $B^\alpha f = \Omega_\alpha \times f$ in the same manner, i.e., by considering (1.2) for $(n+1)/2 < \operatorname{Re}\alpha < n$ and setting

$$\begin{aligned} B^\alpha f(x) &= \int_{\mathbb{R}^n} \Omega_\alpha^k(y) \left[\left(\frac{1}{s}\frac{d}{ds}\right)^k e^{is\operatorname{Im}\langle x,y\rangle} f(x-sy)\right]_{s=1} d\mu(y) \\ &= \int_{\mathbb{R}^n} \Omega_\alpha^k(y)\, e^{it\operatorname{Im}\langle x,y\rangle} \left[\left(\frac{i}{s}\operatorname{Im}\langle x,y\rangle I + \frac{1}{s}\frac{d}{ds}\right)^k f(x-sy)\right]_{s=1} d\mu(y) \end{aligned}$$

for $f \in S$, $\max\{0,(n+1)/2-k\} \le \operatorname{Re}\alpha \le (n+1)/2-k+1$.

In the rest of this section we study the behaviour of some special functions. Introduce

$$\Omega_{\alpha,0}(x) = \left(F^{-1}a(|\xi|)|\xi|^{-\alpha}e^{i|\xi|}\right)(x), \quad \Omega_{\alpha,\infty}(x) = \left(F^{-1}(1-a(|\xi|))|\xi|^{-\alpha}e^{i|\xi|}\right)(x).$$

The function $\Omega_{\alpha,0}(x)$ preserves the local singularities of the kernel $\Omega_\alpha(x)$ and is rapidly decreasing at infinity. More precisely, due to the result of [11], the function

$\Omega_{\alpha,0}(x)$ is smooth on $\mathbb{R}^n \setminus S^{n-1}$, $\Omega_{\alpha,0}(x) \sim \Omega_\alpha(x)$ when $|x| \to 1$ (see above for the asymptotic of $\Omega_\alpha(x)$ when $|x| \to 1$), and for any $M > 0$,

$$\Omega_{\alpha,0}(x) \le c\,|x|^{-M}, \quad |x| \longrightarrow \infty.$$

For $(n+1)/2 < \operatorname{Re}\alpha < n$ or $\operatorname{Re}\alpha = (n+1)/2,\ \operatorname{Im}\alpha \neq 0$ the function $\Omega_{\alpha,0}(x)$ is bounded.

Actually, the above was established in [11] for $\alpha \in \mathbb{R}$ only. But it is not hard to extend it to complex α, taking into account properties of the function $\Omega_\alpha(x)$ and of the function $\Omega_{\alpha,\infty}(x)$ which we now describe.

Contrary to the above situation, the function $\Omega_{\alpha,\infty}(x)$ is smooth at the finite points but preserves the singularity at infinity:

$$\Omega_{\alpha,\infty}(x) \sim c\,|x|^{\operatorname{Re}\alpha - n}, \quad |x| \longrightarrow \infty. \tag{2.6}$$

Moreover, any derivative of the kernel $\Omega_{\alpha,\infty}(x)$ has the same properties: it is smooth at finite points and its asymptotic at infinity is obtained by formal differentiation of (2.6). Let us prove the last assertion, following ideas of [11]. Consider $\mathcal{D}^m = \frac{\partial^m}{\partial x_{k_1}\partial x_{k_2}\ldots\partial x_{k_m}}$, where $1 \le k_m \le n$ are not necessarily distinct. Let us expand ([20, p. 139], [17])

$$(-i)^m \xi_{k_1}\xi_{k_2}\ldots\xi_{k_m} = \sum_{j=1}^{l} |\xi|^{2j} P_j(\xi),$$

where $l = [m/2]$ and $P_j(\xi)$ is a homogeneous (of order zero) harmonic polynomial of degree $m-2j$, $\xi \in \mathbb{R}^n$. We obviously have

$$\mathcal{D}^m\ \Omega_{\alpha,\infty}(x) = \sum_{j=1}^{l} \left(F^{-1} e^{i|\xi|} |\xi|^{2j-\alpha}\ a(|\xi|)\ P_j(\xi)\right)(x). \tag{2.7}$$

Using the rule for the Fourier transform of a product of a radial function with a solid spherical harmonic ([20], p. 158) we have

$$\left(F^{-1} e^{i|\xi|} |\xi|^{2j-\alpha}\ a(|\xi|)\ P_j(\xi)\right)(x) = \omega_\alpha^j(|x|)\ P_j(x),$$

where $\omega_\alpha^j(r)$ is

$$(2\pi)^{1-n} i^{2j-m}\ r^{\alpha-n-2m+2j} \int_0^\infty t^{n/2+m-\alpha}\ a(t/r)\ e^{it/r}\ \mathcal{J}_{n/2+m-2j-1}(t)\ dt.$$

Up to a constant multiplier, this expression can be viewed as a radial function of $n+2m-4j$ variables, being the inverse Fourier transform of the function $e^{i|\eta|}|\eta|^{2j-\alpha} a(|\eta|)$, $\eta \in \mathbb{R}^{n+2m-4j}$, i.e.,

$$\omega_\alpha^j(r) = c\ \left(F^{-1} |\eta|^{2j-\alpha} e^{i|\eta|} a(|\eta|)\right)(|\xi|)\Big|_{|\xi|=r}.$$

For this reason and by (2.6),

$$\omega_\alpha^j(|x|) \sim c\,|x|^{\operatorname{Re}\alpha-n-2m+2j}, \quad |x| \longrightarrow \infty,$$

which together with (2.7) proves our claim.

3. Boundedness of convolution operators

Theorem 3.1. *Let* $0 \le \operatorname{Re}\alpha < n$, $1 \le p \le q \le \infty$. *The operator* A^α *is bounded*

1. *from* L_p *to* L_q, $1 < p \le q < \infty$, *if and only if* $1/q \le 1/p - \operatorname{Re}\alpha/n$ *and either*
$$\frac{1}{p} + \frac{1}{q} \le 1, \quad \frac{1}{p} - \frac{n}{q} \le \operatorname{Re}\alpha - \frac{n-1}{2},$$
or
$$\frac{1}{p} + \frac{1}{q} \ge 1, \quad \frac{n}{p} - \frac{1}{q} \le \operatorname{Re}\alpha + \frac{n-1}{2}.$$
2. *from* L_1 *to* L_q, $1 \le q < \infty$, *if and only if*
$$(n+1)/2 - \operatorname{Re}\alpha < 1/q < (n - \operatorname{Re}\alpha)/n.$$
3. *from* L_1 *to* L_∞ *if and only if* $\operatorname{Re}\alpha > (n+1)/2$ *or both* $\operatorname{Re}\alpha = (n+1)/2$ *and* $\operatorname{Im}\alpha \ne 0$.
4. *from* L_p *to* L_∞, $1 < p < \infty$, *if and only if* $\operatorname{Re}\alpha/n < 1/p < \operatorname{Re}\alpha - (n-1)/2$.

The operator A^α *is of weak* $(1, 2/(n+1-2\operatorname{Re}\alpha))$ *type for* $n/2 \le \operatorname{Re}\alpha < (n+1)/2$, *i.e.,*
$$\operatorname{mes}\{x : |A^\alpha f(x)| > \lambda\} \le c\, \lambda^{-q}\, \|f\|_1^q, \quad 1/q = (n+1)/2 - \operatorname{Re}\alpha. \tag{3.1}$$
It is also of weak $(1, n/(n - \operatorname{Re}\alpha))$ *type for* $n/2 \le \operatorname{Re}\alpha < n$.

Proof. Claims 1–4 were proved in [7] for $\alpha \in \mathbb{R}$, $(n-1)/2 < \operatorname{Re}\alpha < n$ and were extended to all $\alpha \in \mathbb{C}$, $0 \le \operatorname{Re}\alpha < n$ in paper [5]. To complete these results, we prove the weak estimates.

Represent $A^\alpha = A_0^\alpha + A_\infty^\alpha$, where $A_0^\alpha f = \Omega_{\alpha,0} * f$ and $A_\infty^\alpha f = \Omega_{\alpha,\infty} * f$. Due to the behaviour of $\Omega_{\alpha,\infty}(x)$ and by standard Marcinkiewicz arguments, the operator A_∞^α is of weak $(1, n/(n - \operatorname{Re}\alpha))$ type for $0 \le \operatorname{Re}\alpha < n$. Consequently, the latest applies to A^α when $n/2 \le \operatorname{Re}\alpha < n$, as can be seen by routine estimates.

In the same way, having established weak $(1, 2/(n+1-2\operatorname{Re}\alpha))$ boundedness of A_0^α for $(n-1)/2 < \operatorname{Re}\alpha < (n+1)/2$, the same will be true for A^α when $n/2 \le \operatorname{Re}\alpha < (n+1)/2$. To this end it suffices to show (3.1) for the operator $f \mapsto \omega_\alpha * f$ under the condition $\|f\|_1 = 1$ and with 2λ instead of λ on the left side. Here
$$\omega_\alpha(x) = \begin{cases} |1 - |x||^{\operatorname{Re}\alpha - (n+1)/2} & \text{for } \ |x| < 2, \\ 0 & \text{otherwise.} \end{cases}$$
We have
$$\begin{aligned} \operatorname{mes}\{x : |(\omega_\alpha * f)(x)| > 2\lambda\} \ \le\ & \operatorname{mes}\{x : |(\omega_{\alpha,1} * f)(x)| > \lambda\} \\ & + \operatorname{mes}\{x : |(\omega_{\alpha,2} * f)(x)| > \lambda\} \end{aligned}$$
where we denoted
$$\omega_{\alpha,1}(x) = \begin{cases} \omega_{\alpha,1}(x) & \text{for } \ |1 - |x|| < \delta, \\ 0 & \text{otherwise,} \end{cases} \qquad \omega_{\alpha,2}(x) = \omega_\alpha(x) - \omega_{\alpha,1}(x).$$
We are going to choose δ below. First,
$$\operatorname{mes}\{x : |(\omega_{\alpha,1} * f)(x)| > \lambda\} \le 1/\lambda\ \|\omega_{\alpha,1}\|_1 \|f\|_1 = 1/\lambda\ \|\omega_{\alpha,1}\|_1,$$

and

$$\begin{aligned}\|\omega_{\alpha,1}\|_1 &= \int_{\substack{||1-|x||<\delta \\ |x|<2}} |1-|x||^{\mathrm{Re}\alpha-(n+1)/2} d\mu(x) \\ &= |S^{n-1}| \int_{\substack{|1-r|<\delta \\ 0<r<2}} r^{n-1} |1-r|^{\mathrm{Re}\alpha-(n+1)/2} dr \\ &\le 2^n |S^{n-1}| \int_0^{\min\{\delta,2\}} r^{\mathrm{Re}\alpha-(n+1)/2} dr = c_{n,\alpha} \left[\min\{\delta,2\}\right]^{\mathrm{Re}\alpha-(n-1)/2}.\end{aligned}$$

Next,

$$\|\omega_{\alpha,2} * f\|_\infty \le \|\omega_{\alpha,2}\|_\infty \|f\|_1 = \|\omega_{\alpha,2}\|_\infty = \delta^{\mathrm{Re}\alpha-(n+1)/2}.$$

Thus, setting $\delta^{\mathrm{Re}\alpha-(n+1)/2} = \lambda$, we have

$$\mathrm{mes}\{x : |(\omega_{\alpha,2} * f)(x)| > \lambda\} = 0$$

and finally,

$$\begin{aligned}\mathrm{mes}\{x : |(\omega_\alpha * f)(x)| > 2\lambda\} &= \mathrm{mes}\{x : |(\omega_{\alpha,1} * f)(x)| > \lambda\} \\ &\le 2c_{n,\alpha}\, \lambda^{-1/((n+1)/2-\mathrm{Re}\alpha)} = c\,\lambda^{-q}\, \|f\|_1^q,\end{aligned}$$

where $1/q = (n+1)/2 - \mathrm{Re}\,\alpha$, $(n-1)/2 < \mathrm{Re}\,\alpha < (n+1)/2$ and c depends only on n, α. □

4. Boundedness of twisted convolution operators

First we take into account the following general, simple result.

Lemma 4.1. *Let a twisted convolution operator $Tf = \Omega \times f$ be bounded from L_p to L_q, where $p > q$. Then $T = 0$ for $p < \infty$, and in the case $p = \infty$ the restriction of T to L_∞^0 is the null operator.*

Proof. We follow the method of Hörmander ([4]). Let $\tau_h f(x) = f(x-h)$. While operators like T do not commute with translations, they do satisfy the identity

$$\tau_h T f(x) = \left(T e^{-\mathrm{Im}\langle h,\cdot\rangle} \tau_h f\right)(x). \tag{4.1}$$

Now, for $f \in L_p$, $1 \le p < \infty$,

$$\begin{aligned}\|f + \tau_h f\|_p &\longrightarrow 2^{1/p}\, \|f\|_p, \quad h \to \infty, \\ \|f + e^{-\mathrm{Im}\langle h,\cdot\rangle} \tau_h f\|_p &\longrightarrow 2^{1/p}\, \|f\|_p, \quad h \to \infty.\end{aligned}$$

This is obvious for compactly supported functions $f \in C_0^\infty$ and extends to all $f \in L_p$ by the density of C_0^∞ in L_p, $1 \le p < \infty$. Suppose that

$$\|Tf\|_q \le c_{p,q}\, \|f\|_p, \qquad f \in L_p, \qquad 1 \le q < p < \infty. \tag{4.2}$$

Replacing the operator T with $\tau_{-h} T \tau_h$ we get an operator of the same type and with the same norm. Due to (4.1),

$$\|Tf + \tau_h Tf\|_q = \|T(f + \tau_h f)\|_q \le c_{p,q} \|f + e^{-\mathrm{Im}\langle h,\cdot\rangle} \tau_h f\|_p.$$

Letting $h \to \infty$, we get

$$\|Tf\|_q \le 2^{1/p-1/q} c_{p,q} \|f\|_p.$$

For $p > q$ this estimate is better than (4.2), which contradicts our assumption. The same arguments apply to $f \in L_\infty^0$ when $p = \infty$. □

Having established this, we restrict ourselves to $1 \le p \le q \le \infty$, as in the case of translation invariant operators.

Theorem 4.2. *Let $0 \le \operatorname{Re}\alpha < n$. The twisted convolution operator $B^\alpha f = \Omega_\alpha \times f$ is bounded from L_p to L_q, $1 \le p \le q \le \infty$,*

1. *if*

$$\frac{\operatorname{Re}\alpha}{n} \le \frac{1}{p} + \frac{1}{q} \le 1, \quad \frac{1}{p} - \frac{n}{q} \le \operatorname{Re}\alpha - \frac{n-1}{2},$$

unless

$$1/p = \operatorname{Re}\alpha/n, \quad 1/q = 0, \quad n/2 \le \operatorname{Re}\alpha < n$$

or

$$1/p = \operatorname{Re}\alpha - (n-1)/2, \quad 1/q = 0, \quad n/2 \le \operatorname{Re}\alpha < (n+1)/2$$

2. *if*

$$1 \le \frac{1}{p} + \frac{1}{q} \le 2 - \frac{\operatorname{Re}\alpha}{n}, \quad \frac{n}{p} - \frac{1}{q} \le \operatorname{Re}\alpha + \frac{n-1}{2}$$

unless

$$1/p = 1, \quad 1/q = 1 - \operatorname{Re}\alpha/n, \quad n/2 \le \operatorname{Re}\alpha < n$$

or

$$1/p = 1, \quad 1/q = (n+1)/2 - \operatorname{Re}\alpha, \quad n/2 \le \operatorname{Re}\alpha < (n+1)/2.$$

The operator B^α is of weak $(1, 1/((n+1)/2 - \operatorname{Re}\alpha))$ type for $n/2 \le \operatorname{Re}\alpha \le (n+1)/2$, and it is of weak $(1, n/(n - \operatorname{Re}\alpha))$ type for $n/2 \le \operatorname{Re}\alpha < n$.

Proof. The proof of this theorem is quite long and technical. Full details will be given elsewhere. Here we only prove the boundedness of B^α on L_2. This proof is a modification of arguments of Stein and Phong [15] for our twisted case and is of particular interest.

It turns out that the operator $B_0^\alpha f = \Omega_{\alpha,0} \times f$ differs from an operator with compact support by a very nice operator, which is bounded from L_p to L_q for any $1 \le p \le q$. Due to Cowling [1] the twisted convolution operator with compactly supported kernel is bounded on L_p, $1 \le p \le \infty$, if and only if so is the usual convolution operator (with the same kernel). Therefore, we have only to prove the L_2 boundedness for the operator $B_\infty^\alpha f = \Omega_{\alpha,\infty} \times f$.

The case $0 \le \operatorname{Re}\alpha < n/2$ follows from a well-known result (see [19, p. 556], [22, p. 14]), since $\Omega_{\alpha,\infty} \in L_2$ for such α.

Let $n/2 \le \operatorname{Re}\alpha < n$, $\Omega^{\varepsilon}_{\alpha,\infty}(x) = (1-a(\varepsilon|x|))\Omega_{\alpha,\infty}(x)$, and $B^{\alpha}_{\infty,\varepsilon} = \Omega^{\varepsilon}_{\alpha,\infty} \times f$. We have

$$\begin{aligned}
&(B^{\alpha}_{\infty,\varepsilon})^* B^{\alpha}_{\infty,\varepsilon} f(y) \\
&= \int_{\mathbb{R}^n} \overline{\Omega^{\varepsilon}_{\alpha,\infty}(x-y)} e^{-i\operatorname{Im}\langle y,x\rangle} d\mu(x) \int_{\mathbb{R}^n} f(u)\Omega^{\varepsilon}_{\alpha,\infty}(x-u) e^{-i\operatorname{Im}\langle u,x\rangle} d\mu(u) \\
&= \int_{\mathbb{R}^n} f(u) d\mu(u) \int_{\mathbb{R}^n} \overline{\Omega^{\varepsilon}_{\alpha,\infty}(x-y)}\Omega^{\varepsilon}_{\alpha,\infty}(x-u) e^{-i\operatorname{Im}\langle y,x\rangle} e^{-i\operatorname{Im}\langle u,x\rangle} d\mu(x) \\
&= \int_{\mathbb{R}^n} f(u) e^{i\operatorname{Im}\langle u,y-u\rangle} d\mu(u) \int_{\mathbb{R}^n} \overline{\Omega^{\varepsilon}_{\alpha,\infty}(x+u-y)}\Omega^{\varepsilon}_{\alpha,\infty}(x) e^{i\operatorname{Im}\langle x,y-u\rangle} d\mu(x).
\end{aligned}$$

The above change of the order of integration is justified by Fubini's theorem, since

$$\int_{\mathbb{R}^n} |f(u)|\ d\mu(u) \int_{\mathbb{R}^n} \left| \overline{\Omega^{\varepsilon}_{\alpha,\infty}(x+u-y)}\ \Omega^{\varepsilon}_{\alpha,\infty}(x) \right| d\mu(x) \le c\|f\|_2$$

for almost all $y \in \mathbb{R}^n$. Here to estimate the inner integral we applied ([17], [15]) the inequality

$$\int_{\mathbb{R}^k} \frac{d\mu(\tau)}{(1+|\tau|)^a(1+|\eta-\tau|)^b} \le c \begin{cases} (1+|\eta|)^{-\min\{a,b,a+b-k\}}, & \text{if } \max\{a,b\} \ne k, \\ (1+|\eta|)^{k-a-b}\ln(2+|\eta|), & \text{if } \max\{a,b\} = k, \end{cases} \tag{4.3}$$

which holds provided $a, b \ge 0$, $a+b > m$. Thus,

$$(B^{\alpha}_{\infty,\varepsilon})^* B^{\alpha}_{\infty,\varepsilon} f(y) = \int_{\mathbb{R}^n} f(u)\ k_{\alpha,\varepsilon}(y-u)\ e^{i\operatorname{Im}\langle u,y\rangle}\ d\mu(u), \quad f \in L_2,$$

where we denote

$$k_{\alpha,\varepsilon}(u) = \int_{\mathbb{R}^n} \overline{\Omega^{\varepsilon}_{\alpha,\infty}(x-u)}\ \Omega^{\varepsilon}_{\alpha,\infty}(x)\ e^{i\operatorname{Im}\langle x,u\rangle}\ d\mu(x). \tag{4.4}$$

Regarding x, u as elements of $\mathbb{C}^{n/2}$, let us write

$$x = (x_1, x_2, \dots, x_n) = (z_1, z_2, \dots, z_{n/2}) \equiv z$$

and, analogously, $u = (w_1, w_2, \dots, w_{n/2}) \equiv w$. Thus, we can introduce

$$\nabla_z = \left\{ \frac{\partial}{\partial \overline{z}_1}, \frac{\partial}{\partial \overline{z}_2}, \dots, \frac{\partial}{\partial \overline{z}_{n/2}} \right\},$$

where, as usual, $\frac{\partial}{\partial \overline{z}_k}$ denotes the anti-holomorphic derivative in z_k. It is not hard to see that

$$\langle \nabla_z \operatorname{Im}\langle x,u\rangle, u\rangle = \frac{i}{2}\langle u,u\rangle = \frac{i}{2}|u|^2.$$

Thus,

$$e^{i\operatorname{Im}\langle x,u\rangle} = -\frac{2}{|u|^2}\langle \nabla_z e^{i\operatorname{Im}\langle x,u\rangle}, u\rangle.$$

Substituting this in (4.4), integrating by parts, and neglecting the boundary terms, we get

$$\begin{aligned} k_{\alpha,\varepsilon}(u) &= -\frac{2}{|u|^2}\int_{\mathbb{R}^n} \overline{\Omega^{\varepsilon}_{\alpha,\infty}(x-u)\Omega^{\varepsilon}_{\alpha,\infty}(x)}\langle \nabla_z e^{i\,\mathrm{Im}\langle x,u\rangle}, u\rangle \, d\mu(x) \qquad (4.5)\\ &= -\frac{2}{|u|^2}\left\langle \int_{\mathbb{R}^n} \overline{\Omega^{\varepsilon}_{\alpha,\infty}(x-u)\Omega^{\varepsilon}_{\alpha,\infty}(x)}\nabla_z e^{i\,\mathrm{Im}\langle x,u\rangle}\, d\mu(x), u\right\rangle \\ &= \frac{2}{|u|^2}\left\langle \int_{\mathbb{R}^n} \left(\nabla_z \overline{\Omega^{\varepsilon}_{\alpha,\infty}(x-u)\Omega^{\varepsilon}_{\alpha,\infty}(x)}\right) e^{i\,\mathrm{Im}\langle x,u\rangle}\, d\mu(x), u\right\rangle \\ &= \sum_{k=1}^{n/2}\frac{2\overline{w}_k}{|u|^2}\int_{\mathbb{R}^n}\left(\frac{\partial}{\partial \overline{z}_k}\overline{\Omega^{\varepsilon}_{\alpha,\infty}(x-u)\Omega^{\varepsilon}_{\alpha,\infty}(x)}\right) e^{i\,\mathrm{Im}\langle x,u\rangle}\, d\mu(x). \end{aligned}$$

Applying the same procedure to each integral in the sum in (4.5), at the step "m" we arrive at

$$\begin{aligned} k_{\alpha,\varepsilon}(u) &= \frac{2^m}{|u|^{2m}} \sum_{\nu_1+\nu_2+\ldots+\nu_{n/2}=m} \overline{w}^{\nu_1}_{k_1}\overline{v}^{\nu_2}_{k_2}\ldots\overline{w}^{\nu_{n/2}}_{k_{n/2}} \\ &\times \int_{\mathbb{R}^n}\left(\frac{\partial^m}{\partial\overline{z}^{\nu_1}_{k_1}\partial\overline{z}^{\nu_2}_{k_2}\ldots\partial\overline{z}^{\nu_{n/2}}_{k_{n/2}}}\,\overline{\Omega^{\varepsilon}_{\alpha,\infty}(x)}\;\Omega^{\varepsilon}_{\alpha,\infty}(x)\right) e^{i\,\mathrm{Im}\langle x,u\rangle}\, d\mu(x). \end{aligned}$$

Obviously, the derivatives of $\Omega^{\varepsilon}_{\alpha,\infty}(x)$ admit the same estimate at infinity as those of $\Omega_{\alpha,\infty}(x)$ (see the claim after formula (2.6)) and, moreover, uniformly in ε. Thus, using (4.3) we obtain

$$|k_{\alpha,\varepsilon}(x)| \le c\,|x|^{-m}\,(1+|x|)^{2\mathrm{Re}\alpha-n-m}, \quad 0\le \mathrm{Re}\,\alpha < n,$$

with c independent of ε. Fix m such that $\mathrm{Re}\,\alpha - n/4 < m < n$. Since $(B^{\alpha}_{\infty,\varepsilon})^* B^{\alpha}_{\infty,\varepsilon}$ is itself a twisted convolution operator and its kernel is in L_2, we immediately obtain the boundedness estimate

$$\|(B^{\alpha}_{\infty,\varepsilon})^* B^{\alpha}_{\infty,\varepsilon} f\|_2 \le \|k_{\alpha,\varepsilon}\|_2\|f\|_2 \le c\|f\|_2.$$

By limiting arguments we obtain the same estimate for $(B^{\alpha}_{\infty})^* B^{\alpha}_{\infty}$ and, hence, the boundedness of B^{α}_{∞} on L_2. □

5. Concluding remarks

One natural extension of the previous theorem is the consideration of $L_p - L_q$ boundedness properties for the general operator

$$T_{\Omega_\alpha} f(x) = \int_{\mathbb{R}^n} \Omega_\alpha(y)\; f(y-x)\; e^{iP(x,y)}\, d\mu(y), \qquad (5.1)$$

where $P(x,y)$ is a real bilinear form. Boundedness in L_p of the operator T_K (K instead of Ω_α in (5.1)) where $K(x)$ is a Calderon-Zygmund type kernel was considered by Phong and Stein in [15]. It is also of particular interest to investigate

$L_p - L_q$ boundedness for such operators T_K. The operators T_{Ω_α} and T_K have similar global properties of their kernels, but these are locally different.

Another extension of Theorem 4.2 can be given considering more general operators B_b^α with kernels $\Omega_{\alpha,b}(x) = (F^{-1}b(\xi)m_\alpha(\xi))(x)$ with some special $b(\xi)$. For example, one may study radial $b(\xi) = b(|\xi|)$ in a Hölder class of power-like stabilizing functions at infinity. Or, one may investigate homogeneous $b(\xi) = b(\xi/|\xi|)$ satisfying the ellipticity condition $\inf_{\xi\neq 0} |b(\xi)| > 0$ and some smoothness condition. Obviously, it does not always make sense to consider an arbitrary characteristic in the symbol: the kernel's properties will be rather different. But it is essential that for the classes mentioned we still have asymptotics and properties of $\Omega_{\alpha,b}(x)$ like the ones of $\Omega_\alpha(x)$ above (see [6], [7] for details and [18] for the definition of the mentioned Hölder class).

In this connection we mention that Theorem 3.1 (the usual convolution case) was proved in [5] (except for the weak estimates) for more general operators $A_b^\alpha f = \Omega_{\alpha,b} * f$ with radial $b(|\xi|)$ satisfying the following: b and $1/b$ are both in M_p^p, $1 < p < \infty$, and in the case $\operatorname{Re}\alpha = (n+1)/2$, $p = 1$, $q = \infty$ additionally b, $1/b \in M_1^1$. The weak estimates proved in Theorem 3.1 apply to such general operators as well, with the only assumption $|b'(t)| \leq c/t, t > 0$, because in this case the corresponding kernels are estimated with $\leq$ instead of $\sim$ in (2.3)–(2.5) (see [6] for details).

The case of homogeneous $b(\xi)$ was studied in [7], where it was shown that the operator A_b^α has exactly the same boundedness properties as A^α, provided

$$(n-1)/2 < \alpha < n, \quad b(\xi) \in C^m(S^{n-1}), \quad m > 3n/2 - 1,$$

and the ellipticity condition is satisfied. This was extended in [14] to $0 < \alpha < n$, where non-elliptic characteristics were considered as well.

5.1. Acknowledgements

The work was supported by CONACyT grant 35521-E (México, D.F.). The first author is on leave from the Rostov-on-Don University, Russia, and he was supported by the *Lefschetz Research Fellowship* at the Center for Research and Advanced Studies (CINVESTAV-IPN), México, D.F.

References

[1] Cowling, M., A remark on twisted convolution, *Rend. Circ. Mat. Palermo* **2** (1981), 203–209.

[2] Cowling, M., Mantero, A.M., Examples of twisted convolution operators, *Lect. Notes Math.* **908** (1982), 210–216.

[3] Gradshteyn, I.S., Ryzhik, I.M., *Tables of Integrals, Series, and Products*, Academic Press, New York-London-Toronto 1980.

[4] Hörmander, L., Estimates for translation invariant operators in L_p spaces, *Acta. Math.* **104** (1960), 93–140.

[5] Karapetyants, A.N., On $L_p - L_q$ boundedness for convolutions with kernels having singularities on a sphere, *Studia Math.* **144** (2001), no. 2, 121–134.

[6] Karapetyants, A.N., Nogin, V.A., Inversion of potential-type operators with kernels possessing singularities on the sphere, *Journal of Contemporary Math. Anal.* **34** (1999), no.1, 55–69.

[7] Karapetyants, A.N., Nogin, V.A., On the $\mathcal{L}$-characteristic of some potential type operators with oscillating symbols and singularities of the kernels on a sphere, *Dokl. Acad. Nauk* **370** (2000), no. 3, 300–302 [Russian].

[8] Mantero, A.M., Asymmetry of twisted convolution operators, *J. Funct. Anal.* **47** (1982), 7–25.

[9] Mauceri, G., Picardello, M., Ricci, F., Twisted convolution, Hardy spaces and Hörmander multipliers, *Rend. Circ. Mat. Palermo* **2** (1981), 191–202.

[10] Miyachi, A., On some estimates for the wave equation in L^p and H^p, *J. Fac. Sci. Univ. Tokyo Sec. IA* **27** (1980), 331–354.

[11] Miyachi, A., On some singular Fourier multipliers, *J. Fac. Sci. Univ. Tokyo Sec IA* **28** (1981), 267–315.

[12] Nogin, V.A., Samko, S.G., Method of approximating inverse operators and its application to inversion of potential type integral transforms, *Integral Transforms and Special Functions* **1** (1999), no. 2, 1–14.

[13] Nogin, V.A., Samko, S.G., Some applications of potentials and approximative inverse operators in multidimentional fractional calculus, *Fract. Calc. Applied Anal.* **2** (1999), no. 2, 205–228.

[14] Nogin, V.A., Chegolin, A.P., On the L-characteristic of some multiplier operators with oscillating symbols in the non-elliptic case, *Fract. Calc. Appl. Anal.* **4** (2001), no. 1, 67–81.

[15] Phong, D.H., Stein, E.M., Hilbert integrals, singular integrals, and Radon transforms. I., *Acta Math.* **157** (1986), 99–157.

[16] Pini, R., A multiplier result for twisted convolution, *Boll. Un. Mat. Ital. B (7)* **6** (1992), no. 1, 67–78.

[17] Samko, S.G., *Hypersingular Integrals and Their Applications*, Taylor & Francis, in the series "Analytical Methods and Special Functions", vol. 5, 2001.

[18] Samko, S.G., The classes $C^\lambda(R^n)$, and multipliers in the space $I^\alpha(L_p)$ of Riesz potentials, *Izv. Severo-Kavkaz. Naučn. Centra Vysh. Skoly, Ser. Estestv. Nauk* **115** (1977), no. 3, 13–17 [Russian].

[19] Stein, E.M., *Harmonic Analysis: Real-Variable Methods, Orthogonality, and Oscillatory Integrals*, Princeton University Press, Princeton 1993.

[20] Stein, E.M., Weiss, G., *Fourier Analysis on Euclidean Spaces*, Princeton University Press, Princeton 1971.

[21] Strichartz, R., Convolutions with kernels having singularities on a sphere, *Trans. Amer. Math. Soc.* **148** (1970), 461–471.

[22] Thangavelu, S., *Lectures on Hermite and Laguerre Expansions*, Mathematical Notes 42, Princeton University Press, Princeton 1993.

Alexey N. Karapetyants
Department of Mathematical Sciences
University of Arkansas
Fayetteville, AR 72701, USA
e-mail: alexey@math.cinvestav.mx

Enrique Ramírez de Arellano
Departamento de Matemáticas
CINVESTAV del I.P.N.
A. P. 14–740, 07000
México, D.F., México
e-mail: eramirez@math.cinvestav.mx

2000 *Mathematics Subject Classification.*
Primary 43A32; Secondary 44A99 .

Received: 7 June 2001.

Operator Theory:
Advances and Applications, Vol. 142, 147–174

Invertibility of Functional Operators with Slowly Oscillating Non-Carleman Shifts

Alexei Karlovich, Yuri Karlovich and Amarino Lebre

Abstract. We prove criteria for the invertibility of the binomial functional operator

$$A = aI - bW_\alpha$$

in the Lebesgue spaces $L^p(0,1)$, $1 < p < \infty$, where a and b are continuous functions on $(0,1)$, I is the identity operator, W_α is the shift operator, $W_\alpha f = f \circ \alpha$, generated by a non-Carleman shift $\alpha : [0,1] \to [0,1]$ which has only two fixed points 0 and 1. We suppose that $\log \alpha'$ is bounded and continuous on $(0,1)$ and that a, b, α' slowly oscillate at 0 and 1. The main difficulty connected with slow oscillation is overcome by using the method of limit operators.

1. Introduction

Let α be an orientation preserving homeomorphism of $[0,1]$ onto itself which has only two fixed points 0 and 1. So, $\alpha(0) = 0$ and $\alpha(1) = 1$, but $\alpha(t) \neq t$ for $t \in \mathbb{I} := (0,1)$. The function α is referred to as a shift function (*shift*).

Denote by $\beta := \alpha_{-1}$ the inverse function to α. Since α and β strictly monotonically increase on $[0,1]$, their derivatives exist and are positive almost everywhere on $\mathbb{I}$. If α and β are absolutely continuous on $[0,1]$ and $\log \alpha' \in L^\infty := L^\infty(\mathbb{I})$, then the shift operator W_α defined by

$$(W_\alpha \varphi)(t) := \varphi[\alpha(t)], \quad t \in \mathbb{I},$$

is bounded in the Lebesgue space $L^p := L^p(\mathbb{I})$ for every $p \in [1, \infty]$. Moreover, it is invertible and its inverse is given by $W_\alpha^{-1} = W_\beta$.

In what follows we always suppose that $p \in (1, \infty)$ and that $q = p/(p-1)$ is the conjugate exponent to p.

In the space L^p we consider the binomial functional operator with shift

$$A := aI - bW_\alpha \tag{1.1}$$

where I the identity operator, and the coefficients a, b are bounded and continuous (BC) functions on $\mathbb{I}$.

The investigation of two- and one-sided invertibility of functional operators (in particular, of (1.1)) in various functional spaces plays an important role in the theory of functional differential operators (see, e.g., [1, 3, 20]), the theory of singular integral operators, convolution type operators, and pseudodifferential operators with shifts and/or oscillating coefficients (see [2, 19, 21, 22] and references therein), the theory of dynamical systems [6], etc.

In the case of "nice" shifts and coefficients the following invertibility criterion was actually proved in [18] (see also [19, Ch. 2, Th. 4′]).

Theorem 1.1. *Suppose a, b are continuous functions on $[0,1]$ and α is an orientation preserving diffeomorphism of $[0,1]$ onto itself. The operator (1.1) is invertible in the Lebesgue space $L^p, p \in (1,\infty)$, if and only if*

$$\text{either} \quad \min_{t\in[0,1]} |a(t)| > 0, \quad |a(j)| > |b(j)|\Big(\alpha'(j)\Big)^{-1/p}, \quad j \in \{0,1\};$$

$$\text{or} \quad \min_{t\in[0,1]} |b(t)| > 0, \quad |a(j)| < |b(j)|\Big(\alpha'(j)\Big)^{-1/p}, \quad j \in \{0,1\}.$$

Under the same assumptions on the smoothness of shifts and coefficients, this theorem was generalized in [13] to the case of an arbitrary nonempty set of periodic points of shifts. Further, the one-sided invertibility of (1.1) in Lebesgue spaces and in more general rearrangement-invariant spaces was studied in [4, 9, 10, 23] (see also the surveys [12, 14]).

In contrast to the classical assumptions [12, 14, 19, 21, 22], we do not assume the existence of one-sided limits of the coefficients and one-sided derivatives of the shift at the fixed points 0 and 1. In this paper we prove invertibility criteria for the operator (1.1) with slowly oscillating data in Lebesgue spaces L^p.

For a function $\varphi \in BC$, consider its oscillation on a set $J \subset \mathbb{I}$:

$$\omega(\varphi, J) := \sup_{x,y\in J} |\varphi(x) - \varphi(y)|.$$

Following [25], a function $\varphi \in BC$ is called a *slowly oscillating function at* 0 if for every $\lambda \in \mathbb{I}$,

$$\lim_{r\to 0} \omega(\varphi, [\lambda r, r]) = 0, \quad \Big(\text{or equivalently,} \quad \lim_{r\to 0} \omega(\varphi, [r/2, r]) = 0\Big).$$

In this case we will write $\varphi \in SO$. A function $\varphi \in BC$ is called slowly oscillating at 1 if the function $t \mapsto \varphi(1-t)$ belongs to SO. We say that a non-Carleman shift α is a *slowly oscillating shift at* $j \in \{0,1\}$ if $\log \alpha' \in BC$ and α' is a slowly oscillating function at j, or, equivalently (see Lemma 3.8), if $\log \alpha' \in BC$ and

$$\lim_{t\to j} (t-j)\frac{d}{dt}\left(\frac{\alpha(t)-\alpha(j)}{t-j}\right) = 0.$$

We say that the operator (1.1) has *slowly oscillating data* $\{a, b; \alpha\}$ *at the endpoint* $j \in \{0,1\}$ if the coefficients a, b are slowly oscillating functions at j and α is a slowly oscillating shift at j.

Operators with slowly oscillating (Carleman) shifts were studied for the first time in [15]. In that paper singular integral operators with piecewise continuous coefficients and slowly oscillating Carleman backward shifts were interpreted as Mellin pseudodifferential operators with slowly oscillating symbols. Further, by applying the method of limit operators (see also [5, 24]), the investigation of Mellin pseudodifferential operators was reduced to the investigation of Mellin convolutions. Studying Mellin pseudodifferential operators requires that the shift and the coefficients of the initial operator to be infinitely differentiable (apart from the fixed points). These assumptions seem to be redundant, and it is natural to exclude them. For this purpose, in this paper we apply the method of limit operators directly to the operator (1.1). This allows us to overcome the main difficulty connected with slow oscillation and to get necessary conditions for the invertibility of operators (1.1) at the fixed points.

The paper is organized as follows. In Section 2 we assume that $a, b \in BC$ and that α' is invertible in BC. Under these assumptions we formulate sufficient conditions for the two- and one-sided invertibility of the operator (1.1). We also prove that under respective assumptions the right (left) invertible operator (1.1) is not invertible because its kernel (cokernel) has dimension infinity. Clearly, these conditions for strict one-sided invertibility give us necessary conditions for two-sided invertibility. But there is a gap between sufficient and necessary conditions for the two-sided invertibility at the fixed points. This gap is connected with the fact that we do not assume the existence of the one-sided limits at 0 and 1 for the coefficients a, b and for the derivative of the shift α.

To fill this gap we have to assume that the operator (1.1) has slowly oscillating data $\{a, b; \alpha\}$ at 0 and 1. In Section 3 we uncover some properties of slowly oscillating functions and shifts. The method of limit operators allows us to reduce the investigation of the invertibility of the operator (1.1) with "bad" (slowly oscillating) data to the investigation of the invertibility of a family of operators (so called limit operators) of the same form but with "nice" coefficients (continuous on $[0, 1]$) and "nice" shifts (diffeomorphisms of $[0, 1]$ onto itself). On the basis of the results from Section 3, in Section 4 we calculate the limit operators for the operator (1.1). We emphasize that these limit operators already satisfy the conditions of Theorem 1.1.

In Section 5, following the scheme of [9] and using the results of Sections 2 and 4, we prove our main result:

Theorem 1.2. *Suppose the operator* (1.1) *has slowly oscillating data* $\{a, b; \alpha\}$ *at the two endpoints* 0 *and* 1. *The operator* (1.1) *is invertible in the Lebesgue space* $L^p, p \in (1, \infty)$, *if and only if*

$$\text{either} \quad \inf_{t \in \mathbb{I}} |a(t)| > 0, \quad \liminf_{t \to j} \left(|a(t)| - |b(t)| \left(\alpha'(t) \right)^{-1/p} \right) > 0, \quad j \in \{0, 1\};$$

$$\text{or} \quad \inf_{t \in \mathbb{I}} |b(t)| > 0, \quad \limsup_{t \to j} \left(|a(t)| - |b(t)| \left(\alpha'(t) \right)^{-1/p} \right) < 0, \quad j \in \{0, 1\}.$$

2. Sufficient conditions for one-sided invertibility

2.1. Sufficient conditions for invertibility

For a Banach algebra $\mathfrak{A}$ denote by $\mathcal{G}\mathfrak{A}$ the group of all invertible elements in $\mathfrak{A}$. We denote by C the set of all continuous functions $f : \mathbb{I} \to \mathbb{C}$. For $f \in C$, put

$$f_*(j) := \liminf_{t\to j} |f(t)|, \quad f^*(j) := \limsup_{t\to j} |f(t)|, \quad j \in \{0, 1\}.$$

Denote by B the set of all bounded functions which are defined everywhere on $\mathbb{I}$ (in contrast to L^∞). Obviously, $BC = C \cap B$. Clearly, if $f \in BC$ then the upper limits $f^*(0)$ and $f^*(1)$ are finite. It is easy to see that a function $f \in BC$ belongs to $\mathcal{G}BC$ if and only if $f(t) \neq 0$ for every $t \in \mathbb{I}$ and the lower limits $f_*(0)$ and $f_*(1)$ are positive.

Fix an arbitrary point $x \in \mathbb{I}$. Let l be a half-open segment with endpoints x and $\alpha(x)$ such that $x \in l$ but $\alpha(x) \notin l$. Notice that either $x < \alpha(x)$ and then 1 is the attracting point of α, or $\alpha(x) < x$ and then 0 is the attracting point of α (see, e.g., [19, Ch. 1, Section 3]). Put $\alpha_0(t) = t$ and $\alpha_n(t) = \alpha[\alpha_{n-1}(t)]$ for $n \in \mathbb{Z}, t \in [0, 1]$. Assume for definiteness that 0 is the repelling point and 1 is the attracting point for the shift α, that is,

$$\lim_{n\to\infty} \alpha_{-n}(t) = 0, \quad \lim_{n\to\infty} \alpha_n(t) = 1, \quad t \in \mathbb{I}.$$

The opposite case, when 1 is the repelling point and 0 is the attracting point, can be considered absolutely analogously.

Lemma 2.1. *Suppose $g \in BC$ and $\alpha' \in \mathcal{G}BC$. The spectral radius of the weighted shift operator gW_α satisfies the estimate*

$$r(gW_\alpha) \leq \max\Big\{M^*(0), M^*(1)\Big\}, \tag{2.1}$$

where $M(t) := g(t)(\alpha'(t))^{-1/p}$.

Proof. This proof is a slight generalization of the proof given in [21, Theorem 37.1], (see also [19, Ch. 2, Lemmas 6–7] and [1, Theorem 2.2]).

Since $\mathbb{I}$ is invariant with respect to the shift α_n for arbitrary $n \in \mathbb{N}$, we have

$$\begin{aligned} \|(gW_\alpha)^n\varphi\|_p &= \left(\int_0^1 \left|\prod_{k=0}^{n-1} g[\alpha_{k-n}(t)]\right|^p |\varphi(t)|^p |\alpha'_{-n}(t)|\,dt\right)^{1/p} \\ &\leq \sup_{t\in\mathbb{I}} \left(\left|\prod_{k=0}^{n-1} g[\alpha_{k-n}(t)]\right| |\alpha'_{-n}(t)|^{1/p}\right) \|\varphi\|_p. \end{aligned} \tag{2.2}$$

Clearly,

$$\left|\prod_{k=0}^{n-1} g[\alpha_{k-n}(t)]\right| |\alpha'_{-n}(t)|^{1/p} = \prod_{k=1}^{n} \frac{|g[\alpha_{-k}(t)]|}{(\alpha'[\alpha_{-k}(t)])^{1/p}}. \tag{2.3}$$

Since g is bounded and α' is bounded away from zero on $\mathbb{I}$, the quantities $M^*(0)$ and $M^*(1)$ are finite. Fix $\varepsilon > 0$. From the definition of $M^*(j), j \in \{0,1\}$, it follows that there exist points t_1 and t_2 such that

$$\frac{|g(t)|}{(\alpha'(t))^{1/p}} < M^*(0) + \varepsilon,\ t \in (0, t_1), \quad \frac{|g(t)|}{(\alpha'(t))^{1/p}} < M^*(1) + \varepsilon,\ t \in (t_2, 1). \quad (2.4)$$

Since 1 is the attracting point of the shift α, there exists a number $k_0 \in \mathbb{N}$ such that $\alpha_{k_0}(t_1) \in (t_2, 1)$. Put

$$L := \sup_{t\in\mathbb{I}} \frac{|g(t)|}{(\alpha'(t))^{1/p}}, \quad Q := \sup_{t\in(0,t_1)\cup(\alpha_{k_0}(t_1),1)} \frac{|g(t)|}{(\alpha'(t))^{1/p}}.$$

Then we have for every $t \in \mathbb{I}$,

$$\prod_{k=1}^{n} \frac{|g[\alpha_{-k}(t)]|}{(\alpha'[\alpha_{-k}(t)])^{1/p}} \le L^{k_0} Q^{n-k_0}. \quad (2.5)$$

On the other hand, we see from (2.4) that

$$Q \le \max\Big\{M^*(0), M^*(1)\Big\} + \varepsilon. \quad (2.6)$$

Combining (2.2), (2.3), (2.5), (2.6), we obtain

$$\|(gW_\alpha)^n\|_{\mathfrak{B}(L^p)} \le L^{k_0} \Big(\max\Big\{M^*(0), M^*(1)\Big\} + \varepsilon\Big)^{n-k_0}.$$

Applying the spectral radius formula to the latter inequality, we get

$$r(gW_\alpha) \le \max\Big\{M^*(0), M^*(1)\Big\} + \varepsilon.$$

Since ε is arbitrary, this inequality gives us (2.1). □

For further generalizations of the formula for the spectral radius of a weighted shift operator see [1, Ch. 1, Section 5].

From Lemma 2.1, using standard arguments (see, e.g., [10, Section 3]), we get the following.

Lemma 2.2. *Suppose $a, b \in BC$ and $\alpha' \in \mathcal{G}BC$.*

(a) *If $b \in \mathcal{G}BC$ and*

$$\max\Big\{\Big(\frac{a}{b}(\alpha'_{-1})^{-1/p}\Big)^*(0), \Big(\frac{a}{b}(\alpha'_{-1})^{-1/p}\Big)^*(1)\Big\} < 1,$$

then the operator (1.1) *is invertible, and*

$$A^{-1} = -W_\alpha^{-1} \sum_{n=0}^{\infty} (b^{-1} a W_\alpha^{-1})^n b^{-1} I.$$

(b) *If* $a \in \mathcal{G}BC$ *and*

$$\max\left\{\left(\frac{b}{a}(\alpha')^{-1/p}\right)^*(0), \left(\frac{b}{a}(\alpha')^{-1/p}\right)^*(1)\right\} < 1,$$

then the operator (1.1) *is invertible, and*

$$A^{-1} = \sum_{n=0}^{\infty}(a^{-1}bW_\alpha)^n a^{-1}I.$$

2.2. Sufficient conditions for right invertibility

Put

$$\gamma_\pm := \bigcup_{k=1}^{\infty} \alpha_{\pm k}(l).$$

Consider functions $\chi_0, \chi_1 \in C[0,1]$ such that $\chi_0(t) + \chi_1(t) = 1$ for $t \in [0,1]$ and $\chi_0(t) = 1$ for $t \in \gamma_-$, $\chi_1(t) = 1$ for $t \in \gamma_+$.

Using slight modifications of arguments in [9, Lemma 3.5] (see also [11, Lemma 4], [17]), from Lemma 2.2 we can get the following.

Lemma 2.3. *If* $a, b, \alpha' \in \mathcal{G}BC$ *and*

$$\max\left\{\left(\frac{a}{b}(\alpha'_{-1})^{-1/p}\right)^*(1), \left(\frac{b}{a}(\alpha')^{-1/p}\right)^*(0)\right\} < 1, \tag{2.7}$$

then the operator (1.1) *is right invertible. One of its right inverses is*

$$A^{(r)} = \sum_{n=0}^{\infty}(a^{-1}bW_\alpha)^n a^{-1}\chi_0 I - \sum_{n=1}^{\infty}(W_\alpha^{-1}b^{-1}aI)^n a^{-1}\chi_1 I.$$

Lemma 2.4. *Under the assumptions of Lemma* 2.3, $\dim \operatorname{Ker} A = \infty$.

We literally repeat the proof of [9, Lemma 3.6] (see also [11, Lemma 3]).

Corollary 2.5. *If* $a, b, \alpha' \in \mathcal{G}BC$ *and inequality* (2.7) *holds, then the operator* (1.1) *is right invertible but not invertible.*

Corollary 2.5 immediately follows from Lemmas 2.3 and 2.4.

2.3. Sufficient conditions for left invertibility

By analogy with Lemma 2.3 (see also [9, Lemma 3.8], [11, Lemma 4], [17]) it is easy to prove the following.

Lemma 2.6. *If* $a, b, \alpha' \in \mathcal{G}BC$ *and*

$$\max\left\{\left(\frac{a}{b}(\alpha'_{-1})^{-1/p}\right)^*(0), \left(\frac{b}{a}(\alpha')^{-1/p}\right)^*(1)\right\} < 1, \tag{2.8}$$

then the operator (1.1) *is left invertible. One of its left inverses is*

$$A^{(l)} = \sum_{n=0}^{\infty}\chi_1(a^{-1}bW_\alpha)^n a^{-1}I - \sum_{n=1}^{\infty}\chi_0(W_\alpha^{-1}b^{-1}aI)^n a^{-1}I.$$

To get more information on the cokernel of the operator (1.1), we need the following auxiliary proposition.

Proposition 2.7. *Suppose $g, \alpha' \in \mathcal{G}BC$.*

(a) *If* $\left(\frac{1}{g}(\alpha'_{-1})^{-1/p}\right)^*(0) < 1$, *then*

$$\lim_{n\to+\infty}\inf_{t\in l}\left(\left|\prod_{k=0}^{n-1} g[\alpha_{k-n}(t)]\right| |\alpha'_{-n}(t)|^{1/p}\right) = +\infty. \tag{2.9}$$

(b) *If* $\left(g\cdot(\alpha')^{-1/p}\right)^*(1) < 1$, *then*

$$\lim_{n\to+\infty}\inf_{t\in l}\left(\left|\prod_{k=1}^{n} g^{-1}[\alpha_{n-k}(t)]\right| |\alpha'_{n}(t)|^{1/p}\right) = +\infty.$$

Proof. (a) The hypothesis is equivalent to the inequality

$$\left((g\circ\alpha)\cdot(\alpha')^{-1/p}\right)_*(0) > 1.$$

Fix $\varepsilon > 0$ such that

$$\left((g\circ\alpha)\cdot(\alpha')^{-1/p}\right)_*(0) > 1+\varepsilon.$$

From the definition of the lower limit it follows that there exists a point $t_0 \in \mathbb{I}$ such that

$$\frac{|g[\alpha(t)]|}{(\alpha'(t))^{1/p}} > 1+\varepsilon \quad \text{for all} \quad t\in(0,t_0).$$

Since zero is the repelling point of the shift α, there exists a number $k_0 \in \mathbb{N}$ such that $\alpha_{-k}(l) \subset (0,t_0)$ for all $k > k_0$. Hence,

$$\frac{|g[\alpha_{1-k}(t)]|}{(\alpha'[\alpha_{-k}(t)])^{1/p}} > 1+\varepsilon \quad \text{for all} \quad t\in l \quad \text{and} \quad k>k_0. \tag{2.10}$$

On the other hand, in view of (2.3),

$$\left|\prod_{k=0}^{n-1} g[\alpha_{k-n}(t)]\right| |\alpha'_{-n}(t)|^{1/p} = \left(\prod_{k=1}^{n}\frac{|g[\alpha_{1-k}(t)]|}{(\alpha'[\alpha_{-k}(t)])^{1/p}}\right)\cdot\left|\frac{g[\alpha_{-n}(t)]}{g(t)}\right| \tag{2.11}$$

Set

$$M_0 := \inf_{t\in\mathbb{I}}\frac{|g[\alpha(t)]|}{(\alpha'(t))^{1/p}} > 0, \quad N := \inf_{t\in\mathbb{I}}|g(t)|/\sup_{t\in\mathbb{I}}|g(t)| > 0. \tag{2.12}$$

Combining (2.10)–(2.12), we get for sufficiently large n,

$$\begin{aligned}\inf_{t\in l}\left(\left|\prod_{k=0}^{n-1} g[\alpha_{k-n}(t)]\right| |\alpha'_{-n}(t)|^{1/p}\right) &\geq N\cdot\prod_{k=1}^{n}\left(\inf_{t\in l}\frac{|g[\alpha_{1-k}(t)]|}{(\alpha'[\alpha_{-k}(t)])^{1/p}}\right)\\ &\geq NM_0^{k_0}(1+\varepsilon)^{n-k_0}.\end{aligned}$$

The latter inequality implies (2.9). Statement (b) is proved by analogy. □

Lemma 2.8. *Under the assumptions of Lemma* 2.6, $\dim\operatorname{Coker} A = \infty$.

Proof. The idea of the proof is borrowed from [11, Lemma 3]. Let $d \in L^p$ and $\operatorname{supp} d \subset l$. Let us prove that if $d(t) \neq 0$ a. e. on l, then $d \notin \operatorname{Im} A$.

Assume the contrary, that is, assume the equation

$$(Af)(t) = a(t)f(t) - b(t)f[\alpha(t)] = d(t), \quad t \in \mathbb{I}. \tag{2.13}$$

has a solution $f \in L^p$. Since $\operatorname{supp} d \subset l$, we have for every $t \in l$ and $n \in \mathbb{N}$,

$$f[\alpha_{-n}(t)] = \left(\prod_{k=1}^{n} g[\alpha_{-k}(t)]\right) f(t), \tag{2.14}$$

$$f[\alpha_n(t)] = \left(\prod_{k=0}^{n-1} g^{-1}[\alpha_k(t)]\right)\left(f(t) - \frac{d(t)}{a(t)}\right), \tag{2.15}$$

where $g := b/a$. From (2.14) we see that

$$f(t) = \left(\prod_{k=1}^{n} g[\alpha_{n-k}(t)]\right) f[\alpha_n(t)] = \left(\prod_{k=0}^{n-1} g[\alpha_k(t)]\right) f[\alpha_n(t)], \quad t \in \alpha_{-n}(l).$$

Let χ_u stand for the characteristic function of a set $u \subset \mathbb{I}$. For every $n \in \mathbb{N}$,

$$\begin{aligned} \|f\|_p &\geq \|f\chi_{\alpha_{-n}(l)}\|_p = \left(\int_{\alpha_{-n}(l)} \left|\prod_{k=0}^{n-1} g[\alpha_k(t)]\right|^p |f[\alpha_n(t)]|^p dt\right)^{1/p} \\ &= \left(\int_l \left|\prod_{k=0}^{n-1} g[\alpha_{k-n}(t)]\right|^p |f(t)|^p |\alpha'_{-n}(t)| dt\right)^{1/p} \\ &\geq \inf_{t\in l}\left(\left|\prod_{k=0}^{n-1} g[\alpha_{k-n}(t)]\right| |\alpha'_{-n}(t)|^{1/p}\right) \|f\chi_l\|_p. \end{aligned}$$

In view of Proposition 2.7(a), $f \notin L^p$ whenever $f(t) \neq 0$ a. e. on l.

Analogously, from (2.15) we see that

$$f(t) = \left(\prod_{k=1}^{n} g^{-1}[\alpha_{-k}(t)]\right) \varphi[\alpha_{-n}(t)], \quad t \in \alpha_n(l),$$

where $\varphi := f - d/a$. Then for every $n \in \mathbb{N}$,

$$\|f\|_p \geq \inf_{t\in l}\left(\left|\prod_{k=1}^{n} g^{-1}[\alpha_{n-k}(t)]\right| |\alpha'_n(t)|^{1/p}\right) \|\varphi\chi_l\|_p.$$

In view of Proposition 2.7(b), $f \notin L^p$ whenever $\varphi(t) \neq 0$ a. e. on l. Thus, if (2.13) holds and $d(t) \neq 0$ a. e. on l, then $f \notin L^p$, and we are led to a contradiction.

Since $d \notin \operatorname{Im} A$ is arbitrary, it is easily seen that $\dim \operatorname{Coker} A = \infty$. □

Corollary 2.9. *If* $a, b, \alpha' \in \mathcal{G}BC$ *and inequality* (2.8) *holds, then the operator* (1.1) *is left invertible but not invertible.*

Corollary 2.9 immediately follows from Lemmas 2.6 and 2.8.

3. Slowly oscillating functions and shifts

3.1. Some properties of slowly oscillating functions

Let C^1 be the set of all continuously differentiable functions on $\mathbb{I}$.

Below we formulate three sufficient conditions which guarantee slow oscillation.

Proposition 3.1. *Suppose $\varphi \in SO$ and $\psi \in BC$. If*

$$\lim_{t\to 0}\Big(\varphi(t)-\psi(t)\Big)=0,$$

then $\psi \in SO$.

The proof is straightforward.

Proposition 3.2. *If $\varphi \in BC \cap C^1$ and*

$$\lim_{t\to 0} t\varphi'(t)=0, \tag{3.1}$$

then $\varphi \in SO$.

Proof. Fix $\varepsilon > 0$. From (3.1) we infer that there exists a $\delta(\varepsilon) > 0$ such that

$$|t\varphi'(t)| < \varepsilon \quad \text{whenever} \quad t \in (0, \delta(\varepsilon)). \tag{3.2}$$

Fix $\lambda \in \mathbb{I}$ and assume that $r \in (0, \delta(\varepsilon))$. Consider $x, y \in [\lambda r, r]$ and assume for definiteness that $x \leq y$. Then since $\varphi \in C^1$, we obtain

$$|\varphi(x)-\varphi(y)| \leq \int_x^y |t\varphi'(t)|\frac{dt}{t} \leq \left(\sup_{t\in(0,r]} |t\varphi'(t)|\right)\log\frac{1}{\lambda}. \tag{3.3}$$

From (3.2) and (3.3) we get for $r \in (0, \delta(\varepsilon))$,

$$\omega(\varphi,[\lambda r, r]) \leq \left(\sup_{t\in(0,r]} |t\varphi'(t)|\right)\log\frac{1}{\lambda} \leq \varepsilon \log\frac{1}{\lambda}.$$

Hence, $\varphi \in SO$. □

Proposition 3.3. *Let $\varphi \in BC \cap C^1$ and $\psi(t) := t\varphi'(t)$, $t \in \mathbb{I}$. If φ and ψ slowly oscillate at zero, then*

$$\lim_{t\to 0}\psi(t)=0. \tag{3.4}$$

Proof. First let φ and hence ψ be real-valued functions. Contrary to (3.4) assume that there are a constant $c > 0$ and a sequence $r_n \to 0$ such that

$$\text{either} \quad \psi(r_n) \geq 2c > 0 \qquad \text{or} \qquad \psi(r_n) \leq -2c < 0 \tag{3.5}$$

for $n \in \mathbb{N}$. Since ψ slowly oscillates at zero, there exists an $n_0 \in \mathbb{N}$ such that $\omega(\psi, [r_n/2, r_n]) \leq c$ for all $n > n_0$. Hence, by (3.5),

$$\text{either} \quad \inf\Big\{\psi(t) : t \in [r_n/2, r_n]\Big\} \geq c \quad \text{or} \quad \sup\Big\{\psi(t) : t \in [r_n/2, r_n]\Big\} \leq -c.$$

Then

$$\text{either}\quad \varphi(r_n)-\varphi(r_n/2)=\int_{r_n/2}^{r_n}\psi(t)\frac{dt}{t}\geq c\log 2\quad\text{for all}\quad n>n_0,$$

$$\text{or}\quad \varphi(r_n)-\varphi(r_n/2)=\int_{r_n/2}^{r_n}\psi(t)\frac{dt}{t}\leq -c\log 2\quad\text{for all}\quad n>n_0.$$

In both cases $\varphi\notin SO$, and this contradiction proves (3.4).

Passage to the real and imaginary parts of φ and ψ completes the proof. □

Proposition 3.4. *If $\varphi\in SO$, then the function*

$$\zeta(t):=\frac{1}{t}\int_0^t\varphi(s)ds,\quad t\in\mathbb{I},$$

belongs to SO too. Moreover, $\lim_{t\to 0}(\varphi(t)-\zeta(t))=0$.

Proof. It follows from [25, Lemmas 3 and 4] that $\zeta\in SO$. Since $t\zeta'(t)=\varphi(t)-\zeta(t)$ for $t\in\mathbb{I}$, the function $\psi: t\mapsto t\zeta'(t)$ belongs to SO too. It remains to apply Proposition 3.3 with $\varphi=\zeta$. □

We say that $h=\{h_n\}_{n=1}^{\infty}\subset\mathbb{I}$ is a *test sequence (of numbers)* if $\lim_{n\to\infty}h_n=0$.

Lemma 3.5. *Suppose that the continuous functions $\mathcal{F}:\mathbb{I}\times\mathbb{I}\to\mathbb{I}$ and $f_j:\mathbb{I}\to\mathbb{I}$, $j=1,2$, satisfy*

$$xf_1(y)\leq\mathcal{F}(x,y)\leq xf_2(y),\quad x,y\in\mathbb{I}.\tag{3.6}$$

Let $\varphi:\mathbb{I}\to\mathbb{C}$ be a slowly oscillating function at zero. If for some test sequence $h=\{h_n\}_{n=1}^{\infty}$ the limit

$$\lim_{n\to\infty}\varphi(h_n)=:\varphi_h\tag{3.7}$$

exists, then for every $y\in\mathbb{I}$ the limit $\lim_{n\to\infty}\varphi(\mathcal{F}(h_n,y))$ also exists. Moreover,

$$\lim_{n\to\infty}\varphi(\mathcal{F}(h_n,y))=\varphi_h,\tag{3.8}$$

and the convergence is uniform on every segment $J\subset\mathbb{I}$.

Proof. Fix $\varepsilon>0$. Since $\varphi\in SO$, there exists a $\delta=\delta(\varepsilon)\in(0,1)$ such that

$$\omega(\varphi,[\lambda r,r])<\frac{\varepsilon}{2}\quad\text{for all}\quad r\in(0,\delta).\tag{3.9}$$

On the other hand, (3.7) implies the existence of a number $n_1(\varepsilon)\in\mathbb{N}$ such that

$$|\varphi(h_n)-\varphi_h|<\frac{\varepsilon}{2}\quad\text{for all}\quad n\geq n_1(\varepsilon).\tag{3.10}$$

For a given segment $J\subset\mathbb{I}$, put $\lambda:=\min_{y\in J}f_1(y)\in\mathbb{I}$ and set $r_n=h_n$. Then from the property $f_2(y)<1$ and from (3.6) we get

$$\lambda r_n\leq h_n\leq r_n,\quad \lambda r_n\leq\mathcal{F}(h_n,y)\leq r_n\quad\text{for all}\quad y\in J.\tag{3.11}$$

Since $h = \{h_n\}_{n=1}^{\infty}$ is a test sequence, there exists a number $n_2(\varepsilon) \geq n_1(\varepsilon)$ such that $r_n \in (0, \delta)$ for all $n \geq n_2(\varepsilon)$. Hence, from (3.9) and (3.11) we deduce that for every $n \geq n_2(\varepsilon)$,

$$|\varphi(\mathcal{F}(h_n, y)) - \varphi(h_n)| \leq \omega(\varphi, [\lambda r_n, r_n]) < \frac{\varepsilon}{2} \quad \text{for all} \quad y \in J. \tag{3.12}$$

Combining (3.10) and (3.12), we conclude that for every $y \in J$ and every $n \geq n_2(\varepsilon)$,

$$|\varphi(\mathcal{F}(h_n, y)) - \varphi_h| \leq |\varphi(\mathcal{F}(h_n, y)) - \varphi(h_n)| + |\varphi(h_n) - \varphi_h| < \frac{\varepsilon}{2} + \frac{\varepsilon}{2} = \varepsilon,$$

that is, (3.8) holds and the convergence is uniform on $J \subset \mathbb{I}$. □

3.2. Some properties of slowly oscillating shifts

For a shift α and $j \in \{0, 1\}$, put

$$\widetilde{\alpha}_j(t) := \frac{\alpha(t) - \alpha(j)}{t - j}, \quad t \in \mathbb{I}.$$

Proposition 3.6. *If $\alpha' \in \mathcal{G}B$ and $j \in \{0, 1\}$, then the following properties hold:*

(a) *for every $t \in \mathbb{I}$,*

$$m_\alpha t \leq \alpha(t) \leq M_\alpha t, \quad m_\alpha(1 - t) \leq 1 - \alpha(t) \leq M_\alpha(1 - t), \tag{3.13}$$

where

$$0 < m_\alpha := \inf_{t \in \mathbb{I}} \alpha'(t), \quad M_\alpha := \sup_{t \in \mathbb{I}} \alpha'(t) < +\infty; \tag{3.14}$$

(b) *the functions $\widetilde{\alpha}_j$ belong to $\mathcal{G}BC$;*

(c) *$\widetilde{\alpha}_j$ slowly oscillates at j if and only if $\widetilde{\beta}_j$ slowly oscillates at j.*

Proof. Statement (a) follows from the mean-value theorem and the fact that $\alpha(0) = 0, \alpha(1) = 1$. Statement (b) follows from statement (a).

(c) Consider the case $j = 0$. Since $\alpha' \in \mathcal{G}B$ if and only if $\beta' = \alpha'_{-1} \in \mathcal{G}B$, we obtain, in view of statement (b), that $\widetilde{\beta}_0 \in \mathcal{G}BC$. Further, if $s, t \in \mathbb{I}$, then

$$|\widetilde{\beta}_0(s) - \widetilde{\beta}_0(t)| = \left|\frac{\beta(s)}{s}\right| \cdot \left|\frac{\beta(t)}{t}\right| \cdot \left|\frac{s}{\beta(s)} - \frac{t}{\beta(t)}\right| \leq L \left|\frac{s}{\beta(s)} - \frac{t}{\beta(t)}\right|, \tag{3.15}$$

where

$$L := \left(\sup_{t \in \mathbb{I}} |\widetilde{\beta}_0(t)|\right)^2 < +\infty.$$

Let $\lambda \in \mathbb{I}$ and suppose r is sufficiently small. Taking into account that α is a homeomorphism, we have

$$\sup_{s,t \in [\lambda r, r]} \left|\frac{s}{\beta(s)} - \frac{t}{\beta(t)}\right| = \sup_{x,y \in [\beta(\lambda r), \beta(r)]} \left|\frac{\alpha(x)}{x} - \frac{\alpha(y)}{y}\right|. \tag{3.16}$$

From statement (a) we get

$$[\beta(\lambda r), \beta(r)] \subset [m_\beta \lambda r, M_\beta r] =: [\lambda_1 r_1, r_1], \tag{3.17}$$

where m_β and M_β are defined by (3.14), $\lambda_1 := \frac{m_\beta}{M_\beta}\lambda \in (0,\lambda) \subset \mathbb{I}$, $r_1 := M_\beta r$. From (3.15)–(3.17) we deduce that

$$\omega(\widetilde{\beta}_0, [\lambda r, r]) \le L\left(\sup_{x,y\in[\lambda_1 r_1, r_1]} |\widetilde{\alpha}_0(x) - \widetilde{\alpha}_0(y)|\right) = L\omega(\widetilde{\alpha}_0, [\lambda_1 r_1, r_1]). \tag{3.18}$$

If $\widetilde{\alpha}_0$ slowly oscillates at zero, then $\omega(\widetilde{\alpha}_0, [\lambda_1 r_1, r_1]) \to 0$ as $r_1 \to 0$. In this case (3.18) implies that $r = r_1/M_\beta \to 0$ and $\omega(\widetilde{\beta}_0, [\lambda r, r]) \to 0$, i.e., $\widetilde{\beta}_0$ slowly oscillates at zero. For the functions $\widetilde{\alpha}_1$ and $\widetilde{\beta}_1$ statement (c) is proved by analogy. □

Lemma 3.7. *Suppose $\alpha' \in BC$ and $j \in \{0,1\}$. If α' is a slowly oscillating function at j, then $\widetilde{\alpha}_j$ is a slowly oscillating function at j too.*

Proof. Consider the case $j = 1$. If α' slowly oscillates at 1, then $\alpha'(1-t)$ slowly oscillates at zero. Hence, by Proposition 3.4, the function

$$\frac{1-\alpha(1-t)}{t} = \frac{1}{t}\int_0^t \alpha'(1-s)ds$$

belongs to SO together with $\alpha'(1-t)$. This is equivalent to the slow oscillation of $\widetilde{\alpha}_1$ at 1. Applying Proposition 3.4 with $\varphi = \alpha'$, we manage the case $j = 0$. □

One can construct an example of a shift α such that $\log \alpha' \in BC$ and $\widetilde{\alpha}_j$ is a slowly oscillating function at j, but α' does not slowly oscillate at j.

The next lemma gives us equivalent characterizations of slowly oscillating shifts.

Lemma 3.8. *Suppose* $\log \alpha' \in BC$ *and* $j \in \{0,1\}$. *The following statements are equivalent:*

(a) *α' is a slowly oscillating function at j;*

(b) *β' is a slowly oscillating function at j;*

(c) $\lim_{t\to j}(t-j)\widetilde{\alpha}_j'(t) = 0$;

(d) $\lim_{t\to j}(t-j)\widetilde{\beta}_j'(t) = 0$.

Proof. The equivalence (a) ⇔ (b) is proved similarly to Proposition 3.6(c) making use of

$$|\beta'(s) - \beta'(t)| = \frac{|\alpha'(\beta(s)) - \alpha'(\beta(t))|}{\alpha'(\beta(s))\alpha'(\beta(t))}, \quad s,t \in \mathbb{I}.$$

Let us prove the equivalence (a) ⇔ (c) in the case $j = 1$. Taking into account Lemma 3.7, we deduce that $\alpha', \widetilde{\alpha}_1$ slowly oscillate at 1. Therefore, the function

$$t\gamma'(t) = \alpha'(1-t) - \gamma(t), \quad \text{where} \quad \gamma(t) := (1-\alpha(1-t))/t, \tag{3.19}$$

slowly oscillates at zero. Due to Proposition 3.6(b), $\widetilde{\alpha}_1 \in \mathcal{G}BC$. Thus, by Proposition 3.3,

$$\lim_{t\to 1}(t-1)\widetilde{\alpha}_1'(t) = \lim_{t\to 0} t\gamma'(t) = 0. \tag{3.20}$$

Conversely, if (3.20) holds, then by Proposition 3.2, γ slowly oscillates at zero. Hence, from (3.19)–(3.20) and Proposition 3.1 we deduce that $\alpha'(1-t)$ slowly oscillates at zero. This means that α' slowly oscillates at 1.

The case $j=0$ and the equivalence (b) $\Leftrightarrow$ (d) are treated analogously. □

Lemma 3.7 and the second assertion of Proposition 3.4 imply the following.

Corollary 3.9. *Suppose α is a slowly oscillating shift at $j \in \{0,1\}$.*

(a) *If for some sequence $\{h_n\}_{n=1}^{\infty} \subset \mathbb{I}$ which tends to j one of the limits*

$$\lim_{n\to\infty} \widetilde{\alpha}_j(h_n), \qquad \lim_{n\to\infty} \alpha'(h_n)$$

exists, then the other limit exists too. Moreover,

$$\lim_{n\to\infty} \widetilde{\alpha}_j(h_n) = \lim_{n\to\infty} \alpha'(h_n).$$

(b)
$$\liminf_{t\to j} \widetilde{\alpha}_j(t) = \liminf_{t\to j} \alpha'(t), \quad \limsup_{t\to j} \widetilde{\alpha}_j(t) = \limsup_{t\to j} \alpha'(t).$$

Modifying Example 2.3 in [15] one can construct an example of a slowly oscillating non-Carleman shift with

$$\liminf_{t\to j} \alpha'(t) < \limsup_{t\to j} \alpha'(t), \quad j \in \{0,1\}.$$

4. Limit operators for a functional operator

4.1. Abstract approach

Let X be a Banach space and X^* be its dual space. Let $\mathfrak{B}(X)$ denote the Banach algebra of all bounded linear operators on X. We say that an operator $V \in \mathfrak{B}(X)$ is a *pseudoisometry* if $V \in \mathcal{G}\mathfrak{B}(X)$ and $\|V\|_{\mathfrak{B}(X)} = 1/\|V^{-1}\|_{\mathfrak{B}(X)}$. Any sequence $\mathcal{V} = \{V_n\}_{n=1}^{\infty}$ of pseudoisometries is referred to as a *test sequence (of operators)*. Fix some set $\mathbb{V} \subset \mathfrak{B}(X)$ of pseudoisometries.

Let $A \in \mathfrak{B}(X)$ and $\mathcal{V} = \{V_n\}_{n=1}^{\infty} \subset \mathbb{V}$. If the strong limits

$$A_{\mathcal{V}} := s-\lim_{n\to\infty} V_n^{-1} A V_n \quad \text{in } \mathfrak{B}(X), \quad A_{\mathcal{V}^*} := s-\lim_{n\to\infty} (V_n^{-1} A V_n)^* \quad \text{in } \mathfrak{B}(X^*) \quad (4.1)$$

exist, then always $(A_{\mathcal{V}})^* = A_{\mathcal{V}^*}$, and we will refer to the operator $A_{\mathcal{V}}$ as the *limit operator* for the operator A with respect to the test sequence $\mathcal{V}$. Note that usually the limit operator $A_{\mathcal{V}}$ is defined independently of the existence of the strong limit $A_{\mathcal{V}^*}$ (see, e.g., [5], [24]), while we need the existence of both limits (4.1) for our purposes. If the limit operator $A_{\mathcal{V}}$ exists, then it is uniquely determined by A and $\mathcal{V}$, which justifies the notation $A_{\mathcal{V}}$. Further, if $A_{\mathcal{V}}$ is the limit operator of A with respect to $\mathcal{V}$ and if $\mathcal{W}$ is a subsequence of $\mathcal{V}$, then the limit operator $A_{\mathcal{W}}$ of A with respect to $\mathcal{W}$ also exists and coincides with $A_{\mathcal{V}}$.

In the next proposition we summarize some properties of limit operators.

Proposition 4.1. *Let $\mathcal{V} \subset \mathbb{V}$ be a test sequence. If $A, B \in \mathfrak{B}(X)$, $\alpha \in \mathbb{C}$, and the limit operators $A_\mathcal{V}, B_\mathcal{V}$ exist, then $(\alpha A)_\mathcal{V}, (A+B)_\mathcal{V}, (AB)_\mathcal{V}$ also exist and*

$$(\alpha A)_\mathcal{V} = \alpha A_\mathcal{V}, \quad (A+B)_\mathcal{V} = A_\mathcal{V} + B_\mathcal{V}, \quad (AB)_\mathcal{V} = A_\mathcal{V} B_\mathcal{V}.$$

The proof is straightforward (see also [5], [24] and references therein).

The following lemma is a useful tool for the investigation of the invertibility in Banach algebras of operators.

Lemma 4.2. *If an operator $A \in \mathfrak{B}(X)$ is invertible, then for every test sequence $\mathcal{V} = \{V_n\}_{n=1}^\infty \subset \mathbb{V}$ the limit operator $A_\mathcal{V}$ is also invertible.*

This lemma follows from Lemma [8, Ch. 3, Lemma 1.1] due to the existence of the two limits (4.1) and the equality $(A_\mathcal{V})^* = A_{\mathcal{V}^*}$.

4.2. Realization

We will construct a set of pseudoisometries on the space L^p. First, let us start from the simpler case of the half-line $\mathbb{R}_+ := (0, \infty)$. Let $\overline{\mathbb{R}}_+ := [0, \infty]$. For $x \in \mathbb{R}_+$, the multiplicative shift operator $\widetilde{V}_x$ is defined on the space $L^p(\mathbb{R}_+)$ by

$$(\widetilde{V}_x f)(y) := f(y/x), \quad y \in \mathbb{R}_+.$$

Clearly, $\widetilde{V}_x \in \mathcal{G}\mathfrak{B}(L^p(\mathbb{R}_+))$ and $\widetilde{V}_x^{-1} = \widetilde{V}_{1/x}$. Moreover, $\|\widetilde{V}_x\|_{\mathfrak{B}(L^p(\mathbb{R}_+))} = x^{1/p}$. Hence, $\widetilde{V}_x$ is a pseudoisometry for every $x \in \mathbb{R}_+$.

Consider a function $\theta : \overline{\mathbb{R}}_+ \to [0,1]$ such that $\theta(0) = 0$, $\theta(\infty) = 1$, and the derivative θ' exists and is positive for every $x \in \mathbb{R}_+$. Then the transformation θ preserves the natural orientation and is invertible. Denote by θ_{-1} its inverse, which maps $[0,1]$ onto $\overline{\mathbb{R}}_+$.

It is easy to see that the transformation $G : L^p \to L^p(\mathbb{R}_+)$ defined by

$$(G\varphi)(y) := c(y)\varphi(\theta(y)), \quad c(y) := (\theta'(y))^{1/p}, \quad y \in \mathbb{R}_+, \tag{4.2}$$

is an isometric isomorphism with the inverse $G^{-1} : L^p(\mathbb{R}_+) \to L^p$ given by

$$(G^{-1}\psi)(t) := \frac{\psi(\theta_{-1}(t))}{c(\theta_{-1}(t))}, \quad t \in \mathbb{I}. \tag{4.3}$$

Hence,

$$V_x := G^{-1}\widetilde{V}_x G \in \mathcal{G}\mathfrak{B}(L^p), \quad x \in \mathbb{R}_+, \tag{4.4}$$

and

$$V_x^{-1} = V_{1/x}, \quad \|V_x\|_{\mathfrak{B}(L^p)} = \|\widetilde{V}_x\|_{\mathfrak{B}(L^p(\mathbb{R}_+))} = x^{1/p}.$$

So, V_x is a pseudoisometry on L^p. Consider the set of pseudoisometries

$$\mathbb{V} := \Big\{ V_x \in \mathfrak{B}(L^p),\ x \in \mathbb{I} \Big\}.$$

Clearly, we can identify any test sequence $h = \{h_n\}_{n=1}^\infty \subset \mathbb{I}$ with the sequence of pseudoisometries $\mathcal{V}_h = \{V_{h_n}\}_{n=1}^\infty \subset \mathbb{V}$.

For every $x \in \mathbb{R}_+$, put $\gamma_x(y) := y/x$, $y \in \overline{\mathbb{R}}_+$, and

$$w_x := \theta \circ \gamma_x \circ \theta_{-1} \ :\ [0,1] \to [0,1]. \tag{4.5}$$

Clearly, w_x is a homeomorphism with the inverse

$$(w_x)_{-1} = \theta \circ \gamma_{1/x} \circ \theta_{-1} = w_{1/x}. \tag{4.6}$$

Put

$$d_x := \frac{c \circ \theta_{-1} \circ w_x}{c \circ \theta_{-1}}, \qquad W_x := W_{w_x}. \tag{4.7}$$

Proposition 4.3. *For every $x \in \mathbb{R}_+$, we have*

$$V_x = d_x W_x = W_x \frac{1}{d_{1/x}} I. \tag{4.8}$$

Proof. Let $\varphi \in L^p$. From (4.2)–(4.5) we get

$$V_x\varphi = G^{-1}\widetilde{V}_x G\varphi = \left(\frac{c \circ \gamma_x \circ \theta_{-1}}{c \circ \theta_{-1}}\right) \cdot (\varphi \circ \theta \circ \gamma_x \circ \theta_{-1}) = d_x W_x \varphi.$$

Using (4.6) and (4.7) we immediately obtain $d_x \circ w_{1/x} = 1/d_{1/x}$, which gives the second part of (4.8). □

Proposition 4.4.

(a) *If $a \in L^\infty$, then for every $x \in \mathbb{R}_+$,*

$$V_x^{-1} a V_x = (a \circ w_{1/x})I, \quad (V_x^{-1} a V_x)^* = (\overline{a} \circ w_{1/x})I. \tag{4.9}$$

(b) *If $\alpha' \in \mathcal{G}L^\infty$, then for every $x \in \mathbb{R}_+$,*

$$V_x^{-1} W_\alpha V_x = \frac{d_{1/x}}{d_{1/x} \circ \alpha_x} W_{\alpha_x}, \quad (V_x^{-1} W_\alpha V_x)^* = \beta_x' W_{\beta_x} \frac{d_{1/x}}{d_{1/x} \circ \alpha_x} I, \tag{4.10}$$

where

$$\alpha_x := w_x \circ \alpha \circ w_{1/x}, \quad \beta_x := w_x \circ \beta \circ w_{1/x}. \tag{4.11}$$

Proof. (a) If $a \in L^\infty$, then $aI \in \mathfrak{B}(L^p)$. Hence, $V_x^{-1} a V_x \in \mathfrak{B}(L^p)$ for every $x \in \mathbb{R}_+$. From (4.8) we get

$$V_x^{-1} a V_x = d_{1/x} W_{1/x} a W_x \frac{1}{d_{1/x}} I = d_{1/x} (a \circ w_{1/x}) \frac{1}{d_{1/x}} I = (a \circ w_{1/x}) I.$$

Further,

$$(V_x^{-1} a V_x)^* = \Big((a \circ w_{1/x})I\Big)^* = (\overline{a} \circ w_{1/x})I,$$

and so statement (a) is proved.

(b) Clearly, $W_{1/x} W_\alpha W_x = W_{\alpha_x}$. Using this identity and (4.8), we get the first equality in (4.10):

$$V_x^{-1} W_\alpha V_x = d_{1/x} W_{1/x} W_\alpha W_x \frac{1}{d_{1/x}} I = d_{1/x} W_{\alpha_x} \frac{1}{d_{1/x}} I = \frac{d_{1/x}}{d_{1/x} \circ \alpha_x} W_{\alpha_x}.$$

Then passing to adjoint operators in that formula, taking into account that $d_{1/x}$ is real-valued, and using the identity $(W_\alpha)^* = \beta' W_\beta$, $\beta = \alpha_{-1}$, we obtain the second equality in (4.10). Statement (b) and Proposition 4.4 are proved. □

In what follows we take the concrete function

$$\theta(y) := \frac{y}{1+y}, \quad y \in \overline{\mathbb{R}}_+. \tag{4.12}$$

Then

$$\theta_{-1}(t) = \frac{t}{1-t}, \quad t \in [0,1], \tag{4.13}$$

and for every $x \in \mathbb{R}_+$,

$$w_x(t) = \frac{\gamma_x(\theta_{-1}(t))}{1+\gamma_x(\theta_{-1}(t))} = \frac{\theta_{-1}(t)}{x+\theta_{-1}(t)} = \frac{t}{x-xt+t}, \quad t \in [0,1]. \tag{4.14}$$

Analogously,

$$w_{1/x}(t) = \frac{xt}{1+xt-t} =: F(x,t), \quad t \in [0,1]. \tag{4.15}$$

Clearly,

$$xt \le F(x,t) \le \frac{xt}{1-t}, \quad x, t \in \mathbb{I}. \tag{4.16}$$

Lemma 4.5. *If $\alpha' \in \mathcal{G}B$, then for the functions* (4.11) *we have*

$$\sup_{x\in\mathbb{R}_+}\sup_{t\in\mathbb{I}} \alpha'_x(t) < +\infty, \qquad \sup_{x\in\mathbb{R}_+}\sup_{t\in\mathbb{I}} \beta'_x(t) < +\infty.$$

Proof. Let $x \in \mathbb{R}_+$. Clearly, for every $t \in \mathbb{I}$,

$$\alpha'_x(t) = w'_x[\alpha(w_{1/x}(t))] \cdot \alpha'(w_{1/x}(t)) \cdot w'_{1/x}(t) = \alpha'(w_{1/x}(t)) \frac{w'_x[\alpha(w_{1/x}(t))]}{w'_x(w_{1/x}(t))}.$$

Since $\alpha' \in \mathcal{G}B$, we have for every $x \in \mathbb{R}_+$,

$$\alpha'_x(t) \le M_\alpha \sup_{\tau\in\mathbb{I}} \frac{w'_x[\alpha(\tau)]}{w'_x(\tau)}, \quad t \in \mathbb{I}, \tag{4.17}$$

where M_α is defined in (3.14). From (4.14) we get

$$w'_x(t) = \frac{x}{(x-xt+t)^2}, \quad t \in \mathbb{I}.$$

Then for $x \in \mathbb{R}_+$ and $\tau \in \mathbb{I}$ we obtain

$$\frac{w'_x[\alpha(\tau)]}{w'_x(\tau)} = \left(\frac{(1-\tau)x+\tau}{(1-\alpha(\tau))x+\alpha(\tau)}\right)^2. \tag{4.18}$$

From Proposition 3.6(a) we get

$$(1-\alpha(\tau))x + \alpha(\tau) \ge m_\alpha(1-\tau)x + m_\alpha\tau. \tag{4.19}$$

Combining (4.17)–(4.19), we arrive at the estimate

$$\sup_{x\in\mathbb{R}_+}\sup_{t\in\mathbb{I}} \alpha'_x(t) \le \frac{M_\alpha}{m_\alpha^2} < +\infty.$$

Since $\alpha' \in \mathcal{G}B$ implies $\beta' \in \mathcal{G}B$, from the estimate just proved we obtain

$$\sup_{x\in\mathbb{R}_+}\sup_{t\in\mathbb{I}} \beta'_x(t) \le \frac{M_\beta}{m_\beta^2} < +\infty.$$

The lemma is proved. □

4.3. Limit operators for the multiplication operator

We shall use the symbol $\rightrightarrows$ to indicate uniform convergence of sequences in a given segment. We shall use J to denote an arbitrary segment in $\mathbb{I}$.

Lemma 4.6. *Suppose a sequence $\varphi_n \in BC$ converges pointwise to a function $\varphi \in BC$. If this convergence is uniform on every segment $J \subset \mathbb{I}$, then*

$$s-\lim_{n\to\infty} \varphi_n I = \varphi I.$$

The proof is standard.

Theorem 4.7. *If $a \in SO$ and for some test sequence $h = \{h_n\}_{n=1}^{\infty}$ the limit*

$$\lim_{n\to\infty} a(h_n) =: a_h \tag{4.20}$$

exists, then for the multiplication operator aI the limit operator $(aI)_{\mathcal{V}_h}$ with respect to the test sequence of operators

$$\mathcal{V}_h := \left\{ V_{h_n} \in \mathfrak{B}(L^p) \;:\; \lim_{n\to\infty} h_n = 0, \quad \lim_{n\to\infty} a(h_n) = a_h \right\} \subset \mathbb{V} \tag{4.21}$$

exists and this limit operator is given by $(aI)_{\mathcal{V}_h} = a_h I$.

Proof. From Proposition 4.4(a) and (4.15) we get

$$V_{h_n}^{-1} a V_{h_n} = a(F(h_n,\cdot))I, \quad (V_{h_n}^{-1} a V_{h_n})^* = \overline{a(F(h_n,\cdot))}I. \tag{4.22}$$

Since $a \in SO$, the function $\overline{a}$ is slowly oscillating too. Taking into account (4.20) and (4.16), from Lemma 3.5 we obtain

$$a(F(h_n,t)) \rightrightarrows a_h, \quad \overline{a(F(h_n,t))} \rightrightarrows \overline{a_h}, \quad t \in J. \tag{4.23}$$

From (4.22)–(4.23) and Lemma 4.6 we get

$$s-\lim_{n\to\infty} V_{h_n}^{-1} a V_{h_n} = a_h I \in \mathfrak{B}(L^p), \quad s-\lim_{n\to\infty} (V_{h_n}^{-1} a V_{h_n})^* = \overline{a_h} I \in \mathfrak{B}(L^q).$$

Hence, $a_h I$ is the limit operator for aI with respect to the test sequence (4.21). □

4.4. Limit operators for the shift operator

In view of Proposition 4.4(b), to calculate a limit operator for the shift operator W_α, we need a result on the strong convergence of shift operators. The following lemma can be easily obtained from [7, Theorem 1].

Lemma 4.8. *Suppose $\alpha^{(n)} : [0,1] \to [0,1]$ are orientation preserving homeomorphisms such that $\alpha^{(n)}$ and $\alpha_{-1}^{(n)}$ are absolutely continuous on $[0,1]$ and $\log(\alpha^{(n)})' \in L^\infty$ for every $n \in \mathbb{N} \cup \{0\}$. If*

(i) *for some $A > 0$ and all $n \in \mathbb{N} \cup \{0\}$,*

$$\left\| \left(\alpha_{-1}^{(n)} \right)' \right\|_\infty \le A,$$

(ii) *$\alpha^{(n)} \to \alpha^{(0)}$ almost everywhere on $[0,1]$ as $n \to \infty$,* *then the sequence of shift operators* $W_{\alpha^{(n)}} \in \mathfrak{B}(L^p)$ *strongly converges to the shift operator* $W_{\alpha^{(0)}} \in \mathfrak{B}(L^p)$.

Using (4.11) and (4.14)–(4.15), we get for $x \in \mathbb{I}$ and $t \in [0,1]$,

$$\alpha_x(t) = \frac{\alpha[F(x,t)]}{x - x\alpha[F(x,t)] + \alpha[F(x,t)]}, \tag{4.24}$$

$$\beta_x(t) = \frac{\beta[F(x,t)]}{x - x\beta[F(x,t)] + \beta[F(x,t)]}. \tag{4.25}$$

Further, from (4.2) and (4.12)–(4.13) we obtain

$$(c \circ \theta_{-1})(t) = (1 + \theta_{-1}(t))^{-2/p} = \left(1 + \frac{t}{1-t}\right)^{-2/p} = (1-t)^{2/p}, \quad t \in [0,1].$$

Hence, from (4.7) and (4.15) it follows that

$$d_{1/x}(t) = \frac{(c \circ \theta_{-1} \circ w_{1/x})(t)}{(c \circ \theta_{-1})(t)} = (1 + xt - t)^{-2/p},$$

and then

$$f_x(t) := \left(\frac{d_{1/x}}{d_{1/x} \circ \alpha_x}\right)(t) = \left(\frac{1 + x\alpha_x(t) - \alpha_x(t)}{1 + xt - t}\right)^{2/p}, \quad t \in [0,1). \tag{4.26}$$

In what follows we shall denote $\widetilde{\alpha}(t) := \widetilde{\alpha}_0(t) = \alpha(t)/t$, $\widetilde{\beta}(t) := \beta(t)/t$ for $t \in \mathbb{I}$.

Now we apply Lemma 3.5 to shift functions.

Lemma 4.9. *Suppose* $\widetilde{\alpha} \in SO$. *If for some test sequence* $h = \{h_n\}_{n=1}^{\infty}$ *the limit*

$$\lim_{n\to\infty} \widetilde{\alpha}(h_n) =: \varkappa_h \tag{4.27}$$

exists, then we have the following:

(a) *for every* $t \in [0,1]$, *the sequence* $\alpha_{h_n}(t)$ *converges to the function*

$$\alpha_h(t) := \frac{\varkappa_h t}{1 + \varkappa_h t - t};$$

(b) *for every* $t \in [0,1)$, *the sequence* $f_{h_n}(t)$ *converges to the function*

$$f_h(t) := (1 + \varkappa_h t - t)^{-2/p}.$$

In (a)–(b) *the convergence is uniform with respect to* t *on every segment* $J \subset \mathbb{I}$.

Proof. Statement (a) is trivial for the fixed points of α (= endpoints). Consider the case of inner points of $[0,1]$. Due to (4.24),

$$\alpha_{h_n}(t) = \frac{\alpha[F(h_n,t)]/h_n}{1 - h_n\Big(\alpha[F(h_n,t)]/h_n\Big) + \alpha[F(h_n,t)]/h_n}, \quad n \in \mathbb{N}, \quad t \in \mathbb{I}. \tag{4.28}$$

Taking into account (4.16) and (4.27), we deduce from Lemma 3.5 that

$$\frac{\alpha[F(h_n,t)]}{F(h_n,t)} = \widetilde{\alpha}[F(h_n,t)] \rightrightarrows \varkappa_h, \quad t \in J. \tag{4.29}$$

On the other hand,

$$\frac{F(h_n,t)}{h_n} = \frac{t}{1+h_n t - t} \rightrightarrows \frac{t}{1-t}, \quad t \in J. \tag{4.30}$$

From (4.29) and (4.30) we get

$$\frac{\alpha[F(h_n,t)]}{h_n} \rightrightarrows \frac{\varkappa_h t}{1-t}, \quad t \in J. \tag{4.31}$$

Since, by (4.31), the denominator of the fraction in (4.28) uniformly converges on J to the function $1+\varkappa_h t/(1-t)$ and since it is uniformly bounded away from zero on J, we deduce from (4.28) and (4.31) that

$$\alpha_{h_n}(t) \rightrightarrows \frac{\varkappa_h t/(1-t)}{1+\varkappa_h t/(1-t)} = \alpha_h(t), \quad t \in J. \tag{4.32}$$

Statement (a) is proved.

(b) According to (4.26),

$$f_{h_n}(t) = \left(\frac{1+h_n\alpha_{h_n}(t) - \alpha_{h_n}(t)}{1+h_n t - t}\right)^{2/p}, \quad t \in [0,1). \tag{4.33}$$

Since $\alpha_{h_n}(0) = 0$, we get $f_{h_n}(0) = f_h(0) = 1$. As the sequence $1 + h_n t - t$ is uniformly bounded away from zero for all $t \in J$ and all sufficiently large n, we infer from (4.32)–(4.33) that

$$\frac{1+h_n\alpha_{h_n}(t) - \alpha_{h_n}(t)}{1+h_n t - t} \rightrightarrows \frac{1-\alpha_h(t)}{1-t} = \frac{1}{1+\varkappa_h t - t}, \quad t \in J. \tag{4.34}$$

Statement (b) directly follows from (4.33) and (4.34). □

Theorem 4.10. *Suppose $\widetilde{\alpha} \in SO$ and $\alpha' \in \mathcal{G}B$. If for some test sequence $h = \{h_n\}_{n=1}^{\infty}$ the limit (4.27) exists, then*

$$s\text{-}\lim_{n\to\infty} V_{h_n}^{-1} W_\alpha V_{h_n} = f_h W_{\alpha_h}. \tag{4.35}$$

Proof. From the first equality in (4.10) and (4.26) we deduce that

$$V_{h_n}^{-1} W_\alpha V_{h_n} = f_{h_n} W_{\alpha_{h_n}}. \tag{4.36}$$

From Lemma 4.9(b) and Lemma 4.6 we get

$$s\text{-}\lim_{n\to\infty} f_{h_n} I = f_h I. \tag{4.37}$$

Since $\widetilde{\alpha} \in SO$, we infer from Lemma 4.9(a) that

$$\lim_{n\to\infty} \alpha_{h_n}(t) = \alpha_h(t), \quad t \in [0,1]. \tag{4.38}$$

Since $\alpha' \in \mathcal{G}B$, from Lemma 4.5 we deduce that for some $L > 0$,

$$\sup_{t\in\mathbb{I}} \beta'_{h_n}(t) \le L \quad \text{for all} \quad n \in \mathbb{N}. \tag{4.39}$$

Further, for $\beta_h = (\alpha_h)_{-1}$, we have

$$\sup_{t\in\mathbb{I}} \beta_h'(t) = \sup_{t\in\mathbb{I}} \frac{\varkappa_h}{(\varkappa_h - \varkappa_h t + t)^2} \le \varkappa_h \Big(\min\{1, \varkappa_h\}\Big)^{-2}. \tag{4.40}$$

From (4.38)–(4.40) and Lemma 4.8 we obtain

$$s-\lim_{n\to\infty} W_{\alpha_{h_n}} = W_{\alpha_h}. \tag{4.41}$$

Combining (4.36)–(4.37) and (4.41), we arrive at (4.35). □

Under the assumptions of Theorem 4.10 we cannot prove the existence of the strong limit

$$s-\lim_{n\to\infty} (V_{h_n}^{-1} W_\alpha V_{h_n})^*.$$

To prove this we need more restrictions on the shift function α. More precisely, we have to assume that $\alpha' \in SO$ (see Lemma 4.12 and Theorem 4.13). As mentioned before, this implies that $\widetilde{\alpha} \in SO$, but in general $\alpha' \notin SO$ if $\widetilde{\alpha} \in SO$.

Lemma 4.11. *Suppose $\widetilde{\alpha} \in SO$ and $\alpha' \in \mathcal{G}B$. If for some test sequence $h = \{h_n\}_{n=1}^\infty$ the limit (4.27) exists, then for every $t \in [0,1]$ the sequence $\beta_{h_n}(t)$ converges to the function*

$$\beta_h(t) := \frac{t}{\varkappa_h - \varkappa_h t + t}. \tag{4.42}$$

Moreover, this convergence is uniform with respect to t on every segment $J \subset \mathbb{I}$.

Proof. Lemma 4.11 is reduced to Lemma 4.9(a). Since $\alpha' \in \mathcal{G}B$, from Proposition 3.6(b) it follows that $\widetilde{\alpha} \in \mathcal{G}BC$. Hence $\varkappa_h > 0$ and (4.27) implies that

$$\lim_{n\to\infty} \widetilde{\beta}(\alpha(h_n)) = \lim_{n\to\infty} \frac{\beta(\alpha(h_n))}{\alpha(h_n)} = \lim_{n\to\infty} \frac{h_n}{\alpha(h_n)} = \frac{1}{\varkappa_h}. \tag{4.43}$$

Due to Proposition 3.6(a),

$$m_\beta \alpha(h_n) \le \beta(\alpha(h_n)) \le M_\beta \alpha(h_n). \tag{4.44}$$

By Proposition 3.6(c), $\widetilde{\beta} \in SO$. Taking into account that $\widetilde{\beta}(h_n) = \widetilde{\beta}\Big[\beta(\alpha(h_n))\Big]$, we deduce from (4.43)–(4.44) and Lemma 3.5 with $\mathcal{F}(x,y) = \beta(x)$ and the test sequence $h = \{\alpha(h_n)\}_{n=1}^\infty$ that

$$\lim_{n\to\infty} \widetilde{\beta}(h_n) = \lim_{n\to\infty} \widetilde{\beta}(\alpha(h_n)) = \frac{1}{\varkappa_h}. \tag{4.45}$$

Now Lemma 4.11 follows from Lemma 4.9(a) with β in place of α. □

Lemma 4.12. *Suppose α is a slowly oscillating shift at zero. If for some test sequence $h = \{h_n\}_{n=1}^\infty$ the limit (4.27) exists, then for every $t \in \mathbb{I}$ the sequence of the derivatives $\beta_{h_n}'(t)$ converges to the derivative $\beta_h'(t)$. Moreover, this convergence is uniform with respect to t on every segment $J \subset \mathbb{I}$.*

Proof. This lemma is proved by analogy with Lemmas 4.9 and 4.11. By Lemma 3.8, $\beta' \in SO$. From Lemma 3.7 it follows that $\widetilde{\beta} \in SO$. Then, by (4.27), for the test sequence $\{\alpha(h_n)\}_{n=1}^{\infty}$ we have (4.43), which implies (4.45) (see the proof of Lemma 4.11). Furthermore, since β is a slowly oscillating shift at zero, we derive from Corollary 3.9 and (4.45) that

$$\lim_{n\to\infty} \beta'(h_n) = \lim_{n\to\infty} \widetilde{\beta}(h_n) = \frac{1}{\varkappa_h}. \tag{4.46}$$

Then we deduce from (4.46), (4.16), and Lemma 3.5 that

$$\widetilde{\beta}[F(h_n,t)] \rightrightarrows \frac{1}{\varkappa_h}, \quad \beta'[F(h_n,t)] \rightrightarrows \frac{1}{\varkappa_h} \quad \text{for every} \quad J \subset \mathbb{I}. \tag{4.47}$$

From (4.15) and (4.25) it follows that

$$\beta'_{h_n}(t) = \frac{\beta'[F(h_n,t)]}{t^2\Big(h_n/F(h_n,t) + (1-h_n)\widetilde{\beta}[F(h_n,t)]\Big)^2}. \tag{4.48}$$

Combining (4.30), (4.42), (4.47), and (4.48), we infer that

$$\beta'_{h_n}(t) \rightrightarrows \frac{1/\varkappa_h}{t^2\Big((1-t)/t + 1/\varkappa_h\Big)^2} = \frac{\varkappa_h}{(\varkappa_h - \varkappa_h t + t)^2} = \beta'_h(t)$$

on every $J \subset \mathbb{I}$. □

On the basis of Lemmas 4.8–4.9, Theorem 4.10, and Lemmas 4.11–4.12 we can calculate the limit operators for the shift operator W_α.

Theorem 4.13. *Suppose α is a slowly oscillating shift at zero. If for some test sequence $h = \{h_n\}_{n=1}^{\infty}$ the limit*

$$\lim_{n\to\infty} \alpha'(h_n) =: \varkappa_h \tag{4.49}$$

exists, then for the shift operator W_α the limit operator $(W_\alpha)_{\mathcal{V}_h}$ with respect to the test sequence of operators

$$\mathcal{V}_h := \Big\{V_{h_n} \in \mathfrak{B}(L^p) \ : \ \lim_{n\to\infty} h_n = 0, \ \lim_{n\to\infty} \alpha'(h_n) = \varkappa_h\Big\} \subset \mathbb{V} \tag{4.50}$$

exists and this limit operator is given by $(W_\alpha)_{\mathcal{V}_h} = f_h W_{\alpha_h}$.

Proof. From Lemma 3.7 it follows that $\widetilde{\alpha} \in SO$. Corollary 3.9 implies that

$$\lim_{n\to\infty} \widetilde{\alpha}(h_n) = \lim_{n\to\infty} \alpha'(h_n) = \varkappa_h.$$

Then due to Theorem 4.10 we obtain (4.35). From the second equality in (4.10) and from (4.26) we deduce that

$$(V_{h_n}^{-1} W_\alpha V_{h_n})^* = \beta'_{h_n} W_{\beta_{h_n}} f_{h_n} I. \tag{4.51}$$

From Lemmas 4.9(b), 4.12, and 4.6 we infer that

$$s\!-\!\lim_{n\to\infty} f_{h_n} I = f_h I \quad \Big(\in \mathfrak{B}(L^q)\Big), \quad s\!-\!\lim_{n\to\infty} \beta'_{h_n} I = \beta'_h I \quad \Big(\in \mathfrak{B}(L^q)\Big). \tag{4.52}$$

As in the proof of Theorem 4.10, from Lemmas 4.5, 4.8, and 4.11 one can get

$$s-\lim_{n\to\infty} W_{\beta_{h_n}} = W_{\beta_h} \quad \Big(\in \mathfrak{B}(L^q)\Big). \tag{4.53}$$

Combining (4.51)–(4.53), we obtain

$$s-\lim_{n\to\infty} (V_{h_n}^{-1} W_\alpha V_{h_n})^* = s-\lim_{n\to\infty} \beta'_{h_n} W_{\beta_{h_n}} f_{h_n} I = \beta'_h W_{\beta_h} f_h I. \tag{4.54}$$

According to (4.35) and (4.54), $f_h W_{\alpha_h}$ is the limit operator for W_α with respect to the test sequence (4.50). □

4.5. Limit operators for the functional operator

Combining Proposition 4.1, Theorems 4.7 and 4.13, we get the following.

Theorem 4.14. *Suppose the operator* (1.1) *has slowly oscillating data* $\{a, b; \alpha\}$ *at zero. If for some test sequence* $h = \{h_n\}_{n=1}^\infty$ *the limits*

$$\lim_{n\to\infty} a(h_n) =: a_h, \quad \lim_{n\to\infty} b(h_n) =: b_h, \quad \lim_{n\to\infty} \alpha'(h_n) =: \varkappa_h$$

exist, then for the operator (1.1) *the limit operator* $A_{\mathcal{V}_h}$ *with respect to the test sequence of operators*

$$\mathcal{V}_h := \left\{ V_{h_n} \in \mathfrak{B}(L^p) \ : \ \begin{array}{ll} \lim_{n\to\infty} h_n = 0, & \lim_{n\to\infty} a(h_n) = a_h, \\ \lim_{n\to\infty} b(h_n) = b_h, & \lim_{n\to\infty} \alpha'(h_n) = \varkappa_h \end{array} \right\} \subset \mathbb{V}$$

exists and this limit operator is given by $A_{\mathcal{V}_h} = a_h I - b_h f_h W_{\alpha_h}$, *where*

$$f_h(t) := (1 + \varkappa_h t - t)^{-2/p}, \quad \alpha_h(t) := \frac{\varkappa_h t}{1 + \varkappa_h t - t}, \quad t \in [0, 1]. \tag{4.55}$$

5. Necessary conditions for invertibility

5.1. Necessary conditions at the fixed points

For the operator (1.1) with $a, b \in BC$ and $\alpha' \in \mathcal{G}BC$, define the real-valued function

$$\eta(t) := |a(t)| - |b(t)| \Big(\alpha'(t)\Big)^{-1/p}, \quad t \in \mathbb{I},$$

and the real numbers

$$\eta_0(j) := \limsup_{t\to j} \eta(t), \quad \eta_1(j) := \liminf_{t\to j} \eta(t), \quad j \in \{0, 1\}.$$

The results of Section 4 and Theorem 1.1 imply the following.

Theorem 5.1. *Suppose the operator* (1.1) *has slowly oscillating data* $\{a, b; \alpha\}$ *at the endpoint* $j \in \{0, 1\}$. *If the operator* (1.1) *is invertible, then* $0 \notin [\eta_1(j), \eta_0(j)]$.

Proof. Let us start with the case $j = 0$. From the definition of the operator (1.1) with slowly oscillating data at zero it follows that $\eta \in SO$. As $\eta \in BC$, for every $\delta \in [\eta_1(0), \eta_0(0)]$ there exists some test sequence $g = \{g_k\}_{k=1}^\infty$ such that

$$\lim_{k\to\infty} \eta(g_k) = \delta.$$

Since the functions $a, b, \alpha' \in SO$ are bounded, from the bounded sequences

$$\{a(g_k)\}_{k=1}^\infty, \quad \{b(g_k)\}_{k=1}^\infty, \quad \{\alpha'(g_k)\}_{k=1}^\infty$$

we can consecutively extract convergent subsequences

$$\{a(g_{k_n})\}_{n=1}^\infty, \quad \{b(g_{k_n})\}_{n=1}^\infty, \quad \{\alpha'(g_{k_n})\}_{n=1}^\infty.$$

Put $h_n = g_{k_n}$, $n \in \mathbb{N}$. Clearly, $h = \{h_n\}_{n=1}^\infty$ is a subsequence of the test sequence $g = \{g_k\}_{k=1}^\infty$ and the limits

$$\lim_{n\to\infty} a(h_n) =: a_h, \quad \lim_{n\to\infty} b(h_n) =: b_h, \quad \lim_{n\to\infty} \alpha'(h_n) =: \varkappa_h \tag{5.1}$$

exist. Moreover,

$$\delta = |a_h| - |b_h| {\varkappa_h}^{-1/p}.$$

Thus, by Theorem 4.14, for the functional operator (1.1) the limit operator $A_h := A_{\mathcal{V}_h} = a_h I - b_h f_h W_{\alpha_h}$ exists, where a_h and b_h are given by (5.1) and α_h and f_h are defined in (4.55). If the operator (1.1) is invertible, then the operator A_h is invertible too, due to Lemma 4.2. In that case, from Theorem 1.1 we deduce that

$$|a_h| - |b_h f_h(0)| \Big(\alpha_h'(0)\Big)^{-1/p} = |a_h| - |b_h| \varkappa_h^{-1/p} = \delta \neq 0.$$

Since $\delta \in [\eta_1(0), \eta_0(0)]$ is arbitrary, we have proved that $0 \notin [\eta_1(0), \eta_0(0)]$.

It remains to prove the statement in the case $j = 1$. Consider the reflection operator $R \in \mathfrak{B}(L^p)$ defined by

$$(R\varphi)(t) := \varphi(1-t), \quad t \in \mathbb{I}.$$

Clearly, R is involutive and hence, invertible. From the invertibility of $aI - bW_\alpha$ and the obvious equality

$$R(aI - bW_\alpha)R = \widehat{a}I - \widehat{b}W_{\widehat{\alpha}} =: \widehat{A},$$

where

$$\widehat{a}(t) := a(1-t), \quad \widehat{b}(t) := b(1-t), \quad \widehat{\alpha}(t) := 1 - \alpha(1-t),$$

it follows that $\widehat{A}$ is invertible too. Since $\widehat{a}, \widehat{b} \in SO$ and

$$\lim_{t\to 1} (1-t) \frac{d}{dt} \left(\frac{1-\alpha(t)}{1-t} \right) = \lim_{t\to 0} t \frac{d}{dt} \left(\frac{\widehat{\alpha}(t)}{t} \right) = 0,$$

it follows that the operator $\widehat{A}$ has slowly oscillating data at zero. From the part just proved we get

$$0 \notin \left[\liminf_{t\to 0} \widehat{\eta}(t), \limsup_{t\to 0} \widehat{\eta}(t) \right] = [\eta_1(1), \eta_0(1)],$$

where

$$\widehat{\eta}(t) := \eta(1-t) = |\widehat{a}(t)| - |\widehat{b}(t)|\Big(\widehat{\alpha}'(t)\Big)^{-1/p}, \quad t \in \mathbb{I},$$

which completes the proof. □

The following proposition fixes relations between η_0, η_1 and characteristics from Section 2.

Proposition 5.2. *Suppose $\alpha' \in \mathcal{G}BC$ and $a, b \in BC$.*

(a) *If $a \in \mathcal{G}BC$ and $\eta_1(j) > 0$, then*

$$\left(\frac{b}{a}(\alpha')^{-1/p}\right)^*(j) < 1. \tag{5.2}$$

(b) *If $b \in \mathcal{G}BC$, α' slowly oscillates at the endpoint $j \in \{0,1\}$, and $\eta_0(j) < 0$, then*

$$\left(\frac{a}{b}(\alpha'_{-1})^{-1/p}\right)^*(j) < 1. \tag{5.3}$$

Proof. (a) If $\eta_1(j) > 0$, then there is an $\varepsilon > 0$ such that for every t in a neighborhood of j,

$$|a(t)| > |b(t)|\Big(\alpha'(t)\Big)^{-1/p} + \varepsilon \quad \text{or, equivalently,} \quad \frac{|b(t)|}{|a(t)|}\Big(\alpha'(t)\Big)^{-1/p} < 1 - \frac{\varepsilon}{|a(t)|}.$$

Hence

$$\left(\frac{b}{a}(\alpha')^{-1/p}\right)^*(j) \le 1 - \varepsilon\left(\inf_{t\in\mathbb{I}} \frac{1}{|a(t)|}\right) < 1.$$

(b) If $\eta_0(j) < 0$, then analogously to (a),

$$\left(\frac{a}{b}(\alpha'_{-1}\circ\alpha)^{-1/p}\right)^*(j) = \left(\frac{a}{b}(\alpha')^{1/p}\right)^*(j) < 1, \quad j \in \{0,1\}. \tag{5.4}$$

By Lemma 3.8, the function $\alpha'_{-1} = \beta'$ slowly oscillates at j together with α'. Then taking into account Proposition 3.6(a) one can prove, by analogy with Lemma 3.5, that

$$\left(\frac{a}{b}(\alpha'_{-1}\circ\alpha)^{-1/p}\right)^*(j) = \left(\frac{a}{b}(\alpha'_{-1})^{-1/p}\right)^*(j), \quad j \in \{0,1\}. \tag{5.5}$$

Finally, (5.3) follows from (5.4) and (5.5). □

Now we are in a position to prove necessary conditions for the invertibility of the operator (1.1) at the endpoints (= fixed points of the shift α).

Theorem 5.3. *Suppose the operator* (1.1) *has slowly oscillating data $\{a, b; \alpha\}$ at both endpoints* 0 *and* 1. *If the operator* (1.1) *is invertible, then*

$$\textit{either} \quad \eta_1(j) > 0, \quad j \in \{0,1\} \qquad \textit{or} \quad \eta_0(j) < 0, \quad j \in \{0,1\}. \tag{5.6}$$

Proof. This statement is proved by analogy with [9, Lemma 5.1]. Since $\mathcal{G}\mathfrak{B}(L^p)$ is open in the norm topology, we can without loss of generality suppose that

$a, b \in \mathcal{G}BC$. Assume the contrary. Then one of the following four conditions is fulfilled.

1. $\eta_0(0) < 0 < \eta_1(1)$. In view of Proposition 5.2, these conditions imply that inequality (2.8) is satisfied. By Corollary 2.9, the operator (1.1) is not invertible, and we arrive at a contradiction.
2. $\eta_0(1) < 0 < \eta_1(0)$. Analogously, by Proposition 5.2, these conditions imply that inequality (2.7) is fulfilled. We get a contradiction again, due to Corollary 2.5.
3. $\eta_1(0) \le 0 \le \eta_0(0)$. In this case we immediately obtain a contradiction from Theorem 5.1.
4. $\eta_1(1) \le 0 \le \eta_0(1)$. As in the previous case, we conclude that this situation is impossible by virtue of Theorem 5.1.

Thus, in each case we establish a contradiction, and this proves the theorem. □

5.2. Necessary conditions at inner points

Under assumption (5.6) we put

$$\kappa_- := \begin{cases} 0, & \eta_1(0) > 0, \\ 1, & \eta_0(0) < 0, \end{cases} \qquad \kappa_+ := \begin{cases} 0, & \eta_1(1) > 0, \\ 1, & \eta_0(1) < 0. \end{cases}$$

Let $n \in \mathbb{N}$. Define the six projections $\Pi_n^\pm, \Pi_n^0, \widetilde{\Pi}_n^\pm, \widetilde{\Pi}_n^0$ as the operators of multiplication by the characteristic functions of the sets

$$\gamma_n^\pm := \bigcup_{k=n+1}^{\infty} \alpha_{\pm k}(l), \quad \gamma_n^0 := \bigcup_{k=-n}^{n} \alpha_k(l), \quad \widetilde{\gamma}_n^\pm := \bigcup_{k=n+1\pm\kappa_\pm}^{\infty} \alpha_{\pm k}(l), \quad \widetilde{\gamma}_n^0 := \bigcup_{k=-n+\kappa_-}^{n+\kappa_+} \alpha_k(l).$$

The space L^p decomposes into the direct sums

$$L^p = \Pi_n^- L^p \dotplus \Pi_n^0 L^p \dotplus \Pi_n^+ L^p, \tag{5.7}$$

$$L^p = \widetilde{\Pi}_n^- L^p \dotplus \widetilde{\Pi}_n^0 L^p \dotplus \widetilde{\Pi}_n^+ L^p. \tag{5.8}$$

We consider $A : L^p \to L^p$ as an operator acting from the direct sum (5.8) into the direct sum (5.7). This operator can be represented as the 3×3 operator matrix

$$A = \begin{bmatrix} \Pi_n^- A \widetilde{\Pi}_n^- & \Pi_n^- A \widetilde{\Pi}_n^0 & \Pi_n^- A \widetilde{\Pi}_n^+ \\ \Pi_n^0 A \widetilde{\Pi}_n^- & \Pi_n^0 A \widetilde{\Pi}_n^0 & \Pi_n^0 A \widetilde{\Pi}_n^+ \\ \Pi_n^+ A \widetilde{\Pi}_n^- & \Pi_n^+ A \widetilde{\Pi}_n^0 & \Pi_n^+ A \widetilde{\Pi}_n^+ \end{bmatrix}. \tag{5.9}$$

Lemma 5.4.

(a) *If $\eta_0(1)\eta_1(1) > 0$, then the operator $\Pi_n^+ A \widetilde{\Pi}_n^+ : \widetilde{\Pi}_n^+ L^p \to \Pi_n^+ L^p$ is invertible for every sufficiently large n.*

(b) *If $\eta_0(0)\eta_1(0) > 0$, then the operator $\Pi_n^- A \widetilde{\Pi}_n^- : \widetilde{\Pi}_n^- L^p \to \Pi_n^- L^p$ is invertible for every sufficiently large n.*

The statement is proved by the literal repetition of the proof in [9, Lemma 6.2]. In this proof we essentially use Lemma 2.2 and Proposition 5.2.

As usual, denote by $[L^p(l)]^r$ the Banach space of vectors with r components from $L^p(l)$, equipped with the Euclidean norm. Consider the operators

$$\sigma_n : \Pi_n^0 L^p \to [L^p(l)]^{2n+1}, \quad \widetilde{\sigma}_n : \widetilde{\Pi}_n^0 L^p \to [L^p(l)]^{2n+1}$$

defined by

$$(\sigma_n\varphi)(t) = \{\varphi[\alpha_k(t)]\}_{k=-n}^{n}, \quad (\widetilde{\sigma}_n\varphi)(t) = \{\varphi[\alpha_k(t)]\}_{k=-n+\kappa_-}^{n+\kappa_+}, \quad t \in l. \tag{5.10}$$

It is easily seen that the operators σ_n and $\widetilde{\sigma}_n$ are invertible. Their inverses are given by

$$\begin{aligned} \sigma_n^{-1}\Big(\{\psi_k(t)\}_{k=-n}^{n}\Big) &= \psi_k[\alpha_{-k}(t)], && t \in \alpha_k(l), \quad k \in \{-n, \dots, n\}, \\ \widetilde{\sigma}_n^{-1}\Big(\{\psi_k(t)\}_{k=-n+\kappa_-}^{n+\kappa_+}\Big) &= \psi_k[\alpha_{-k}(t)], && t \in \alpha_k(l), \quad k \in \{-n+\kappa_-, \dots, n+\kappa_+\}, \end{aligned}$$

respectively. Then

$$A_n = \sigma_n \Pi_n^0 A \widetilde{\Pi}_n^0 \widetilde{\sigma}_n^{-1} : [L^p(l)]^{2n+1} \to [L^p(l)]^{2n+1} \tag{5.11}$$

is the operator of multiplication by the square matrix function

$$\mathcal{A}_n(t) = \Big(a[\alpha_k(t)]\delta_{k,j} - b[\alpha_k(t)]\delta_{k,j-1}\Big)_{k=-n,\dots,n;\, j=-n+\kappa_-,\dots,n+\kappa_+}, \quad t \in \bar{l}, \tag{5.12}$$

with entries continuous on $\bar{l}$, where $\delta_{k,j}$ is the Kronecker delta.

Lemma 5.5. *Under assumption* (5.6) *the operator* (1.1) *is invertible if and only if the operator* (5.11) *is invertible for every sufficiently large* n.

Using Lemma 5.4 this statement can be proved by analogy with [16, Lemma 4].

Proposition 5.6.

(a) *If* $\eta_0(0) < 0$ *and* $\eta_0(1) < 0$, *then* $\det \mathcal{A}_n(t) \neq 0$ *for every* $t \in \bar{l}$ *if and only if* $b[\alpha_k(t)] \neq 0$ *for every* $k \in \{-n, \dots, n\}$ *and* $t \in \bar{l}$.

(b) *If* $\eta_1(0) > 0$ *and* $\eta_1(1) > 0$, *then* $\det \mathcal{A}_n(t) \neq 0$ *for every* $t \in \bar{l}$ *if and only if* $a[\alpha_k(t)] \neq 0$ *for every* $k \in \{-n, \dots, n\}$ *and* $t \in \bar{l}$.

This statement is obvious.

5.3. Proof of Theorem 1.2

Now we are able to prove the main result of this paper. We argue by analogy with the proof of [9, Theorem 6.8].

Proof. Sufficiency follows from Lemma 2.2 and Proposition 5.2.

Necessity. Suppose that the operator (1.1) is invertible. Then, by Theorem 5.3, either $\eta_1(0) > 0$ and $\eta_1(1) > 0$, or $\eta_0(0) < 0$ and $\eta_0(1) < 0$. In view of Lemma 5.5, the operator (5.11) and, hence, the operator $\mathcal{A}_n(\cdot)I$ are invertible for every sufficiently large n.

If $\eta_0(j) < 0, j \in \{0, 1\}$, then, by Proposition 5.6(a), $b[\alpha_k(t)] \neq 0$ for every $k \in \{-n, \dots, n\}$, $t \in \bar{l}$ and every sufficiently large n. Hence, $b(t) \neq 0$ for all $t \in \mathbb{I}$. Besides, from the definition of η_0 we see that $b_*(j) > 0, j \in \{0, 1\}$. Thus, $b \in \mathcal{G}BC$.

Analogously, using Proposition 5.6(b), one can prove that $a \in \mathcal{G}BC$ in the case $\eta_1(j) > 0, j \in \{0, 1\}$, which completes the proof of this theorem. □

Acknowledgements

The first author is partially supported by F.C.T. (Portugal) grant PRAXIS XXI/BPD/ 22006/99; the second author is partially supported by CONACYT (México) grant, Cátedra Patrimonial, nivel II, No. 990017-R2000; all the authors are partially supported by F.C.T. (Portugal) under Project No. POCTI/34222/99-FEDER.

References

[1] Antonevich, A.B., *Linear Functional Equations. Operator Approach*, Birkhäuser Verlag, Basel, Berlin 1995.

[2] Antonevich, A., Belousov, M., Lebedev, A., *Functional Differential Equations: II. C^*-applications. Parts 1–2*, Longman, Harlow 1998

[3] Antonevich, A., Lebedev, A., *Functional-Differential Equations: I. C^*-theory*, Longman Scientific & Technical, Harlow 1994.

[4] Aslanov, V., Karlovich, Yu.I., One-sided invertibility of functional operators in reflexive Orlicz spaces, *Dokl. Akad. Nauk AzSSR* **45** (1989), no. 11–12, 3–7 [in Russian].

[5] Böttcher, A., Karlovich, Yu.I., Rabinovich, V.S., The method of limit operators for one-dimensional singular integrals with slowly oscillating data, *J. Operator Theory* **43** (2000), 171–198.

[6] Chicone, C., Latushkin, Yu., *Evolution Semigroups in Dynamical Systems and Differential Equations*, AMS, Providence, RI 1999.

[7] Drakhlin, M.E., Stepanov, E., Weak convergence of inner superposition operators, *Proc. Amer. Math. Soc.* **126** (1998), 173–179.

[8] Gohberg, I., Feldman, I.A., *Convolution Equations and Projection Methods for Their Solution*, AMS, Providence, RI 1974.

[9] Karlovich, A.Yu., Criteria for one-sided invertibility of a functional operator in rearrangement-invariant spaces of fundamental type, *Mathematische Nachrichten* **229** (2001), 91–118.

[10] Karlovich, A.Yu., Karlovich, Yu.I., One-sided invertibility of binomial functional operators with a shift on rearrangement-invariant spaces, *Integr. Equat. Oper. Theory* **42** (2002), 201–228.

[11] Karlovich, Yu.I., On the invertibility of functional operators with non-Carleman shift in Hölder spaces, *Diff. Uravn.* **20** (1984), 2165–2169 [in Russian].

[12] ———, The continuous invertibility of functional operators in Banach spaces, *Dissertationes Mathematicae* **340** (1995), 115–136.

[13] Karlovich, Yu.I., Kravchenko, V.G., A Noether theory for a singular integral operator with a shift having periodic points, *Soviet Math. Dokl.* **17** (1976), 1547–1551.

[14] Karlovich, Yu.I., Kravchenko, V.G., Litvinchuk, G.S., Invertibility of functional operators in Banach spaces, *Functional-Differential Equations, Perm. Politekh. Inst., Perm* (1990), 18–58 [in Russian].

[15] Karlovich, Yu.I., Lebre, A.B., Algebra of singular integral operators with a Carleman backward slowly oscillating shift, *Integr. Equat. Oper. Theory* **41** (2001), 288–323.

[16] Karlovich, Yu.I., Mardiev, R., One-sided Invertibility of Functional Operators and the $n(d)$-normality of Singular Integral Operators with Translation in Hölder Spaces, *Differential Equations* **24** (1988), 350–359.

[17] Kitover, A.K., Spectrum of automorphisms with weight, and the Kamowitz-Scheinberg theorem, *Funct. Anal. Appl.* **13** (1979), 57–58.

[18] Kravchenko, V.G., On a Singular Integral Operator with a Shift, *Soviet Math. Dokl.* **15** (1974), 690–694.

[19] Kravchenko, V.G., Litvinchuk, G.S., *Introduction to the Theory of Singular Integral Operators with Shift*, Series: Mathematics and its applications **289**, Kluwer Academic Publishers, Dordrecht, Boston, London 1994.

[20] Kurbatov, V.G., *Functional Differential Operators and Equations*, Series: Mathematics and its applications **473**, Kluwer Academic Publishers, Dordrecht, Boston, London 1999.

[21] Litvinchuk, G.S., *Boundary Value Problems and Singular Integral Equations with Shift*, Nauka, Moscow 1977 [in Russian].

[22] ______, *Solvability Theory of Boundary Value Problems and Singular Integral Equations with Shift*, Series: Mathematics and its applications **523**, Kluwer Academic Publishers, Dordrecht, Boston, London 2000.

[23] Mardiev, R., A criterion for the semi-Noetherian property of one class of singular integral operators with a non-Carleman shift, *Dokl. Akad. Nauk UzSSR* **2** (1985), 5–7 [in Russian].

[24] Rabinovich, V.S., Roch, S., Silbermann, B., Fredholm theory and finite section method for band-dominated operators, *Integr. Equat. Oper. Theory* **30** (1998), 452–495.

[25] Sarason, D., Toeplitz operators with piecewise quasicontinuous symbols, *Indiana Univ. Math. J.* **26** (1977), 817–838.

A.Yu. Karlovich and A. B. Lebre
Departamento de Matemática
Instituto Superior Técnico
Av. Rovisco Pais
1049 – 001 Lisboa, Portugal
e-mail: akarlov@math.ist.utl.pt
e-mail: alebre@math.ist.utl.pt

Yu. I. Karlovich
Departamento de Matemáticas
CINVESTAV del I.P.N.
Apartado Postal 14–740
07000 México, D.F., México
e-mail: karlovich@servm.uaem.mx

2000 *Mathematics Subject Classification.*
Primary 39B32; Secondary 47A10, 47B38.

Received: 7 June 2001.

Operator Theory:
Advances and Applications, Vol. 142, 175–188

Hadamard-Type Integral Equations and Fractional Calculus Operators

Anatoly A. Kilbas

Abstract. The paper is devoted to the study of the integral equation

$$\frac{1}{\Gamma(\alpha)}\int_a^x \left(\frac{u}{x}\right)^\mu \left(\log\frac{x}{u}\right)^{\alpha-1} f(u)\frac{du}{u} = g(x) \quad (0 < a < x < b)$$

with real μ and $\alpha > 0$ on a finite segment $[a, b]$ of the real line. We prove conditions for the existence of a solution $f(x)$ of this equation in the space $X_\mu(a, b)$ of Lebesgue measurable functions f on (a, b) such that $\int_a^b |u^{\mu-1}f(u)|du < \infty$. Explicit formulas for the solution $f(x)$ are established. We also describe properties of the Hadamard-type fractional integrals defined by the left-hand side of the above equation and of the corresponding fractional derivatives.

1. Introduction

The purpose of this paper is to solve the integral equation

$$\frac{1}{\Gamma(\alpha)}\int_a^x \left(\frac{u}{x}\right)^\mu \left(\log\frac{x}{u}\right)^{\alpha-1} f(u)\frac{du}{u} = g(x) \quad (0 < a < x < b), \tag{1.1}$$

where Γ is the Euler Gamma-function [2, Section 1.1], $\mu \in \mathbf{R} = (-\infty, \infty)$, $\alpha > 0$ and $[a, b]$ is a finite segement of the real line. The left-hand side defines an operator $\mathcal{J}^\alpha_{a+,\mu}$ by

$$(\mathcal{J}^\alpha_{a+,\mu}f)(x) = \frac{1}{\Gamma(\alpha)}\int_a^x \left(\frac{u}{x}\right)^\mu \left(\log\frac{x}{u}\right)^{\alpha-1} f(u)\frac{du}{u} \quad (x > a > 0). \tag{1.2}$$

When $\mu = 0$, the integral takes the form

$$(\mathcal{J}^\alpha_{a+}f)(x) = \frac{1}{\Gamma(\alpha)}\int_a^x \left(\log\frac{x}{u}\right)^{\alpha-1} f(u)\frac{du}{u} \quad (x > a > 0), \tag{1.3}$$

and this integral was introduced by Hadamard [3] in the case $a = 0$ and is therefore often referred to as the Hadamard fractional integral of order α (see [4, Section 18.3 and Section 23.1, notes to Section 18.3]). We call the more general integral in (1.2) a Hadamard-type fractional integral of order α, we refer to the equation in (1.1) as a Hadamard-type integral equation, and its particular case for $\mu = 0$,

$$\frac{1}{\Gamma(\alpha)}\int_a^x \left(\log\frac{x}{u}\right)^{\alpha-1} f(u)\frac{du}{u} = g(x) \quad (0 < a < x < b), \tag{1.4}$$

will be called the Hadamard integral equation.

We study equation (1.1) in the space $X_\mu(a,b)$ ($\mu \in \mathbf{R}$) of those Lebesgue measurable functions f on $[a,b]$ for which $x^{\mu-1}f(x)$ is absolutely integrable:

$$X_\mu(a,b) = \left\{ f : [a,b] \to \mathbf{C} : \quad \int_a^b |u^{\mu-1} f(u)| du < \infty \right\}, \quad X(a,b) \equiv X_0(a,b). \quad (1.5)$$

We prove necessary and sufficient conditions for the solvability of equation (1.1) in $X_\mu(a,b)$ in terms of the function $\mathcal{J}^{n-\alpha}_{a+,\mu} g$ ($n-1<\alpha \leq n$, $n \in \mathbf{N} = \{1,2,,\ldots\}$) in the weighted space $AC^n_\mu[a,b]$ of functions $h(x)$ such that $x^\mu h(x)$ has $\delta = x(d/dx)$ derivatives up to order $n-1$ on $[a,b]$ with $\delta^{n-1}[x^\mu h(x)] \in AC[a,b]$:

$$AC^n_\mu[a,b] = \left\{ h : [a,b] \to \mathbf{C} : \quad \left(x \frac{x}{dx} \right)^{n-1} [x^\mu h(x)] \in AC[a,b] \right\}. \quad (1.6)$$

Here $AC[a,b]$ is the set of absolutely continuous functions on $[a,b]$, which coincides with the space of primitives of Lebesgue measurable functions:

$$h \in AC[a,b] \Leftrightarrow h(x) = c + \int_a^x \psi(t)dt, \ \psi(t) \in L_(a,b); \quad (1.7)$$

see, for example, [4, (1.4)].

We construct the solution of equation (1.1) in the explicit form

$$f(x) = x^{-\mu} \delta^n x^\mu \left(\mathcal{J}^{n-\alpha}_{a+,\mu} g \right)(x), \ \delta = x \frac{d}{dx} \ (n-1<\alpha \leq n, \ n \in \mathbf{N}); \quad (1.8)$$

in particular, we have

$$f(x) = \delta^n \left(\mathcal{J}^{n-\alpha}_{a+,\mu} g \right)(x), \ \delta = x \frac{d}{dx} \ (n-1<\alpha \leq n, \ n \in \mathbf{N}), \quad (1.9)$$

for the Hadamard equation (1.4), where

$$(\mathcal{J}^0_{a+,\mu} f)(x) = f(x). \quad (1.10)$$

As we have mentioned, the solvability conditions obtained for the equation (1.1) with noninteger α ($n-1<\alpha<n$, $n \in \mathbf{N}$) are expressed in terms of the auxiliary function $\mathcal{J}^{n-\alpha}_{a+,\mu} g = \mathcal{J}^{1-\{\alpha\}}_{a+,\mu} g$. In the case $0<\alpha<1$ we establish sufficient conditions for the solvability of (1.1) in terms of the function g itself and give an alternative form for the explicit solution (1.8).

Our results obtained for the Hadamard-type integral equation (1.1) are analogues of those given in [5] (see also [1]) for the Abel integral equation

$$\frac{1}{\Gamma(\alpha)} \int_a^x \frac{f(t)dt}{(x-t)^{1-\alpha}} = g(x) \quad (a<x<b, \ n-1<\alpha \leq n, \ n \in \mathbf{N}) \quad (1.11)$$

in the space $L(a,b)$ of summable functions on a finite interval $(a,b) \subset \mathbf{R}$. For this equation, we have the following result proved in [5].

Proposition 1.1. *Let $n-1<\alpha\leq n$, $n\in\mathbf{N}$. The Abel integral equation (1.11) is solvable in $L(a,b)$ if and only if the following conditions hold:*

1) $$(I_{a+}^{n-\alpha}g)(x)\in AC^n[a,b],$$

2) $$(I_{a+}^{n-\alpha}g)(a)=(I_{a+}^{n-\alpha}g)'(a)=\cdots=(I_{a+}^{n-\alpha}g)^{(n-1)}(a)=0.$$

If these conditions are satisfied, then the equation has a unique solution, which can be given by

$$f(x)=(D_{a+}^{\alpha}g)(x)\equiv D^n(I_{a+}^{n-\alpha}g)(x),\ D=\frac{d}{dx},\tag{1.12}$$

where

$$(I_{a+}^{\alpha}g)(x)=\frac{1}{\Gamma(\alpha)}\int_a^x\frac{g(t)dt}{(x-t)^{1-\alpha}}\ (\alpha>0),\ (I_{a+}^{0}g)(x)=g(x).\tag{1.13}$$

We note that our results show that the weighted spaces $X_\mu(a,b)$ and $AC_\mu^n[a,b]$ play the same role for the Hadamard-type equation (1.1) as the spaces $L(a,b)$ and $AC^n[a,b]$ play for the Abel equation (1.11) and that the derivative of the form $\delta=x(d/dx)$ is the natural substitute of the usual derivative $D=d/dx$ when passing from (1.11) to (1.1).

The right-hand sides $D_{a+}^{\alpha}g$ of (1.12) and $I_{a+}^{\alpha}g$ of (1.13) are known as Riemann-Liouville fractional derivatives and integrals of order $\alpha\geq 0$, respectively [4, (2.32), (2.17) and (2.40)], while the particular case of the right-hand side of (1.9) given by

$$(\mathcal{D}_{a+}^{\alpha}y)(x)=\left(x\frac{d}{dx}\right)^n\left(\mathcal{J}_{a+}^{n-\alpha}g\right)(x)\ (a>0;\ n-1<\alpha\leq n,\ n\in\mathbf{N}),\tag{1.14}$$

is known as the Hadamard fractional derivative of order α [4, (18.54)]. Thus the right-hand side of (1.8) may be called a Hadamard-type fractional derivative of order α:

$$(\mathcal{D}_{a+,\mu}^{\alpha}g)(x)=x^{-\mu}\delta^n x^{\mu}\left(\mathcal{J}_{0+,\mu}^{n-\alpha}g\right)(x),\ \delta=x\frac{d}{dx}\tag{1.15}$$

$$(a>0;\ n-1<\alpha\leq n,\ n\in\mathbf{N},\ \mu\in\mathbf{R}).$$

We describe some properties of the Hadamard-type fractional integrals (1.2) and derivatives (1.15) similar to those for the Riemann-Liouville fractional integrals (1.13) and derivatives (1.12) on a finite interval (a,b) of the real line $\mathbf{R}$.

The paper is organized as follows. Section 2 deals with the boundedness of the Hadamard-type fractional integral (1.1) in the space $X_\mu(a,b)$ and the semigroup property for such an integral and for the Hadamard-type fractional derivative (1.15). Section 3 is devoted to the solvability of equation (1.1) with $\alpha>0$ in the space $X_\mu(a,b)$, and Section 4 is concerned with its particular case $0<\alpha<1$. Conditions for the existence of the Hadamard-type fractional derivative (1.15) and for the validity of its reciprocal property with the Hadamard-type fractional integral (1.2) are given in Section 5. The corresponding assertions for the Hadamard integral equation (1.4) and for the Hadamard fractional integrals (1.3) and fractional derivatives (1.14) are also presented in Sections 3, 4 and 2, 5, respectively.

2. Hadamard-type fractional integrals and derivatives in $X_\mu(a,b)$

First we prove that the Hadamard-type fractional integral $\mathcal{J}^\alpha_{a+,\mu}f$ is defined for $f \in X_\mu(a,b)$. We denote by $\|f\|_{X_\mu}$ the norm of f in $X_\mu(a,b)$:

$$\|f\|_{X_\mu} = \int_a^b u^{\mu-1}|f(u)|du. \tag{2.1}$$

Lemma 2.1. *If $\alpha > 0$, $\mu \in \mathbf{R}$, and $a > 0$, then the operator $\mathcal{J}^\alpha_{a+,\mu}$ is bounded in $X_\mu(a,b)$:*

$$\|\mathcal{J}^\alpha_{a+,\mu}f\|_{X_\mu} \leq K\|f\|_{X_\mu}, \; K = \frac{1}{\Gamma(\alpha+1)}\left[\log\left(\frac{b}{a}\right)\right]^\alpha. \tag{2.2}$$

Proof. Since $f \in X_\mu(a,b)$ and thus $t^{\mu-1}f(t) \in L(a,b)$, we can apply Fubini's theorem and Dirichlet's relation [4, Theorem 1.1 and (1.32)] to obtain

$$\|\mathcal{J}^\alpha_{a+,\mu}f\|_{X_\mu} \leq \frac{1}{\Gamma(\alpha)}\int_a^b u^{\mu-1}|f(u)|du\int_u^b\left(\log\frac{x}{u}\right)^{\alpha-1}\frac{dx}{x}.$$

Making the change of variable $x/u = t$ we get

$$\begin{aligned}\|\mathcal{J}^\alpha_{a+,\mu}f\|_{X_\mu} &\leq \frac{1}{\Gamma(\alpha)}\int_a^b u^{\mu-1}|f(u)|du\int_1^{b/u}(\log t)^{\alpha-1}\frac{dt}{t} \\ &= \frac{1}{\Gamma(\alpha+1)}\int_a^b u^{\mu-1}|f(u)|\left[\log\left(\frac{b}{u}\right)\right]^\alpha du \leq K\int_a^b u^{\mu-1}|f(u)|du,\end{aligned}$$

where K is given by (2.2). This and (2.1) yield (2.2). □

Now we prove the semigroup property for the Hadamard-type fractional integrals (1.2).

Lemma 2.2. *If $\alpha > 0$, $\beta > 0$, $\mu \in \mathbf{R}$, $a > 0$, and $f \in X_\mu(a,b)$, then the semigroup property holds:*

$$\mathcal{J}^\alpha_{a+,\mu}\mathcal{J}^\beta_{a+,\mu}f = \mathcal{J}^{\alpha+\beta}_{a+,\mu}f. \tag{2.3}$$

Proof. Applying Fubini's theorem we find

$$\begin{aligned}(\mathcal{J}^\alpha_{a+,\mu}\mathcal{J}^\beta_{a+,\mu}f)(x) &= \frac{1}{\Gamma(\alpha)}\int_a^x\left(\frac{u}{x}\right)^\mu\left(\log\frac{x}{u}\right)^{\alpha-1}\frac{du}{u} \\ &\times\frac{1}{\Gamma(\beta)}\int_a^u\left(\frac{t}{u}\right)^\mu\left(\log\frac{u}{t}\right)^{\beta-1}f(t)\frac{dt}{t} \\ &= \frac{x^{-\mu}}{\Gamma(\alpha)\Gamma(\beta)}\int_a^x t^{\mu-1}f(t)dt\int_t^x\left(\log\frac{x}{u}\right)^{\alpha-1}\left(\log\frac{u}{t}\right)^{\beta-1}\frac{du}{u}.\end{aligned} \tag{2.4}$$

The inner integral is evaluated by the change of variable $\tau = \log(u/t)/\log(x/t)$ and by using the formulas [2, 1.5(1) and 1.5(5)] for the Beta-function:

$$\begin{aligned}\int_t^x \left(\log\frac{x}{u}\right)^{\alpha-1}\left(\log\frac{u}{t}\right)^{\beta-1}\frac{du}{u} &= \int_0^1 \left(\tau\log\frac{x}{t}\right)^{\beta-1}\left[(1-\tau)\log\frac{x}{t}\right]^{\alpha-1}\log\frac{x}{t}d\tau \\ &= \left(\log\frac{x}{t}\right)^{\alpha+\beta-1} B(\beta,\alpha) \\ &= \left(\log\frac{x}{t}\right)^{\alpha+\beta-1}\frac{\Gamma(\alpha)\Gamma(\beta)}{\Gamma(\alpha+\beta)}.\end{aligned}$$

Substituting this relation into (2.4) we get

$$(\mathcal{J}^{\alpha}_{a+,\mu}\mathcal{J}^{\beta}_{a+,\mu}f)(x) = \frac{1}{\Gamma(\alpha+\beta)}\int_a^x \left(\frac{u}{x}\right)^{\mu}\left(\log\frac{x}{u}\right)^{\alpha+\beta-1}\frac{du}{u},$$

which, together with (1.2), yields (2.3). □

Next we consider the composition of the operators of Hadamard-type fractional differentiation (1.15) and fractional integration (1.2).

Lemma 2.3. *If $\alpha > \beta > 0$ and $\mu \in \mathbf{R}$, then for $f \in X_\mu(a,b)$,*

$$\mathcal{D}^{\beta}_{a+,\mu}\mathcal{J}^{\alpha}_{a+,\mu}f = \mathcal{J}^{\alpha-\beta}_{a+,\mu}f. \tag{2.5}$$

In particular, if $\beta = m \in \mathbf{N}$, then

$$\mathcal{D}^{m}_{a+,\mu}\mathcal{J}^{\alpha}_{a+,\mu}f = \mathcal{J}^{\alpha-m}_{a+,\mu}f. \tag{2.6}$$

Proof. Let $m-1 < \beta \leq m$ ($m \in \mathbf{N}$). If $\beta = m$, then by (1.15)

$$(\mathcal{D}^{m}_{a+,\mu}y)(x) = x^{-\mu}\left(x\frac{d}{dx}\right)^{m} x^{\mu}y(x) \tag{2.7}$$

and hence

$$(\mathcal{D}^{m}_{a+,\mu}\mathcal{J}^{\alpha}_{a+,\mu}f)(x) = x^{-\mu}\left(x\frac{d}{dx}\right)^{m-1} x\frac{d}{dx}\frac{1}{\Gamma(\alpha)}\int_a^x u^{\mu}\left(\log\frac{x}{u}\right)^{\alpha-1} f(u)\frac{du}{u}.$$

Applying the formula for differentiation under the integral sign and using relation [2, 1.2(1)] for the Gamma-function

$$\Gamma(z+1) = z\Gamma(z), \tag{2.8}$$

and (1.2) we obtain

$$\begin{aligned}(\mathcal{D}^{m}_{a+,\mu}\mathcal{J}^{\alpha}_{a+,\mu}f)(x) &= x^{-\mu}\left(x\frac{d}{dx}\right)^{m-1}\frac{x}{\Gamma(\alpha)}\int_a^x u^{\mu}\frac{\partial}{\partial x}\left(\log\frac{x}{u}\right)^{\alpha-1} f(u)\frac{du}{u} \\ &= x^{-\mu}\left(x\frac{d}{dx}\right)^{m-1}\frac{1}{\Gamma(\alpha-1)}\int_a^x u^{\mu}\frac{\partial}{\partial x}\left(\log\frac{x}{u}\right)^{\alpha-2} f(u)\frac{du}{u} \\ &= x^{-\mu}\left(x\frac{d}{dx}\right)^{m-1} x^{\mu}(\mathcal{J}^{\alpha-1}_{a+,\mu}f)(x).\end{aligned}$$

Repeating this procedure k $(1 \leq k \leq m)$ times we get

$$(\mathcal{D}^m_{a+,\mu}\mathcal{J}^\alpha_{a+,\mu}f)(x) = x^{-\mu}\left(x\frac{d}{dx}\right)^{m-k} x^\mu(\mathcal{J}^{\alpha-k}_{a+,\mu}f)(x),$$

and (2.6) follows for $k = m$.

If $m-1 < \beta < m$, then (2.5) follows from (2.3) and (2.6):

$$\mathcal{D}^\beta_{a+,\mu}\mathcal{J}^\alpha_{a+,\mu}f = (\mathcal{D}^m_{a+,\mu}\mathcal{I}^{m-\beta}_{a+,\mu}\mathcal{J}^\alpha_{a+,\mu}f = \mathcal{D}^m_{a+,\mu}\mathcal{J}^{m+\alpha-\beta}_{a+,\mu}f = \mathcal{J}^{\alpha-\beta}_{a+,\mu}f.$$

□

Calculation of the Hadamard-type integral (1.2) for a power function gives the so-called incomplete Gamma-function $\gamma(\nu, x)$ defined for $\nu > 0$ and $x \geq 0$ by [2, 6.9(22)]

$$\gamma(\nu, x) = \int_0^x t^{\nu-1}e^{-t}dt. \tag{2.9}$$

Lemma 2.4. *Let $\alpha > 0$, $a \geq 0$, $\mu \in \mathbf{R}$, and $\omega \in \mathbf{R}$ be such that $\mu + \omega > 0$. Then*

$$(\mathcal{J}^\alpha_{a+,\mu}t^\omega)(x) = \frac{\gamma\left[\alpha, (\omega+\mu)\log(x/a)\right]}{\Gamma(\alpha)}(\mu+\omega)^{-\alpha}x^\omega. \tag{2.10}$$

In particular, for $\mu = 0$ and $\omega > 0$,

$$(\mathcal{J}^\alpha_{a+}t^\omega)(x) = \frac{\gamma\left[\alpha, \omega\log(x/a)\right]}{\Gamma(\alpha)}\omega^{-\alpha}x^\omega. \tag{2.11}$$

Proof. By (1.2), making the change of variable $\log(x/u) = \tau$ and $\tau(\mu+\omega) = y$ and using (2.9) we obtain

$$\begin{aligned}
(\mathcal{J}^\alpha_{a+,\mu}t^\omega)(x) &= \frac{x^{\omega-1}}{\Gamma(\alpha)}\int_a^x \left(\frac{u}{x}\right)^{\mu+\omega-1}\left(\log\frac{x}{u}\right)^{\alpha-1}du \\
&= \frac{x^\omega}{\Gamma(\alpha)}\int_0^{\log(x/a)} e^{-\tau(\mu+\omega)}\tau^{\alpha-1}d\tau \\
&= \frac{x^\omega}{\Gamma(\alpha)}(\mu+\omega)^{-\alpha}\int_0^{(\mu+\omega)\log(x/a)} y^{\alpha-1}e^{-y}dy \\
&= \frac{\gamma\left[\alpha, (\omega+\mu)\log(x/a)\right]}{\Gamma(\alpha)}(\mu+\omega)^{-\alpha}x^\omega,
\end{aligned}$$

and (2.10) is proved. □

Corollary 2.5. *Let $\alpha > 0$, $\mu \in \mathbf{R}$, and $\omega \in \mathbf{R}$ be such that $\mu + \omega > 0$. Then*

$$(\mathcal{J}^\alpha_{0+,\mu}t^\omega)(x) = (\mu+\omega)^{-\alpha}x^\omega, \ (\mathcal{D}^\alpha_{0+,\mu}t^\omega)(x) = (\mu+\omega)^\alpha x^\omega. \tag{2.12}$$

In particular, for $\mu = 0$ and $\omega > 0$,

$$(\mathcal{J}^\alpha_{0+}t^\omega)(x) = \omega^{-\alpha}x^\omega, \ (\mathcal{D}^\alpha_{0+}t^\omega)(x) = \omega^\alpha x^\omega. \tag{2.13}$$

Proof. The first relation in (2.12) follows from (2.10) in conjunction with the relation

$$\gamma(\nu,\infty)=\Gamma(\nu). \tag{2.14}$$

The formula (1.15) with $a=0$ together with the first relation in (2.12) yields the second one. □

3. Hadamard-type integral equations in $X_\mu(a,b)$

A criterion for the solvability of the Hadamard-type integral equation (1.1) is given by the following result.

Theorem 3.1. *Let $n-1<\alpha\leq n$, $n\in\mathbf{N}$, $a>0$, and $\mu\in\mathbf{R}$. The Hadamard-type integral equation (1.1) is solvable in the space $X_\mu(a,b)$ if and only if the following conditions hold:*

1)
$$(\mathcal{J}^{n-\alpha}_{a+,\mu}g)(x)\in AC^n_\mu[a,b], \tag{3.1}$$

2)
$$\lim_{x\to a+}\left(x\frac{d}{dx}\right)^k x^\mu(\mathcal{J}^{n-\alpha}_{a+,\mu}g)(x)=0\quad(k=0,1,\ldots,n-1). \tag{3.2}$$

If these conditions are satisfied, then the equation has a unique solution, which can be given by

$$f(x)=x^{-\mu}\left(x\frac{x}{dx}\right)^n x^\mu\frac{1}{\Gamma(n-\alpha)}\int_a^x\left(\frac{u}{x}\right)^\mu\left(\log\frac{x}{u}\right)^{n-\alpha-1}g(u)\frac{du}{u}, \tag{3.3}$$

where $(\mathcal{J}^{n-\alpha}_{a+,\mu}g)(x)$ is defined by (1.2) and (1.10).

Proof. First we prove the necessity. Let the equation (1.1) be solvable in $X_\mu(a,b)$. Applying the operator $\mathcal{J}^{n-\alpha}_{a+,\mu}$ to both sides of (1.1) and using relation (2.3) with $\beta=\alpha$ and α replaced by $n-\alpha$, we get

$$(\mathcal{J}^n_{a+,\mu}f)(x)=(\mathcal{J}^{n-\alpha}_{a+,\mu}g)(x). \tag{3.4}$$

Applying the operator $\mathcal{D}^k_{a+,\mu}$ $(k=1,2,\ldots,n-1)$ to (3.4) and using relation (2.6) of Lemma 2.3, we find

$$(\mathcal{J}^{n-k}_{a+,\mu}f)(x)=\mathcal{D}^k_{a+,\mu}(\mathcal{J}^{n-\alpha}_{a+,\mu}g)(x)\quad(k=0,1,\ldots,n-1). \tag{3.5}$$

In particular, for $k=n-1$,

$$(\mathcal{J}^1_{a+,\mu}f)(x)=\mathcal{D}^{n-1}_{a+,\mu}(\mathcal{J}^{n-\alpha}_{a+,\mu}g)(x), \tag{3.6}$$

and hence from (1.2) and (2.7) we get

$$\int_a^x u^{\mu-1}f(u)du=\left(x\frac{d}{dx}\right)^{n-1}x^\mu(\mathcal{J}^{n-\alpha}_{a+,\mu}g)(x). \tag{3.7}$$

After differentiation we obtain

$$f(x)=x^{-\mu}\left(x\frac{d}{dx}\right)^n x^\mu(\mathcal{J}^{n-\alpha}_{a+,\mu}g)(x),$$

which in conjunction with (1.2) yields (3.3).

Thus if equation (1.1) is solvable in the space $X_\mu(a,b)$, then its solution is necessarily given by (3.3) and therefore it is unique.

According to (1.2) and (1.15), relation (3.5) can be rewritten in the form

$$\frac{1}{\Gamma(n-k)}\int_a^x \left(\log\frac{x}{u}\right)^{n-k-1} u^{\mu-1}f(u)du = \delta^k x^\mu (\mathcal{J}^{n-\alpha}_{a+,\mu}g)(x) \tag{3.8}$$

$$(k=0,1,\ldots,n-1).$$

From (3.8) we see that $\delta^k x^\mu(\mathcal{J}^{n-\alpha}_{a+,\mu}g)(x) \in C[a,b]$ for $k=0,1,2\ldots,n-2$, while (3.7) yields $\delta^{n-1}x^\mu(\mathcal{J}^{n-\alpha}_{a+,\mu}g)(x) \in AC[a,b]$ in accordance with the condition $x^{\mu-1}f(x) \in L(a,b)$ and (1.7). By (1.6), this means that $\mathcal{J}^{n-\alpha}_{a+,\mu}g \in AC^n_\mu[a,b]$ and hence (3.1) holds. Relation (3.2) follows from (3.8) by passing to the limit $x \to a+$, and thus the necessity is proved.

Now we prove the sufficiency. Since $(\mathcal{J}^{n-\alpha}_{a+,\mu}g)(x) \in AC^n_\mu[a,b]$, we have

$$\delta^{n-1}x^\mu(\mathcal{J}^{n-\alpha}_{a+,\mu}g)(x) \in AC[a,b]$$

and hence

$$\frac{d}{dx}\left[\delta^{n-1}x^\mu(\mathcal{J}^{n-\alpha}_{a+,\mu}g)(x)\right] \in L(a,b),\ x^{1-\mu}\frac{d}{dx}\left[\delta^{n-1}x^\mu(\mathcal{J}^{n-\alpha}_{a+,\mu}g)(x)\right] \in X_\mu(a,b).$$

Therefore the function $f(x)$ given by (3.3) exists almost everywhere on $[a,b]$ and belongs to $X_\mu(a,b)$. Let us show that it is indeed a solution of (1.1). For this purpose we substitute (3.3) into the left-hand side of (1.1) and denote the result by $h(x)$, i.e.,

$$\frac{1}{\Gamma(\alpha)}\int_a^x \left(\frac{u}{x}\right)^\mu \left(\log\frac{x}{u}\right)^{\alpha-1} \left[u^{-\mu}\delta^n u^\mu \left(\mathcal{J}^{n-\alpha}_{a+,\mu}g\right)(u)\right]\frac{du}{u} = h(x). \tag{3.9}$$

We shall show that $h(x)=g(x)$, which proves the theorem. (3.9) is an equation of the type (1.1) with respect to $u^{-\mu}\delta^n u^\mu\left(\mathcal{J}^{n-\alpha}_{a+,\mu}g\right)(u)$. It is certainly solvable since it is a merely a notation. So by (3.3) we have

$$x^{-\mu}\delta^n x^\mu\left(\mathcal{J}^{n-\alpha}_{a+,\mu}g\right)(x) = x^{-\mu}\delta^n x^\mu\left(\mathcal{J}^{n-\alpha}_{a+,\mu}h\right)(x),$$

that is,

$$\frac{d}{dx}\delta^{n-1}x^\mu\left(\mathcal{J}^{n-\alpha}_{a+,\mu}g\right)(x) = \frac{d}{dx}\delta^{n-1}x^\mu\left(\mathcal{J}^{n-\alpha}_{a+,\mu}h\right)(x).$$

The functions $\delta^{n-1}x^\mu\left(\mathcal{J}^{n-\alpha}_{a+,\mu}g\right)(x)$ and $\delta^{n-1}x^\mu\left(\mathcal{J}^{n-\alpha}_{a+,\mu}h\right)(x)$ are absolutely continuous on $[a,b]$, the first by assumption and the second by virtue of (3.7) with $h(x)$ in the right-hand side. Hence

$$\delta^{n-1}x^\mu\left(\mathcal{J}^{n-\alpha}_{a+,\mu}g\right)(x) - \delta^{n-1}x^\mu\left(\mathcal{J}^{n-\alpha}_{a+,\mu}h\right)(x) = c$$

with some constant c. We have $\lim_{x\to a+}\delta^{n-1}x^\mu\left(\mathcal{J}^{n-\alpha}_{a+,\mu}g\right)(x)=0$ due to condition (3.2) and $\lim_{x\to a+}\delta^{n-1}x^\mu\left(\mathcal{J}^{n-\alpha}_{a+,\mu}h\right)(x)=0$ because (3.9) is a solvable equation. Hence $c=0$, and thus

$$\delta^{n-1}x^\mu\left(\mathcal{J}^{n-\alpha}_{a+,\mu}g\right)(x) = \delta^{n-1}x^\mu\left(\mathcal{J}^{n-\alpha}_{a+,\mu}h\right)(x).$$

Similar arguments yield the relations

$$\delta^{n-k} x^{\mu}\left(\mathcal{J}_{a+,\mu}^{n-\alpha} g\right)(x) = \delta^{n-k} x^{\mu}\left(\mathcal{J}_{a+,\mu}^{n-\alpha} h\right)(x)$$

for $k = 1, 2, \ldots, n$. When $k = n$, we have

$$x^{\mu}\left(\mathcal{J}_{a+,\mu}^{n-\alpha} g\right)(x) = x^{\mu}\left(\mathcal{J}_{a+,\mu}^{n-\alpha} h\right)(x)$$

or

$$\frac{1}{\Gamma(n-\alpha)} \int_a^x \left(\frac{u}{x}\right)^{\mu} \left(\log \frac{x}{u}\right)^{n-\alpha-1} [g(u) - h(u)] \frac{du}{u} = 0.$$

The latter is equation of the form (1.1). The uniqueness of its solution leads to the relation $g(x) - h(x) = 0$. This completes the proof of theorem. □

Corollary 3.2. *Let $n - 1 < \alpha \leq n$, $n \in \mathbf{N}$, and $a > 0$. The Hadamard integral equation (1.4) is solvable in the space $X(a,b)$ if and only if the following conditions hold:*

1) $$(\mathcal{J}_{a+}^{n-\alpha} g)(x) \in AC_0^n[a,b], \tag{3.10}$$

2) $$\lim_{x \to a+} \left(x \frac{d}{dx}\right)^k (\mathcal{J}_{a+}^{n-\alpha} g)(x) = 0 \quad (k = 0, 1, \ldots, n-1). \tag{3.11}$$

If these conditions are satisfied, then the equation has a unique solution, which can be given by

$$f(x) = \left(x \frac{d}{dx}\right)^n \frac{1}{\Gamma(n-\alpha)} \int_a^x \left(\log \frac{x}{u}\right)^{n-\alpha-1} g(u) \frac{du}{u}. \tag{3.12}$$

Remark 3.3. By analogy with Abel's equation (1.1), whose solution (1.12) is presented by the Riemann-Liouville fractional derivative (1.15), the solution (3.3) of equation (1.1) is given by the Hadamard-type fractional derivative (1.15). But the conditions for the solvability of these equations are expressed by different spaces of functions. The spaces $L(a,b)$ and $AC^n[a,b]$ are suitable for the former equation, while the weighted spaces $X_{\mu}[a,b]$ and $AC_{\mu}^n[a,b]$ are convenient for the latter one.

4. Hadamard-type integral equations in $X_{\mu}(a,b)$ in the case $0 < \alpha < 1$

The results of Theorem 3.1 and Corollary 3.2 can be simplified in the case of equation (1.1) with $0 < \alpha < 1$.

Theorem 4.1. *The Hadamard-type integral equation (1.1) with $0 < \alpha < 1$ is solvable in the space $X_{\mu}(a,b)$ if and only if the following conditions hold:*

$$x^{\mu}(\mathcal{J}_{a+,\mu}^{1-\alpha} g)(x) \in AC[a,b], \quad \lim_{x \to a+} x^{\mu}(\mathcal{J}_{a+,\mu}^{1-\alpha} g)(x) = 0. \tag{4.1}$$

If these conditions are satisfied, then the equation has a unique solution, which can be given by

$$f(x) = x^{1-\mu} \frac{d}{dx} x^{\mu} \frac{1}{\Gamma(1-\alpha)} \int_a^x \left(\frac{u}{x}\right)^{\mu} \left(\log \frac{x}{u}\right)^{-\alpha} g(u) \frac{du}{u}. \tag{4.2}$$

Corollary 4.2. *The Hadamard integral equation (1.4) with $0 < \alpha < 1$ is solvable in the space $X(a,b)$ if and only if the following conditions hold:*

$$(\mathcal{J}_{a+}^{1-\alpha} g)(x) \in AC[a,b], \quad \lim_{x\to a+} (\mathcal{J}_{a+}^{1-\alpha} g)(x) = 0. \tag{4.3}$$

When these conditions are satisfied, the equation has a unique solution, namely

$$f(x) = x\frac{d}{dx}\frac{1}{\Gamma(1-\alpha)}\int_a^x \left(\log\frac{x}{u}\right)^{-\alpha} g(u)\frac{du}{u}. \tag{4.4}$$

A criterion for the solvability of the Hadamard-type equation (1.1) with $0 < \alpha < 1$ was given in Theorem 4.1 in terms of the function $(\mathcal{J}_{a+,\mu}^{1-\alpha} g)(x)$. Now we give a sufficient solvability condition in terms of the function $g(x)$ itself. For this we need an auxiliary result.

Lemma 4.3. *If $x^\mu g(x) \in AC[a,b]$ and $0 < \alpha < 1$, then $x^\mu (J_{a+,\mu}^{1-\alpha} g)(x) \in AC[a,b]$ and*

$$x^\mu (J_{a+,\mu}^{1-\alpha} g)(x) = \frac{1}{\Gamma(2-\alpha)}\left[g_0(a)\left(\log\frac{x}{a}\right)^{1-\alpha} + \int_a^x \left(\log\frac{x}{t}\right)^{1-\alpha} g_0'(t)dt\right], \tag{4.5}$$

where

$$g_0(x) = x^\mu g(x). \tag{4.6}$$

Proof. By (4.6) and the hypothesis of the theorem, $g_0(x) \in AC[a,b]$ and hence, by (1.7),

$$g_0(t) = g_0(a) + \int_a^u g_0'(t)dt, \; g_0'(t) \in L(a,b).$$

Substituting this equality into $x^\mu (J_{a+,\mu}^{1-\alpha} g)(x)$ we get

$$\begin{aligned} x^\mu (J_{a+,\mu}^{1-\alpha} g)(x) &= \frac{g_0(a)}{\Gamma(1-\alpha)}\int_a^x \left(\log\frac{x}{u}\right)^{-\alpha}\frac{du}{u} \\ &+ \frac{1}{\Gamma(1-\alpha)}\int_a^x \left(\log\frac{x}{u}\right)^{-\alpha}\frac{du}{u}\int_a^u g_0'(t)dt. \end{aligned} \tag{4.7}$$

The first integral can be evaluated directly:

$$\int_a^x \left(\log\frac{x}{u}\right)^{-\alpha}\frac{du}{u} = -\int_a^x \left(\log\frac{x}{u}\right)^{-\alpha} d\left(\log\frac{x}{u}\right) = \frac{1}{(1-\alpha)}\left(\log\frac{x}{a}\right)^{1-\alpha}. \tag{4.8}$$

Changing the order of integration and evaluating the inner integral, we obtain

$$\begin{aligned} \int_a^x \left(\log\frac{x}{u}\right)^{-\alpha}\frac{du}{u}\int_a^u g_0'(t)dt &= \int_a^x g_0'(t)dt\int_t^x \left(\log\frac{x}{u}\right)^{-\alpha}\frac{du}{u} \\ &= \frac{1}{(1-\alpha)}\int_a^x \left(\log\frac{x}{t}\right)^{1-\alpha} g_0'(t)dt. \end{aligned} \tag{4.9}$$

Substituting (4.8) and (4.9) into (4.7) and taking (2.8) into account we arrive at (4.5).

Now we prove that $x^{\mu}(J^{1-\alpha}_{a+,\mu}g)(x) \in AC[a,b]$. The first function in (4.5) is absolutely continuous on $[a,b]$ because

$$\left(\log\frac{x}{a}\right)^{1-\alpha} = (1-\alpha)\int_a^x \left(\log\frac{t}{a}\right)^{-\alpha}\frac{dt}{t}$$

and $[\log(t/a)]^{-\alpha}/t \in L(a,b)$:

$$\int_a^b \left(\log\frac{t}{a}\right)^{-\alpha}\frac{dt}{t} = \frac{1}{(1-\alpha)}\left(\log\frac{b}{a}\right)^{1-\alpha} < \infty.$$

It can be directly verified that

$$\int_a^x \left(\log\frac{x}{u}\right)^{-\alpha}\frac{du}{u}\int_a^u g_0'(t)dt = \int_a^x \left[\int_a^t \frac{1}{t}\left(\log\frac{t}{s}\right)^{-\alpha} g_0'(s)ds\right]dt \tag{4.10}$$

and $(1/t)\int_a^t [\log(t/s)]^{-\alpha} g_0'(s)ds \in L(a,b)$:

$$\begin{aligned}\int_a^b \left|\frac{1}{t}\int_a^t \left(\log\frac{t}{s}\right)^{-\alpha} g_0'(s)ds\right| dt &\leq \int_a^b |g_0'(s)|ds \int_s^b \left(\log\frac{t}{s}\right)^{-\alpha} d\left(\log\frac{t}{s}\right)\\ &= K\int_a^b |g_0'(s)|ds < \infty,\end{aligned}$$

with $K = \frac{1}{(1-\alpha)}\left(\log\frac{b}{a}\right)^{1-\alpha}$, because $g_0'(s) \in L(a,b)$. This completes the proof of Lemma 4.3. □

The following result presents sufficient conditions for the solvability of the Hadamard-type integral equation (1.1) with $0 < \alpha < 1$ in terms of the function g.

Theorem 4.4. *If $x^{\mu}g(x) \in AC[a,b]$, then the Hadamard-type integral equation (1.1) with $0 < \alpha < 1$ is solvable in $X_{\mu}(a,b)$ and its solution (4.2) may be represented in the form*

$$f(x) = \frac{x^{-\mu}}{\Gamma(1-\alpha)}\left[g_0(a)\left(\log\frac{x}{a}\right)^{-\alpha} + \int_a^x \left(\log\frac{x}{t}\right)^{-\alpha} g_0'(t)dt\right], \tag{4.11}$$

where $g_0(x)$ is given by (4.6).

Proof. If $x^{\mu}g(x) \in AC[a,b]$, then, by Lemma 4.3, $x^{\mu}(J^{1-\alpha}_{a+,\mu}g)(x) \in AC[a,b]$. Hence, (4.5) and (4.10) yield $\lim_{x\to a}\left[x^{\mu}(J^{1-\alpha}_{a+,\mu}g)(x)\right] = 0$. Thus, the solvability conditions in (4.1) are satisfied and we can apply Theorem 4.1. According to this theorem the unique solution of equation (1.1) is given by (4.2). Since

$$f(x) = x^{1-\mu}\frac{d}{dx}\left[x^{\mu}(J^{1-\alpha}_{a+,\mu}g)(x)\right]$$

and $x^{\mu}(J^{1-\alpha}_{a+,\mu}g)(x)$ is given by (4.5), differentiation of (4.5) yields (4.11). □

Corollary 4.5. *If $g(x) \in AC[a,b]$, then the Hadamard integral equation (1.4) with $0<\alpha<1$ is solvable in $X(a,b)$ and its solution (4.4) may be represented in the form*

$$f(x)=\frac{1}{\Gamma(1-\alpha)}\left[g(a)\left(\log\frac{x}{a}\right)^{-\alpha}+\int_a^x\left(\log\frac{x}{t}\right)^{-\alpha}g'(t)dt\right]. \tag{4.12}$$

Remark 4.6. Corollary 4.5 is the analogue of the corresponding result for the Abel equation (1.11) [4, Corollary of Lemma 2.1]: if $f(x) \in AC[a,b]$, then (1.11) with $0<\alpha<1$ is solvable in $L(a,b)$ and its solution (1.12) can be represented in the form

$$f(x)=\frac{1}{\Gamma(1-\alpha)}\left[\frac{f(a)}{(x-t)^{\alpha}}+\int_a^x\frac{f'(t)}{(x-t)^{\alpha}}\right]. \tag{4.13}$$

5. The existence of Hadamard-type fractional derivatives and Hadamard-type differentiation and integration as reciprocal operations

The proof of the necessity part of Theorem 3.1 yields conditions for the existence of the Hadamard-type fractional derivative (1.15).

Theorem 5.1. *Let $\alpha>0$, $n-1<\alpha\leq n$ $(n\in\mathbf{N})$, $a>0$, $\mu\in\mathbf{R}$ and let $g(x)$ be such that $(\mathcal{J}^{n-\alpha}_{a+,\mu}g)(x)\in AC^n_{\mu}[a,b]$. Then the function $(\mathcal{D}^{\alpha}_{a+,\mu}g)(x)$ given by (1.15) exists almost everywhere.*

In particular, if $0<\alpha<1$ and $(\mathcal{J}^{1-\alpha}_{a+,\mu}g)(x)\in AC[a,b]$, then

$$(\mathcal{D}^{\alpha}_{a+,\mu}g)(x)=x^{1-\mu}\frac{d}{dx}x^{\mu}\int_a^x\left(\frac{u}{x}\right)^{\mu}\left(\log\frac{x}{u}\right)^{-\alpha}g(u)\frac{du}{u} \tag{5.1}$$

exists almost everywhere.

Corollary 5.2. *Let $\alpha>0$, $n-1<\alpha\leq n$ $(n\in\mathbf{N})$, $a>0$ and let $g(x)$ be such that $(\mathcal{J}^{n-\alpha}_{a+}g)(x)\in AC^n[a,b]$. Then the function $(\mathcal{D}^{\alpha}_{a+}g)(x)$ given by (1.14) exists almost everywhere.*

In particular, if $0<\alpha<1$ and $(\mathcal{J}^{1-\alpha}_{a+}g)(x)\in AC[a,b]$, then

$$(\mathcal{D}^{\alpha}_{a+}g)(x)=x\frac{d}{dx}\int_a^x\left(\log\frac{x}{u}\right)^{-\alpha}g(u)\frac{du}{u} \tag{5.2}$$

exists almost everywhere.

In the case $0<\alpha<1$ Theorem 4.4 provides us with simpler conditions for the existence of $\mathcal{D}^{\alpha}_{a+,\mu}g$ and a representation different from (5.1).

Theorem 5.3. *If $0<\alpha<1$, $a>0$, $\mu\in\mathbf{R}$, and $x^{\mu}g(x)\in AC[a,b]$, then $(\mathcal{D}^{\alpha}_{a+,\mu}g)(x)$ exists almost everywhere and can be represented in the form*

$$(\mathcal{D}^{\alpha}_{a+,\mu}g)(x)=\frac{x^{-\mu}}{\Gamma(1-\alpha)}\left[g_0(a)\left(\log\frac{x}{a}\right)^{-\alpha}+\int_a^x\left(\log\frac{x}{t}\right)^{-\alpha}g_0'(t)dt\right], \tag{5.3}$$

where $g_0(x)=x^{\mu}g(x)$.

Corollary 5.4. *If $0 < \alpha < 1$, $a > 0$, and $g(x) \in AC[a,b]$, then $(\mathcal{D}^{\alpha}_{a+}g)(x)$ exists almost everywhere and can be represented in the form*

$$(\mathcal{D}^{\alpha}_{a+}g)(x) = \frac{1}{\Gamma(1-\alpha)}\left[g(a)\left(\log\frac{x}{a}\right)^{-\alpha} + \int_a^x \left(\log\frac{x}{t}\right)^{-\alpha} g'(t)dt\right]. \tag{5.4}$$

It is known [4, Section 2.6] that the Riemann-Liouville fractional differentiation D^{α}_{a+} and integration I^{α}_{a+}, given (1.12) and (1.13) are reciprocal operations if the former is applied first, i.e., $D^{\alpha}_{a+}I^{\alpha}_{a+}g = g$ for $g(x) \in L(a,b)$, while $I^{\alpha}_{a+}D^{\alpha}_{a+}f \neq f$ in general. Conditions for the validity of the second reciprocal relation, $I^{\alpha}_{a+}D^{\alpha}_{a+}f = f$, were given in [4, Theorem 2.4]. Theorem 3.1 yields a similar result for the Hadamard-type fractional differentiation (1.15) and integration (1.2).

Theorem 5.5. *Let $\alpha > 0$, $n-1 < \alpha \leq n$ $(n \in \mathbf{N})$, and $\mu \in \mathbf{R}$. Then the following statements are true for the Hadamard-type fractional operators $\mathcal{D}^{\alpha}_{a+,\mu}$ and $\mathcal{J}^{\alpha}_{a+,\mu}$ given by (1.15) and (1.2).*

(a) *If $f \in X_{\mu}(a,b)$, then*

$$\mathcal{D}^{\alpha}_{a+,\mu}\mathcal{J}^{\alpha}_{a+,\mu}f = f. \tag{5.5}$$

(b) *If $g(x)$ satisfies the conditions in (3.1) and (3.2), then*

$$\mathcal{J}^{\alpha}_{a+,\mu}\mathcal{D}^{\alpha}_{a+,\mu}g = g. \tag{5.6}$$

(c) *If $0 < \alpha < 1$ and $x^{\mu}g(x) \in AC[a,b]$, then (5.6) is valid.*

Proof. Let $f \in X_{\mu}(a,b)$ and $n-1 < \alpha \leq n$ $(n \in \mathbf{N})$. Using (1.15) and (1.2) and applying relation (2.3) with $\beta = \alpha$ and α replaced by $n - \alpha$, we get

$$\begin{aligned}
\left(\mathcal{D}^{\alpha}_{a+,\mu}\mathcal{J}^{\alpha}_{a+,\mu}f\right)(x) &= x^{-\mu}\left(x\frac{x}{dx}\right)^{n} x^{\mu}\left(\mathcal{J}^{n-\alpha}_{a+,\mu}\mathcal{J}^{\alpha}_{a+,\mu}f\right)(x) \\
&= x^{-\mu}\left(x\frac{x}{dx}\right)^{n} x^{\mu}\left(\mathcal{J}^{n}_{a+,\mu}f\right)(x) \\
&= x^{-\mu}\left(x\frac{x}{dx}\right)^{n-1} x\frac{d}{dx}\frac{1}{(n-1)!}\int_a^x \left(\log\frac{x}{u}\right)^{n-1} u^{\mu}f(u)\frac{du}{u}.
\end{aligned}$$

Taking differentiation under the integral sign we obtain

$$\left(\mathcal{D}^{\alpha}_{a+,\mu}\mathcal{J}^{\alpha}_{a+,\mu}f\right)(x) = x^{-\mu}\left(x\frac{x}{dx}\right)^{n-1}\frac{1}{(n-2)!}\int_a^x \left(\log\frac{x}{u}\right)^{n-2} u^{\mu}f(u)\frac{du}{u}.$$

Repeating this procedure n times similarly to the proof of Lemma 2.3 we arrive at (5.5), and thus (a) is proved.

Let now $g(x)$ satisfy conditions (3.1) and (3.2). Then, in accordance with Theorem 3.1, $f(x) = (\mathcal{D}^{\alpha}_{a+,\mu}g)(x)$ is the unique solution of the equation (1.1):

$$\frac{1}{\gamma(\alpha)}\int_a^x \left(\frac{u}{x}\right)^{\mu}\left(\log\frac{x}{u}\right)^{\alpha-1}(\mathcal{D}^{\alpha}_{a+,\mu}g)(u)\frac{du}{u} = g(x),$$

which yields (5.6) and proves (b). Assertion (c) follows from Theorem 4.4. Thus Theorem 5.5 is proved. □

Corollary 5.6. *Let $\alpha > 0$ and $n-1 < \alpha \leq n$ $(n \in \mathbf{N})$. Then the following statements are true for the Hadamard fractional operators $\mathcal{D}^{\alpha}_{a+}$ and $\mathcal{J}^{\alpha}_{a+}$ given by (1.14) and (1.3).*

(a) *If $f \in X(a,b)$, then*

$$\mathcal{D}^{\alpha}_{a+}\mathcal{J}^{\alpha}_{a+}f = f. \tag{5.7}$$

(b) *If $g(x)$ satisfies the conditions in (3.11) and (3.12), then*

$$\mathcal{J}^{\alpha}_{a+}\mathcal{D}^{\alpha}_{a+}g = g. \tag{5.8}$$

(c) *If $0 < \alpha < 1$ and $g(x) \in AC[a,b]$, then (5.8) is valid.*

Acknowledgements

The author was partly supported by the Belarusian Fundamental Research Fund.

References

[1] Dzherbashyan, M.M., *Integral Transforms and Representation of Functions in Complex Domain* (Russian), Nauka, Moscow 1968.

[2] Erdelyi, A., Magnus, W., Oberhettinger, F, Tricomi, F.G., *Higher Transcendental Functions, Vol. 1,* McGraw-Hill Coop., New York 1953.

[3] Hadamard, J., Essai sur l'étude des fonctions données par leur développement de Taylor, *J. Mat. Pure et Appl., Sér. 4,* **8** (1892), 101–186.

[4] Samko, S.G., Kilbas, A.A., Marichev, O.I., *Fractional Integrals and Derivatives. Theory and Applications*, Gordon and Breach, Yverdon et alibi 1993.

[5] Tamarkin, J.D., On integrable solution of Abel's integral equatuion, *Ann. Math., Ser 2,* **31** (1930), 219–229.

Anatoly A. Kilbas
Department of Mathematics and Mechanics
Belarusian State University
220050 Minsk, Belarus
e-mail: `kilbas@bsu.by`

2000 *Mathematics Subject Classification.*
Primary 45E10; Secondary 26A33.

Received: 28 May 2001.

Operator Theory:
Advances and Applications, Vol. 142, 189–211

Factorization of Singular Integral Operators with a Carleman Shift via Factorization of Matrix Functions

Viktor Kravchenko, Amarino Lebre and Juan Rodríguez

Abstract. This paper is devoted to singular integral operators with a linear fractional Carleman shift of arbitrary order preserving the orientation on the unit circle. The main goal is to obtain a special factorization of the operator with the help of a factorization of a related matrix function in a suitable algebra. This factorization allows us to characterize the kernel and the range of the operator under consideration, similarly to the case of singular integral operators without shift.

1. Introduction

Let $\mathbb{T}$ denote the unit circle with the usual anti-clockwise orientation, and denote by $\mathbb{T}_+$ and $\mathbb{T}_-$ the interior and exterior (which we assume to include the point at infinity) of $\mathbb{T}$, respectively. It is well known that the singular operator with Cauchy kernel, S, defined almost everywhere on $\mathbb{T}$ by

$$S\varphi(t) = \frac{1}{\pi i} \int_{\mathbb{T}} \frac{1}{\tau - t}\, \varphi(\tau)\, d\tau,$$

where the integral is understood in the sense of its principal value, represents a bounded linear operator in $L_p(\mathbb{T})$ for all $p \in (1,\infty)$. A very useful property of this operator is that $S^2 = I$, I denoting the identity operator in $L_p(\mathbb{T})$, $p \in (1,\infty)$. This permits us to introduce a pair of complementary projection operators in $L_p(\mathbb{T})$ by $P_\pm = (I \pm S)/2$ and to decompose $L_p(\mathbb{T}) = L_p^+(\mathbb{T}) \oplus \overset{\circ}{L}{}_p^-(\mathbb{T})$ with $L_p^+(\mathbb{T}) = \operatorname{im} P_+$ and $\overset{\circ}{L}{}_p^-(\mathbb{T}) = \operatorname{im} P_-$. We also set $L_p^-(\mathbb{T}) = \overset{\circ}{L}{}_p^-(\mathbb{T}) \oplus \mathbb{C}$.

We consider a shift function α on $\mathbb{T}$, i.e., α is a homeomorphism of $\mathbb{T}$ onto itself which is differentiable on $\mathbb{T}$ and whose derivative does not vanish there. The shift is said to be of Carleman type if it satisfies the Carleman condition for some $n \in \mathbb{N}$, that is,

$$\alpha_n(t) = t \quad \forall t \in \mathbb{T},$$

where $\alpha_1 = \alpha, \alpha_k = \alpha \circ \alpha_{k-1}, k = 2, \ldots, n$. The least value of $n \in \mathbb{N}$ for which the above relation holds is then called the order of the shift. Two cases can occur: the shift α either preserves or changes the given orientation on $\mathbb{T}$. In the first case we refer to α as an orientation preserving or forward shift and in the other case α is said to be an orientation changing or backward shift. It is known that there are no Carleman backward shifts of order greater than two (see [KL]). Below we will take a linear fractional shift for α.

Together with a linear fractional Carleman shift function we consider a weighted shift operator $U : L_p(\mathbb{T}) \to L_p(\mathbb{T})$, $p \in (1, \infty)$, defined by

$$(Uf)(t) = u(t)\, f(\alpha(t)), \quad t \in \mathbb{T},$$

where the function u, depending on α, is chosen in such a way that the following properties hold for U:

$$\text{(i)} \;\; U^n = I, \qquad\qquad \text{(ii)} \;\; US = \pm\, S\,U, \tag{1.1}$$

where n is the order of the shift α and in (ii) the sign is taken to be $+$ or $-$ in dependence on whether α is a forward or backward shift on $\mathbb{T}$, respectively.

An operator in $L_p(\mathbb{T})$ of the form

$$A = \sum_{j=0}^{n-1} a_j\, U^j, \tag{1.2}$$

where each $a_j, j = 0, \ldots, n-1$, is a function defined on $\mathbb{T}$, is called a functional operator and the functions a_j are referred to as the coefficients of the operator A.

In this paper we consider singular integral operators of the form

$$T_{A,B} = A\,P_+ + B\,P_- \tag{1.3}$$

where A and B are functional operators with $L_\infty(\mathbb{T})$ coefficients. Similarly to what happens with singular integral operators without shift (see the next section for details) and under certain conditions which guarantee that $T_{A,B}$ is a Fredholm operator in $L_p(\mathbb{T})$, $p \in (1, \infty)$, we are mainly interested in establishing that the operator $T_{A,B}$ can be factorized as a product of three operators of the same nature, two of them being invertible and the remaining one being simpler than the original operator, which makes it possible to characterize its kernel and range. Problems of this type were studied in the works [DK, KSh, KLL, KLR].

As is known (see [KL]), for a forward Carleman shift of order n the invertibility of the functional operator (1.2) with continuous coefficients is equivalent to the invertibility of the matrix function

$$\mathcal{A} = \begin{pmatrix} a_0 & a_1 & a_2 & \ldots & a_{n-1} \\ a_{n-1}(\alpha) & a_0(\alpha) & a_1(\alpha) & \ldots & a_{n-2}(\alpha) \\ a_{n-2}(\alpha_2) & a_{n-1}(\alpha_2) & a_0(\alpha_2) & \ldots & a_{n-3}(\alpha_2) \\ \ldots & \ldots & \ldots & \ldots & \ldots \\ a_1(\alpha_{n-1}) & a_2(\alpha_{n-1}) & a_3(\alpha_{n-1}) & \ldots & a_0(\alpha_{n-1}) \end{pmatrix} \tag{1.4}$$

(here and in what follows $x(\alpha)$ means the composition $x \circ \alpha$), and the Fredholm theory in $L_p(\mathbb{T})$, $p \in (1, \infty)$, for the operator (1.3) is analogous to that of the operator

$$S_{\mathcal{A},\mathcal{B}} = \mathcal{A}\,P_+ + \mathcal{B}\,P_- \tag{1.5}$$

with matrix coefficients of the form (1.4). For this reason, in the works [DK]–[KSh] the Fredholm characteristics of (1.3) were studied using a classical factorization of the matrix function $\mathcal{C} = \mathcal{A}^{-1}\mathcal{B}$ as a main tool to obtain the required factorization of that operator. In particular, in the paper [KLR] devoted to the case of a Carleman linear fractional shift of order 2 preserving the orientation on $\mathbb{T}$, the results clearly exhibit a natural connection between the factorization of $T_{A,B}$ and a factorization of the matrix function $\mathcal{C}$ in a subalgebra of 2×2 matrix functions related to the shift. This, in particular, implies that each factor of such a factorization of $\mathcal{C}$ must belong to the subalgebra under consideration. Therefore, the central factor of that factorization of $\mathcal{C}$ cannot be taken to be a diagonal matrix as it is in the classical factorization, $\Lambda = \operatorname{diag}\{t^{\kappa_1}, t^{\kappa_2}\}$.

In the case of a linear fractional shift changing the orientation on $\mathbb{T}$ it is not possible to find a subalgebra of 2×2 matrix functions which permit us to obtain a similar result for the factorization of $T_{A,B}$. This fact can be explained as follows: contrary to the orientation preserving case, when $P_\pm U = U P_\pm$, in the orientation changing case we have

$$P_+U = UP_-,$$

and thus the coefficients of the paired operator $T_{A,B}$ cannot be considered separately.

As is well known, from a classical factorization of the matrix function $\mathcal{C}$ in $L_p(\mathbb{T})$ (below we recall this concept), say $\mathcal{C} = \mathcal{C}_+\Lambda\mathcal{C}_-$, one can obtain a factorization of the operator in (1.5) in the form

$$S_{\mathcal{A},\mathcal{B}} = \mathcal{A}\mathcal{C}_+ \left(P_+ + \Lambda P_-\right)\left(\mathcal{C}_+^{-1}P_+ + \mathcal{C}_-P_-\right). \tag{1.6}$$

Let us associate to the operator $S_{\mathcal{A},\mathcal{B}}$ the block diagonal matrix function

$$\mathcal{M} = \begin{pmatrix} \mathcal{A} & 0 \\ 0 & \mathcal{B} \end{pmatrix}. \tag{1.7}$$

Then corresponding to the factorization of the operator $S_{\mathcal{A},\mathcal{B}}$ in (1.6), we have a factorization of $\mathcal{M}$ in the form

$$\mathcal{M} = \begin{pmatrix} \mathcal{A}\mathcal{C}_+ & 0 \\ 0 & \mathcal{A}\mathcal{C}_+ \end{pmatrix}\begin{pmatrix} \mathcal{I} & 0 \\ 0 & \Lambda \end{pmatrix}\begin{pmatrix} \mathcal{C}_+^{-1} & 0 \\ 0 & \mathcal{C}_- \end{pmatrix} \tag{1.8}$$

and, reciprocally, starting from the above factorization of $\mathcal{M}$ one can easily obtain a factorization of the operator $S_{A,B}$ in the form (1.6).

This observation gives us the basic idea of our approach to investigate singular integral operators with a linear fractional Carleman shift, that is, paired operators of the form (1.3) with functional coefficients of the form (1.2). It happens that the factorization of this type of operators is strongly related to the factorization (1.8) of the block diagonal matrices (1.7) in the subalgebra

$$\mathfrak{V}_1 = \left\{\mathcal{M} \in L_\infty(\mathbb{T})^{2n\times 2n} : \mathcal{M} = \begin{pmatrix} \mathcal{E}^{-1} & 0 \\ 0 & \mathcal{E}^{-1} \end{pmatrix}\mathcal{M}(\alpha)\begin{pmatrix} \mathcal{E} & 0 \\ 0 & \mathcal{E} \end{pmatrix}\right\} \tag{1.9}$$

with $\mathcal{E}^n = \mathcal{I}_n$ in the case of a forward shift of order $n \geq 2$ and in the subalgebra

$$\mathfrak{V}_2 = \left\{ \mathcal{M} \in L_\infty(\mathbb{T})^{4\times 4} : \mathcal{M} = \begin{pmatrix} 0 & \mathcal{E} \\ \mathcal{E} & 0 \end{pmatrix} \mathcal{M}(\alpha) \begin{pmatrix} 0 & \mathcal{E} \\ \mathcal{E} & 0 \end{pmatrix} \right\}$$

with $\mathcal{E}^2 = \mathcal{I}_2$ in the case of a backward shift.

It should also be mentioned that in the course of our investigations a paper by T. Ehrhardt [E], dedicated to the study of Toeplitz + Hankel operators, came to our knowledge. In this paper the author presents an interesting method for the factorization of the operators under consideration, which consists in a successive factorization of matrix functions (in two steps, called anti-symmetric and asymmetric factorizations). We note that this work is devoted to the particular case of the backward shift (flip) on the unit circle $\alpha(t) = t^{-1}$, $t \in \mathbb{T}$, but this is not essential, since the same kind of ideas can be used to deal with more general cases, in the same way as our ideas could serve to deal with this particular case.

Due to space limitations, in this paper we consider only the case of a forward linear fractional Carleman shift leaving the case of a backward linear fractional Carleman shift to a separate publication.

2. Preliminaries

On the unit circle we consider a linear fractional transformation of the general form

$$\alpha(t) = \frac{at+b}{ct+d},$$

which we suppose to be one to one ($ad - bc \neq 0$), non-trivial ($c \neq 0$), and well defined on $\mathbb{T}$ ($d/c \notin \mathbb{T}$). In addition we require that α maps $\mathbb{T}$ onto itself. Then an easy calculation shows that α is either the so-called flip, $\alpha(t) = 1/t$, or that it can be written in the form

$$\alpha(t) = \theta\,\frac{t-\beta}{\overline{\beta}t-1}, \quad \theta \in \mathbb{T}, \ \beta \in \mathbb{C} \setminus \mathbb{T}. \tag{2.1}$$

Note that the flip can be considered as the limit case of the above expression as β tends to infinity along the imaginary axis and $\theta = 1$. We are interested in those linear fractional homeomorphisms of $\mathbb{T}$ onto itself which satisfy a Carleman condition of order $n \in \mathbb{N}$.

We remark that a shift has fixed points on $\mathbb{T}$ only if it is backward. If α is an orientation preserving shift and we consider it as a function defined on $\mathbb{C} \setminus \{1/\overline{\beta}\}$, then it has two fixed points, one in $\mathbb{T}_+$ and the other in $\mathbb{T}_-$. These fixed points, say $t_+ \in \mathbb{T}_+$ and $t_- \in \mathbb{T}_-$, are the roots of the polynomial $p(t) = \overline{\beta}\,t^2 - (1+\theta)\,t + \theta\,\beta$ and are connected by the relation

$$t_+ = \frac{1}{\overline{t_-}}. \tag{2.2}$$

It is easy to characterize linear fractional Carleman shifts of order $n = 2$. Indeed, a shift of the form (2.1) is of this type if and only if $\theta = 1$. The choice of the constant β in (2.1) determines the nature of the shift α: if $|\beta| < 1$ then α is orientation preserving and if $|\beta| > 1$ then α is an orientation changing shift. Of course, the flip operator is an orientation changing shift.

As mentioned before, there are no Carleman backward shifts of order greater than two. Therefore it remains to characterize the linear fractional Carleman forward shifts of order $n > 2$. We claim that α given by (2.1) is such a shift if and only if there exists a primitive n-root of the unit, that is, an ω satisfying

$$\omega^n = 1, \quad \omega^m \neq 1 \ \forall m \in \mathbb{N} : m < n, \tag{2.3}$$

such that in (2.1)

$$\theta = \frac{\omega\, t_- - t_+}{\omega\, t_+ - t_-}, \quad \beta = \frac{(\omega - 1)\, t_-\, t_+}{\omega\, t_- - t_+}, \tag{2.4}$$

where $t_\pm \in \mathbb{T}_\pm$, the fixed points of α, are connected by (2.2).

In fact, if ω is a primitive n-root of the unit, θ and β are given by (2.4), and α is defined by (2.1), then one can establish by induction (or see [F]) that, for every $t \in \mathbb{T}$,

$$\alpha_k(t) = \theta_k\, \frac{t - \beta_k}{\overline{\beta_k}\, t - 1}, \quad k \in \mathbb{N}, \tag{2.5}$$

with

$$\theta_k = \frac{\omega^k\, t_- - t_+}{\omega^k\, t_+ - t_-}, \quad \beta_k = \frac{(\omega^k - 1)\, t_-\, t_+}{\omega^k\, t_- - t_+}, \quad k \in \mathbb{N}. \tag{2.6}$$

Therefore $\theta_n = -1$, $\beta_n = 0$ and thus $\alpha_n(t) = t, t \in \mathbb{T}$, whereas $\alpha_k(t) \not\equiv t$ for $k < n$. Conversely, if α is defined by (2.1) with fixed points $t_\pm$ and n is the least natural number such that $\alpha_n(t) = t, t \in \mathbb{T}$, then, setting $\omega = \frac{\theta\, t_- - t_+}{\theta\, t_+ - t_-}$, it follows that θ is as in (2.4) and, since $1 + \theta = \overline{\beta}\,(t_+ + t_-)$, one concludes that β is also as in (2.4). Using again the formulas established in the first part, one concludes that the requirement that α be a Carleman shift of order n implies that ω is a primitive n-root of the unit.

Suppose that α is a linear fractional forward Carleman shift. Denoting by $\operatorname{ind}_{\mathbb{T}} \alpha$ the winding number of α on $\mathbb{T}$, we have $\operatorname{ind}_{\mathbb{T}} \alpha = 1$. Thus α can be factorized as

$$\alpha = \alpha_+\, t\, \alpha_-, \tag{2.7}$$

where $\alpha_+^{\pm 1} \in R_+(\mathbb{T})$, $\alpha_-^{\pm 1} \in R_-(\mathbb{T})$. Here, with $R(\mathbb{T})$ denoting the algebra of all rational functions without poles on $\mathbb{T}$, $R_\pm(\mathbb{T})$ denote the subalgebras consisting of the functions with poles outside $\mathbb{T}_\pm$. For further reference we fix this factorization. If α is an n-order forward shift then we normalize the above factorization by $\alpha_-(t_-) = 1$, yielding

$$\alpha_+(t) = \frac{\omega\,(t_- - t_+)}{(\omega - 1)\, t + t_- - \omega\, t_+}, \qquad \alpha_-(t) = \frac{(\omega\, t_- - t_+)\, t + (1 - \omega)\, t_-\, t_+}{\omega\,(t_- - t_+)\, t}. \tag{2.8}$$

With this choice one can easily establish that $\alpha_+(t_+) = \omega$ and

$$\prod_{k=0}^{n-1} \alpha_+ \circ \alpha_k = 1, \tag{2.9}$$

with the convention that $\alpha_0(t) = \alpha_n(t) = t$, $t \in \mathbb{T}$.

Together with the linear fractional forward shift function (2.1), we consider the weighted shift operator $U : L_p(\mathbb{T}) \to L_p(\mathbb{T})$, $p \in (1, \infty)$, defined by

$$(Uf)(t) = \omega\, \alpha_+(t)\, f(\alpha(t))$$

where ω is the primitive n-root of the unit as in (2.3). It is easily seen that with this definition the shift operator U satisfies the properties in (1.1).

Let us now recall some results from the general theory of singular integral operators without shift. The main tool for studying singular integral operators with matrix coefficients is the notion of factorization. For $p \in (1, \infty)$, set $q = p/(p-1)$.

Definition 2.1. A matrix function $\mathcal{C} \in L_\infty(\mathbb{T})^{n\times n}$ is said to admit a generalized factorization in (or relative to) $L_p(\mathbb{T})^n$ if

(a) it can be represented in the form

$$\mathcal{C} = \mathcal{C}_+ \,\Lambda\, \mathcal{C}_-$$

where $\mathcal{C}_+ \in L_p^+(\mathbb{T})^{n\times n}, \mathcal{C}_+^{-1} \in L_q^+(\mathbb{T})^{n\times n}, \mathcal{C}_- \in L_q^-(\mathbb{T})^{n\times n}, \mathcal{C}_-^{-1} \in L_p^-(\mathbb{T})^{n\times n}$, and $\Lambda = \operatorname{diag}\{t^{\kappa_1}, \ldots, t^{\kappa_n}\}$, $\kappa_1 \geq \cdots \geq \kappa_n$ being integers,

(b) the operator $B = \mathcal{C}_+ P_- \mathcal{C}_+^{-1}$ is bounded in $L_p(\mathbb{T})^n$.

The numbers κ_j, $j = 1, \ldots, n$, being uniquely determined by the matrix function $\mathcal{C}$, are called the partial indices of $\mathcal{C}$. Sometimes it is convenient to consider the number $\ell \leq n$ of partial indices which are pairwise distinct, and we will write $\Lambda = \operatorname{diag}\{t^{\kappa_1} \mathcal{I}_{d_1}, \ldots, t^{\kappa_\ell} \mathcal{I}_{d_\ell}\}$, where d_i is the multiplicity of the partial index κ_i, $i = 1, \ldots, \ell$.

It should be noted that a continuous matrix function admits a generalized factorization in $L_p(\mathbb{T})^n$, $p \in (1, \infty)$, if and only if it is non-singular on $\mathbb{T}$ and the outer factors of any factorization of it belong to every space $L_r(\mathbb{T})$, $r \in (1, \infty)$, although they are not necessarily continuous (see, for instance, [LS]). For this reason, if $\mathcal{C} \in C(\mathbb{T})^{n\times n}$, we simply speak of a factorization of $\mathcal{C}$ instead of a factorization of $\mathcal{C}$ in $L_p(\mathbb{T})$. In addition, if $\mathcal{C} \in \mathbf{B}^{n\times n}$, where $\mathbf{B}$ is a decomposable algebra of continuous functions on the unit circle, $\mathbf{B} = \mathbf{B}_+ \oplus \overset{\circ}{\mathbf{B}}_-$, with $\mathbf{B}_+ = P_+\mathbf{B}$ and $\overset{\circ}{\mathbf{B}}_- = P_-\mathbf{B}$, then $\mathcal{C}_\pm \in \mathbf{B}_\pm^{n\times n}$ where $\mathbf{B}_- = \overset{\circ}{\mathbf{B}}_- \oplus \mathbb{C}$. Clearly, this applies to the algebra $\mathbf{B} = R(\mathbb{T})$ of rational functions without poles on $\mathbb{T}$.

Let $\mathcal{A}, \mathcal{B} \in L_\infty(\mathbb{T})^{n\times n}$ and consider, in $L_p(\mathbb{T})^n$, the operator

$$S_{\mathcal{A},\mathcal{B}} = \mathcal{A}\, P_+ + \mathcal{B}\, P_-.$$

The following statement is known (see [LS]).

Theorem 2.2. *Let $p \in (1,\infty)$. Then $S_{\mathcal{A},\mathcal{B}}$ is a Fredholm operator in $L_p(\mathbb{T})^n$ if and only if the following conditions hold:*

(i) $\mathcal{A}, \mathcal{B} \in \mathcal{G}L_\infty(\mathbb{T})^{n\times n}$ *(the group of invertible elements in $L_\infty(\mathbb{T})^{n\times n}$),*

(ii) $\mathcal{C} = \mathcal{A}^{-1}\,\mathcal{B}$ *admits a generalized factorization in $L_p(\mathbb{T})^n$.*

In the case where $\mathcal{C} = \mathcal{A}^{-1}\,\mathcal{B}$ admits a generalized factorization in $L_p(\mathbb{T})^n$, say $\mathcal{C} = \mathcal{C}_+\,\Lambda\,\mathcal{C}_-$, with bounded outer factors, this bounded factorization induces a factorization of $S_{\mathcal{A},\mathcal{B}}$ in the form (cf. (1.6))

$$S_{\mathcal{A},\mathcal{B}} = \mathcal{A}\mathcal{C}_+\,S_{I,\Lambda}\,S_{\mathcal{C}_+^{-1},\mathcal{C}_-}. \tag{2.10}$$

Note that the left operator on the right-hand side of the above representation is just a multiplication operator, corresponding to a singular integral operator with equal coefficients, $\mathcal{A}\mathcal{C}_+ = S_{\mathcal{A}\mathcal{C}_+,\mathcal{A}\mathcal{C}_+}$. This operator and the right operator are invertible in $L_p(\mathbb{T})^n$. The Fredholm characteristics of $S_{\mathcal{A},\mathcal{B}}$ coincide with those of $S_{I,\Lambda}$. A very special case occurs if all the partial indices are equal to zero ($\Lambda = I$). In this case the factorization is said to be canonical and the operator $S_{\mathcal{A},\mathcal{B}}$ is invertible. If the factorization is not bounded then similar results hold, but then it is necessary to make use of condition (b) of Definition 2.1 in its full power.

3. Analysis of the case of a forward shift

In this section we assume that α is a linear fractional forward Carleman shift of an arbitrary order $n \in \mathbb{N}\setminus\{1\}$. According to the results of the previous section, this means that α can be written in the form

$$\alpha(t) = \frac{(\omega\,t_- - t_+)\,t + (1-\omega)\,t_-\,t_+}{(\omega-1)\,t + t_- - \omega\,t_+}, \tag{3.1}$$

where ω is a primitive n-root of the unit and the fixed points t_-, t_+ of α are connected by the relation $t_+\,\overline{t_-} = 1$.

We consider singular integral operators with shift $T_{A,B}$ of the form (1.3), with functional coefficients A, B of the form (1.2). Our goal is to obtain a factorization of $T_{A,B}$ as described in the introduction, thus generalizing the results obtained in [KLR] for the case $n = 2$ to a shift of an arbitrary order $n \geq 2$. For this reason, we sometimes refer to that paper for proofs which are similar.

Let us denote by $\mathbf{F}_n$ the algebra of all functional operators acting in $L_p(\mathbb{T})$, $p \in (1,\infty)$, of the form (1.2) with coefficients in $L_\infty(\mathbb{T})$, and let $\mathbf{L}_\alpha^{n\times n}$ denote the subalgebra of $L_\infty(\mathbb{T})^{n\times n}$ consisting of matrix functions of the form (1.4) with $a_j \in L_\infty(\mathbb{T})$, $j = 0,\dots,n-1$. It is easily seen that the subalgebra $\mathbf{L}_\alpha^{n\times n}$ is characterized by the following property.

Proposition 3.1. $A \in \mathbf{L}_\alpha^{n\times n}$ *if and only if* $A \in L_\infty(\mathbb{T})^{n\times n}$ *and*

$$\mathcal{A} = \mathcal{E}^{-1}\,\mathcal{A}(\alpha)\,\mathcal{E}, \tag{3.2}$$

where $\mathcal{E}$ denotes the permutation matrix

$$\mathcal{E} = \begin{pmatrix} 0 & 1 & 0 & \cdots & 0 \\ 0 & 0 & 1 & \cdots & 0 \\ \vdots & \vdots & \vdots & \ddots & \vdots \\ 0 & 0 & 0 & \cdots & 1 \\ 1 & 0 & 0 & \cdots & 0 \end{pmatrix}. \tag{3.3}$$

We define $\pi : \mathbf{F}_n \to \mathbf{L}_\alpha^{n\times n}$ as the map that associates the matrix function (1.4) to the functional operator A in (1.2), that is,

$$A = \sum_{j=0}^{n-1} a_j\, U^j \mapsto \mathcal{A} \text{ given by (1.4).}$$

Let us emphasize that

(i) π is an algebraic isomorphism,

which can be deduced by standard linear algebra considerations. Let $\mathbf{F}_n^\pm$ be the subalgebras of $\mathbf{F}_n$ that are constituted by the functional operators with coefficients in $L_\infty(\mathbb{T})^\pm$ (the subalgebras of $L_\infty(\mathbb{T})$ consisting of the functions which admit analytic extension into $\mathbb{T}_\pm$ and are bounded at ∞ in the case of $\mathbb{T}_-$). Another remarkable property of the map π is that

(ii) $\pi(\mathbf{F}_n^\pm) = \mathbf{L}_\alpha^{n\times n} \cap L_\infty^\pm(\mathbb{T})^{n\times n}$,

which is a consequence of the commutation relation $P_\pm\, U = U\, P_\pm$.

These properties of π will permit us, by analogy with the case $n = 2$, to define a concept of factorization for matrix functions in the algebra $\mathbf{L}_\alpha^{n\times n}$, so-called α-factorization, which leads to a factorization of the corresponding singular integral operator with shift satisfying the properties mentioned in the introduction, in the same way as the classical factorization of a matrix function allows us to factorize the associated singular integral operator (without shift) as described at the end of the previous section. Before introducing this new concept we establish some auxiliary results connected with the properties of the factors in a classical factorization of a matrix function belonging to $\mathbf{L}_\alpha^{n\times n}$.

3.1. Auxiliary results

The following result describes the dependence of the outer factors of a factorization of a matrix function in the class $\mathbf{L}_\alpha^{n\times n}$ on the shift α.

Proposition 3.2. *If $\mathcal{A} \in \mathbf{L}_\alpha^{n\times n}$ admits a factorization in $L_p(\mathbb{T})^n$, $p \in (1,\infty)$, say $\mathcal{A} = \mathcal{A}_+\,\Lambda\,\mathcal{A}_-$ with $\Lambda = \operatorname{diag}\{t^{\kappa_1}\mathcal{I}_{d_1},\dots,t^{\kappa_\ell}\mathcal{I}_{d_\ell}\}$, then the outer factors satisfy the identities*

$$\mathcal{A}_+ = \mathcal{E}^{-1}\mathcal{A}_+(\alpha)\,\Lambda_+\mathcal{H}, \tag{3.4}$$

$$\mathcal{A}_- = \Lambda^{-1}\mathcal{H}^{-1}\Lambda\,\Lambda_-\mathcal{A}_-(\alpha)\,\mathcal{E}, \tag{3.5}$$

where $\mathcal{E}$ denotes the matrix (3.3),

$$\Lambda_\pm = \operatorname{diag}\{\alpha_\pm^{\kappa_1}\mathcal{I}_{d_1},\dots,\alpha_\pm^{\kappa_\ell}\mathcal{I}_{d_\ell}\},$$

and $\mathcal{H}$ is a polynomial matrix function with an upper triangular block structure,

$$\mathcal{H} = \begin{pmatrix} \mathcal{H}_1 & * & * \\ 0 & \ddots & * \\ 0 & 0 & \mathcal{H}_l \end{pmatrix} \tag{3.6}$$

where each block $\mathcal{H}_i$, $i = 1, \dots \ell$, on the main diagonal is a non-singular constant matrix of order d_i, the multiplicity of the partial index κ_i.

Proof. Taking into account that a matrix function $\mathcal{A} \in \mathbf{L}_\alpha^{n\times n}$ satisfies (3.2), we obtain another factorization of $\mathcal{A}$ in $L_p(\mathbb{T})^n$ in the form

$$\mathcal{A} = \left(\mathcal{E}^{-1}\mathcal{A}_+ (\alpha)\,\Lambda_+\right) \Lambda \left(\Lambda_- \mathcal{A}_- (\alpha)\,\mathcal{E}\right).$$

All the other conclusions are consequences of known results on the connection of two different factorizations of the same matrix function (see [LS]). □

The upper triangular block matrix $\mathcal{H}$ in this proposition plays an important role in what follows. We collect some of its properties.

Proposition 3.3. *Under the conditions of the previous proposition,*

(i) $\prod_{j=0}^{n-1} (\Lambda_+\mathcal{H})\,(\alpha_j) = \mathcal{I}_n$,

(ii) *the set $\sigma(\mathcal{H})$ of all eigenvalues of the matrix functions* (3.6) *coincides with the set of all n-roots of the unit,*

(iii) *if $\mathcal{H}_i$ is one of the blocks on the main diagonal of $\mathcal{H}$, then the set $\sigma(\mathcal{H}_i)$ is uniquely determined by the matrix function $\mathcal{A}$ and thus independent of the factors of a factorization of $\mathcal{A}$.*

Proof. (i) From (3.4) we have $\Lambda_+\mathcal{H} = \mathcal{A}_+^{-1}(\alpha)\,\mathcal{E}\,\mathcal{A}_+$. By successive composition with the shift α, the result follows immediately.

(ii) Evaluating (3.5) at the fixed point $t_- \in \mathbb{T}_-$ of the shift α, we get

$$\mathcal{A}_-^{-1}(t_-)\,\Lambda^{-1}(t_-)\,\mathcal{H}(t_-)\,\Lambda(t_-)\,\mathcal{A}_-(t_-) = \mathcal{E}^{-1},$$

which means that $\mathcal{H}(t_-)$ and $\mathcal{E}^{-1}$ are similar matrices. Therefore they possess the same set of eigenvalues. Since $\sigma(\mathcal{H}) = \sigma(\mathcal{H}(t_-))$, the result follows.

(iii) If $\mathcal{A} = \mathcal{A}'_+\,\Lambda\,\mathcal{A}'_-$ and $\mathcal{A} = \mathcal{A}''_+\,\Lambda\,\mathcal{A}''_-$ are two different factorizations of $A \in \mathbf{L}_\alpha^{n\times n}$, then there exists a polynomial matrix function $\widetilde{\mathcal{H}}$ with the structure (3.6) such that

$$\mathcal{A}''_- = \Lambda^{-1}\widetilde{\mathcal{H}}^{-1}\Lambda\,\mathcal{A}'_-.$$

By Proposition 3.2, there exist two polynomial matrix functions $\mathcal{H}', \mathcal{H}''$ with the same block structure such that

$$\begin{aligned} \mathcal{A}'_- &= \Lambda^{-1}\,(\mathcal{H}')^{-1}\,\Lambda\,\Lambda_-\,\mathcal{A}'_-(\alpha)\,\mathcal{E}^{-1}, \\ \mathcal{A}''_- &= \Lambda^{-1}\,(\mathcal{H}'')^{-1}\,\Lambda\,\Lambda_-\mathcal{A}''_-(\alpha)\,\mathcal{E}^{-1}. \end{aligned}$$

Combining these relations at the fixed point $t_- \in \mathbb{T}_-$ of the shift α, we find that

$$\mathcal{H}''(t_-) = \left(\widetilde{\mathcal{H}}(t_-)\right)^{-1} \mathcal{H}'(t_-) \widetilde{\mathcal{H}}(t_-),$$

and therefore $\mathcal{H}'(t_-)$ and $\mathcal{H}''(t_-)$ as well as their blocks $\mathcal{H}'_i$ and $\mathcal{H}''_i$, $i = 1, \dots, \ell$, are similar. Consequently the blocks $\mathcal{H}'_i$ and $\mathcal{H}''_i$, $i = 1, \dots, \ell$, possess the same set of eigenvalues. □

According to this proposition, to each partial index κ_i with multiplicity d_i of a matrix function $\mathcal{A} \in \mathbf{L}_\alpha^{n\times n}$, taken from the set of ℓ distinct partial indices, we may associate a set of d_i distinct n-roots of the unit, which are the eigenvalues of the block matrix $\mathcal{H}_i$. In other words, the correspondence

$$\kappa_i \mapsto \sigma(\mathcal{H}_i)$$

is bijective. This permits us to define an order relation in the set of the n-roots of the unit

$$D_{\mathcal{A}} = \{\varepsilon_1, \dots, \varepsilon_n\}. \tag{3.7}$$

Namely, we consider the ordering

$$\varepsilon_{i_1} \in \mathcal{H}_{j_1}, \ \ \varepsilon_{i_2} \in \mathcal{H}_{j_2}, \ \ j_1 > j_2 \ \ \Rightarrow \ \ i_1 > i_2, \tag{3.8}$$

in $D_{\mathcal{A}}$, where $\mathcal{H}_i$, $i = 1, \dots, \ell$, are the blocks on the main diagonal of (3.6). Obviously, if all the partial indices are different, then the elements of $D_{\mathcal{A}}$ can be ordered in a unique way according to the previous rule, since in that case all the blocks $\mathcal{H}_i$, $i = 1, \dots, n$, have dimension 1. In all the other cases, and without loss of generality, we consider one of the orders that the previous rule determines.

Proposition 3.4. *Suppose that $\mathcal{A} \in \mathbf{L}_\alpha^{n\times n}$ admits a factorization in $L_p(\mathbb{T})^n$, $p \in (1, \infty)$, and let $D_{\mathcal{A}}$ be the set* (3.7)*. Then the correspondence*

$$\theta : D_{\mathcal{A}} \to D_{\mathcal{A}}, \ \ \theta(\varepsilon_i) = \varepsilon_i\, \omega^{\kappa_i}, \ \ i = 1, \dots, n, \tag{3.9}$$

is bijective. Here ω is the primitive n-root of the unit determined by the shift α.

Proof. From (3.4) we have $\Lambda_+\mathcal{H} = \mathcal{A}_+^{-1}(\alpha)\,\mathcal{E}\,\mathcal{A}_+$. Evaluating this expression at the fixed point $t_+ \in \mathbb{T}_+$ of α, one concludes that the matrices $\Lambda_+\mathcal{H}(t_+)$ and $\mathcal{E}$ are similar. Therefore, the eigenvalues of $\Lambda_+\mathcal{H}(t_+)$ are all the n-roots of the unit. Since $\alpha_+(t_+) = \omega$, it follows that

$$\sigma[\Lambda_+\mathcal{H}(t_+)] = D_{\mathcal{A}} = \{\varepsilon_j \in D_{\mathcal{A}} : \varepsilon_j = \varepsilon_i\, \omega^{\kappa_i}, \ \ i = 1, \dots, n\}. \qquad \square$$

Our final result in this part serves as a preparation for the concept of α-factorization and it is connected with the possibility of exhibiting some "simple" elements in the class $\mathbf{L}_\alpha^{n\times n}$. To formulate it we need some more notation.

Let $\theta : D_{\mathcal{A}} \to D_{\mathcal{A}}$ be the map defined in Proposition 3.4 and define a permutation matrix $\mathcal{N} = (n_{ij}) \in \mathbb{C}^{n\times n}$ by

$$n_{ij} = \begin{cases} 1 & \text{if } \theta(\varepsilon_j) = \varepsilon_i, \\ 0 & \text{otherwise.} \end{cases} \tag{3.10}$$

Clearly,

$$\mathcal{N}\,\operatorname{diag}\{\theta(\varepsilon_1),\dots,\theta(\varepsilon_n)\}\,\mathcal{N}^{-1}=\operatorname{diag}\{\varepsilon_1,\dots,\varepsilon_n\}.$$

Further, among the diagonalizing matrices for the permutation matrix $\mathcal{E}$ defined in (3.3), we choose a unitary matrix $\mathcal{S}$ such that

$$\mathcal{S}\,\mathcal{E}\,\mathcal{S}^{-1}=\operatorname{diag}\{\varepsilon_1,\dots,\varepsilon_n\},$$

or, more precisely,

$$\mathcal{S}=\frac{1}{\sqrt{n}}\begin{pmatrix} 1 & \varepsilon_1^{-1} & \dots & \varepsilon_1^{-n+1}\\ 1 & \varepsilon_2^{-1} & \dots & \varepsilon_2^{-n+1}\\ \vdots & \vdots & \ddots & \vdots\\ 1 & \varepsilon_n^{-1} & \dots & \varepsilon_n^{-n+1}\end{pmatrix}. \tag{3.11}$$

Proposition 3.5. *Let $\kappa_1,\dots,\kappa_n$ be integers, let ω be the primitive n-root of the unit associated with the shift α, and let $\chi\in L_\infty(\mathbb{T})$ be such that*

$$\chi(\alpha)=\omega\,\chi\,.$$

Then the matrix function

$$\mathcal{R}=\mathcal{S}^{-1}\,\mathcal{N}\,\operatorname{diag}\{\chi^{\kappa_1},\dots,\chi^{\kappa_n}\}\,\mathcal{S}$$

belongs to $\mathbf{L}_\alpha^{n\times n}$.

Proof. We only have to prove that $\mathcal{R}=\mathcal{E}^{-1}\,\mathcal{R}(\alpha)\,\mathcal{E}$. Using the definitions of $\mathcal{N}$ and $\mathcal{S}$, we obtain successively

$$\begin{aligned}\mathcal{E}^{-1}\,\mathcal{R}(\alpha)\,\mathcal{E} &= \mathcal{E}^{-1}\,\mathcal{S}^{-1}\,\mathcal{N}\,\operatorname{diag}\{\omega^{\kappa_1}\chi^{\kappa_1},\dots,\omega^{\kappa_1}\chi^{\kappa_n}\}\,\mathcal{S}\,\mathcal{E}\\ &= \mathcal{E}^{-1}\,\mathcal{S}^{-1}\,\mathcal{N}\operatorname{diag}\{\theta(\varepsilon_1)\chi^{\kappa_1},\dots,\theta(\varepsilon_n)\chi^{\kappa_n}\}\,\mathcal{S}\\ &= \mathcal{E}^{-1}\,\mathcal{S}^{-1}\,\operatorname{diag}\{\varepsilon_1,\dots,\varepsilon_n\}\,\mathcal{N}\,\operatorname{diag}\{\chi^{\kappa_1},\dots,\chi^{\kappa_n}\}\,\mathcal{S}\\ &= \mathcal{S}^{-1}\,\mathcal{N}\,\operatorname{diag}\{\chi^{\kappa_1},\dots,\chi^{\kappa_n}\}\,\mathcal{S}=\mathcal{R}.\end{aligned}$$

□

3.2. α-factorization in the algebra $\mathbf{L}_\alpha^{n\times n}$

We now introduce a factorization concept (α-factorization) that is closely connected with the factorization in the subalgebra $\mathfrak{V}_1$ (see (1.9)) and thus provides an adequate tool for treating singular integral operators with a linear fractional forward Carleman shift of order $n\geq 2$. By $\mathbf{N}_\alpha^{n\times n}$ we denote the algebra of all (not necessarily bounded) $n\times n$ matrix functions defined on $\mathbb{T}$ which satisfy (3.2).

Definition 3.6. A function $\mathcal{A}\in\mathbf{L}_\alpha^{n\times n}$ is said to admit an *α-factorization in $L_p(\mathbb{T})^n$*, $p\in(1,\infty)$, if it can be represented in the form

$$\mathcal{A}=\mathcal{A}_{+,\alpha}\,\mathcal{R}\,\mathcal{A}_{-,\alpha},$$

where

(i) the factors $\mathcal{A}_{\pm,\alpha}$ are in $\mathbf{N}_\alpha^{n\times n}$ and satisfy all the properties of the outer factors in a factorization in $L_p(\mathbb{T})^n$ (see Definition 2.1),

(ii) $\mathcal{R}$ is of the form

$$\mathcal{R}=\mathcal{S}^{-1}\,\mathcal{N}\,\operatorname{diag}\{\chi^{\kappa_1},\dots,\chi^{\kappa_n}\}\,\mathcal{S}, \tag{3.12}$$

where $\kappa_1, \dots, \kappa_n$ are integers such that $\kappa_1 \geq \dots \geq \kappa_n$, $\mathcal{S}$ and $\mathcal{N}$ are the constant non-singular matrices given by (3.10) and (3.11), respectively, and $\mathcal{X}$ is a rational function satisfying

$$\mathcal{X}(\alpha) = \omega \mathcal{X} \qquad \text{and} \qquad \operatorname{ind}_{\mathbb{T}} \mathcal{X} = 1. \tag{3.13}$$

Here ω is the primitive n-root determined by the shift α and $\operatorname{ind}_{\mathbb{T}} \mathcal{X}$ denotes the winding number of the function $\mathcal{X}$ on $\mathbb{T}$.

Let us note that as a consequence of Proposition 3.5 the matrix function $\mathcal{R}$ belongs to the algebra $\mathbf{L}_\alpha^{n\times n}$. Therefore each factor in an α-factorization belongs to $\mathbf{N}_\alpha^{n\times n}$. This fact is of crucial importance for the development of the theory. Also note that the requirement that $\mathcal{X}$ be rational together with the condition $\operatorname{ind}_{\mathbb{T}} \mathcal{X} = 1$ immediately implies that any α-factorization of $\mathcal{A} \in \mathbf{L}_\alpha^{n\times n}$ gives rise to a factorization of $\mathcal{A}$ in $L_p(\mathbb{T})$. In fact, starting from a factorization $\mathcal{X} = \mathcal{X}_+ t \mathcal{X}_-$ of the rational function $\mathcal{X}$, setting $Z_+ = \mathcal{S}^{-1} \mathcal{N} \operatorname{diag}\{\mathcal{X}_+^{\kappa_1}, \dots, \mathcal{X}_+^{\kappa_n}\}$ and $Z_- = \operatorname{diag}\{\mathcal{X}_-^{\kappa_1}, \dots, \mathcal{X}_-^{\kappa_n}\}\, \mathcal{S}$, we obtain a factorization of $\mathcal{A}$ in $L_p(\mathbb{T})^n$ in the form

$$\mathcal{A} = \mathcal{A}_{+,\alpha}\, Z_+ \operatorname{diag}\{t^{\kappa_1}, \dots, t^{\kappa_n}\}\, Z_-\, \mathcal{A}_{-,\alpha}.$$

The operators B_1 and B_2 acting in $L_p(\mathbb{T})^n$ by the rules $B_1 \varphi = \mathcal{A}_{+,\alpha} P_- \mathcal{A}_{+,\alpha}^{-1} \varphi$ and $B_2 \varphi = \mathcal{A}_{+,\alpha} Z_+ P_- Z_+^{-1} \mathcal{A}_{+,\alpha}^{-1} \varphi$ are simultaneously bounded or not bounded, since they differ by a compact operator in $L_p(\mathbb{T})^n$. Hence, the numbers $\kappa_1, \dots, \kappa_n$ are nothing but the partial indices of $\mathcal{A}$.

As in the usual concept of factorization we may choose the rational function $\mathcal{X}$ as simple as possible. However, the definition given above does not depend on this choice. This conclusion can be obtained as follows. Suppose $\mathcal{A} = \mathcal{A}_{+,\alpha} \mathcal{R} \mathcal{A}_{-,\alpha}$ is an α-factorization in $L_p(\mathbb{T})^n$ of $\mathcal{A} \in \mathbf{L}_\alpha^{n\times n}$ for a given function $\mathcal{X}$ satisfying (3.13). Further, suppose that $\hat{\mathcal{X}}$ is another rational function satisfying (3.13) and let $\hat{\mathcal{R}}$ be the function defined by (3.12) upon the substitution of $\mathcal{X}$ by $\hat{\mathcal{X}}$. Then, making use of the two factorizations $\mathcal{X} = \mathcal{X}_+ t \mathcal{X}_-$, $\hat{\mathcal{X}} = \hat{\mathcal{X}}_+ t \hat{\mathcal{X}}_-$, setting $r_+ = \mathcal{X}_+/\hat{\mathcal{X}}_+$, $r_- = \mathcal{X}_-/\hat{\mathcal{X}}_-$ and $\mathcal{Q}_+ = \mathcal{S}^{-1} \mathcal{N} \operatorname{diag}\{r_+^{\kappa_1}, \dots, r_+^{\kappa_n}\} \mathcal{N}^{-1} \mathcal{S}$, $\mathcal{Q}_- = \mathcal{S}^{-1} \operatorname{diag}\{r_-^{\kappa_1}, \dots, r_-^{\kappa_n}\}\, \mathcal{S}$, we get another α-factorization of $\mathcal{A}$ in $L_p(\mathbb{T})^n$ in the form $\mathcal{A} = \mathcal{A}_{+,\alpha} \mathcal{Q}_+ \hat{\mathcal{R}} \mathcal{Q}_- \mathcal{A}_{-,\alpha}$, since $\mathcal{Q}_+ \in R_+(\mathbb{T})^{n\times n}$ and $\mathcal{Q}_- \in R_-(\mathbb{T})^{n\times n}$.

The following result, which will be used later, can be easily proved.

Lemma 3.7. *If $\mathcal{X}$ is a function satisfying the two conditions of* (3.13) *and $\mathcal{X} = \mathcal{X}_+ t \mathcal{X}_-$ is a factorization of it, then*

$$\mathcal{X}_+(\alpha_m) = \omega^m \mathcal{X}_+ \prod_{j=0}^{m-1} \alpha_+^{-1}(\alpha_j), \quad \mathcal{X}_-(\alpha_m) = \mathcal{X}_- \prod_{j=0}^{m-1} \alpha_-^{-1}(\alpha_j)$$

for all $m = 1, \dots, n-1$.

For the sake of definiteness, let us now present a concrete example of a function $\mathcal{X}$ satisfying the two conditions of Definition 3.6.

Proposition 3.8. *The rational function*

$$\mathcal{X} = \sum_{j=0}^{n-1} \omega^{-j}\, \alpha_j, \tag{3.14}$$

satisfies both conditions of (3.13).

Proof. It is straightforward to verify that $\mathcal{X}(\alpha) = \omega\mathcal{X}$. Now we prove that $\mathrm{ind}_{\mathbb{T}}\, \mathcal{X} = 1$. To this end, we first identify the zeros and the poles of $\mathcal{X}$. Note that $\mathcal{X}$ is the quotient of a polynomial of degree n by a polynomial of degree $n-1$. The poles of $\mathcal{X}$ are the points $1/\beta_k$, $k = 1, \ldots, n-1$, where β_k is defined by (2.6). It is easily seen that these poles belong to $\mathbb{T}_-$. On the other hand, since ω is an n-root of the unit, we immediately conclude that t_- and t_+, the fixed points of α, are zeros of $\mathcal{X}$. We prove that $t_+ \in \mathbb{T}_+$ is a zero of the first order and that $t_- \in \mathbb{T}_-$ is a zero of order $n-1$ of $\mathcal{X}$, which implies that $\mathrm{ind}_{\mathbb{T}}\, \mathcal{X} = 1$.

Evaluating the derivative of α at the fixed points, we get $\alpha'(t_\pm) = \omega^{\pm 1}$. Since $\mathcal{X}'(t_+) = n$, it follows that t_+ is a simple zero of $\mathcal{X}$. On the other hand, by successive differentiation of the relation $\mathcal{X}(\alpha) = \omega\,\mathcal{X}$ we get

$$\alpha'\,\mathcal{X}'(\alpha) = w\,\mathcal{X}', \quad \chi^{(m)}(\alpha)\,(\alpha')^m + \sum_{s=1}^{m-1} \chi^{(s)}(\alpha)\, f_s = \omega\chi^{(m)}, \quad m \geq 2,$$

where f_s, $s = 1, \ldots, m-1$, is a function analytic in a neighbourhood of the point t_-. As ω is a primitive n-root of the unit, we first conclude that $\mathcal{X}'(t_-) = 0$ and then, by recursion, we establish that $\mathcal{X}^{(m)}(t_-) = 0$ for all $m = 1, \ldots, n-1$, which is the desired conclusion. □

Now we state and prove the main result of this part, which reveals the generality of the concept of α-factorization.

Theorem 3.9. *Let $\mathcal{A} \in \mathbf{L}_\alpha^{n\times n}$ and $p \in (1, \infty)$. Then the following assertions are equivalent:*

(i) *$\mathcal{A}$ admits a factorization in $L_p(\mathbb{T})^n$,*

(ii) *$\mathcal{A}$ admits an α-factorization in $L_p(\mathbb{T})^n$.*

Proof. The proof of (ii) ⇒ (i) was already done after Definition 3.6. So we are left to prove that (i) ⇒ (ii).

Suppose that

$$\mathcal{A} = \mathcal{A}_+\, \Lambda\, \mathcal{A}_-, \quad \Lambda = \mathrm{diag}\{t^{\kappa_1}, \ldots, t^{\kappa_n}\}, \tag{3.15}$$

is a factorization of $\mathcal{A} \in \mathbf{L}_\alpha^{n\times n}$ in $L_p(\mathbb{T})^n$. If this is a canonical factorization ($\Lambda = \mathcal{I}$), then it is also a canonical α-factorization. Now suppose that it is a non-canonical factorization ($\Lambda \neq \mathcal{I}$). Then, by Proposition 3.2, we know that the outer factors satisfy relations (3.4) and (3.5), for some polynomial matrix $\mathcal{H}$ with the block structure (3.6). Let $D_\mathcal{A}$ be the set of the n-roots of the unit (which is the set of eigenvalues of $\mathcal{H}$) ordered as in (3.8), and define the matrices $\mathcal{N}$ and $\mathcal{S}$ by (3.10) and (3.11).

Now, let $\mathcal{R}$ be of the form (3.12) for some rational function $\mathcal{X}$ satisfying conditions (3.13). According to Proposition 3.5, we have $\mathcal{R} \in \mathbf{L}_\alpha^{n\times n}$. Using the procedure described in the proof of (ii) $\Rightarrow$ (i), we can get a factorization of $\mathcal{R}$ in $L_p(\mathbb{T})^n$, say

$$\mathcal{R} = \widetilde{\mathcal{R}}_+ \Lambda \widetilde{\mathcal{R}}_-,$$

where $\widetilde{\mathcal{R}}_+ = \mathcal{S}^{-1}\mathcal{N}\Lambda_{\mathcal{X}_+}$ and $\widetilde{\mathcal{R}}_- = \Lambda_{\mathcal{X}_-}\mathcal{S}$ with $\Lambda_{\mathcal{X}_\pm} = \mathrm{diag}\{\mathcal{X}_\pm^{\kappa_1}, \ldots, \mathcal{X}_\pm^{\kappa_n}\}$, and then we straightforwardly verify that

$$\widetilde{\mathcal{R}}_+ = \mathcal{E}^{-1}\widetilde{\mathcal{R}}_+(\alpha)\Lambda_+ \mathcal{D}_\varepsilon \quad , \quad \widetilde{\mathcal{R}}_- = \mathcal{D}_\varepsilon \Lambda_- \widetilde{\mathcal{R}}_-(\alpha)\,\mathcal{E}, \tag{3.16}$$

with $\mathcal{D}_\varepsilon = \mathrm{diag}\{\varepsilon_1, \ldots, \varepsilon_n\}$.

We claim that if we succeed to find another factorization of $\mathcal{R}$, say $\mathcal{R} = \mathcal{R}_+ \Lambda \mathcal{R}_-$, in such a way that

$$\mathcal{R}_+ = \mathcal{E}^{-1}\mathcal{R}_+(\alpha)\,\Lambda_+\mathcal{H}, \tag{3.17}$$

then we get an α-factorization of $\mathcal{A}$ in $L_p(\mathbb{T})^n$: $\mathcal{A} = \mathcal{A}_{+,\alpha}\,\mathcal{R}\,\mathcal{A}_{-,\alpha}$, with $\mathcal{A}_{+,\alpha} = \mathcal{A}_+\mathcal{R}_+^{-1}$, $\mathcal{A}_{-,\alpha} = \mathcal{R}_-^{-1}\mathcal{A}_-$. In fact, in such a case we have

$$\mathcal{A}_{+,\alpha} = \mathcal{E}^{-1}\mathcal{A}_+(\alpha)\Lambda_+\mathcal{H}\mathcal{H}^{-1}\Lambda_+^{-1}\mathcal{R}_+^{-1}(\alpha)\mathcal{E} = \mathcal{E}^{-1}(\mathcal{A}_+\mathcal{R}_+^{-1})(\alpha)\mathcal{E} = \mathcal{E}^{-1}\mathcal{A}_{+,\alpha}(\alpha)\mathcal{E}$$

and so $\mathcal{A}_{+,\alpha} \in \mathbf{N}_\alpha^{n\times n}$. Note that if (3.17) holds, then $\mathcal{R}_- = \Lambda^{-1}\mathcal{H}^{-1}\Lambda\Lambda_-\mathcal{A}_-(\alpha)\,\mathcal{E}$ also holds, by Proposition 3.2. Thus the verification of the fact that $\mathcal{A}_{-,\alpha} \in \mathbf{N}_\alpha^{n\times n}$ is similar to the previous one. The remaining properties of the α-factorization are consequences of the fact that $\mathcal{R}$ is a rational function and the analyticity properties of the factors in its factorization.

Using the results of [LS] on the connection between two factorizations in $L_p(\mathbb{T})^2$ of the same matrix function, to establish the existence of the matrix function $\mathcal{R}_+$ with the required property, we have to show that there exists a polynomial matrix function $\mathcal{Q}$ with the block structure (3.6) such that $\mathcal{R}_+ = \widetilde{\mathcal{R}}_+\,\mathcal{Q}^{-1}$ satisfies equation (3.17). Making use of (3.16), we find that such $\mathcal{Q}$ exists if and only if it satisfies the relation

$$\Lambda_+\,\mathcal{H}\,\mathcal{Q} = \mathcal{Q}(\alpha)\,\Lambda_+\,\mathcal{D}_\varepsilon.$$

If we represent the matrix function $\mathcal{Q}$ in terms of its columns, $\mathcal{Q} = \mathrm{col}(q_1, \ldots, q_n)$, then each column q_i must satisfy the property

$$\Lambda_+\,\mathcal{H}\,q_i = \varepsilon_i\,\alpha_+^{\kappa_i}q_i(\alpha).$$

Now, let W denote the unweighted shift operator on $L_p(\mathbb{T})^n$, $W\varphi = \varphi \circ \alpha$, let W^{-1} be its inverse and, for each $i = 1, \ldots, n$, let V_i be the operator defined by $V_i\varphi = W^{-1}\Lambda_+\,\mathcal{H}\,\alpha_+^{-\kappa_i}\varphi$. Then the last equation takes the form

$$(I - \varepsilon_i^{-1}\,V_i)\,q_i = 0. \tag{3.18}$$

Note that, as a consequence of (2.9) and Proposition 3.3(i), the shift operator V_i satisfies the identity

$$V_i^{\,n} = I$$

and, therefore, we can introduce a projection operator Π_i in $L_p(\mathbb{T})^n$ by

$$\Pi_i = \frac{1}{n}\sum_{j=0}^{n-1} \varepsilon_i^{-j}\, V_i^j .$$

Since $(I - \varepsilon_i^{-1} V_i)\,\Pi_i = 0$, if we choose $q_i \in \operatorname{im}\Pi_i$, we get a solution of (3.18). We shall take q_i in a special form.

Let $\mathcal{H}_s \in \mathbb{C}^{d_s \times d_s}$, $s = 1, \ldots, \ell$, be one of the constant blocks on the main diagonal of $\mathcal{H}$, let ε_k be one of its eigenvalues and u_k a corresponding eigenvector: $\mathcal{H}_s\, u_k = \varepsilon_k\, u_k$. Let $\mathcal{B}_s \in \mathbb{C}^{d_s \times d_s}$ be a diagonalizing matrix for $\mathcal{H}_s$. Then

$$\text{block– diag}\{\mathcal{H}_1, \ldots, \mathcal{H}_\ell\} = \mathcal{B}\,\text{diag}\{\varepsilon_1, \ldots, \varepsilon_n\}\,\mathcal{B}^{-1},$$

where $\mathcal{B} = \text{block– diag}\{\mathcal{B}_1, \ldots, \mathcal{B}_\ell\}$. Set

$$q_i = \Pi_i\,(0, \ldots, 0, u_i, 0, \ldots, 0)^T,$$

where the vector on the right-hand side is represented by blocks. Then, taking into account the block structure of $\mathcal{H}$, we conclude that the vector q_i has the block representation

$$q_i = (*, \ldots, *, u_i, 0, \ldots, 0)^T,$$

where the components of each block $*$ are polynomials in α_+^{-1}, thus being themselves polynomials. With this choice for its columns, we find that $\mathcal{Q}$ has the required structure. □

The results derived above allow us to extract an important consequence for the partial indices of a matrix function belonging to $\mathbf{L}_\alpha^{n\times n}$.

Corollary 3.10. *Suppose $\mathcal{A} \in \mathbf{L}_\alpha^{n\times n}$ admits a factorization in $L_p(\mathbb{T})^n$, $p \in (1,\infty)$. Then the difference between any two of the partial indices of $\mathcal{A}$ is a multiple of n.*

Proof. Let $\kappa_1 \geq \cdots \geq \kappa_n$ be the partial indices of $\mathcal{A}$. From Proposition 3.4 we know that the map $\theta : D_{\mathcal{A}} \to D_{\mathcal{A}}$ is bijective. Thus

$$\theta\,(\varepsilon_i) = \theta\,(\varepsilon_j) \Rightarrow \varepsilon_i = \varepsilon_j$$

and, consequently,

$$\varepsilon_i = \varepsilon_i \omega^{\kappa_i - \kappa_j},$$

and the result follows. □

3.3. Operator factorization and Fredholm characteristics of $T_{A,B}$

Now we use the concept of α-factorization to obtain a factorization of a singular integral operator with a fractional linear forward Carleman shift of order $n \geq 2$. This involves the assumption that the coefficients (which are functional operators) of the singular integral operator with the shift under consideration are invertible. So we may restrict ourselves to operators of the form

$$T_A = P_+ + A\,P_- \quad , \quad A \in \mathbf{F}_n, \tag{3.19}$$

in the space $L_p(\mathbb{T})$, $p \in (1,\infty)$.

In order to avoid undesirable technical difficulties which are not essential in the development of the theory, in this section we make the following assumption concerning all the factorizations of matrix functions under consideration: we suppose that these factorizations are bounded in the sense introduced at the end of Section 2

Theorem 3.11. *Let $A \in \mathbf{F}_n$ be a functional operator such that $\mathcal{A} = \pi(A)$ admits a bounded α-factorization in $L_p(\mathbb{T})^n$. Then the operator T_A can be factorized as*

$$T_A = A_+ T_R \,(A_+^{-1} P_+ + A_- P_-), \tag{3.20}$$

where $A_\pm = \pi^{-1}(\mathcal{A}_{\pm,\alpha})$ and $R = \pi^{-1}(\mathcal{R})$.

Proof. As in the case $n = 2$ considered in [KLR], the above representation of T_A is a consequence of the properties of the map π noted in the beginning of this section. In fact, if $\mathcal{A} = \mathcal{A}_{+,\alpha} \mathcal{R} \mathcal{A}_{-,\alpha}$ is an α-factorization of $\mathcal{A} = \pi(A)$ with bounded factors, then the functional operator A is factorized as $A = A_+ R A_-$ where $A_\pm = \pi^{-1}(\mathcal{A}_{\pm,\alpha}) \in \mathbf{F}_n^\pm$ and $R = \pi^{-1}(\mathcal{R}) \in \mathbf{F}_n$ has rational coefficients. The invariance of the subalgebras $\mathbf{F}_n^\pm$ under the map π implies that $P_\mp A_\pm P_\pm = 0$ and $P_\mp A_\pm^{-1} P_\pm = 0$. Consequently,

$$\begin{aligned} T_A &= P_+ + AP_- = A_+(P_+ + RP_-)(A_+^{-1}P_+ + A_- P_-) \\ &= A_+ T_R \,(A_+^{-1}P_+ + A_- P_-). \end{aligned}$$

□

Suppose that the conditions of the previous theorem are fulfilled. Since the left and right operators on the right-hand side of (3.20) are invertible in $L_p(\mathbb{T})$, the Fredholm characteristics of T_A and of T_R coincide. Therefore, to proceed, we exploit the properties of the latter operator.

First of all note that

$$R = \sum_{j=1}^{n} r_{1j}\, U^{j-1},$$

where r_{1j}, $j = 1, \ldots, n$, are the entries in the first line of the matrix $\mathcal{R}$. Also according to Definition 3.6, we have

$$r_{1j} = \frac{1}{n} \sum_{i=1}^{n} \varepsilon_i^{1-j} \chi^{\kappa_i},$$

where $\varepsilon_1, \ldots, \varepsilon_n$ are the roots of the unit, $\kappa_1 \geq \cdots \geq \kappa_n$ are the partial indices of $\mathcal{A}$, and χ is a rational function satisfying (3.13).

Associated with the set of roots of the unit, we introduce a set of projection operators in $L_p(\mathbb{T})$ by

$$Q_{\varepsilon_i} = \frac{1}{n} \sum_{k=0}^{n-1} \varepsilon_i^{-k} U^k, \quad i = 1, \ldots, n.$$

so that $\sum_{i=1}^{n} Q_{\varepsilon_i} = I$ and $Q_{\varepsilon_i} Q_{\varepsilon_j} = 0$ for $i \neq j$, which can be verified directly. Then

$$R = \sum_{i=1}^{n} \mathcal{X}^{\kappa_i} Q_{\varepsilon_i}.$$

Now, for each $i = 1, \ldots, n$, let $\hat{\varepsilon}_i = \theta(\varepsilon_i)$, where θ is the bijective map (3.9). Since $\mathcal{X}$ satisfies $\mathcal{X}(\alpha) = \omega \mathcal{X}$, one easily checks that $\mathcal{X}^{\kappa_i} Q_{\varepsilon_i} = Q_{\hat{\varepsilon}_i} \mathcal{X}^{\kappa_i}$, $i = 1, \ldots, n$. Therefore R can alternatively be represented as

$$R = \sum_{i=1}^{n} Q_{\hat{\varepsilon}_i} \mathcal{X}^{\kappa_i} \tag{3.21}$$

and, by the properties of the projections Q_{ε_i}, we also have

$$R^{-1} = \sum_{i=1}^{n} \mathcal{X}^{-\kappa_i} Q_{\hat{\varepsilon}_i} = \sum_{i=1}^{n} Q_{\varepsilon_i} \mathcal{X}^{-\kappa_i}.$$

These representations allow us to obtain a generalized inverse of T_A and to derive sufficient conditions for its one-sided invertibility.

Theorem 3.12. *Suppose the conditions of Theorem* 3.11 *are fulfilled and let* $\kappa_1 \geq \cdots \geq \kappa_n$ *be the partial indices of* $\mathcal{A} = \pi(A)$. *Let* $\mathcal{R}$ *be the matrix function given by* (3.12)–(3.14) *and put* $R = \pi^{-1}(\mathcal{R})$. *Then* T_A *is generalized invertible and a generalized inverse of it is given by*

$$T_A^{(-1)} = \left(A_+ P_+ + A_-^{-1} P_-\right) R T_{R^{-1}} R^{-1} A_+^{-1}.$$

Moreover, we have

(i) *T_A is a left invertible operator if $\kappa_1 \leq 0$,*

(ii) *T_A is a right invertible operator if $\kappa_n \geq 0$,*

and, therefore, T_A is an invertible operator if $\kappa_1 = \cdots \kappa_n = 0$. In each of these cases, the one-sided or two-sided inverse of T_A is given by the expression above.

Proof. We follow the lines of the proof of the corresponding result for the case $n = 2$. Taking into account (3.20) and the invertibility of the outer operators there, we have to prove that a generalized inverse of T_R is given by

$$T_R^{(-1)} = R\, T_{R^{-1}}\, R^{-1} = (R P_+ + P_-) R^{-1}, \tag{3.22}$$

which is a right inverse of T_R if $\kappa_n \geq 0$ and a left inverse of T_R if $\kappa_1 \leq 0$. Note that with the choice for the function $\mathcal{X}$ made in the definition of R, we have $P_- \mathcal{X} P_+ = 0$. Consequently,

$$P_- R P_+ = 0 \text{ if } \kappa_n \geq 0 \quad \text{and} \quad P_- R^{-1} P_+ = 0 \text{ if } \kappa_1 \leq 0.$$

A straightforward computation shows that the operator $T_R^{(-1)}$ is a left inverse of T_R if $\kappa_1 \leq 0$ and a right inverse of T_R if $\kappa_n \geq 0$. Therefore, it remains to prove that $T_R^{(-1)}$, defined in (3.22), is a generalized inverse of T_R if not all the partial indices have the same sign.

For each $\kappa \in \mathbb{Z}$, consider the scalar singular integral operators without shift $T_\kappa = P_+ + \chi^\kappa P_-$ and $\Lambda_\kappa = \chi^\kappa P_+ + P_-$ in the space $L_p(\mathbb{T})$. A well-known fact is that Λ_ε is left or right invertible according to whether κ is non-positive or non-negative, respectively, and that, moreover, a one-sided inverse is the operator $\Lambda_\kappa^{(-1)} = \chi^{-\kappa} P_+ + P_-$. Thus $\Lambda_\kappa \Lambda_\kappa^{(-1)} \Lambda_\kappa = \Lambda_\kappa$ and $T_\kappa \Lambda_\kappa^{(-1)} \Lambda_\kappa = T_\kappa$ for $\kappa \in \mathbb{Z}$.

From the representations of R and R^{-1} given before this theorem, we get

$$T_R = \sum_{j=1}^{n} T_{\kappa_i} Q_{\varepsilon_i} = \sum_{j=1}^{n} Q_{\hat{\varepsilon}_i} T_{\kappa_i} \;\; ; \;\; T_{R^{-1}} = \sum_{j=1}^{n} T_{-\kappa_i} Q_{\hat{\varepsilon}_i} = \sum_{j=1}^{n} Q_{\varepsilon_i} T_{-\kappa_i}. \tag{3.23}$$

It is easily seen that (the sums taken from 1 to n)

$$(R P_+ + P_-) R^{-1} T_R = \sum_i \Lambda_{\kappa_i} \Lambda_{\kappa_i}^{(-1)} Q_{\hat{\varepsilon}_i}$$

and that $Z \stackrel{\text{def}}{=} T_R (R P_+ + P_-) R^{-1} T_R = \sum_{j,i} Q_{\hat{\varepsilon}_j} T_{\kappa_j} \Lambda_{\kappa_i} \Lambda_{\kappa_i}^{(-1)} Q_{\hat{\varepsilon}_i}$.

A further computation shows that

$$Z = T_R + \sum_{j \neq i} Q_{\hat{\varepsilon}_j} T_{\kappa_j} (\Lambda_{\kappa_i} \Lambda_{\kappa_i}^{(-1)} - I) Q_{\hat{\varepsilon}_i}.$$

To finish the proof it remains to show that the sum on the right-hand side of this expression vanishes. We show that each term in that sum vanishes. This is clear if $\kappa_i \leq 0$, since in such a case Λ_{κ_i} is right invertible with right inverse $\Lambda_{\kappa_i}^{(-1)}$. On the hand, if $\kappa_i > 0$, from

$$T_{\kappa_j} (\Lambda_{\kappa_i} \Lambda_{\kappa_i}^{(-1)} - I) = (\chi^{\kappa_j} - P_+ \chi^{\kappa_i}) P_- \chi^{-\kappa_i} P_+$$

and the commutation relation $\chi^{\kappa_i} Q_{\varepsilon_i} = Q_{\hat{\varepsilon}_i} \chi^{\kappa_i}$, $i = 1, \ldots, n$, it follows that, for any $j, i = 1, \ldots, n$ with $j \neq i$, we have

$$Q_{\hat{\varepsilon}_j} T_{\kappa_j} (\Lambda_{\kappa_i} \Lambda_{\kappa_i}^{(-1)} - I) Q_{\hat{\varepsilon}_i} = 0.$$

Thus $Z = T_R$ and hence $R T_{R^{-1}} R^{-1}$ is a generalized inverse of T_R, as claimed. □

It should be remarked that the sufficient conditions for the invertibility of the operator T_A given in the previous theorem are not necessary. This fact was already noticed in [KLR] for the case $n = 2$ and exhibits a new feature of singular integral operators with shift which is in contrast to the well-known results for singular integral operators without shift, for which the invertibility is equivalent to the condition that all partial indices of the corresponding matrix function be equal to zero.

Our next aim is to characterize the kernel and the range of the operator T_A through these characteristics of T_R. It is clear from (3.23) that this will be accomplished if we succeed to describe the kernel and the range of an operator of the form $Q_{\hat{\varepsilon}} T_\kappa$ with $\varepsilon \in D_A$ and $\kappa \in \mathbb{Z}$, considered as an operator into $\operatorname{im} Q_{\hat{\varepsilon}}$. This will be done in the next proposition and to formulate the result we need to introduce some notation.

Let $\varepsilon \in D_{\mathcal{A}}$ and $\kappa \in \mathbb{Z}$. Introduce a new projection operator in $L_p(\mathbb{T})$ by

$$Q_{\varepsilon}^{(\kappa)} = \frac{1}{n} \sum_{j=0}^{n-1} \varepsilon^{-j} V_{\kappa}^{j}, \quad V_{\kappa} = \alpha_{+}^{-\kappa} U. \tag{3.24}$$

Further, let $Z_{\varepsilon} = \left\{m \in \mathbb{Z} : \left[\frac{s+1-m}{n}\right] = \frac{s+1-m}{n}\ ,\ \varepsilon = \omega^{s}\right\}$, where ω is the primitive root of the unit associated with the shift α, define $\Gamma_{\varepsilon,0} = \emptyset$ and

$$\Gamma_{\varepsilon,\kappa} = \begin{cases} Z_{\varepsilon} \cap \{0, \ldots, \kappa - 1\} & \text{if } \kappa > 0 \\ Z_{\varepsilon} \cap \{\kappa, \ldots, -1\} & \text{if } \kappa < 0 \end{cases} \tag{3.25}$$

and set

$$\gamma(\varepsilon, \kappa) = \# \Gamma_{\varepsilon,\kappa}. \tag{3.26}$$

Proposition 3.13. *Let $\varepsilon \in D_{\mathcal{A}}$, $\hat{\varepsilon} = \theta(\varepsilon)$, and $\kappa \in \mathbb{Z}$. For $p \in (1, \infty)$, consider the operator $Q_{\hat{\varepsilon}} T_{\kappa} : L_p(\mathbb{T}) \to \operatorname{im} Q_{\hat{\varepsilon}}$. Then*

(i) $\ker[Q_{\hat{\varepsilon}} T_{\kappa}] = \{0\}$ *if* $\kappa \leq 0$ *;* $\operatorname{im}[Q_{\hat{\varepsilon}} T_{\kappa}] = \operatorname{im} Q_{\hat{\varepsilon}}$ *if* $\kappa \geq 0$,

(ii) $\dim \ker[Q_{\hat{\varepsilon}} T_{\kappa}] = \gamma(\varepsilon, \kappa)$ *if* $\kappa > 0$ *;* $\operatorname{codim} \operatorname{im}[Q_{\hat{\varepsilon}} T_{\kappa}] = \gamma(\varepsilon, \kappa)$ *if* $\kappa < 0$.

(iii) *Moreover, if $\kappa > 0$, then*

$$\ker[Q_{\hat{\varepsilon}} T_{\kappa}] = \operatorname{span}\left\{(\chi_{+}^{\kappa} - t^{-\kappa} \chi_{-}^{-\kappa})\, p_m : m \in \Gamma_{\varepsilon,\kappa}\right\}$$

and, if $\kappa < 0$, then

$$\operatorname{im}[Q_{\hat{\varepsilon}} T_{\kappa}] = \left\{f \in \operatorname{im} Q_{\hat{\varepsilon}} : \int_{\mathbb{T}} f(t)\, \chi_{+}^{-\kappa}(t)\, p_{m-\kappa}(t) = 0,\ \ m \in \Gamma_{\varepsilon,\kappa}\right\}.$$

Here $t_{\pm} \in \mathbb{T}_{\pm}$ are the fixed points of α, $\chi = \chi_{+} t \chi_{-}$ is a factorization of χ, and

$$p_m(t) = (t - t_{+})^{\kappa - 1 - m} (t_{-} - t)^{m}, \quad m = 0, 1, \ldots, |\kappa| - 1. \tag{3.27}$$

Proof. For any $\varphi \in L_p(\mathbb{T})$ write $\varphi = \varphi^{+} + \varphi^{-}$ with $\varphi^{+} \in L_p^{+}(\mathbb{T})$ and $\varphi^{-} \in \overset{\circ}{L}{}_p^{-}(\mathbb{T})$.

First we prove all the assertions concerning the kernel of $Q_{\hat{\varepsilon}} T_{\kappa}$. In view of the relation $\chi^{\kappa} Q_{\varepsilon} = Q_{\hat{\varepsilon}} \chi^{\kappa}$, one establishes that $\varphi \in \ker[Q_{\hat{\varepsilon}} T_{\kappa}]$ if and only if φ is a solution to the Riemann-Hilbert problem

$$Q_{\hat{\varepsilon}} \varphi^{+} = -\chi^{\kappa} Q_{\varepsilon} \varphi^{-}.$$

The solutions to this problem are very well known (see [GK]). If $\kappa \leq 0$, then the problem has only the trivial solution, and so $Q_{\hat{\varepsilon}} T_{\kappa}$ is injective. If $\kappa > 0$, which we suppose from now on, then its solutions can be determined in terms of a factorization of χ, say $\chi = \chi_{+} t \chi_{-}$, in the form

$$Q_{\hat{\varepsilon}} \varphi^{+} = \chi_{+}^{\kappa} p, \quad Q_{\varepsilon} \varphi^{-} = -t^{-\kappa} \chi_{-}^{-\kappa} p,$$

where $p \in P^{\kappa-1}$ (the space of all polynomials with degree at most $\kappa - 1$) is such that $\chi_{+}^{\kappa} p \in \operatorname{im} Q_{\hat{\varepsilon}}$. Using the properties of the factors in a factorization of χ given in Lemma 3.7, we have

$$Q_{\hat{\varepsilon}} \chi_{+}^{\kappa} = \chi_{+}^{\kappa} Q_{\varepsilon}^{(\kappa)}, \quad Q_{\varepsilon} (t^{-\kappa} \chi_{-}^{-\kappa}) = t^{-\kappa} \chi_{-}^{-\kappa} Q_{\varepsilon}^{(\kappa)},$$

where $Q_\varepsilon^{(\kappa)}$ is defined in (3.24). Therefore, $\varphi \in \ker Q_{\hat\varepsilon} T_\kappa$ if and only if

$$\varphi = (\chi_+^{\kappa} - t^{-\kappa}\chi_-^{-\kappa})\, p, \;\; p \in \operatorname{im} Q_\varepsilon^{(\kappa)} \cap P^{\kappa-1}. \tag{3.28}$$

Now, note that

$$p \in \operatorname{im} Q_\varepsilon^{(\kappa)} \text{ if and only if } V_\kappa\, p = \varepsilon\, p,$$

where V_κ is also given in (3.24). In fact, if $V_\kappa\, p = \varepsilon\, p$, then from the definition of $Q_\varepsilon^{(\kappa)}$ it follows that $p \in \operatorname{im} Q_\varepsilon^{(\kappa)}$, and the converse is a consequence of the equality $(I - \varepsilon^{-1} V_\kappa)\, Q_\varepsilon^{(\kappa)} = 0$.

Consider the following basis of $P^{\kappa-1}$:

$$p_m(t) = (t - t_+)^{\kappa-1-m}\,(t_- - t)^m, \;\; m = 0, 1, \ldots, \kappa - 1.$$

Then making use of the equalities

$$\alpha - t_+ = \alpha_+\,(t - t_+), \;\; t_- - \alpha = \omega^{-1}\,\alpha_+(t_- - t),$$

which can be easily verified from the factorization of α considered (see (2.8)), we arrive at

$$V_\kappa\, p_m = \alpha_+^{-\kappa} U\, p_m = \omega^{1-m}\, p_m, \;\; m = 0, 1, \ldots, \kappa - 1,$$

from which we conclude that the polynomials p_m such that $\omega^{1-m} = \varepsilon$ constitute a basis for the subspace $\operatorname{im} Q_\varepsilon^{(\kappa)} \cap P^{\kappa-1}$. Since $\varepsilon = \omega^s$ for some integer s, it is now easy to see that this set of polynomials can be characterized as in the assertion, and this together with (3.28) concludes the proof of the first part.

Now we prove the assertions concerning the range of $Q_{\hat\varepsilon} T_\kappa$. This will be done by considering the transpose operator to $T_\kappa Q_{\hat\varepsilon}$, which can be viewed as an operator from $\operatorname{im} Q'_{\hat\varepsilon}$ into $L_q(\mathbb{T})$, q being the conjugate to p. We start by identifying the transpose operator U' to U. Using the Riesz representation theorem and taking into account that $\alpha' = \omega^{-1}\,\alpha_+^2$, we find that

$$U'\,\psi(t) = \omega^{-1}\alpha_+^{-1}(\alpha_{n-1}(\tau))\,\psi(\alpha_{n-1}(\tau)),$$

which means that U' coincides with the inverse of U regarded as an operator acting in $L_q(\mathbb{T})$.

For the transpose operators to T_κ and $Q_{\hat\varepsilon}$ we have

$$T_\kappa{}' = P_+ + P_-\overline{\chi}^{\,\kappa}, \;\; Q_{\hat\varepsilon}{}' = \sum_{j=1}^{n} \hat\varepsilon^j\,(U')^j.$$

Since we are considering $Q_{\hat\varepsilon} T_\kappa$ as an operator into $\operatorname{im} Q_{\hat\varepsilon}$, its transpose operator coincides with $T_\kappa{}'|\operatorname{im} Q'_{\hat\varepsilon}$. A function $\psi \in \operatorname{im} Q_{\hat\varepsilon}{}'$ belongs to $\ker T_\kappa{}'$ if and only if $\psi^+ = 0$ and

$$P_-\overline{\chi}^{\,\kappa}\,\psi^- = 0.$$

As $\operatorname{ind}_{\mathbb{T}} \overline{\chi} = -1$, if $\kappa \geq 0$, this equation has only the trivial solution and so $T_\kappa{}'$ is injective in $\operatorname{im} Q_{\hat\varepsilon}{}'$ and, consequently, $\operatorname{im} Q_{\hat\varepsilon} T_\kappa = \operatorname{im} Q_{\hat\varepsilon}$.

Now suppose that $\kappa < 0$. Then the solutions of the last equation can be obtained using the factorization $\overline{X} = \overline{X}_- t^{-1} \overline{X}_+$ similarly to what we have done before, yielding

$$\psi^- = Q_{\hat{\varepsilon}}{}'(\overline{X}_+^{-\kappa} t^\kappa p),$$

where $p \in P^{|\kappa|-1}$. Set $R^{|\kappa|-1} = \{t^{-|\kappa|} p : p \in P^{|\kappa|-1}\}$. Taking into account that $\overline{X}_+^\kappa Q_{\hat{\varepsilon}}{}' = (Q_\varepsilon^{(\kappa)})' \overline{X}_+^\kappa I$, we conclude that a function $\psi \in \operatorname{im} Q_{\hat{\varepsilon}}{}'$ belongs to $\ker T_\kappa{}'$ if and only if

$$\psi = \overline{X}_+^{-\kappa} q, \quad q \in (Q_\varepsilon^{(\kappa)})' \cap R^{|\kappa|-1}.$$

We have

$$(Q_\varepsilon^{(\kappa)})' = \frac{1}{n} \sum_{j=0}^{n-1} \varepsilon^j (V_\kappa{}')^j, \quad V_\kappa{}' = U' \,\overline{\alpha}_+{}^{-\kappa} I,$$

and, as in the first part, we establish that $q \in \operatorname{im}(Q_\varepsilon^{(\kappa)})'$ if and only if $\varepsilon V_\kappa' q = q$.

Consider again the set (3.27). As we saw before, for each $m = 0, \ldots, -\kappa - 1$, we have $U p_m = \omega^{1-m} \alpha_+^{-\kappa} p_m$. Thus, $U(t^\kappa p_m) = \omega^{1-m} \alpha_-^\kappa t^\kappa p_m$, or

$$\begin{aligned} t^\kappa p_m &= \omega^{1-m} U^{-1} [\alpha_-^\kappa (t^\kappa p_m)] \\ &= \omega^{1-m-\kappa} U^{-1} [\overline{\alpha}_+^{-\kappa} (t^{-\kappa} p_m)] = \omega^{1-m-\kappa} V_\kappa'(t^\kappa p_m), \end{aligned}$$

where we have used the easily verified relation $\alpha_- = \omega^{-1} \overline{\alpha}_+^{-1}$. Therefore, a basis for $(Q_\varepsilon^{(\kappa)})' \cap R^{|\kappa|-1}$ is the set of functions $t^{-\kappa} p_m$, where $m = 0, \ldots, -\kappa - 1$ is such that $\omega^{1-m-\kappa} = \varepsilon$. Equivalently, a basis for $(Q_\varepsilon^{(\kappa)})' \cap R^{|\kappa|-1}$ is the set of functions $t^{-\kappa} p_{m-\kappa}$ where $m = \kappa, \ldots, -1$ is such that $\omega^{1-m} = \varepsilon$. It is easily verified that this last condition determines the set $\Gamma_{\varepsilon,\kappa}$. Consequently, $\ker T_\kappa{}'|_{\operatorname{im} Q_{\hat{\varepsilon}}'} = \{\overline{X}_+^{-\kappa} t^\kappa p_{m-\kappa} : m \in \Gamma_{\varepsilon,\kappa}\}$.

Since $\operatorname{codim} \operatorname{im}[Q_{\hat{\varepsilon}} T_\kappa] = \dim \ker T_\kappa{}'|_{\operatorname{im} Q_{\hat{\varepsilon}}'}$, it follows that $\operatorname{codim} \operatorname{im}[Q_{\hat{\varepsilon}} T_\kappa] = \gamma(\varepsilon, \kappa)$. On the other hand, a function f is in $\operatorname{im}[Q_{\hat{\varepsilon}} T_\kappa]$ if and only if

$$(f, \psi) = \int_{\mathbb{T}} f(t) \overline{\psi(t)} \, |dt| = 0$$

for all $\psi \in \ker T_\kappa{}'|_{\operatorname{im} Q_{\hat{\varepsilon}}'}$. Taking into account that $dt = i\, t\, |dt|$ and that, for all $m = 0, \ldots, -\kappa - 1$ and all $t \in \mathbb{T}$, $\overline{p_m(t)} = c_m t^{1+\kappa} p_m(t)$ with $c_m \in \mathbb{C} \setminus \{0\}$, it follows that $f \in \operatorname{im}[Q_{\hat{\varepsilon}} T_\kappa]$ if and only if

$$\int_{\mathbb{T}} f(t) \, X_+^{-\kappa} p_{m-\kappa} \, dt = 0 \quad , \quad m \in \Gamma_{\varepsilon,\kappa},$$

and this completes the proof. □

Gathering the results of Theorem 3.11, representation (3.23) and Proposition 3.13, we are now able to characterize the kernel and the range of the operator T_A.

Theorem 3.14. *Suppose the conditions of Theorem* 3.11 *are fulfilled, let* $\kappa_1 \geq \cdots \geq \kappa_n$ *be the partial indices of* $\mathcal{A} = \pi(A)$, *and let* $D_{\mathcal{A}} = \{\varepsilon_1, \ldots, \varepsilon_n\}$ *be the set of* n*-roots of the unit. Then*

$$\dim\ker T_A = \sum_{\kappa_i > 0} \gamma(\varepsilon_i, \kappa_i), \quad \operatorname{codim}\, \operatorname{im} T_A = \sum_{\kappa_i < 0} \gamma(\varepsilon_i, \kappa_i), \tag{3.29}$$

where $\gamma(\varepsilon_i, \kappa_i)$ *is defined by* (3.25)-(3.26). *Moreover, if* T_A *is non-injective then*

$$\ker T_A = \operatorname{span}\left\{B(\chi_+^{\kappa_i} - t^{-\kappa_i}\chi_-^{-\kappa_i})\, p_{m_i} : \kappa_i > 0\,,\ m_i \in \Gamma_{\varepsilon_i,\kappa_i}\right\},$$

and if T_A *is non-surjective then*

$$\operatorname{im} T_A = \left\{ f \in L_p(\mathbb{T}) : \int_{\mathbb{T}} f(t)\, \chi_+^{-\kappa_i}(t)\, p_{m_i - \kappa_i}(t) = 0,\ \kappa_i < 0\,,\ m_i \in \Gamma_{\varepsilon_i,\kappa_i} \right\},$$

where $B = A_+ P_+ + A_-^{-1} P_-$ *and the polynomials* p_{m_i} *are given by* (3.27).

Acknowledgements

All the authors are members of Applied Mathematics Centre-IST, which is supported by FCT, the Portuguese Foundation for Science and Technology.

References

[DK] Drekova, G.V. and Kravchenko, V.G., Dimension and structure of the kernel and cokernel of a singular integral operator with linear-fractional Carleman shift and with conjugation, *Soviet Math. Dokl.* **42:3** (1991), 743–746.

[E] Ehrhardt, T., Invertibility Theory for Toeplitz + Hankel operators and singular integral operators with flip. Preprint, 2000.

[F] Ford, L.R., *Automorphic Functions*, Chelsea Publishing Company, New York 1957.

[GK] Gohberg, I. and Krupnik, N.Ja., *One-Dimensional Linear Singular Integral Operators*, Vols. 53 and 54 of Operator Theory: Advances and Applications, Birkhäuser Verlag 1992.

[KLL] Kravchenko, V.G., Lebre, A.B. and Litvinchuk, G.S., Spectrum problems for singular integral operators with Carleman shift, *Math. Nachr.* **226** (2001), 129–151.

[KLR] Kravchenko, V.G., Lebre, A.B. and Rodríguez, J.S., Factorization of singular integral operators with a Carleman shift and spectral problems. To appear in *J. Integral Equations Appl.*.

[KL] Kravchenko, V.G. and Litvinchuk, G.S., *Introduction to the Theory of Singular Integral Operators with Shift*, Vol. 289 of Mathematics and its Applications, Kluwer Academic Publishers 1994.

[KSh] Kravchenko, V.G. and Shaev, A., The theory of solvability of singular integral equations with a linear fractional Carleman shift, *Soviet Math. Dokl.* **43:1** (1991), 73–77.

[LS] Litvinchuk, G.S. and Spitkovskii, I.M., *Factorization of Measurable Matrix Functions*, Birkhäuser 1987.

Viktor G. Kravchenko
Área Dep. de Matemática
Universidade do Algarve
Campus de Gambelas
8000-810 Faro, Portugal
e-mail: vkravch@ualg.pt

Amarino B. Lebre
Departamento de Matemática
Instituto Superior Técnico
Av. Rovisco Pais
1049-001 Lisboa, Portugal
e-mail: alebre@math.ist.utl.pt

Juan S. Rodríguez
Área Dep. de Matemática
Universidade do Algarve
Campus de Gambelas
8000-810 Faro, Portugal
e-mail: jsanchez@ualg.pt

2000 *Mathematics Subject Classification.*
Primary 47G10; Secondary 47A68, 47A10.

Received: 31 May 2001 .

Operator Theory:
Advances and Applications, Vol. 142, 213–221

Single Elements of Operator Algebras

William E. Longstaff

Abstract. An element s of an (abstract) algebra $\mathcal{A}$ is a *single element* of $\mathcal{A}$ if $asb = 0$ and $a, b \in \mathcal{A}$ implies that $as = 0$ or $sb = 0$. Such elements were first used by operator theorists in 1961 when J.R. Ringrose proved that algebraic isomorphisms between certain nest algebras were spatial. They have been successfully used since, towards a similar end, namely, in proving the (quasi-)spatiality of algebraic isomorphisms between reflexive operator algebras. These results, and others concerning the structure and the rank of single elements in operator algebras of various types, are reviewed in this article.

1. Introduction

Definition 1.1. An element s of an (abstract) algebra $\mathcal{A}$ is called a single element of $\mathcal{A}$ if

$$a\,s\,b = 0 \quad \text{with} \quad a, b \in \mathcal{A} \quad \text{implies that} \quad a\,s = 0 \quad \text{or} \quad s\,b = 0.$$

The above definition makes sense in any (associative, multiplicative) semigroup with a zero element. However, with the exception of [20], the notion of 'single element' has not been studied much in this broader context. Indeed, almost all of the investigation has been done in an operator-theoretic context. In such a context single elements arise naturally. For example, every rank 1 operator is a single element of any algebra containing it. For this reason it is tempting to think of single elements as 'abstract rank 1 operators' (relative to the algebra in question). Since singleness is a purely algebraic notion, single elements will be preserved under algebraic isomorphisms (that is, multiplicative linear bijections) between algebras. In particular, the image of a rank 1 operator under such an isomorphism will always be single, though not necessarily of rank 1 (as we shall see).

Throughout what follows all scalars are assumed complex (unless otherwise specified). We let X denote a complex Banach space with (topological) dual X^*. By 'operator' we mean 'bounded linear transformation' and by 'subspace' we mean 'closed linear manifold'. We let $\mathcal{B}(X)$ denote the set of operators acting on X. A subspace lattice on X is, by definition, a family $\mathcal{L}$ of subspaces of X which contains the trivial subspaces (0) and X, and is closed under arbitrary intersections and arbitrary closed linear spans, that is, for every family $\{L_\gamma\}_{\gamma\in\Gamma}$ of elements of $\mathcal{L}$ both $\bigcap_{\gamma\in\Gamma} L_\gamma$ and $\bigvee_{\gamma\in\Gamma} L_\gamma$ belong to $\mathcal{L}$. For any subspace lattice $\mathcal{L}$ and any element $L \in \mathcal{L}$ we let L_- be the element of $\mathcal{L}$ given by $L_- = \vee\{M \in \mathcal{L} : L \not\subseteq M\}$ (taking $(0)_-$ to be (0)).

Single elements have been used to study reflexive, generally non-self-adjoint, operator algebras. These arise as follows. If $\mathcal{L}$ is a subspace lattice on X, we put

$$\operatorname{Alg}\mathcal{L} = \{T \in \mathcal{B}(X) : T(L) \subseteq L \quad \text{for all} \quad L \in \mathcal{L}\}.$$

(So $\operatorname{Alg}\mathcal{L}$ is the set of operators which leave every member of $\mathcal{L}$ invariant.) An operator algebra $\mathcal{A} \subseteq \mathcal{B}(X)$ is reflexive if it equals $\operatorname{Alg}\mathcal{L}$ for some subspace lattice $\mathcal{L}$. (The definition of reflexivity was first given in [23].)

Clearly, the lattice and topological properties of the lattice $\mathcal{L}$ are reflected in algebraic and topological properties of the algebra $\operatorname{Alg}\mathcal{L}$. Over the past 40 years or so the following types of subspace lattices $\mathcal{L}$ have been investigated:

- nest algebras. Here $\mathcal{L}$ is a nest, that is, it is totally ordered by inclusion.
- ABSL algebras. Here $\mathcal{L}$ is an atomic Boolean subspace lattice (ABSL), that is, it is complemented, distributive and atomic. (An element $K \in \mathcal{L}$ is an atom of $\mathcal{L}$ if $(0) \subseteq L \subseteq K$ with $L \in \mathcal{L}$ implies that $L = (0)$ or K. A subspace lattice $\mathcal{L}$ is atomic if every non-zero element is the closed linear span of the atoms which it contains.)
- CSL algebras. Here $\mathcal{L}$ is a commutative subspace lattice (CSL) on a Hilbert space, that is, a subspace lattice with the property that the orthogonal projections onto the members of $\mathcal{L}$ pairwise commute. Such subspace lattices are necessarily distributive.
- JSL algebras. Here $\mathcal{L}$ is a $\mathcal{J}$-lattice (JSL), that is, a subspace lattice satisfying
  (1) $\vee\{K : K \in \mathcal{J}(\mathcal{L})\} = X$,
  (2) $\cap\{K_- : K \in \mathcal{J}(\mathcal{L})\} = (0)$,
  (3) $K \vee K_- = X$ for every $K \in \mathcal{J}(\mathcal{L})$,
  (4) $K \cap K_- = (0)$ for every $K \in \mathcal{J}(\mathcal{L})$,

  where, by definition, $\mathcal{J}(\mathcal{L}) = \{K \in \mathcal{L} : K \neq (0) \text{ and } K_- \neq X\}$ is the set of $\mathcal{J}$-elements of $\mathcal{L}$. Every ABSL is a JSL (with $\mathcal{J}(\mathcal{L})$ its set of atoms [13]).
- CDSL algebras. Here $\mathcal{L}$ is a completely distributive subspace lattice, that is, a subspace lattice satisfying $L = \cap\{M_- : M \in \mathcal{L} \text{ and } M \not\subseteq L\}$. (This is not the original definition but it is equivalent to it; see [13].) Every nest, every finite CSL, and every ABSL is a CDSL.

2. Results

The notion of 'single element' was first isolated and used by J. R. Ringrose in 1961 [24]. He investigated certain kinds of nest algebras and showed that in them the rank 1 operators could be characterized algebraically: $S \in \operatorname{Alg}\mathcal{L}$ has rank 1 if and only if

$$S \neq 0 \text{ and } ASB = 0 \text{ with } A, B \in \operatorname{Alg}\mathcal{L} \text{ implies that } AS = 0 \text{ or } SB = 0$$

(that is, if and only if S is a single element of $\operatorname{Alg}\mathcal{L}$). This algebraic characterization of rank 1 operators means that such operators are preserved under algebraic isomorphisms, a necessary condition for spatiality.

Later, in 1966 [25], Ringrose showed that every algebraic isomorphism $\phi : \operatorname{Alg}\mathcal{L}_1 \to \operatorname{Alg}\mathcal{L}_2$, where $\operatorname{Alg}\mathcal{L}_1$ and $\operatorname{Alg}\mathcal{L}_2$ are nest algebras on Hilbert spaces, is spatial in the sense that there exists a bicontinuous bijection S such that $\phi(A) = SAS^{-1}$ for all $A \in \operatorname{Alg}\mathcal{L}_1$.

If $f \in X$ and $e^* \in X^*$ are non-zero vectors, the operator $e^* \otimes f$ acting according to $(e^* \otimes f)(x) = e^*(x)f, \quad x \in X$, is a rank 1 operator and all rank 1 operators arise in this way. Note that every rank 1 operator is a single element of every algebra containing it:

Let $A(e^* \otimes f)B = 0$ with $A, B \in \mathcal{B}(X)$. Then $e^*(Bx)Af = 0$ for all $x \in X$. Thus $Af = 0$ or $e^*(Bx) = 0$ for all $x \in X$. If the former is true $A(e^* \otimes f) = 0$; if the latter $(e^* \otimes f)B = 0$.

So single elements are, in a sense, abstract rank 1 operators relative to the algebra in question.

In 1971, J.A. Erdos [3] gave the first definition of 'single element'.

Theorem 2.1 (J.A. Erdos [3]). *Let $\mathcal{A}$ be a C^* algebra and let $s \in \mathcal{A}$. There exists a faithful representation ϕ of $\mathcal{A}$ on a Hilbert space such that $\phi(s)$ has rank 1 if and only if s is a non-zero single element.*

This Banach algebra line of investigation was pursued in [1, 2, 4, 6, 7, 8].

In 1982 M.S. Lambrou [11] considered $\operatorname{Alg}\mathcal{L}$ where $\mathcal{L}$ is an atomic Boolean subspace lattice (ABSL) on Banach space. For these he showed that every non-zero single element has rank 1 and that every algebraic isomorphism $\phi : \operatorname{Alg}\mathcal{L}_1 \to \operatorname{Alg}\mathcal{L}_2$ is quasi-spatial (and not necessarily spatial) in the sense that there exists a not necessarily bounded, closed, injective, densely defined linear transformation S with dense range, and with domain invariant under $\operatorname{Alg}\mathcal{L}_1$ such that

$$\phi(A)Sx = SAx \quad \text{for all} \quad A \in \operatorname{Alg}\mathcal{L}_1 \text{ and for all } x \in \operatorname{Dom}(S).$$

F. Gilfeather and R.L. Moore [5, 19] considered $\operatorname{Alg}\mathcal{L}$ where $\mathcal{L}$ is a completely distributive commutative subspace lattice (CDCSL) on separable Hilbert space. For such subspace lattices they showed that an algebraic isomorphism $\phi : \operatorname{Alg}\mathcal{L}_1 \to \operatorname{Alg}\mathcal{L}_2$ need not preserve rank, but if it does, it is quasi-spatial. Whether or not such a mapping must be spatial is still an open question. Gilfeather and Moore also showed that single elements can have ranks greater than 1.

Example 2.2 (F. Gilfeather and R.L. Moore [5, 19]). Let $\mathcal{L}$ be the subspace lattice on $\mathbb{C}^4$ given by

$$\mathcal{L} = \{(0), \langle e_1\rangle, \langle e_2\rangle, \langle e_1, e_2\rangle, \langle e_1, e_2, e_3\rangle, \langle e_1, e_2, e_4\rangle, \mathbb{C}^4\}$$

where $\langle e, f, g, \ldots \rangle$ denotes the linear span of $e, f, g, \ldots$. Here

$$\operatorname{Alg}\mathcal{L} = \{\begin{bmatrix} a & 0 & x & y \\ 0 & b & z & t \\ 0 & 0 & c & 0 \\ 0 & 0 & 0 & d \end{bmatrix} : a, b, c, d, x, y, z, t \in \mathbb{C}\}$$

and the mapping

$$\begin{bmatrix} a & 0 & x & y \\ 0 & b & z & t \\ 0 & 0 & c & 0 \\ 0 & 0 & 0 & d \end{bmatrix} \mapsto \begin{bmatrix} a & 0 & x & y \\ 0 & b & z & -t \\ 0 & 0 & c & 0 \\ 0 & 0 & 0 & d \end{bmatrix}$$

defines an algebraic automorphism ϕ of $\operatorname{Alg}\mathcal{L}$.

The element

$$R = \begin{bmatrix} 0 & 0 & 1 & 1 \\ 0 & 0 & 1 & 1 \\ 0 & 0 & 0 & 0 \\ 0 & 0 & 0 & 0 \end{bmatrix}$$

has rank 1, and its image

$$\phi(R) = \begin{bmatrix} 0 & 0 & 1 & 1 \\ 0 & 0 & 1 & -1 \\ 0 & 0 & 0 & 0 \\ 0 & 0 & 0 & 0 \end{bmatrix}$$

is a single element of rank 2.

As we have already noted, a subspace lattice which is a nest, an ABSL or a finite CSL is completely distributive. In 1983 [19] R.L. Moore asked if there was a lattice-theoretic necessary and sufficient condition for non-zero single elements to have rank 1 in $\operatorname{Alg}\mathcal{L}$, when $\mathcal{L}$ is completely distributive and commutative. This question was answered affirmatively in 1997. We shall come to this answer soon.

If $\mathcal{L}_i$, $i = 1, 2$ are subspace lattices on the Banach spaces X_i, $i = 1, 2$, respectively, their ordinal sum $\mathcal{L}_1 + \mathcal{L}_2$ is the subspace lattice on $X_1 \oplus X_2$ given by

$$\mathcal{L}_1 + \mathcal{L}_2 = \{L \oplus (0) : L \in \mathcal{L}_1\} \cup \{X_1 \oplus M : M \in \mathcal{L}_2\}.$$

In 1990 M.S. Lambrou showed that single elements could have infinite rank.

Theorem 2.3 (M.S. Lambrou [12]). *Let $\mathcal{L}$ be the completely distributive subspace lattice consisting of the ordinal sum of an ABSL with m atoms and an ABSL with n atoms, where m and n are possibly infinite. Then the maximum rank of a single element of* $\operatorname{Alg}\mathcal{L}$ *is* $\min\{m, n\}$.

Pentagon subspace lattices were studied in [10]. A pentagon subspace lattice $\mathcal{P}$ on X is a set of five subspaces $\mathcal{P} = \{(0), K, L, M, X\}$ where $L \subset M$, $K \cap M = (0)$ and $K \vee L = X$. Since $M \cap (K \vee L) \neq (M \cap K) \vee (M \cap L)$ such subspace lattices are non-modular so non-distributive. For every pentagon $\mathcal{P}$ it was shown that every non-zero single element of $\operatorname{Alg}\mathcal{P}$ has rank 1. It was also shown that every algebraic isomorphism $\phi : \operatorname{Alg}\mathcal{P}_1 \to \operatorname{Alg}\mathcal{P}_2$, where $\mathcal{P}_1$ and $\mathcal{P}_2$ are pentagons on Banach spaces, is quasi-spatial. It is not known if such mappings must be spatial.

Oreste Panaia carried all of these investigations forward in a significant way in his Ph.D. thesis [21] in 1995. In this thesis the class of $\mathcal{J}$-subspace lattices (JSL's) on Banach spaces was introduced, and it was shown that

1. in the Alg of every JSL, non-zero single elements have rank 1,
2. every algebraic isomorphism $\phi : \operatorname{Alg}\mathcal{J}_1 \to \operatorname{Alg}\mathcal{J}_2$, where $\mathcal{J}_1$ and $\mathcal{J}_2$ are JSL's on reflexive Banach spaces, is quasi-spatial,
3. if $\mathcal{L}_1$ and $\mathcal{L}_2$ are finite distributive subspace lattices (not necessarily commutative) on reflexive Banach spaces, every algebraic isomorphism $\phi : \operatorname{Alg}\mathcal{L}_1 \to \operatorname{Alg}\mathcal{L}_2$ which preserves rank is quasi-spatial (see also [22]; more recent results concerning JSL's can be found in [15, 18]).

This work also led to an affirmative answer to the question raised by Moore.

Theorem 2.4 (W.E. Longstaff, Oreste Panaia [14]). *Let $\mathcal{L}$ be a completely distributive subspace lattice on a reflexive Banach space X. Let $n \in \mathbb{Z}^+ \cup \{\infty\}$. There exists a single element of rank n in* $\operatorname{Alg}\mathcal{L}$ *if and only if there exist families $\{M_i\}_1^n, \{K_i\}_1^n$ of $\mathcal{J}$-elements satisfying*

(1) $M_i \cap M_j = (0), i \neq j,$

(2) $K_{i\_} \vee K_{j\_} = X, i \neq j,$

(3) $\bigvee_1^n M_i \subseteq \bigcap_1^n K_j.$

Corollary 2.5. *In any completely distributive subspace lattice $\mathcal{L}$, every non-zero single element has rank* 1 *if and only if there do not exist $\mathcal{J}$-elements M_1, M_2, K_1, K_2 satisfying*

$$M_1 \cap M_2 = (0), \quad K_{1\_} \vee K_{2\_} = X, \quad M_1 \vee M_2 \subseteq K_1 \cap K_2.$$

It easily follows that if $\mathcal{L}$ is a nest or an ABSL, every non-zero single element has rank 1, that is, the results of Ringrose and Lambrou follow.

Our final results are taken from two papers by W.E. Longstaff and Oreste Panaia [16, 17]. They concern finite CSL algebras. The single elements of such algebras can be fully described.

Finite CSL's arise in the following way.

Let n be a positive integer and let $\preceq$ be a partial order on $\{1, 2, \ldots, n\}$. Call a subset $\mathcal{F}$ of $\{1, 2, \ldots, n\}$ *hereditary* if $i \preceq j \in \mathcal{F} \Rightarrow i \in \mathcal{F}$. The set of hereditary subsets is a finite distributive lattice under the inclusion order. Next, let $\{H_i\}_1^n$ be a family of non-zero separable Hilbert spaces and put $H = \bigoplus_1^n H_i$. The set of subspaces

$$\mathcal{L} = \{\bigoplus_{i \in \mathcal{F}} : \mathcal{F} \text{ is a hereditary subset of } \{1, 2, \ldots, n\}\}$$

is a finite CSL on H. All finite CSL's on separable Hilbert space arise in this way.

For this $\mathcal{L}$

$$\operatorname{Alg}\mathcal{L} = \{(T_{i,j}) \in \mathcal{B}(H) : T_{i,j} = 0 \text{ if } i \npreceq j\}$$

that is, the elements of $\operatorname{Alg}\mathcal{L}$ are those $n \times n$ operator-entried matrices having 0's in those positions (i,j) where $i \npreceq j$. In particular, if $\{e_1, e_2, \ldots, e_n\}$ is an orthonormal basis for H and $\preceq$ is a partial order on $\{1, 2, \ldots, n\}$, the matrix algebra

$$\mathcal{A}_n(\preceq) = \{(a_{i,j}) : a_{i,j} = 0 \text{ if } i \npreceq j\}$$

is the Alg of a finite CSL on H (take $H_i = \langle e_i \rangle$ for every i). The algebras of the type $\mathcal{A}_n(\preceq)$ are called matrix incidence algebras. (These algebras are of interest in combinatorics.) The 0, 1-matrix whose (i,j)-entry is 1 precisely when $i \preceq j$ is called the incidence matrix of the partial order.

Example 2.6.

1. When the partial order is equality, $\mathcal{A}_n(=)$ is the algebra of all $n \times n$ diagonal matrices and the corresponding CSL is an ABSL with n (one-dimensional) atoms.
2. When the partial order is the usual order, $\mathcal{A}_n(\leq)$ is the algebra of all upper triangular $n \times n$ matrices, and the corresponding CSL is a nest with $n+1$ elements.
3. When the partial order has incidence matrix

$$\begin{bmatrix} 1 & 0 & 1 & 1 \\ 0 & 1 & 1 & 1 \\ 0 & 0 & 1 & 0 \\ 0 & 0 & 0 & 1 \end{bmatrix}$$

we have

$$\mathcal{A}_n(\preceq) = \left\{ \begin{bmatrix} * & 0 & * & * \\ 0 & * & * & * \\ 0 & 0 & * & 0 \\ 0 & 0 & 0 & * \end{bmatrix} \right\}$$

where the $*$'s denote arbitrary elements of $\mathbb{C}$, and the corresponding CSL is the ordinal sum of two 2-atom ABSL's. This is the example of Moore and Gilfeather mentioned earlier.

In Example 2.6(3) direct computation shows that every matrix of the form

$$\begin{bmatrix} 0 & 0 & x & y \\ 0 & 0 & z & t \\ 0 & 0 & 0 & 0 \\ 0 & 0 & 0 & 0 \end{bmatrix} \qquad \text{with } x, y, z, t \neq 0$$

is a single element and every single element has rank ≤ 2. For most algebras, similar direct computations are too difficult. The algebra $\mathcal{A}_n(\preceq)$ is completely determined by the partial order $\preceq$, so its single elements must be, and are, describable in terms of the partial order. As we shall see, this leads to a description of the single elements of finite CSL algebras, and also of the single elements of $\operatorname{Alg}\mathcal{D}$, where $\mathcal{D}$ is a finite distributive subspace lattice on a finite-dimensional space.

Theorem 2.7. *Let n be a positive integer and let $\preceq$ be a partial order on $\{1,2,\dots,n\}$. The matrix $S=(s_{i,j})\in\mathcal{A}_n(\preceq)$ is a single element of $\mathcal{A}_n(\preceq)$ if and only if*

(i) *$r_i\neq 0$ and $c_j\neq 0$ implies that $s_{i,j}\neq 0$,*

(ii) *$i\preceq j_1$ and $i\preceq j_2$ for some i implies that r_{j_1} and r_{j_2} are linearly dependent,*

(iii) *$i_1\preceq j$ and $i_2\preceq j$ for some j implies that c_{i_1} and c_{i_2} are linearly dependent.*

Here r_i denotes the ith row of S and c_j denotes the jth column of S.

This theorem holds for matrix incidence algebras over arbitrary fields. From (i), if S is single, then there exist subsets $I,J\subseteq\{1,2,\dots,n\}$ such that $s_{i,j}\neq 0$ if and only if $(i,j)\in I\times J$ (so single elements 'live on grids'). It follows that every single element in any incidence algebra over $\mathbb{Z}_2$ has rank 1.

Corollary 2.8. *Let X, respectively Y, be a set of minimal, respectively maximal, elements of the partially ordered set $(\{1,2,\dots,n\},\preceq)$ satisfying $x\preceq y$ for all $x\in X, y\in Y$. Every $n\times n$ matrix S satisfying $s_{i,j}\neq 0$ if and only if $(i,j)\in X\times Y$ is a single element of $\mathcal{A}_n(\preceq)$. If m is the largest positive integer for which there exist such sets X,Y with $|X|=|Y|=m$, then m is the maximum rank that a single element of $\mathcal{A}_n(\preceq)$ can have.*

Corollary 2.9. *In $\mathcal{A}_n(\preceq)$ all non-zero single elements have rank 1 if and only if there do not exist distinct minimal elements x_1,x_2 and distinct maximal elements y_1,y_2 such that $x_i\preceq y_j$ for all i,j.*

Applications

1. By Corollary 2.9, it is now obvious that, in $\mathcal{A}_n(=)$, all non-zero single elements have rank 1. This also holds for $\mathcal{A}_n(\leq)$ since in the case there is a unique minimal element.
2. When the partial order has incidence matrix

$$\begin{bmatrix}1&0&1&1\\0&1&1&1\\0&0&1&0\\0&0&0&1\end{bmatrix}$$

it is clear, by Corollary 2.8, that single elements have rank at most 2 in $\mathcal{A}_n(\preceq)$.

More generally, if the partial order (on $\{1,2,3,\dots,m+n\}$) has incidence matrix $\begin{bmatrix}I_m&J\\0&I_n\end{bmatrix}$ where I_m and I_n are the identity matrices of sizes m and n, respectively, and J is the $m\times n$ matrix with every entry equal to 1, single elements of $\mathcal{A}_n(\preceq)$ have rank at most $\min\{m,n\}$.

Finally, for finite CSL algebras, that is, algebras of the form $\operatorname{Alg}\mathcal{L}$ where $\mathcal{L}$ is a finite commutative subspace lattice, the single elements arise as follows.

Theorem 2.10. *Let $H_i, i = 1, 2, \ldots, n$ be Hilbert spaces. Let $H = \bigoplus_{i=1}^n H_i$ and let $\preceq$ be a partial order on the set $\{1, 2, \ldots, n\}$. Let $f_i \in H_i, e_j \in H_j$ be non-zero vectors. Then $e_j \otimes f_i \in \mathcal{B}(H_j, H_i)$ and the $n \times n$ operator-entried matrix $(e_j \otimes f_i)$ (with (i, j)-element equal to $e_j \otimes f_i$) belongs to $\mathcal{B}(H)$. Let $S = (s_{i,j})$ be a single element of $\mathcal{A}_n(\preceq)$. Then the Hadamard (that is, entry-wise) product matrix $S \otimes_h (e_j \otimes f_i)$ is a single element of* $\operatorname{Alg}\mathcal{L}$ *with the same rank as S. All single elements of* $\operatorname{Alg}\mathcal{L}$ *arise in this way.*

The preceding theorem leads to a description of the single elements of reflexive algebras of the form $\operatorname{Alg}\mathcal{D}$, where $\mathcal{D}$ is a finite distributive subspace lattice on a finite-dimensional space. We may assume that the underlying space is a Hilbert space H. Then [9] there exists an invertible operator $T \in \mathcal{B}(H)$ and a finite CSL $\mathcal{L}$ on H such that $\mathcal{D} = T\mathcal{L} = \{T(L) : L \in \mathcal{L}\}$. We have $\operatorname{Alg}\mathcal{D} = T(\operatorname{Alg}\mathcal{L})T^{-1}$ and the set of non-zero single elements of $\operatorname{Alg}\mathcal{D}$ is $\{TST^{-1} : S$ is a non-zero single element of $\operatorname{Alg}\mathcal{L}\}$.

References

[1] Dalla, L., Giotopoulos, S., Katseli, N., Skeletons of the unit ball of a C^*-algebra, *Math. Balkanica (N. S.)* **(1) 1** (1987), 83–88.

[2] Dalla, L., Giotopoulos, S., Katseli, N., The socle and finite-dimensionality of a semiprime Banach algebra, *Studia Math.* **(2) 92** (1989), 201–204.

[3] Erdos, J.A., On certain elements of C^* algebras, *Illinois J. Math.* **15** (1971), 682–693.

[4] Erdos, J.A., Giotopoulos, S., Lambrou, M.S., Rank one elements of Banach algebras, *Mathematika* **24** (1977), 178–181.

[5] Gilfeather, F., Moore, R.L., Isomorphisms of certain CSL algebras, *J. Funct. Anal.* **(2) 67** (1986), 264–291.

[6] Giotopoulos, S., A note on annihilator Banach algebras, *Math. Proc. Cambridge Philos. Soc.* **97** (1985), 101–106.

[7] Giotopoulos, S., C_p-classes of operators in C^*-algebras, *J. Austral. Math. Soc. Ser. A* **(2) 40** (1986), 153–160.

[8] Giotopoulos, S., Katseli, N., Modular annihilator algebras, *Period. Math. Hungar.* **(2) 22** (1991), 91–95.

[9] Harrison, K.J., Longstaff, W.E., Automorphic images of commutative subspace lattices, *Trans. Amer. Math. Soc.* **(1) 296** (1986), 217–228.

[10] Katavolos, A., Lambrou, M.S., Longstaff, W.E., Pentagon subspace lattices on Banach spaces, *J. Op. Theory* **46** (2001), 355–380.

[11] Lambrou, M.S., Automatic continuity and implementation of homomorphisms, *U. Crete (manuscript)* (1982).

[12] Lambrou, M.S., On the ranks of operators in reflexive algebras, *Lin. Alg. & Applic.* **142** (1990), 211–235.

[13] Longstaff, W.E., Strongly reflexive lattices, *J. London Math. Soc.* **(2) 11** (1975), 491–498.

[14] Longstaff, W.E., Panaia, O., On the ranks of single elements of reflexive operator algebras, *Proc. Amer. Math. Soc.* **(10) 125** (1997), 2875–2882.

[15] Longstaff, W.E., Panaia, O., $\mathcal{J}$-subspace lattices and subspace M-bases, *Studia Math.* **(3) 139** (2000), 197–212.

[16] Longstaff, W.E., Panaia, O., Single elements of matrix incidence algebras, *Lin. Alg. & Applic.* **318** (2000), 117–126.

[17] Longstaff, W.E., Panaia, O., Single elements of finite CSL algebras, *Proc. Amer. Math. Soc.* **(4) 129** (2000), 1021–1029.

[18] Longstaff, W.E., Nation, J.B., Panaia, O., Abstract refexive sublattices and completely distributive collapsibility, *Bull. Aust. Math. Soc.* **58** (1998), 245–260.

[19] Moore, R.L., Isomorphisms of certain CSL algebras, *U. Alabama (manuscript)* (1983), .

[20] Munn, W.D., Brandt congruences on inverse semigroups, *Proc. London Math. Soc.* **(3) 14** (1964), 154–164.

[21] Panaia, O., Quasi-spatiality of isomorphisms for certain classes of operator algebras, *Ph. D. dissert., U. West. Aust.* (1995).

[22] Panaia, O., Algebraic isomorphisms and finite distributive subspace lattices, *J. London Math. Soc.* **(2) 59** (1999), 1033–1048.

[23] Radjavi, H., Rosenthal, P., Invariant subspaces and weakly closed algebras, *Bull. Amer. Math. Soc.* **(5) 74** (1968), 1013–1014.

[24] Ringrose, J.R., Algebraic isomorphisms between ordered bases, *Amer. J. Math.* **83** (1961), 463–478.

[25] Ringrose, J.R., On some algebras of operators II, *Proc. London Math. Soc.* **(3) 16** (1966), 385–402.

William E. Longstaff
Department of Mathematics & Statistics
The University of Western Australia
35 Stirling Highway, Crawley
Western Australia 6009, Australia
e-mail: `longstaf@maths.uwa.edu.au`

2000 *Mathematics Subject Classification.*
Primary 47L35; Secondary 47C05.

Received: 29 January 2001 .

Operator Theory:
Advances and Applications, Vol. 142, 223–251

Skew Incidence Half-Plane Diffraction Problems: Particular Aspects of the Factorization Problem

Ernst Lüneburg and Hamit A. Serbest

Abstract. Problems of diffraction at half-planes or half-plane junctions with different impedance boundary conditions for normal incidence depend upon the polarization of the incident plane electromagnetic wave: either the electric vector or the magnetic vector is parallel to the discontinuity of the boundary condition (TE and TM mode). For isotropic boundary conditions both cases can be treated independently from each other. For normal incidence but arbitrary polarization the diffraction problem can be solved by splitting up the polarization of the incident wave into the sum of the TE and the TM modes which can often be solved by the Wiener-Hopf technique.

For skew (oblique) incidence this splitting up into independent TE and TM modes is no longer possible. The Wiener-Hopf technique then leads to systems with 2×2 or 4×4 Fourier symbol matrices. These can be decoupled by polynomial similarity transformations to single equations or 2×2 block diagonal matrices everywhere in the complex plane except at points of the singular set. Traditionally for about 40 years these constraints have been neglected. Then the final solution of the diffraction problem involves unphysical "leaky wave" type poles that are eliminated by fixing undetermined constants in the classical Wiener-Hopf solution by the residue calculus technique. The concept of restricted range of the analytic family of polynomial transformations that are used to reduce the Fourier symbol matrix functions to simpler forms is shown to be closely related to the pole elimination procedure. As a representative example, diffraction by a two-face impedance half-plane for arbitrary angles of incidence is treated in some detail. Other examples are outlined.

1. Introduction

About five years ago one of the authors presented a review paper with the title "The Sommerfeld problem: Methods, generalizations and frustrations", at the Sommerfeld '96 Workshop, Freudenstadt, [28]. The present contribution may be considered as a continuation of that paper dealing with some particular aspects of diffraction of electromagnetic waves by half-planes and half-plane structures.

It is well known that diffraction of a plane electromagnetic wave at a completely conducting half-plane for so-called normal incidence, when the wave vector

$\mathbf{k}^i$ of the incident wave is perpendicular to the diffraction edge, can be decoupled by splitting up the electric vector of the incident plane wave into two components, one perpendicular to the scattering plane spanned by the normal $\mathbf{n}$ of the half-plane and $\mathbf{k}^i$ (E polarization, TE case) and one inside the scattering plane (H polarization, TM case). This leads to two 2-dimensional half-plane boundary values problems with Dirichlet and Neumann boundary conditions that can be solved by the Sommerfeld method, the Wiener-Hopf technique and by some other ad hoc approaches.

The extension of the problem of diffraction by a half-plane for skew (oblique) incidence in the scalar acoustic case with soft and hard boundary conditions is straightforward: only the wave number k has to be replaced by the reduced wave number. For electromagnetic waves the situation is more complicated due to the transversality of the electric vector. It was shown by P.C. Clemmow [9] in 1951 that a solution of the three-dimensional Sommerfeld half-plane problem for skew incidence can be found from the corresponding two-dimensional normal incidence cases with Dirichlet and Neumann conditions. Clemmow's method consists in splitting up the incident electromagnetic wave field into two parts generated by an electric and a magnetic Hertz vector parallel to the diffraction edge.

Over the last decades the influence of material properties on edge diffraction phenomena has been studied extensively by solving boundary value problems for half-planes with different impedance boundary conditions (Leontovich [25]). The problem of diffraction of plane electromagnetic waves for perpendicular (normal) incidence with respect to the edge of a wedge with different impedance boundary conditions on both faces was first solved in 1958 by G.D. Maliuzhinetz [31]. The solution includes the two-face impedance half-plane problem as a special case for vanishing wedge angle. The extension of the wedge problem with arbitrary wedge angle to arbitrary oblique angle of incidence has not found a general closed-form solution yet. The corresponding two-face impedance half-plane problem was solved by Bucci and Franceschetti [2] in 1976 by an extension of Maliuzhinetz's original method; cf. also Syed and Volakis [55], Senior and Legault [52] and Osipov and Norris [37].

The solution of the impedance half-plane problem by the Wiener-Hopf technique was first given by Senior [43] in 1952 for normal incidence and equal impedances on both sides. As for the Sommerfeld half-plane, for normal incidence the electric vector can be resolved into one component parallel to the edge (E polarization) and one component perpendicular to it (H polarization), and this decomposition remains valid after applying the boundary conditions. This decomposition into independent transverse field pairs, however, is no longer valid for arbitrary skew incidence. In 1959 Senior [44] extended his previous work to arbitrary angles of incidence. One objective of Senior's work was to find out if the method proposed by Clemmow in 1951 for Dirichlet and Neumann boundary conditions could be extended to more general boundary conditions. Senior concluded that a splitting up into two 2-dimensional decoupled parts corresponding to normal incidence (with reduced wave number of course) does not seem to be possible.

It was, however, shown shortly afterward by Williams [57] in 1960 that such a procedure is indeed possible by introducing certain combinations of tangential derivatives of the tangential electric and magnetic field components that later on were recognized as the perpendicular (normal) electric and magnetic field components (see also Senior [45]). Due to the obvious advantage of dealing with decoupled boundary conditions, the use of normal field components has become the standard approach for a large variety of isotropic half-plane and half-plane junction diffraction problems for skew incidence with different boundary conditions for the last 40 years: Senior [44, 46, 47, 48, 49, 50, 51], Rojas [41, 42], Uzgören et al [56], Büyükaksoy et al. [3, 4], Ly et al [30], Nepa et al. [35], Passow [38] and others.

The higher-order singularities of the normal components close to the diffraction edge lead to constants in the classical Wiener-Hopf solution which are determined in such a way that the residues of "unphysical leaky wave poles" that appear in the final solution vanish. It will be shown that the unexpected occurrence of such poles is due to the omission of some subtle mathematical points that require certain properties of families of polynomial matrix transformations.

In the case of different face impedances the Wiener-Hopf formulation for perpendicular incidence leads to a system of two coupled functional equations with a 2×2 matrix function of the Daniele-Khrapkov type. The factorization was first achieved by Hurd [19] in 1976 by introducing the Wiener-Hopf-Hilbert (WHH) technique (see also Rawlins [39], Hurd and Prezeździecki [20], and Spitkovskii and Tashbaev [54]). Hurd's WHH solution agrees with Maliuzhinetz's solution. For arbitrary skew angles of incidence, only a cursory treatment by Senior [48] of 1978 exists. A more detailed treatment will be presented in this contribution. We limit ourselves to this example as being representative for an entire class of similar situations. It will be shown that also the solution of the two-face impedance half-plane problem for oblique incidence can be expressed in terms of the corresponding solutions of the normal incidence case. The final solutions, however, are coupled in a subtle manner giving right to Senior in a certain way.

The Wiener-Hopf formulation of the problem of diffraction of an obliquely incident plane wave at a two-face impedance half-plane for the Fourier transforms of the tangential field components, presented in this contribution, yields another example of a Wiener-Hopf system with a 4×4 Fourier symbol matrix. This matrix-valued function can be block diagonalized by similarity transformation with a particular family of 4×4 polynomial transformation matrices for all values of the complex Fourier transform variable except at two points that belong to the singular set of this mapping and that are related to the obliqueness parameter. At these points the Jordan structure of the symbol matrix changes. These transformation matrices correspond to the introduction of the Fourier transforms of the normal electric and magnetic field components and the same result follows from the direct application of the decoupled normal electric and magnetic field components.

It will be shown that consideration of the range of the mapping of the Fourier transforms of the tangential to the normal field components and vice versa is closely related to this conventional physically based reasoning of the removal of unwanted leaky wave pole contributions. This mapping consists of rational transformations which allow the reduction of the Fourier symbol matrix function to block diagonal form everywhere in the complex Fourier transform plane if at certain exceptional discrete points constraints are imposed on the range of the mapping. This concept of "restricted range" (of rational transformations) is applicable to a variety of other diffraction problems and the (block) diagonalization of their Fourier symbol matrix functions.

The present contribution extends the results of Hurd, Senior, Volakis, Rojas and others on diffraction of electromagnetic waves at half-plane structures for the case of normal incidence to the case of arbitrary incidence. This extension reveals a subtle coupling effect between the independent TE and TM modal forms due to their edge conditions that previously has not been addressed properly.

2. Preliminaries

The first step towards the factorization of a square $n \times n$ matrix $A : \mathcal{C}^n \to \mathcal{C}^n$ is to consider its invariant subspaces. A subspace $\mathcal{M} \subset \mathcal{C}^n$ is said to be invariant for A if $A\mathcal{M} \subset \mathcal{M}$. Elementary invariant subspaces are the kernel and the image (range) of A, Ker$A$ and ImA, and also the root subspaces to the eigenvalue λ of A

$$\mathcal{R}_\lambda(A) = \{\mathbf{x} \in \mathcal{C}^n \,|\, (A - \lambda I)^n = 0\}. \tag{2.1}$$

These root subspaces contain the vectors of any Jordan chain of A corresponding to λ.

If $\lambda_1, \ldots, \lambda_r$ are all the different eigenvalues of a transformation $A : \mathcal{C}^n \to \mathcal{C}^n$ then $\mathcal{C}^n$ decomposes into the direct sum

$$\mathcal{C}^n = \mathcal{R}_{\lambda_1}(A) \oplus \cdots \oplus \mathcal{R}_{\lambda_r}(A). \tag{2.2}$$

The simplest form that can be achieved by similarity transformations is the Jordan form. The reduction to its Jordan form may facilitate significantly the study of invariant subspaces of a transformation A

$$S^{-1}AS = J_{k_1}(\lambda_1) \oplus \cdots \oplus J_{k_p}(\lambda_p). \tag{2.3}$$

This is a direct sum of Jordan blocks of the general form

$$J_{k_j}(\lambda_j) = \begin{bmatrix} \lambda_j & 1 & 0 & \cdots & 0 \\ 0 & \lambda_j & 1 & & \cdot \\ \cdot & & & & \cdot \\ \cdot & & & & 0 \\ & & & & 1 \\ 0 & 0 & & \cdots & \lambda_j \end{bmatrix}. \tag{2.4}$$

The concepts of partial multiplicities and geometric and algebraic multiplicities of an eigenvalue λ of A are intimately related to the Jordan form of A.

The invariant subspaces of the generic $n \times n$ Jordan block $J(\lambda_0)$ are the $n+1$ subspaces $\{0\}$ and $\mathrm{Span}\{\mathbf{e}_1, \ldots, \mathbf{e}_k\}$ for $k = 1, \ldots, n$. The root subspace $\mathcal{R}_\lambda(J)$ is the whole $\mathcal{C}^n$. In particular,

$$\mathrm{Ker} J(\lambda_0) = \begin{cases} \{0\} & \text{if } \lambda_0 \neq 0 \\ \mathrm{Span}\{\mathbf{e}_1\} & \text{if } \lambda_0 = 0 \end{cases}, \tag{2.5}$$

and

$$\mathrm{Im} J(\lambda_0) = \begin{cases} \mathcal{C}^n & \text{if } \lambda_0 \neq 0 \\ \mathrm{Span}\{\mathbf{e}_1, \ldots, \mathbf{e}_{n-1}\} & \text{if } \lambda_0 = 0 \end{cases}. \tag{2.6}$$

Eigenvectors are often not sufficient for the description of all A-invariant subspaces. It is necessary to consider also generalized eigenvectors. A chain of vectors $\mathbf{x}_0, \mathbf{x}_1, \ldots, \mathbf{x}_k$ is called a Jordan chain corresponding to λ_0 if

$$\begin{aligned} A\mathbf{x}_0 = \lambda_0 \mathbf{x}_0 \quad &\to \quad (A - \lambda_0 I)\mathbf{x}_0 = 0 \\ A\mathbf{x}_1 - \lambda_0 \mathbf{x}_1 = \mathbf{x}_0 \quad &\to \quad (A - \lambda_0 I)^2 \mathbf{x}_1 = 0 \\ A\mathbf{x}_2 - \lambda_0 \mathbf{x}_2 = \mathbf{x}_1 \quad &\to \quad (A - \lambda_0 I)^3 \mathbf{x}_2 = 0 \\ \cdots \quad & \quad\quad \cdots \\ A\mathbf{x}_k - \lambda_0 \mathbf{x}_k = \mathbf{x}_{k-1} \quad &\to \quad (A - \lambda_0 I)^{k+1} \mathbf{x}_k = 0. \end{aligned} \tag{2.7}$$

The vectors in a Jordan chain are linearly independent.

A transformation $A : \mathcal{C}^n \to \mathcal{C}^n$ (or an $n \times n$ matrix A) is called diagonable or diagonalizable if there exist eigenvectors $\mathbf{x}_1, \ldots, \mathbf{x}_n$ of A that form a basis in $\mathcal{C}^n$. Equivalently, an $n \times n$ matrix A is diagonable if for some nonsingular matrix S the matrix $S^{-1}AS$ has a diagonal form:

$$S^{-1}AS = \mathrm{diag}\,[\lambda_1\ \lambda_2\ \cdots\ \lambda_n]. \tag{2.8}$$

So a diagonable $n \times n$ matrix has n Jordan blocks in its Jordan form and each block is of size 1 (some or all of the eigenvalues λ_i may coincide).

Let $\mathcal{M} \subset \mathcal{X}$ be invariant with respect to the operator A and let $\mathcal{N}$ be a complementary subspace (not necessarily orthogonal) such that $\mathcal{M} \oplus \mathcal{N} = \mathcal{X}$. In the basis $\{\mathcal{M}, \mathcal{N}\}$ the matrix of A assumes the form

$$A = \begin{bmatrix} A_{11} & A_{12} \\ 0 & A_{22} \end{bmatrix} \quad \text{acting on} \quad \mathcal{X} = \begin{bmatrix} \mathcal{M} \\ \mathcal{N} \end{bmatrix}. \tag{2.9}$$

If $\mathcal{N}$ is invariant with respect to A, the matrix of A assumes the form

$$A = \begin{bmatrix} A_{11} & 0 \\ A_{21} & A_{22} \end{bmatrix} \quad \text{acting on} \quad \mathcal{X} = \begin{bmatrix} \mathcal{M} \\ \mathcal{N} \end{bmatrix}. \tag{2.10}$$

$\mathcal{M}$ is said to be reducing if the subspace $\mathcal{N}$ is also A invariant, so that $A_{12} = A_{21} = 0$. Then the subspaces $\mathcal{M}$ and $\mathcal{N}$ are called reducing subspaces, and they yield a decomposition of A into a direct sum, i.e., a representation as a generalized block diagonal matrix

$$A \leftrightarrow \begin{bmatrix} A_1 & 0 \\ 0 & A_2 \end{bmatrix} \tag{2.11}$$

where A_1 and A_2 are the matrix representations on the complementary invariant subspaces.

This procedure can be continued to a complete disjoint system of invariant reducing subspaces

$$\mathcal{X} = \mathcal{M}_1 \oplus \mathcal{M}_2 \oplus \cdots \oplus \mathcal{M}_r \tag{2.12}$$

and the decomposition of A into a direct sum of operators which are parts of A,

$$A = A_1 \oplus A_2 \oplus \cdots \oplus A_r. \tag{2.13}$$

A reducing subspace $\mathcal{M}$ of A is said to be irreducible with respect to A if there is no nontrivial reducing subspace of $A\mathcal{M}$. A central problem in the investigation of the structure of A is the search for a complete disjoint system of irreducible invariant reducing subspaces of A and the study of the structure of A restricted to these irreducible subspaces. The problem can be solved by the spectral representation and the Jordan representation of A.

We need to extend the analysis to the case where the transformation A depends analytically on a complex parameter, $A = A(\alpha)$, i.e., we consider analytic families of transformations and analytic families of their invariant subspaces.

A very simple example is provided by the Fourier symbol matrix function for diffraction at two right-angle staggered parallel Dirichlet half-planes separated by the plate distance 2b, in which case

$$A(\alpha) = \begin{bmatrix} 1 & \exp[2ib\kappa(\alpha)] \\ \exp[2ib\kappa(\alpha)] & 1 \end{bmatrix} \tag{2.14}$$

with $\kappa(\alpha) = \sqrt{k^2 - \alpha^2}$. A similarity transformation with a constant matrix C leads to

$$CA(\alpha)C^{-1} = \begin{bmatrix} 1 + \exp[2ib\kappa(\alpha)] & 0 \\ 0 & 1 - \exp[2ib\kappa(\alpha)] \end{bmatrix}, C = \frac{1}{\sqrt{2}} \begin{bmatrix} 1 & 1 \\ -1 & 1 \end{bmatrix}. \tag{2.15}$$

Unfortunately the general situation is much more complicated and is far from having a solution even for 2×2 matrix functions. Also the thoroughly studied Daniele-Khrapkov class (Heins [16], Daniele [10], [11], [12], Khrapkov [22], Câmara et al. [7], [8], Meister and Speck [32], [33], dos Santos et al. [13]) cannot be factorized using its spectral properties and Jordan structure.

One says that $A(\alpha)$: $\mathcal{C}^n \to \mathcal{C}^n$ is an analytic family in the complex plane Ω where Ω denotes the entire closed complex α plane $\mathcal{C} \cup \{\infty\}$ if in a neighbourhood U_{α_0} of each point $\alpha_0 \in \mathcal{C}$ the transformation $A(\alpha)$ admits a representation as a power series

$$A(\alpha) = \sum_{j=0}^{\infty} A_j (\alpha - \alpha_0)^j, \quad \alpha \in U_{\alpha_0} \tag{2.16}$$

where $A_0, A_1, \ldots$ are transformations from $\mathcal{C}^n \to \mathcal{C}^n$. If α_0 lies at infinity the transformation $\widetilde{A}(\zeta) = A(1/\zeta)$ at $\zeta = 0$ is considered.

For the definition of analytic families of subspaces we refer to Gohberg et al. [14]. In particular if for every $\alpha_0 \in \Omega$ the vectors $\mathbf{x}_1(\alpha_0), \ldots, \mathbf{x}_p(\alpha_0)$ are linearly independent, then

$$\text{Span}\{\mathbf{x}_1(\alpha), \ldots, \mathbf{x}_p(\alpha)\}, \quad \alpha \in \Omega \tag{2.17}$$

is an analytic family of subspaces. Also $\text{Ker}A(\alpha)$ and $\text{Im}A(\alpha)$ are analytic families of subspaces if $\dim \text{Ker}A(\alpha)$ is constant, i.e., if the rank of $A(\alpha)$ or, equivalently, the dimension of $\text{Im}A(\alpha)$ is constant.

Generally, without the extra assumption that the dimension of $\text{Ker}A(\alpha)$ is independent of α, the families of subspaces $\text{Ker}A(\alpha)$ and $\text{Im}A(\alpha)$, where $A(\alpha)$: $\mathcal{C}^n \to \mathcal{C}^n$ is an analytic family on Ω, are not analytic on Ω.

By definition the singular set $S(A)$ of $A(\alpha)$ is the set of all $\alpha_0 \in \Omega$ for which

$$\text{rank}\, A(\alpha_0) < \max_{\alpha\in\Omega}\text{rank}\, A(\alpha). \tag{2.18}$$

The singular set $S(A)$ is discrete; cf. Gohberg et al. [14].

For a transformation X: $\mathcal{C}^n \to \mathcal{C}^n$ we denote by $\nu(X)$ the number of distinct eigenvalues of X. Obviously, $1 \leq \nu(X) \leq n$. We cite from Gohberg et al. [14]:

Let $A(\alpha) : \mathcal{C}^n \to \mathcal{C}^n$ be an analytic family of transformations on Ω. Then for all $\alpha \in \Omega$ except on a discrete set S_0 we have

$$\nu(A(\alpha)) = \max_{\alpha\in\Omega} \nu(A(\alpha)); \tag{2.19}$$

for $\alpha_0 \in S_0$ we have

$$\nu(A(\alpha_0)) < \max_{\alpha\in\Omega} \nu(A(\alpha)). \tag{2.20}$$

The points from S_0 will be called the *multiple points* of the analytic family of transformations $A(\alpha)$, because at these points the eigenvalues of $A(\alpha)$ attain higher multiplicity than "usual".

Many of the theoretical results we apply for the solution of some problems in diffraction theory in the next sections can be found in two monographs (in particular the second one), namely,

- Baumgärtel, H., *Operator Theory: Advances and Applications, vol. 15, Analytic Perturbation Theory for Matrices and Operators*, Birkhäuser, Boston, Cambridge, Mass., 1985, [1];
- Gohberg, I., P. Lancaster, and L. Rodman, *Invariant Subspaces of Matrices with Applications*, John Wiley, New York, 1986, [14].

The full theory of invariant subspaces and analytical perturbation theory, however, is not needed for the following simple examples.

3. The two-face impedance half-plane problem

Let a two-face impedance half-plane be given by $\{x \geq 0,\, y = 0, -\infty < z < \infty\}$. We assume the following conditions for the electric and magnetic fields ($e^{-i\omega t}$ time dependence). The homogeneous isotropic impedance (Leontovich) boundary conditions (cf. Leontovich [25], Senior and Volakis [53] or Hoppe and Rahmat-Samii [18]), in vectorial form, read

$$\mathbf{E} - (\mathbf{n}\cdot\mathbf{E})\mathbf{n} = \eta^{\pm} Z\, \mathbf{n}\times\mathbf{H} \qquad y = \pm 0,\, x > 0, \tag{3.1}$$

where $\mathbf{n}$ is the outward unit normal. The plus and minus superscripts denote the upper and lower faces of the half-plane and the complementary screen. The $\eta^{\pm}$ are the surface impedances relative to the intrinsic impedance Z of free space.

Due to the edge condition the components E_z and H_z parallel to the edge behave like $A + B\rho^{1/2}$ as $\rho \to 0$, where ρ is the distance from the field point to the edge and A and B are material constants, whereas the field components E_x, E_y and H_x, H_y, perpendicular (normal) to the diffraction edge, grow as $\rho^{-1/2}$ as $\rho \to 0$ and thus have higher-order edge singularities. The Sommerfeld radiation condition ensures that the diffracted fields propagate outward from the diffraction screen and vanish at infinity.

The incident field $\mathbf{E}_i$ is assumed to be a plane electromagnetic wave with wave vector $\mathbf{k} = k\hat{\mathbf{k}}$ ($k = \omega\sqrt{\mu\epsilon}$) and amplitude vector $\mathbf{E}^0$:

$$\mathbf{E}_i = \mathbf{E}^0\, e^{i\mathbf{k}\cdot\mathbf{r}},\ \mathbf{H}_i = \mathbf{H}^0\, e^{i\mathbf{k}\cdot\mathbf{r}} \qquad \mathbf{k} \perp \mathbf{E}^0, \mathbf{H}^0, \tag{3.2}$$

where the unit vector $\hat{\mathbf{k}}$, which specifies the direction of incidence, is given by

$$\hat{\mathbf{k}} = k_x\hat{\mathbf{x}} + k_y\hat{\mathbf{y}} + k_z\hat{\mathbf{z}} = -\sin\theta_i\cos\phi_i\,\hat{\mathbf{x}} - \sin\theta_i\sin\phi_i\,\hat{\mathbf{y}} + \cos\theta_i\,\hat{\mathbf{z}}. \tag{3.3}$$

For the sake of definiteness we assume $0 \leq \phi_i,\, \theta_i \leq \pi$. The z component $k_z = \cos\theta_i$ is called the obliqueness or obliquity parameter. For normal incidence, $\theta_i = \pi/2$, and hence $k_z = 0$.

Maxwell's equations and the zero-divergence relations imply the following scalar boundary conditions NH and NE for the **normal** field components $H_y^{\pm}$ and $E_y^{\pm}$ on the screen:

$$\text{NH} \qquad \frac{\partial H_y^{\pm}}{\partial y} \pm \frac{ik}{\eta^{\pm}} H_y^{\pm} = 0 \qquad x > 0,\ y = \pm 0 \tag{3.4}$$

$$\text{NE} \qquad \frac{\partial E_y^{\pm}}{\partial y} \pm ik\eta^{\pm} E_y^{\pm} = 0 \qquad x > 0,\ y = \pm 0. \tag{3.5}$$

These boundary conditions for the normal field components $H_y^{\pm}$, $E_y^{\pm}$ are linear combinations of the tangential derivatives of the tangential field components $H_{x,z}^{\pm}$, $E_{x,z}^{\pm}$ and are always decoupled independently of the form of the incident wave. This fact was implicitly noted already by Grünberg [15] in 1943 and demonstrated by Williams [57] and Senior [45] in 1960. Their approach, i.e., the use of normal rather than tangential field components, was applied extensively later on by all authors dealing with problems of diffraction of obliquely incident plane waves by half-plane structures and half-plane junctions for a large variety of boundary conditions. The normal field components, however, can be derived from the tangential field components only outside the complex characteristics $\partial/\partial x = \pm i\partial/\partial z$ corresponding to two discrete "transition" points in the complex Fourier transform domain.

For a normally incident plane wave with $k_z = 0$ the electric vector can always be decomposed into the sum of two components, one of which is parallel to the diffraction edge (E polarization) whereas the other is perpendicular to it (**H** polarization). For the E polarization case the field quantities H_z and E_x, E_y vanish, and we obtain

$$E_{||} \qquad \frac{\partial E_z^{\pm}}{\partial y} \pm \frac{ik}{\eta^{\pm}} E_z^{\pm} = 0, \quad \frac{\partial H_y^{\pm}}{\partial y} \pm \frac{ik}{\eta^{\pm}} H_y^{\pm} = 0. \tag{3.6}$$

For the H polarization case the field quantities E_z and H_x, H_y vanish. Hence

$$H_{||} \qquad \frac{\partial H_z^{\pm}}{\partial y} \pm ik\eta^{\pm} H_z^{\pm} = 0, \quad \frac{\partial E_y^{\pm}}{\partial y} \pm ik\eta^{\pm} E_y^{\pm} = 0. \tag{3.7}$$

Thus, for normal incidence, these are two boundary conditions of the same type for E polarization with E_z and H_y and for H polarization with H_z and E_y. It must be realized, however, that there exist different edge conditions for the same type of boundary conditions

$$E_z, H_z \sim \rho^{1/2}, \quad H_y, E_y \sim \rho^{-1/2} \qquad \rho \to 0 \tag{3.8}$$

with corresponding behavior of their Fourier transforms at infinity. The higher-order edge singularity going from E_z to H_y and from H_z to E_y is due to taking the derivative with respect to x. All equations for E_z, H_y, H_z, and E_y are of the common type

$$\frac{\partial \psi^{\pm}}{\partial y} \pm is_{\pm}\psi^{\pm} = 0 \qquad x > 0,\ y = \pm 0. \tag{3.9}$$

This equation was first solved by Hurd [19] in 1976 by the Wiener-Hopf-Hilbert technique assuming that $\psi^{\pm} \sim \rho^{1/2}$. The result can be directly applied to the first of equations $E_{||}$ and $H_{||}$ but to the second ones only after some modification because of the higher-order edge singularity of the normal field components.

The scattered fields $\mathbf{E}_s$ and $\mathbf{H}_s$ can conveniently be expressed in terms of electric and magnetic Hertz vectors. Introducing Fourier representations with the square root function $\Gamma(\alpha)$ defined in the complex α plane as

$$\Gamma(\alpha) = \sqrt{\chi^2 - \alpha^2}, \quad \chi^2 = k^2 - k_z^2, \quad \operatorname{Im}\chi \geq 0 \tag{3.10}$$

such that $\Gamma(0) = +\chi$, the reduced propagation constant, we arrive at the following Wiener-Hopf system for the tangential electric field components

$$\mathcal{G}(\alpha)\boldsymbol{\mathcal{L}}(\alpha) = \boldsymbol{\mathcal{U}}(\alpha) + \boldsymbol{\mathcal{F}}(\alpha) \tag{3.11}$$

with the characteristic 4×4 symbol matrix function

$$\mathcal{G}(\alpha) = \frac{1}{2} \begin{bmatrix} I + \frac{\eta^+}{k\Gamma(\alpha)} M(\alpha) & \eta^+ I + k\Gamma(\alpha) M^{-1}(\alpha) \\ -\left(I + \frac{\eta^-}{k\Gamma(\alpha)} M(\alpha)\right) & \eta^- I + k\Gamma(\alpha) M^{-1}(\alpha) \end{bmatrix} \tag{3.12}$$

and with

$$\boldsymbol{\mathcal{U}}(\alpha) = \begin{bmatrix} \mathbf{U}_1 \\ \mathbf{U}_2 \end{bmatrix}, \ \mathcal{L}(\alpha) = \begin{bmatrix} \mathbf{L}_1 \\ \mathbf{L}_2 \end{bmatrix}, \ \boldsymbol{\mathcal{F}}(\alpha) = \begin{bmatrix} \mathbf{F}_1 \\ \mathbf{F}_2 \end{bmatrix}. \tag{3.13}$$

$\mathcal{G}(\alpha)$ and $\boldsymbol{\mathcal{F}}(\alpha)$ are known whereas $\mathbf{U}_j(\alpha)$ and $\mathbf{L}_j(\alpha)$ $(j = 1, 2)$ are unknown 2-column vectors with elements regular in the upper α half-plane $H^U = \{\alpha \,|\mathrm{Im}\,(\alpha) > -\mathrm{Im}\,(\chi)\}$ and in the lower α half-plane $H^L = \{\alpha \,|\mathrm{Im}\,(\alpha) < \mathrm{Im}\,(k_x)\}$, respectively.

The matrix $M(\alpha)$ is the coupling matrix between the Fourier transforms of the tangential components of the scattered electric and magnetic vectors field and reads

$$M(\alpha) = \begin{bmatrix} \chi^2 & \alpha k_z \\ \alpha k_z & k^2 - \alpha^2 \end{bmatrix}, \ M^{-1}(\alpha) = \frac{1}{k^2\,\Gamma^2(\alpha)} \begin{bmatrix} k^2 - \alpha^2 & -\alpha k_z \\ -\alpha k_z & \chi^2 \end{bmatrix} \tag{3.14}$$

and acts on the two-dimensional complex vector space $\mathcal{C}^2$ of field components. $M(\alpha)$ and $M^{-1}(\alpha)$ are symmetric **polynomial** matrices and form analytic families of $\mathcal{C}^2 \to \mathcal{C}^2$ transformations in the entire complex α plane Ω and in the set $\Omega \setminus \{\pm\chi\}$, punctured at $\alpha = \pm\chi$, respectively. $M(\alpha)$ is singular at the branch points $\pm\chi$ of $\Gamma(\alpha)$ because $\det M(\alpha) = k^2\,\Gamma^2(\alpha)$. The coupling matrix becomes diagonal for $k_z = 0$, implying the independence of the $E_{||}$ and $H_{||}$ cases for normal incidence.

For the eigenvector/eigenvalue analysis of the coupling matrix $M(\alpha)$ we introduce left row eigenvectors of $M(\alpha)$ by

$$\mathbf{z}_1(\alpha)M(\alpha) = \lambda_1(\alpha)\mathbf{z}_1(\alpha) = k^2\mathbf{z}_1(\alpha), \tag{3.15}$$
$$\mathbf{z}_2(\alpha)M(\alpha) = \lambda_2(\alpha)\mathbf{z}_2(\alpha) = \Gamma^2(\alpha)\mathbf{z}_2(\alpha), \tag{3.16}$$

$$\mathbf{z}_1(\alpha) = [\,\alpha,\ k_z],\qquad \lambda_1(\alpha) = k^2, \tag{3.17}$$
$$\mathbf{z}_2(\alpha) = [\,-k_z,\ \alpha\,],\quad \lambda_2(\alpha) = \Gamma^2(\alpha). \tag{3.18}$$

The transformation matrix of the eigenvectors is given by

$$T(\alpha) = \begin{bmatrix} z_1(\alpha) \\ z_2(\alpha) \end{bmatrix} = \begin{bmatrix} \alpha & k_z \\ -k_z & \alpha \end{bmatrix}. \tag{3.19}$$

The eigenvalues are polynomials in α; they coincide for $\alpha = \pm ik_z$: $\lambda_1 = k^2 = \lambda_2(\pm ik_z)$. Thus

$$T(\alpha)M(\alpha) = D(\alpha)\,T(\alpha) \tag{3.20}$$

for all values of α, where

$$D(\alpha) = \begin{bmatrix} \lambda_1(\alpha) & 0 \\ 0 & \lambda_2(\alpha) \end{bmatrix} = \begin{bmatrix} k^2 & 0 \\ 0 & \Gamma^2(\alpha) \end{bmatrix} \tag{3.21}$$

is the matrix of the eigenvalues of $M(\alpha)$. Equation (3.20) is valid for all values of $\alpha \in \Omega$. It is, however, a similarity transformation only for $\det T(\alpha) \neq 0$, i.e., $\alpha \neq \pm ik_z$.

The normalized eigenvectors and eigenvalues of the mapping $T(\alpha)$: $\mathcal{C}^2 \to \mathcal{C}^2$ themselves satisfy the relations

$$T(\alpha)\,\mathbf{x}_i = \gamma_i(\alpha)\,\mathbf{x}_i \qquad i = 1, 2 \tag{3.22}$$

and are given by (an asterisk denotes complex conjugation)

$$\mathbf{x}_1 = \frac{1}{\sqrt{2}}\begin{bmatrix} 1 \\ -i \end{bmatrix}, \quad \mathbf{x}_2 = \frac{1}{\sqrt{2}}\begin{bmatrix} 1 \\ i \end{bmatrix} = \mathbf{x}_1^*, \tag{3.23}$$

$$\left.\begin{aligned} \gamma_1(\alpha) &= \alpha - ik_z \\ \gamma_2(\alpha) &= \alpha + ik_z \end{aligned}\right\} \gamma_1(\alpha)\gamma_2(\alpha) = \alpha^2 + k_z^2. \tag{3.24}$$

The orthogonal eigenvectors $\mathbf{x}_1$ and $\mathbf{x}_2$ are independent of α and span the entire space $\mathcal{C}^2$.

Hence the coupling matrices $M(\alpha)$ can be diagonalized by similarity transformations with the transformation matrices $T(\alpha)$ except for the discrete singular set (cf. Gohberg et al. [14]) defined as

$$S(T) = \{\alpha_s | \operatorname{rank} T(\alpha_s) < \max \operatorname{rank} T(\alpha)\} = \{\alpha_1, \alpha_2\} = \{\pm ik_z\} \tag{3.25}$$

when $T(\alpha)$ is singular. This result can also be found by using Rawlins' criteria ([40], 1992) for polynomial diagonalization.

The points $\alpha_1 = ik_z$ and $\alpha_2 = -ik_z$ of the singular set are related to the obliqueness parameter k_z. They coincide for $k_z = 0$. It is remarkable (cf. [14], 1986) that the singular set $S(T)$ coincides with the set of multiple points S_0 of the matrix $M(\alpha)$ defined as

$$S_0 = \{\alpha_0 | \nu[M(\alpha_0)] < \max \nu[M(\alpha)]\}, \tag{3.26}$$

where $\nu(M)$ denotes the number of distinct eigenvalues of the matrix M. At $\alpha_1 = ik_z$ and at $\alpha_2 = -ik_z \in S(T)$ the matrix $M(\alpha)$ assumes the forms

$$M(\pm ik_z) = \begin{bmatrix} k^2 - k_z^2 & \pm ik_z^2 \\ \\ \pm ik_z^2 & k^2 + k_z^2 \end{bmatrix} = M^\dagger(\mp ik_z), \tag{3.27}$$

where the dagger means Hermitian conjugation. These matrices are nonsingular. Their twofold eigenvalues coincide with k^2 as their common value. The matrix $M(\pm ik_z)$ is degenerated (nonderogatory) since the algebraic multiplicity is 2 whereas the geometric multiplicity is only 1. The only eigenvector of $M(ik_z)$ is $\mathbf{x}_1$ and the only eigenvector of $M(-ik_z)$ is $\mathbf{x}_2$,

$$M(ik_z)\mathbf{x}_1 = k^2\mathbf{x}_1, \quad M(-ik_z)\mathbf{x}_2 = k^2\mathbf{x}_2. \tag{3.28}$$

Hence there are not enough linearly independent eigenvectors to diagonalize this matrix. This implies that the matrix $M(\alpha)$ cannot be diagonalized by (nonsingular) similarity transformations at $\alpha = \pm ik_z$. It can be reduced to its simplest form, which is its Jordan form, by the Jordan chain $\{\mathbf{x}_1, \mathbf{x}_2\}$ where $\mathbf{x}_2$ is a generalized eigenvector of $\mathbf{x}_1$ belonging to the eigenvalue k^2 of $M(ik_z)$ and $\mathbf{x}_1$ is a generalized eigenvector of $\mathbf{x}_2$ belonging to the same eigenvalue k^2 of $M(-ik_z)$,

$$M(ik_z)\mathbf{x}_2 - k^2\mathbf{x}_2 = -2k_z^2\mathbf{x}_1, \tag{3.29}$$

$$M(-ik_z)\mathbf{x}_1 - k^2\mathbf{x}_1 = -2k_z^2\mathbf{x}_2. \tag{3.30}$$

The eigenvector $\mathbf{x}_1$ ($\mathbf{x}_2$) and the generalized eigenvector $\mathbf{x}_2$ ($\mathbf{x}_1$) are always orthogonal. For $k_z = 0$ the right-hand side of the generalized eigenvector equation vanishes, and the generalized eigenvector becomes a second independent true eigenvector.

The Jordan form of the matrices $M(ik_z)$, i.e., the form that these matrices assume in the basis $\{\mathbf{x}_1, \mathbf{x}_2\}$, reads

$$X^{-1}M(ik_z)X = \begin{bmatrix} k^2 & -2k_z^2 \\ 0 & k^2 \end{bmatrix}, \quad X = [\mathbf{x}_1 \;\; \mathbf{x}_2]. \tag{3.31}$$

Taking the complex conjugate leads to the Jordan form of $M(-ik_z)$. Only for $k_z = 0$ this Jordan form is diagonal as a multiple of the unit matrix. Hence the analytic family of transformations $M(\alpha)$: $\mathcal{C}^2 \to \mathcal{C}^2$ does not have a fixed Jordan structure.

At $\alpha \in S(T)$ the matrix $T(\alpha)$ can be written in the outer form as rank 1 matrices,

$$T(ik_z) = 2ik_z\mathbf{x}_2\mathbf{x}_2^\dagger, \quad T(-ik_z) = -2ik_z\mathbf{x}_1\mathbf{x}_1^\dagger. \tag{3.32}$$

For all values of $\alpha \in \Omega$, including the singular set (set of multiple points), the matrices $M(\alpha)$ and $M^{-1}(\alpha)$ can be transformed to

$$T(\alpha)M(\alpha) = D(\alpha)T(\alpha), \quad T(\alpha)M^{-1}(\alpha) = D^{-1}(\alpha)T(\alpha). \tag{3.33}$$

For $M^{-1}(\alpha)$ the branch points $\pm\chi$ have to be excluded. These are similarity transformations, however, only for $\alpha \neq \{\pm ik_z\}$.

The transformation matrices $T(\alpha)$ have the properties

$$\text{Im}T(\alpha) = \begin{cases} \text{span}\,\{\mathbf{x}_1, \mathbf{x}_2\} \\ \text{span}\,\{\mathbf{x}_2\} \\ \text{span}\,\{\mathbf{x}_1\} \end{cases}; \quad \text{dimIm}T(\alpha) = \begin{cases} 2 & \alpha \notin S(T) \\ 1 & \alpha = ik_z \\ 1 & \alpha = -ik_z \end{cases} \tag{3.34}$$

and, correspondingly,

$$\text{Ker}T(\alpha) = \begin{cases} \{0\} \\ \text{span}\,\{\mathbf{x}_1\} \\ \text{span}\,\{\mathbf{x}_2\} \end{cases}; \quad \text{dimKer}T(\alpha) = \begin{cases} 0 & \alpha \notin S(T) \\ 1 & \alpha = ik_z \\ 1 & \alpha = -ik_z \end{cases}, \tag{3.35}$$

where Im denotes the image (range) and Ker denotes the kernel of the transformation. As always, for all $\alpha \in \Omega$

$$\text{dimKer}T(\alpha) + \text{dimIm}T(\alpha) = \text{dim}\mathcal{C}^2 = 2. \tag{3.36}$$

Since the dimensions of $\text{Im}T(\alpha)$ and $\text{Ker}T(\alpha)$ are not constant but depend on α, the families of subspaces $\text{Im}T(\alpha)$ and $\text{Ker}T(\alpha)$ are not analytic on the complex plane but only on the restricted open domain $\Omega \backslash S(T)$ (cf. [14]). This restricted range for the Fourier transformation of the normal components is intimately related to the fact that their boundary conditions are derived from the (unrestricted) boundary conditions for the tangential field components by tangential differentiation. This restricted domain, however, is the domain for inversion or transformation from the normal to the tangential field components that require the full two-dimensional complex vector space $\mathcal{C}^2$.

Banach's theorem states that every bounded linear operator, say $F\colon X \to Y$, that maps one Banach space X onto another Banach space Y and is one-to-one has a bounded inverse. A bounded linear operator is left invertible (by a bounded linear operator) if and only if it is injective and has complemented range; it is right invertible, if and only if it is surjective and has complemented kernel. In the present case the mapping into the space $\mathcal{C}^2$ is not onto (not surjective), and hence the mapping is not invertible. If Banach's theorem, nevertheless, is assumed to be applicable as has been done in this context ever since the first publications by W.E. Williams and T.B.A. Senior in 1959–1960, the solutions obtained must be considered with caution. Indeed, the final expressions for the solution of the boundary value problem contain poles that produce nonphysical waves growing at infinity (leaky waves). These pole contributions can be eliminated by an ad hoc procedure by determining some undetermined constants (that arise in the Wiener-Hopf solution because of the higher-order edge singularities of the normal field components) requesting the vanishing of the corresponding residues by the residue theorem. This means that Sommerfeld's radiation condition has to be invoked again although it was completely incorporated into the analysis right at the beginning by choosing appropriate Green functions. Having incorporated the radiation condition, the remaining steps for obtaining the solution should be dictated by the rules of mathematics and logic only.

The difficulty just pointed out can be overcome by redefining the range of the mapping $T(\alpha)$ at the critical values $\alpha = \pm ik_z \in S(T)$ by restrictions (constraints). This can be done effectively in the following way.

Let $\mathbf{F}(\alpha) \in \mathcal{C}^2$ be an arbitrary two-component analytic vector in the range of $T(\alpha)$ given in the Cartesian basis $\boldsymbol{\mathcal{E}} = \{\mathbf{e}_1, \mathbf{e}_2\}$ and in the orthonormal basis $\mathcal{B} = \{\mathbf{x}_1, \mathbf{x}_2\}$ of (3.23) as

$$\mathbf{F}(\alpha) = f_1(\alpha)\mathbf{e}_1 + f_2(\alpha)\mathbf{e}_2 = g_1(\alpha)\mathbf{x}_1 + g_2(\alpha)\mathbf{x}_2. \tag{3.37}$$

Using the vector representations in the basis $\boldsymbol{\mathcal{E}}$,

$$\mathbf{e}_1 = \begin{bmatrix} 1 \\ 0 \end{bmatrix}, \quad \mathbf{e}_2 = \begin{bmatrix} 0 \\ 1 \end{bmatrix}, \tag{3.38}$$

$$\mathbf{x}_1 = \frac{1}{\sqrt{2}}(\mathbf{e}_1 - i\mathbf{e}_2) = \frac{1}{\sqrt{2}}\begin{bmatrix} 1 \\ -i \end{bmatrix}, \quad \mathbf{x}_2 = \frac{1}{\sqrt{2}}(\mathbf{e}_1 + i\mathbf{e}_2) = \frac{1}{\sqrt{2}}\begin{bmatrix} 1 \\ i \end{bmatrix}, \tag{3.39}$$

a comparison yields

$$g_1(\alpha) = \frac{1}{\sqrt{2}}[f_1(\alpha) + if_2(\alpha)], \quad g_2(\alpha) = \frac{1}{\sqrt{2}}[f_1(\alpha) - if_2(\alpha)]. \tag{3.40}$$

The restricted range of $T(\alpha)$ at $\alpha = ik_z$ is span$\{\mathbf{x}_2\}$; hence to have the full range span $\{\mathbf{x}_1, \mathbf{x}_2\}$ we require $g_1(ik_z) = 0$ implying

$$f_1(ik_z) + if_2(ik_z) = 0 \quad \Rightarrow \quad [1\ ,\ i\,]\,\mathbf{f}(ik_z) = 0. \tag{3.41}$$

Similarly, the restricted range of $T(\alpha)$ for $\alpha = -ik_z$ is span$\{\mathbf{x}_1\}$; hence we demand $g_2(-ik_z) = 0$ implying

$$f_1(-ik_z) - if_2(-ik_z) = 0 \Rightarrow [1, -i\,]\,\mathbf{f}(-ik_z) = 0. \tag{3.42}$$

Now, with these constraints the range of $T(\alpha)$ is span$\{\mathbf{x}_1, \mathbf{x}_2\}$ for all values of α. The subspace of $\mathcal{C}^2$ that consists of the points that satisfy the constraint equations (3.41) and (3.42) will be denoted as $\mathcal{C}_0^2$.

The solution of the Wiener-Hopf system (3.11) requires the factorization of the matrix function $\mathcal{G}(\alpha)$ in the form

$$\mathcal{G}(\alpha) = \mathcal{G}^U(\alpha)\mathcal{G}^L(\alpha), \tag{3.43}$$

where $\mathcal{G}^U(\alpha)$ and $[\mathcal{G}^U(\alpha)]^{-1}$ are holomorphic in the upper α half-plane H^U and $\mathcal{G}^L(\alpha)$ and $[\mathcal{G}^L(\alpha)]^{-1}$ are holomorphic in the lower α half-plane H^L with algebraic increase at infinity. With the known techniques it is not possible to factorize even a 2×2 matrix function in the general case, and much less is known for 4×4 matrices. Since this diffraction problem contains the TE and TM cases for normal incidence with 2×2 symbol matrices of the Daniele-Khrapkov form, the most that can be expected is reduction of the Fourier symbol matrix to block diagonal form.

The application of polynomial similarity transformations uses the fact that the 4×4 matrix function (3.12) is composed of 2×2 unit matrices and the matrix functions $M(\alpha)$ and $M^{-1}(\alpha)$ (matrix polynomials and their inverses), which can be diagonalized by polynomial similarity almost everywhere in the complex α plane. The invariance of the unit matrices with respect to similarity is the main reason for applying similarity transformations rather than using matrix polynomials with constant nonzero determinants in order to reduce the matrix polynomial to its Smith form. Reduction to Smith form, however, has been proposed to play a major role in the 1984 Wiener-Hopf solution of the Daniele-Khrapkov type symbol matrix function for the two-face impedance half-plane problem for normal incidence by Lüneburg and Hurd [27].

The reduction and factorization are done explicitly with respect to a contour like the real axis or with respect to a strip of analyticity like $H = H^L \cap H^U$ in the complex α plane and then extended by analytical continuation to the half-planes H^L and H^U, respectively. It is necessary that the reduction to block diagonal form should be possible in the entire complex plane Ω. In the present case this reduction is not always possible due to the Jordan structure of the symbol matrix. For a general account of the relevant analytic perturbation theory for matrices we refer to Baumgärtel [1].

The matrix function $\mathcal{G}(\alpha)$ of (3.12) can be reduced to block diagonal form with two 2×2 block matrices of the Daniele-Khrapkov form by a polynomial (nonsingular) similarity transformation and a permutation (reordering) similarity transformation only for values of $\alpha \notin S(T)$ or $\alpha \neq \pm ik_z$. However, with the help of relation (3.33), which is valid for all $\alpha \in \Omega$, the matrix function $\mathcal{G}(\alpha)$ can be transformed to the form

$$\mathcal{T}(\alpha)\mathcal{G}(\alpha) =: \widehat{\mathcal{G}}(\alpha)\mathcal{T}(\alpha), \tag{3.44}$$

where $\mathcal{T}(\alpha) = I \otimes T(\alpha) = T(\alpha) \oplus T(\alpha)$. Here the circled times denotes the tensor or Kronecker product and the circled plus the direct sum (see [17]). Explicitly, $\widehat{\mathcal{G}}(\alpha)$ assumes the form of the matrix function $\mathcal{G}(\alpha)$ on the left-hand side of the preceding equation with $M(\alpha)$ replaced by the diagonal matrix $D(\alpha)$ and $M^{-1}(\alpha)$ replaced by $D^{-1}(\alpha)$. $\widehat{\mathcal{G}}(\alpha)$ is a sparse matrix function. The 8 matrix elements in the matrix $\mathcal{G}(\alpha)$ of (3.12), which are proportional to k_z, now vanish

$$\widehat{\mathcal{G}}(\alpha) = \frac{1}{2}\begin{bmatrix} 1 + \frac{\eta^+ k}{\Gamma(\alpha)} & 0 & \eta^+ + \frac{\Gamma(\alpha)}{k} & 0 \\ 0 & 1 + \frac{\eta^+\Gamma(\alpha)}{k} & 0 & \eta^+ + \frac{k}{\Gamma(\alpha)} \\ -(1 + \frac{\eta^- k}{\Gamma(\alpha)}) & 0 & \eta^- + \frac{\Gamma(\alpha)}{k} & 0 \\ 0 & -(1 + \frac{\eta^-\Gamma(\alpha)}{k}) & 0 & \eta^- + \frac{k}{\Gamma(\alpha)} \end{bmatrix}. \tag{3.45}$$

Relation (3.44) is valid for all $\alpha \in \Omega \setminus \{\pm\chi\}$, including the singular set $S(T)$, and defines the matrix function $\widehat{\mathcal{G}}(\alpha)$. It is a (nonsingular) similarity transformation, however, only when $\det T(\alpha) \neq 0$, i.e., for $\alpha \notin S(T)$.

The classical concept of the factorization of Wiener-Hopf symbol matrix functions (cf. for instance Noble [36] or Büyükaksoy and Serbest [5]) involves the reduction of the original matrix to diagonal block matrices of smaller sizes that allow explicit factorizations for all values of the variable α.

Algebraic increase of the factor matrices at infinity requires the determination of coefficients in a scalar or vectorial polynomial ansatz by other reasoning (see Noble [36], Chapter I.1.7 and Rawlins [40]). This occurs for poles in the upper or lower complex half-planes and ideas how to remove these undesired poles can be found in Câmara et al. [6], where the concept of "meromorphic factorization" is used.

The present problem, however, is not directly related to the appearance of poles in one or the other complex half-planes of the Wiener-Hopf factorization problem (although an incorrect treatment implies the occurrence of poles at the final stage of solution) but with the changing Jordan structure of the Fourier symbol matrix function in the complex domain. This kind of problem can occur only in the genuine matrix function case and not with scalar functions. The solution method implies that the factorization of the matrix function cannot be performed per se independently from the Fourier transform function spaces on which the factor matrix functions are acting: a complete function theoretic factorization is only possible if these function spaces are suitably restricted at the discrete singular set of the symbol matrix function in the complex domain.

A further similarity transformation with the permutation (reordering) matrix

$$\mathcal{P} = \begin{bmatrix} 1 & 0 & 0 & 0 \\ 0 & 0 & 1 & 0 \\ 0 & 1 & 0 & 0 \\ 0 & 0 & 0 & 1 \end{bmatrix} = \mathcal{P}^{-1} = \mathcal{P}^T \tag{3.46}$$

reduces the 4×4 matrix $\hat{\mathcal{G}}(\alpha)$ to block diagonal form

$$\widetilde{\mathcal{G}}(\alpha) := \mathcal{P}\widehat{\mathcal{G}}(\alpha)\mathcal{P} = \begin{bmatrix} \widetilde{G}_1(\alpha) & 0 \\ 0 & \widetilde{G}_2(\alpha) \end{bmatrix} \tag{3.47}$$

with the block diagonal matrices $\widetilde{G}_1(\alpha)$ and $\widetilde{G}_2(\alpha)$ given by

$$\widetilde{G}_1(\alpha) = \frac{1}{2} \begin{bmatrix} 1 + \frac{\eta^+ k}{\Gamma(\alpha)} & \eta^+ + \frac{\Gamma(\alpha)}{k} \\ -(1 + \frac{\eta^- k}{\Gamma(\alpha)}) & \eta^- + \frac{\Gamma(\alpha)}{k} \end{bmatrix}, \tag{3.48}$$

$$\widetilde{G}_2(\alpha) = \frac{1}{2} \begin{bmatrix} 1 + \frac{\eta^+ \Gamma(\alpha)}{k} & \eta^+ + \frac{k}{\Gamma(\alpha)} \\ -(1 + \frac{\eta^- \Gamma(\alpha)}{k}) & \eta^- + \frac{k}{\Gamma(\alpha)} \end{bmatrix}. \tag{3.49}$$

The 2×2 matrices $\widetilde{G}_1(\alpha)$ and $\widetilde{G}_2(\alpha)$ depend on the skewness parameter k_z only via the square root function $\Gamma(\alpha)$. They coincide for $\alpha \in S(T)$,

$$\widetilde{G}(\pm ik_z) := \widetilde{G}_1(\pm ik_z) = \widetilde{G}_2(\pm ik_z) = \tfrac{1}{2} \begin{bmatrix} 1 + \eta^+ & 1 + \eta^+ \\ -(1 + \eta^-) & 1 + \eta^- \end{bmatrix}, \tag{3.50}$$

and are completely independent of k_z. This emphasizes again the significance of the parameters $\alpha = \pm ik_z$ of the set $S(T)$ as transition points between the (generalized) transverse magnetic (TM) and transverse electric (TE) modes.

By multiplication from the left and right with appropriate nonsingular constant matrices, both block matrices $\widetilde{G}_1(\alpha)$ and $\widetilde{G}_2(\alpha)$ of (3.48) and (3.49) can be reduced to Hurd's canonical matrix ([19], 1976)

$$H(\alpha) = \frac{1}{2} \begin{bmatrix} s_1 + \Gamma(\alpha) & 1 + s_1/\Gamma(\alpha) \\ -(s_2 + \Gamma(\alpha)) & 1 + s_2/\Gamma(\alpha) \end{bmatrix}. \tag{3.51}$$

Thus the factorization of the matrix $\widetilde{\mathcal{G}}(\alpha)$ is reduced to the factorization of Hurd's canonical matrix $H(\alpha)$. Explicit solutions of this classical problem have been presented by various authors (Hurd [19], Hurd and Prezeździecki [20], Lüneburg and Hurd [27], Meister et al. [34]).

For equal impedances $\eta^+ = \eta^- \equiv \eta$ the generic matrix $H(\alpha)$ assumes a form that can be decoupled elementarily by pre-multiplication with a constant matrix:

$$H(\alpha) = \frac{1}{2}\begin{bmatrix} \eta+\Gamma(\alpha) & 1+\eta/\Gamma(\alpha) \\ -(\eta+\Gamma(\alpha)) & 1+\eta/\Gamma(\alpha) \end{bmatrix}$$

$$= \frac{1}{2}\begin{bmatrix} 1 & 1 \\ -1 & 1 \end{bmatrix}\begin{bmatrix} \eta+\Gamma(\alpha) & 0 \\ 0 & 1+\eta/\Gamma(\alpha) \end{bmatrix}, \tag{3.52}$$

leading to scalar Wiener-Hopf equations (cf. Senior [46, 47]).

The well-known Wiener-Hopf-Hilbert solution of the two-face impedance half-plane problem by Hurd and others relies on the assumption of normal incidence. The present work extends these results to the case of oblique incidence. It will be shown that the Wiener-Hopf equations for the TE and TM modes can be decoupled but that their solutions are coupled via their edge conditions.

Applying relations (3.44) and (3.47), the Wiener-Hopf equation (3.11) assumes the form

$$\widetilde{\mathcal{G}}(\alpha)\widetilde{\mathcal{L}}(\alpha) = \widetilde{\mathcal{U}}(\alpha) + \widetilde{\mathcal{F}}(\alpha), \qquad \alpha \in H, \tag{3.53}$$

for all values of $\alpha \in \Omega$ with

$$\widetilde{\mathcal{L}}(\alpha) = \mathcal{P}\mathcal{T}(\alpha)\mathcal{L}(\alpha) = \begin{bmatrix} \widetilde{\mathbf{L}}_1(\alpha), \\ \widetilde{\mathbf{L}}_2(\alpha) \end{bmatrix} \tag{3.54}$$

and similarly for $\widetilde{\mathcal{U}}(\alpha)$ and $\widetilde{\mathcal{F}}$.

We can assume without restriction of generality that $\mathrm{Im}\,(k_z) \geq 0$. Then the constraints on the solutions take the form

$$\{[1\,, \pm i] \otimes I_2\}\, \widetilde{\mathcal{L}}(\pm ik_z) = 0, \tag{3.55}$$

or more explicitly,

$$\widetilde{\mathbf{L}}_1(\pm ik_z) \pm i\widetilde{\mathbf{L}}_2(\pm ik_z) = 0. \tag{3.56}$$

Due to (3.50) and the fact that these restrictions are automatically satisfied for the excitation terms $\widetilde{\mathcal{F}}(\pm ik_z)$, the term $\widetilde{\mathcal{L}}(-ik_z)$ has been used rather than $\widetilde{\mathcal{U}}(-ik_z)$.

In the present matrix case the direct solution of the system (3.11) is not possible but only the solution of the modified system of equations (3.53). Here a first approach for the determination of the unknown constant four-vector is the conventional approach based on the use of the boundary conditions for the normal electric and magnetic field components and the subsequent elimination of artificial unphysical poles. The occurrence of poles is due to the fact that the normal field components are not completely independent but are coupled by means of range restrictions that involve simultaneously the normal electric and the magnetic field components. The range of the mapping from the tangential electric and magnetic field components to their normal components is not the entire complex domain $\mathcal{C}^2$ but is restricted. Hence this mapping is not (left and right) invertible.

A second approach makes explicit use of the range restrictions and is described in detail in the present contribution. This procedure completely clarifies the phenomenon of the occurrence of unphysical leaky wave poles during some intermediate stage of solution and of their elimination by the residue technique.

In analogy to (3.43), (3.44), and (3.47) we assume

$$\mathcal{PTG}^U(\alpha) = \mathcal{P}\widehat{\mathcal{G}}^U(\alpha)\mathcal{PPT} =: \widetilde{\mathcal{G}}^U(\alpha)\mathcal{PT}, \tag{3.57}$$

$$\mathcal{PTG}^L(\alpha) = \mathcal{P}\widehat{\mathcal{G}}^L(\alpha)\mathcal{PPT} =: \widetilde{\mathcal{G}}^L(\alpha)\mathcal{PT}. \tag{3.58}$$

Thus, for the factorization of the block diagonal matrix $\widetilde{G}(\alpha)$ of the Wiener-Hopf equation (3.53), we assume the form

$$\widetilde{\mathcal{G}}(\alpha) = \widetilde{\mathcal{G}}^U(\alpha)\widetilde{\mathcal{G}}^L(\alpha) \quad \text{with} \quad \widetilde{G}_i(\alpha) = \widetilde{G}_i^U(\alpha)\widetilde{G}_i^L(\alpha) \tag{3.59}$$

for $i = 1, 2$, where the factor matrices are defined by (3.57) and (3.58). This is formally a decomposition of the original 4×4 Wiener-Hopf system acting on $\mathcal{C}^4$ into two disjoint 2×2 systems acting, however, not on two separated direct sum $\mathcal{C}^2$-spaces but on the restricted direct sum space $\mathcal{C}_0^4 := (\mathcal{C}^2 \oplus \mathcal{C}^2)_0$, since the constraints (3.55) or (3.56) involve elements of both decomposing $\mathcal{C}^2$-spaces.

The transformations from $\mathcal{G}^{U,L}(\alpha)$ to $\widetilde{\mathcal{G}}^{U,L}(\alpha)$ are similarity transformations for all $\alpha \notin S(T)$ just as for $\mathcal{G}(\alpha)$ to $\widetilde{\mathcal{G}}(\alpha)$. Referring to (3.50) we furthermore assume that at the points $\pm ik_z$ of the singular set,

$$\widetilde{G}_1^{L,U}(\pm ik_z) = \widetilde{G}_2^{L,U}(\pm ik_z). \tag{3.60}$$

Using this factorization, the Wiener-Hopf equation (3.53) can be written in the form

$$\widetilde{\mathcal{G}}^L(\alpha)\widetilde{\mathcal{L}}(\alpha) = [\widetilde{\mathcal{G}}^U(\alpha)]^{-1}\widetilde{\mathcal{U}}(\alpha) + [\widetilde{\mathcal{G}}^U(\alpha)]^{-1}\widetilde{\mathcal{F}}(\alpha). \tag{3.61}$$

The block diagonal structure of the matrices $\widetilde{\mathcal{G}}^{L,U}(\alpha)$ and of their inverses implies that this equation actually consists of two decoupled 2-component vectorial functional equations

$$\widetilde{G}_i^L(\alpha)\widetilde{\mathbf{L}}_i(\alpha) = [\widetilde{G}_i^U(\alpha)]^{-1}\widetilde{\mathbf{U}}_i(\alpha) + [\widetilde{G}_i^U(\alpha)]^{-1}\widetilde{\mathbf{F}}_i(\alpha) \tag{3.62}$$

for $i = 1, 2$. Additive decomposition, analytical continuation, and application of Liouville's extended theorem (cf. Noble [36]) leads to the solution of the Wiener-Hopf equation in the form

$$\widetilde{\mathcal{L}}(\alpha) = [\widetilde{\mathcal{G}}^L(\alpha)]^{-1}\left\{\frac{[\widetilde{\mathcal{G}}^U(k_x)]^{-1}\widetilde{\mathcal{F}}^0(k_x)}{\alpha - k_x} + \mathcal{Q}\right\}, \tag{3.63}$$

$$\widetilde{\mathcal{U}}(\alpha) = \widetilde{\mathcal{G}}^U(\alpha)\left\{\frac{[\widetilde{\mathcal{G}}^U(k_x)]^{-1}\widetilde{\mathcal{F}}^0(k_x) - [\widetilde{\mathcal{G}}^U(\alpha)]^{-1}\widetilde{\mathcal{F}}^0(\alpha)}{\alpha - k_x} + \mathcal{Q}\right\}. \tag{3.64}$$

The asymptotic behavior of the various terms as $\alpha \to \infty$ in the lower and upper half-planes H^L and H^U implies that the unknown four-vector $\mathcal{Q}$ is constant.

The constraints of (3.55) or (3.56) are used to determine the unknown constant four-vector $\mathcal{Q} = [\mathbf{Q}_1, \mathbf{Q}_2]^T$. Explicitly, we obtain

$$\mathbf{Q}_1 = \frac{1}{k_x^2 + k_z^2} \left\{ k_x [\widetilde{G}_1^U(k_x)]^{-1} \widetilde{\mathbf{F}}_1^0(k_x) - k_z [\widetilde{G}_2^U(k_x)]^{-1} \widetilde{\mathbf{F}}_2^0(k_x) \right\}, \tag{3.65}$$

$$\mathbf{Q}_2 = \frac{1}{k_x^2 + k_z^2} \left\{ k_z [\widetilde{G}_1^U(k_x)]^{-1} \widetilde{\mathbf{F}}_1^0(k_x) + k_x [\widetilde{G}_2^U(k_x)]^{-1} \widetilde{\mathbf{F}}_2^0(k_x) \right\}, \tag{3.66}$$

or together, in compact form,

$$\mathcal{Q} = \frac{1}{k_x^2 + k_y^2} \left[T^T(k_x) \otimes I_2 \right] [\widetilde{\mathcal{G}}^U(k_x)]^{-1} \widetilde{\mathcal{F}}^0(k_x). \tag{3.67}$$

The final solution of the Wiener-Hopf equation reads

$$\begin{aligned} \widetilde{\mathcal{L}}(\alpha) &= \begin{bmatrix} \widetilde{\mathbf{L}}_1(\alpha) \\ \widetilde{\mathbf{L}}_2(\alpha) \end{bmatrix} \\ &= \frac{k}{2\pi i} [\widetilde{\mathcal{G}}^L(\alpha)]^{-1} \left\{ \left[\frac{1}{\alpha - k_x} I + \frac{1}{k_x^2 + k_z^2} T^T(k_x) \right] \otimes I \right\} [\widetilde{\mathcal{G}}^U(k_x)]^{-1} \mathcal{E}^0 \end{aligned} \tag{3.68}$$

with the excitation term

$$\mathcal{E}^0 = \left\{ \begin{bmatrix} 1 \\ 0 \end{bmatrix} \otimes \begin{bmatrix} k_y + \eta^+ \\ k_y - \eta^- \end{bmatrix} \right\} E_y^0 + \left\{ \begin{bmatrix} 0 \\ 1 \end{bmatrix} \otimes \begin{bmatrix} 1 + \eta^+ k_y \\ 1 - \eta^- k_y \end{bmatrix} \right\} Z H_y^0. \tag{3.69}$$

Here the mixed-product property $(A \otimes B)(C \otimes D) = (AC) \otimes (BD)$ (see [17], 1991) has been used repeatedly. The constraints (3.55) or (3.56) are easily verified.

The coupling between the vector components $\widetilde{\mathbf{L}}_1(\alpha)$ and $\widetilde{\mathbf{L}}_2(\alpha)$ is due to the obliqueness parameter k_z in the off-diagonal elements of the matrix

$$T^T(k_x) \otimes I = \begin{bmatrix} k_x I & -k_z I \\ k_z I & k_x I \end{bmatrix}. \tag{3.70}$$

For $k_z = 0$, i.e., for normal incidence, there is no coupling between the electric and magnetic fields. The coupling between electric and magnetic fields for $k_z \neq 0$ is a pure edge diffraction effect.

The special excitation $E_y^0 = 0$ means that the incident electric field vector $\mathbf{E}^0$ is parallel to the half-plane and that the magnetic vector $\mathbf{H}^0$ lies in the plane of incidence spanned by the normal $\mathbf{y}$ to the half-plane and the incident propagation vector $\mathbf{k}$. The excitation $H_z^0 = 0$, on the other hand, implies that the incident magnetic field vector $\mathbf{H}^0$ is parallel to the half-plane and that the electric vector $\mathbf{E}^0$ lies in the scattering plane.

For a detailed account of this problem see also Lüneburg and Serbest [29].

4. Other examples

We present some other examples from the literature that are similar to the two-face impedance half-plane problem for oblique incidence and refer to the short communication by Rawlins [40] on polynomial diagonalization, where several typical examples have been collected.

4.1. Example 1

In their study of diffraction of an obliquely incident plane wave by an anisotropic half-plane, Hurd and Lüneburg [21] arrived at the factorization of the matrix function

$$G(\alpha) = \begin{bmatrix} 1 + \dfrac{k^2-\alpha^2}{k\,\eta\,\Gamma(\alpha)} & -\dfrac{k_z\,\alpha}{k\,\eta\,\Gamma(\alpha)} \\ -\dfrac{k_z\alpha}{k\,\eta\,\Gamma(\alpha)} & 1+\dfrac{\chi^2}{k\,\eta\,\Gamma(\alpha)} \end{bmatrix} = I + \frac{1}{k\,\eta\,\Gamma(\alpha)} \begin{bmatrix} k^2-\alpha^2 & -k_z\,\alpha \\ -k_z\,\alpha & \chi^2 \end{bmatrix} \tag{4.1}$$

with $\Gamma(\alpha) = \sqrt{\chi^2 - \alpha^2}$ and the reduced wavenumber $\chi^2 = k^2 - k_z^2$. The eigenvalues of $G(\alpha)$ are given by

$$\lambda_1(\alpha) = 1 + \frac{\Gamma(\alpha)}{\eta\,k}, \quad \lambda_2(\alpha) = 1 + \frac{k}{\eta\,\Gamma(\alpha)}. \tag{4.2}$$

The eigenvalues coincide for the multiple points $\alpha = \pm ik_z$: $\lambda_1(\pm ik_z) = \lambda_2(\pm ik_z) = 1 + \eta^{-1} \doteq \lambda$. For $\alpha = \pm ik_z$ we have

$$G(\pm ik_z) = \begin{bmatrix} 1 + \dfrac{k^2+k_z^2}{\eta k^2} & \dfrac{\mp ik_z^2}{\eta k^2} \\ \dfrac{\mp ik_z^2}{\eta k^2} & 1 + \dfrac{k^2-k_z^2}{\eta k^2} \end{bmatrix}. \tag{4.3}$$

These matrices are nonsingular but have only one eigenvector corresponding to the only eigenvalue λ.

The points $\{\pm ik_z\}$ form the singular set $S(G)$ since the Jordan structure of $G(\alpha)$ is changing at the points of this set. Introducing the polynomial transformation

$$T(\alpha) = \begin{bmatrix} \alpha & -k_z \\ k_z & \alpha \end{bmatrix}, \quad T^{-1}(\alpha) = \frac{1}{\alpha^2+k_z^2} \begin{bmatrix} \alpha & k_z \\ -k_z & \alpha \end{bmatrix} \quad (\alpha \neq \pm ik_z) \tag{4.4}$$

we obtain

$$G(\alpha)T(\alpha) = T(\alpha)\operatorname{diag}\left[\lambda_1(\alpha), \lambda_2(\alpha)\right] \qquad (\text{all } \alpha). \tag{4.5}$$

This is a similarity transformation for all α outside the singular set.

For $\alpha \notin S(G)$ we introduce

$$R_1(\alpha) = T(\alpha)R_{10}T^{-1}(\alpha) = \frac{1}{\alpha^2 + k_z^2}\begin{bmatrix} \alpha^2 & \alpha k_z \\ \alpha k_z & k_z^3 \end{bmatrix} \tag{4.6}$$

and

$$R_2(\alpha) = T(\alpha)R_{20}T^{-1}(\alpha) = \frac{1}{\alpha^2 + k_z^2}\begin{bmatrix} k_z^2 & -\alpha k_z \\ -\alpha k_z & \alpha^3 \end{bmatrix} \tag{4.7}$$

where S_1 and S_2 are given by equations (4.44) and note that $R_1(\alpha)$ and $R_2(\alpha)$ are complementary projectors. Hence,

$$G(\alpha) = \lambda_1(\alpha)R_1(\alpha) + \lambda_2(\alpha)R_2(\alpha) \quad \text{for} \quad \alpha \notin S(G). \tag{4.8}$$

4.2. Example 2

The problem of diffraction of obliquely incident plane waves by a partially coated dielectric slab leads to the factorization of the following matrix function (cf. Lüneburg et al. [26]):

$$\begin{aligned} G(\alpha) &= G_a(\alpha) + G_b(\alpha) \\ &= \frac{1}{\Gamma_1(\alpha)}\begin{bmatrix} k_1^2 - k_z^2 & \alpha\, k_z \\ \alpha\, k_z & k_1^2 - \alpha^2 \end{bmatrix} + \frac{\mu}{\Gamma_2(\alpha)}\begin{bmatrix} k_2^2 - k_z^2 & \alpha\, k_z \\ \alpha\, k_z & k_2^2 - \alpha^2 \end{bmatrix} \end{aligned} \tag{4.9}$$

where $\Gamma_i(\alpha) = \sqrt{\chi_i^2 - \alpha^2}$, $\chi_i^2 = k_i^2 - k_z^2$ $(i = 1, 2)$. The eigenvalues of $G(\alpha)$ read

$$\lambda_1(\alpha) = \frac{k_1^2}{\Gamma_1(\alpha)} + \frac{\mu\, k_2^2}{\Gamma_2(\alpha)}, \quad \lambda_2(\alpha) = \Gamma_1(\alpha) + \mu\,\Gamma_2(\alpha) \tag{4.10}$$

and coincide for the singular set

$$S(G) = \{\alpha = \pm i k_z\} \tag{4.11}$$

with $\lambda_1(\pm ik_z) = \lambda_2(\pm ik_z) = k_1 + \mu\, k_2$. The sum terms $G_a(\alpha)$ and $G_b(\alpha)$ of $G(\alpha)$ commute and, hence, can simultaneously be transformed by the matrix $T(\alpha)$ of equation (4.4) according to

$$G(\alpha)T(\alpha) = T(\alpha)\mathrm{diag}\,[\lambda_1(\alpha), \lambda_2(\alpha)]. \tag{4.12}$$

This is a similarity transformation for all α outside the singular set $S(G)$.

4.3. Example 3

Lebre [24] (see also Rawlins [40]) considered canonical factorization of the special class of 2×2 matrix functions of the form

$$G(\alpha) - \begin{bmatrix} 1 & a(\alpha)\,\rho^2(\alpha) \\ a(\alpha) & 1 \end{bmatrix} \tag{4.13}$$

where $a(\alpha)$ is an arbitrary analytic function and $\rho(\alpha)$ is a rational polynomial. With

$$m(\alpha) = \begin{bmatrix} 1 & 0 \\ 0 & \rho(\alpha) \end{bmatrix} \quad \text{and} \quad m^{-1}(\alpha) = \begin{bmatrix} 1 & 0 \\ 0 & \rho^{-1}(\alpha) \end{bmatrix} \tag{4.14}$$

we obtain the symmetric form

$$\widetilde{G}(\alpha) = m(\alpha)G(\alpha)m^{-1}(\alpha) = \begin{bmatrix} 1 & a(\alpha)\rho(\alpha) \\ a(\alpha)\rho(\alpha) & 1 \end{bmatrix}. \tag{4.15}$$

The eigenvectors and eigenvalues of $G(\alpha)$ are given as

$$G(\alpha)\mathbf{x}_i(\alpha) = \lambda_i(\alpha)\mathbf{x}_i(\alpha) \quad (i = 1, 2) \tag{4.16}$$

with

$$\mathbf{x}_1(\alpha) = \begin{bmatrix} \rho(\alpha) \\ 1 \end{bmatrix}, \quad \mathbf{x}_2(\alpha) = \begin{bmatrix} -\rho(\alpha) \\ 1 \end{bmatrix} \tag{4.17}$$

and

$$\lambda_1[G(\alpha)] = 1 + a(\alpha)\rho(\alpha), \quad \lambda_2[G(\alpha)] = 1 - a(\alpha)\rho(\alpha) \tag{4.18}$$

Defining the matrix of eigenvectors and its inverse by

$$X(\alpha) = [\mathbf{x}_1(\alpha), \mathbf{x}_2(\alpha)] = \begin{bmatrix} \rho(\alpha) & -\rho(\alpha) \\ 1 & 1 \end{bmatrix}, \tag{4.19}$$

$$X^{-1}(\alpha) = \frac{1}{2\rho(\alpha)} \begin{bmatrix} 1 & \rho(\alpha) \\ -1 & \rho(\alpha) \end{bmatrix} \quad (\rho(\alpha) \neq 0), \tag{4.20}$$

we obtain for $\rho(\alpha) \neq 0$

$$X^{-1}(\alpha)G(\alpha)X(\alpha) = D_1(\alpha) = \begin{bmatrix} 1 + a(\alpha)\rho(\alpha) & 0 \\ 0 & 1 - a(\alpha)\rho(\alpha) \end{bmatrix}. \tag{4.21}$$

For $\rho(\alpha) = 0$ or $\alpha = \alpha_i$ $(i = 1, 2, \ldots)$, i.e., for the discrete singular set $S(G) = \{\alpha_i | \rho(\alpha_i) = 0, i = 1, 2, \ldots\}$, both eigenvalues coincide, $\lambda_1(\alpha_i) = \lambda_2(\alpha_i) =: \lambda = 1$ for all i, and $D_1(\alpha_i) = I$, the unit matrix. The matrices $G(\alpha_i)$ $(i = 1, 2, \ldots)$ assume the form

$$G(\alpha_i) = \begin{bmatrix} 1 & 0 \\ a(\alpha_i) & 1 \end{bmatrix}. \tag{4.22}$$

These matrices are already in their Jordan normal form and indicate that there is only one eigenvector $x_1 = [0\ 1]^T$ that belongs to the double eigenvalue $\lambda = 1$, except when also $a(\alpha_i) = 0$. Hence, the matrix $G(\alpha)$ cannot be diagonalized for all values of the complex variable α, but only for α outside the discrete singular set $S(G)$. For $\alpha \in S(G)$ the underlying space $\mathcal{C}^2$ can be spanned by the eigenvector $\mathbf{e}_2 = [0\ \ 1]^T$ and a generalized eigenvector $\mathbf{e}_1 = [1\ \ 0]^T$, say

$$G|_{\rho=0}\mathbf{e}_1 - \mathbf{e}_1 = a(\alpha_i)\mathbf{e}_2. \tag{4.23}$$

The set $\{\mathbf{e}_2, \mathbf{e}_1\}$ forms a Jordan chain to the eigenvalue λ.

Defining

$$Z(\alpha) = \mathrm{adj}\{X\}(\alpha) = \begin{bmatrix} 1 & \rho(\alpha) \\ -1 & \rho(\alpha) \end{bmatrix} \quad \text{with} \quad Z_0 = Z|_{\rho=0} = \begin{bmatrix} 1 & 0 \\ -1 & 0 \end{bmatrix} \tag{4.24}$$

we obtain

$$Z(\alpha)G(\alpha) = \begin{bmatrix} 1 + a(\alpha)\rho(\alpha) & [1 + a(\alpha)\rho(\alpha)]\rho(\alpha) \\ -[1 - a(\alpha)\rho(\alpha)] & [1 - a(\alpha)\rho(\alpha)]\rho(\alpha) \end{bmatrix} = D_1(\alpha)Z(\alpha). \tag{4.25}$$

This relation remains meaningful even for $\rho(\alpha) = 0$:

$$Z_0 G|_{\rho=0} = \begin{bmatrix} 1 & 0 \\ -1 & 0 \end{bmatrix} = D_1|_{\rho=0} Z_0 = I Z_0 = Z_0. \tag{4.26}$$

It should be noted that the matrix Z_0 is a projector: $Z_0^2 = Z_0$.

The eigenvalues of Z_0 are 0 and 1 and the corresponding eigenvectors are $\mathbf{e}_2$ and $\mathbf{y}$. Hence

$$Z_0 \mathbf{e}_2 = 0 \quad \text{and} \quad Z_0 \mathbf{y} = \mathbf{y}, \quad \mathbf{y} = \begin{bmatrix} 1 \\ -1 \end{bmatrix} = \mathbf{e}_1 - \mathbf{e}_2. \tag{4.27}$$

We note that

$$\operatorname{Im} Z(\alpha) = \begin{cases} \mathcal{C}^2, & \alpha \notin S(G) \\ \operatorname{span}\{\mathbf{y}\}, & \alpha \in S(G) \end{cases} . \tag{4.28}$$

On the singular set, the matrix $G(\alpha_i)$ can be replaced by the matrix $D_1|_{\rho=0} = I$ provided that any vector $\widetilde{\mathbf{L}}(\alpha) \in Z(\alpha)\mathcal{C}^2$ satisfies the following constraints on the singular set $S(G)$:

$$\widetilde{\mathbf{L}}(\alpha) = \begin{bmatrix} \widetilde{L}_1(\alpha) \\ \widetilde{L}_2(\alpha) \end{bmatrix} \quad \Longrightarrow \widetilde{L}_1(\alpha_i) + \widetilde{L}_2(\alpha_i) = 0. \tag{4.29}$$

These constraints guarantee that the vector $\widetilde{\mathbf{L}}(\alpha)$ lies in the range of $Z(\alpha)$ for all values of α. The restricted space is denoted as $\mathcal{C}_0^2$. The restricted range has to be taken into account for the Wiener-Hopf solution with such a symbol matrix.

4.4. Example 4

Following Meister and Speck [32], [33] we consider the diagonal matrix

$$D(\alpha) = \begin{bmatrix} \lambda_1(\alpha) & 0 \\ 0 & \lambda_2(\alpha) \end{bmatrix} \tag{4.30}$$

where $\lambda_1(\alpha)$ and $\lambda_2(\alpha)$ are arbitrary analytic functions. $D(\alpha)$ can be written in the form

$$D(\alpha) = \lambda_1(\alpha) P_{10} + \lambda_2(\alpha) P_{20} \tag{4.31}$$

where

$$P_{10} = \begin{bmatrix} 1 & 0 \\ 0 & 0 \end{bmatrix} \quad \text{and} \quad P_{20} = \begin{bmatrix} 0 & 0 \\ 0 & 1 \end{bmatrix} \tag{4.32}$$

are complementary (singular) projection matrices: $P_{i0}^2 = P_{i0}$ $(i = 1, 2)$, $P_{10} \oplus P_{20} = I$.

We consider the polynomial matrix

$$T(\alpha) = \begin{bmatrix} \alpha & -p \\ p & \alpha \end{bmatrix} \tag{4.33}$$

and perform a similarity transformation for $\alpha \neq \pm ip$:

$$\begin{aligned} G(\alpha) &= T(\alpha)\, D(\alpha)\, T^{-1}(\alpha) \qquad (4.34) \\ &= \frac{1}{\alpha^2+p^2} \begin{bmatrix} \alpha^2\,\lambda_1(\alpha) + p^2\,\lambda_2(\alpha) & \alpha\,[\lambda_1(\alpha)-\lambda_2(\alpha)]\,p \\ \alpha\,[\lambda_1(\alpha)-\lambda_2(\alpha)]\,p & p^2\,\lambda_1(\alpha) + \alpha^2\,\lambda_2(\alpha) \end{bmatrix}. \end{aligned}$$

Since the transpose of T is equal to its adjoint, $T^T = \operatorname{adj} T$, the matrix $G(\alpha)$ is symmetric, $G(\alpha)^T = G(\alpha)$.

According to equation (4.31) the matrix $G(\alpha)$ can be split up as

$$G(\alpha) = \lambda_1(\alpha)\, R_1(\alpha) + \lambda_2(\alpha)\, R_2(\alpha) \qquad (4.35)$$

where

$$R_i = T(\alpha)\, R_{i0} T^{-1}(\alpha) \quad \text{for} \quad i = 1, 2. \qquad (4.36)$$

The matrices $R_1(\alpha)$ and $R_2(\alpha)$, like R_{10} and R_{20}, are symmetric, idempotent and singular $R_i^2(\alpha) = R_i(\alpha)$ $(i = 1, 2)$, and, being similar to R_{i0}, add up to the unit matrix, $R_1(\alpha) \oplus R_2(\alpha) = I$. Furthermore, they are commuting, $R_1\, R_2 = R_2\, R_1$, i.e., they have the same eigenvectors and, hence, can be simultaneously diagonalized.

4.5. Example 5

Meister and Speck [32, 33] considered the matrix

$$G(\alpha) = \frac{1}{k_2^2 t_2^2} \begin{bmatrix} \alpha^2\,\psi - k_2^2\,t_2^2 & \alpha\,\xi_2\,\psi \\ \alpha\,\xi_2\,\psi & \xi_2^2\,\psi - k_2^2\,t_2^2 \end{bmatrix} \qquad (4.37)$$

occurring in elastodynamic problems (see Kupradze [23]) with $\psi = 4\tau - 3k_2^2$ and $\tau = \alpha^2 + \xi_2^2 - t_1 t_2$. Introducing the polynomial matrix

$$T(\alpha) = \begin{bmatrix} \alpha & -\xi_2 \\ \xi_2 & \alpha \end{bmatrix}, \qquad (4.38)$$

for which

$$T^{-1}(\alpha) = \frac{1}{\alpha^2+\xi_2^2} \begin{bmatrix} \alpha & \xi_2 \\ -\xi_2 & \alpha \end{bmatrix}, \qquad (4.39)$$

we obtain for $\alpha \neq \pm i\xi_2$ the diagonal matrix (Jordan form)

$$\begin{aligned} D(\alpha) &= T^{-1}(\alpha)\, G(\alpha)\, T(\alpha) \qquad (4.40) \\ &= \frac{1}{k_2^2 t_2^2} \begin{bmatrix} \alpha^2\,\psi - k_2^2\,t_2^2 + \xi_2^2\,\psi & 0 \\ 0 & -k_2^2\,t_2^2 \end{bmatrix} = \begin{bmatrix} \lambda_1(\alpha) & 0 \\ 0 & \lambda_2(\alpha) \end{bmatrix}. \end{aligned}$$

The matrix $G(\alpha)$ can be decomposed as

$$G(\alpha) = \lambda_1(\alpha) R_1(\alpha) + \lambda_2(\alpha) R_2(\alpha) = \frac{\alpha^2\,\psi - k_2^2\,t_2^2 + \xi_2^2\,\psi}{k_2^2 t_2^2}\, R_1(\alpha) - R_2(\alpha) \qquad (4.41)$$

with

$$R_1(\alpha) = T(\alpha) R_{10} T^{-1}(\alpha) = \frac{1}{\alpha^2+\xi_2^2} \begin{bmatrix} \alpha^2 & \alpha\,\xi_2 \\ \alpha\,\xi_2 & \xi_2^2 \end{bmatrix} \qquad (4.42)$$

and

$$R_2(\alpha) = T(\alpha) R_{20} T^{-1}(\alpha) = \frac{1}{\alpha^2 + \xi_2^2} \begin{bmatrix} \xi_2^2 & -\alpha\,\xi_2 \\ -\alpha\,\xi_2 & \alpha^2 \end{bmatrix}, \tag{4.43}$$

$$R_{10} = \begin{bmatrix} 1 & 0 \\ 0 & 0 \end{bmatrix}, \quad R_{20} = \begin{bmatrix} 0 & 0 \\ 0 & 1 \end{bmatrix}. \tag{4.44}$$

$R_1(\alpha)$ and $R_2(\alpha)$ are complementary projectors (idempotent, singular) and their direct sum is $R_1(\alpha) \oplus R_2(\alpha) = I$.

At the points of the singular set $S(G) = \{\pm i\xi_2\}$ the eigenvalues of the matrices $G(\pm i\xi_2)$ coincide, $\lambda_1 = \lambda_2 = -1$, and there exists only one eigenvector,

$$G(i\xi_2) y_1 = -y_1, \quad G(-i\xi_2) y_2 = -y_2 \tag{4.45}$$

with

$$y_1 = \frac{1}{\sqrt{2}} \begin{bmatrix} 1 \\ -i \end{bmatrix}, \quad y_2 = \frac{1}{\sqrt{2}} \begin{bmatrix} -i \\ 1 \end{bmatrix}. \tag{4.46}$$

For $G(i\xi_2)$ the vector y_2 and for $G(-i\xi_2)$ the vector y_1 are generalized eigenvectors. The Jordan chains $\{y_1, y_2\}$ and $\{y_2, y_1\}$, respectively, span $\mathcal{C}^2$ on the singular set $S(G)$.

5. Concluding remarks

In this paper the Wiener-Hopf solution for the homogeneous isotropic two-face impedance half-plane problem in the oblique incidence case is presented in some detail. In contrast to previous solutions of this problem by Bucci and Franceschetti [2] in 1976, who used the Maliuzhinetz approach ([31], 1958) and ad hoc assumptions for the expressions of the spectral normal field components, and a cursory Wiener-Hopf treatment by Senior [48] in 1978, who used normal field components and the traditional pole elimination procedure, the present contribution is based upon the new rigorous concept of "range restrictions" for the polynomial transformation matrices between the Fourier transforms of the tangential and the normal field components. These constraints on the solution space arise because the Jordan structure of the Fourier symbol matrix function is changing on the singular set in the complex domain. The method is related to the theory of invariant subspaces of analytical families of transformation matrices.

Superficially, the traditional engineering approach of elimination of unphysical leaky wave poles from the solution and the method of range restriction of rational transformations to reduce the Fourier symbol matrix function presented here appear to be equivalent. This is correct, indeed, as far as the final results are concerned. The second method, however, clearly exhibits and illuminates the internal structure and coupling of the two possible TE and TM wave modes. A separation of the original 4×4 Wiener-Hopf system into two 2×2 systems is possible only within a properly restricted four-column complex vector space.

Some other representative examples which exhibit the same behavior of changing the Jordan structure of the Fourier symbol matrix function in the complex plane are quoted from the literature. Their solutions can be obtained in a way similar to the one outlined in the present contribution. The corresponding solutions obtained with the Maliuzhinetz approach need to be critically re-assessed.

References

[1] Baumgärtel, H., *Operator Theory: Advances and Applications, vol. 15, Analytic Perturbation Theory for Matrices and Operators*, Birkhäuser, Boston, Cambridge, Mass. 1985.

[2] Bucci, O.M. and G. Franceschetti, Electromagnetic scattering by a half-plane with two face impedances, *Radio Sci.* **11(1)** (1976), 49–59.

[3] Büyükaksoy, A., E. Erdogan, and A.H. Serbest, Diffraction of an obliquely incident plane wave by a thin dielectric half-plane, *Bull. Tech. Univ. Istanbul* **41** (1988), 655–664.

[4] Büyükaksoy, A., G. Uzgören, and A.H. Serbest, Diffraction of an obliquely incident plane wave by the discontinuity of a two part thin dielectric plane, *Int. J. Eng. Sci.* **27** (1989), 701–710.

[5] Büyükaksoy, A. and A.H. Serbest, Wiener-Hopf Factorization Methods and Applications to some Diffraction Problems, *M. Hashimoto, M. Idemen, and O. A. Tretyakov (Eds): Analytical and Numerical Methods in Electromagnetic Wave Theory, Science House Co.,* Tokyo (1993), 257–315.

[6] Câmara, M.C., A.B. Lebre, and F.-O. Speck, Meromorphic factorization, partial index estimates and elastodynamic diffraction problems, *Math. Nachr.* **157** (1992), 291–317.

[7] Câmara, M.C., A.F. dos Santos, and M.A. Bastos, Generalised factorization for Daniele-Khrapkov matrix functions: Partial indices, *J. Math. Anal. Appl.* **190** (1995), 142–164.

[8] ———, Generalised factorization for Daniele-Khrapkov matrix functions: Explicit formulas, *J. Math. Anal. Appl.* **190** (1995), 295–328.

[9] Clemmow, P.C., *The Plane Wave Spectrum Representation of Electromagnetic Fields*, Pergamon, New York 1966.

[10] Daniele, V.G., On the factorization of Wiener-Hopf matrices in problems solvable with Hurd's method, *IEEE Trans. Antennas Propag.* **AP-26** (1978), 614–616.

[11] ———, On the solutions of vector Wiener-Hopf equations occurring in scattering problems, *Radio Sci.* **19(5)** (1984), 1173–1178.

[12] ———, On the solution of two coupled Wiener-Hopf equations, *SIAM J. Appl. Math.* **44** (1984), 667–680.

[13] dos Santos, A.F., A.B. Lebre, and F.S. Teixeira, The diffraction problem for a half-plane with different face impedances: Revisited, *J. Math. Anal. Appl.* **140** (1989), 485–509.

[14] Gohberg, I., P. Lancaster, and L. Rodman, *Invariant Subspaces of Matrices with Applications*, John Wiley, New York 1986.

[15] Grünberg, G., Suggestion for a theory of coastal refraction, *Phys. Rev.* **63** (1943), 185–189.

[16] Heins, A.A., Systems of Wiener-Hopf integral equations and their applications to some boundary value problems in electromagnetic theory, *Proc. Symp. Appl. Math. II, McGraw-Hill Book Company, New York* (1950), 76–81.

[17] Horn, R.A. and C.R. Johnson, *Topics in Matrix Analysis*, Cambridge Univ. Press, New York 1991.

[18] Hoppe, D.J. and Y. Rahmat-Samii, *Impedance Boundary Conditions in Electromagnetics*, Taylor and Francis, Philadelphia, Pa. 1995.

[19] Hurd, R.A., The Wiener-Hopf-Hilbert method for diffraction problems, *Can. J. Phys.* **54** (1976), 775–780.

[20] Hurd, R.A. and S. Przeździecki, Diffraction by a half-plane with different face impedances: A re-examination, *Can. J. Phys.* **59** (1981), 1337–1347.

[21] Hurd, R.A. and E. Lüneburg, Diffraction by an anisotropic impedance half-plane, *Can. J. Phys.* **63** (1985), 1135–1140.

[22] Khrapkov, A.A., Certain cases of the elastic equilibrium of an infinite wedge with a non-symmetric notch at the vertex subjected to concentrated forces, *Prikl. Mat. Mekh.* **35** (1971), 625–637.

[23] Kupradze, V.D. (ed.), *Three-Dimensional Problems of the Mathematical Theory of Elasticity and Thermoelasticity*, North-Holland, Amsterdam 1979.

[24] Lebre, A.B., Factorization in the Wiener algebra of a class of 2 × 2 matrix functions, *Integral Equations and Operator Theory* **12** (1989), 408–423.

[25] Leontovich, M.A., Investigations on radiowave propagation, Part II, *Printing House of the Academy of Sciences* Moscow (1948), 5–12.

[26] Lüneburg, E., R.A. Hurd, and E. Meister, *Diffraction by a partially coated dielectric half-plane*, Preprint No. 1156, Techn. Hochschule Darmstadt, 1988.

[27] Lüneburg, E. and R.A. Hurd, On the diffraction problem of a half-plane with different face impedances, *Can. J. Phys.* **62** (1984), 353–360.

[28] Lüneburg, E., The Sommerfeld problem: Methods, generalizations and frustrations, *Proceedings of the Sommerfeld'96 Workshop, Freudenstadt, 30 September–4 October 1996, Peter Lang, Frankfurt* (1997), 145–162.

[29] Lüneburg, E. and H. Serbest, Diffraction of an obliquely incident plane wave by a two-face impedance half-plane: Wiener-Hopf approach, *Radio Sci.* **35(6)** (2000), 1361–1374.

[30] Ly, H.C., R.G. Rojas, and P.H. Pathak, EM plane wave diffraction by a planar junction or two thin material half-planes: Oblique incidence, *IEEE Trans. Antennas Propag.* **41(4)** (1993), 429–440.

[31] Maliuzhinetz, G.D., Excitation, reflection and emission of surface waves from a wedge with given face impedances, *Sov. Phys. Dokl., Engl. Transl.* **3** (1958), 752–755.

[32] Meister, E. and F.-O. Speck, Wiener-Hopf factorization of certain non-rational matrix functions in mathematical physics, *The Gohberg Anniversary Collection, Vol. II, H. Dym, S. Goldberg, M.A. Kashoek, P. Lancaster (eds.), Proc. Conf. Calgary 1988, Birkhäuser, Basel* (1989), 385–394.

[33] ———, The explicit solution of elastodynamical diffraction problems by symbol factorization, *Z. Anal. Anw.* **8(4)** (1989), 307–328.

[34] Meister, E., P.A. Santos, and F.S. Teixeira, A Sommerfeld-type diffraction problem with second-order boundary conditions, *Z. Angew. Math. Mech.* **72** (1992), 621–630.

[35] Nepa, P., G. Manara, and A. Armogida, Plane wave scattering by edges in unidirectionally conducting screens, *Radio Science* **35(6)** (2000), 1265–1278.

[36] Noble, B., *Methods Based on the Wiener-Hopf Technique for the Solution of Partial Differential Equations*, Pergamon Press, London 1958.

[37] Osipov, A.V. and A.N. Norris, The Malyuzhinets theory for scattering from wedge boundaries, *Wave Motion* **29(4)** (1999), 313–340.

[38] Passow, A., *Sommerfeld-Halbebenenproblem mit elektromagnetisch anisotropen Leontovich-Randbedingungen, Dissertation D 17, TU Darmstadt, 141 pp.*, Shaker Verlag, Aachen 1999.

[39] Rawlins, A.D., The solution of a mixed boundary value problem in the theory of diffraction by a semi-infinite plane, *Proc. R. Soc. London, Ser. A* **346** (1975), 469–484.

[40] ———, A note on polynomial diagonalization and Wiener-Hopf factorization of 2×2 matrices, *Z. Angew. Math. Mech.* **72** (1992), 373–375.

[41] Rojas, R.G., Wiener-Hopf analysis of the EM diffraction by an impedance discontinuity in a planar surface and an impedance half-plane, *IEEE Trans. Antennas Propag.* **AP-36** (1988), 71–83.

[42] ———, Electromagnetic diffraction for an obliquely incident plane wave field by a wedge with impedance faces, *IEEE Trans. Antennas Propag.* **AP-36** (1988), 956–970.

[43] Senior, T.B.A., Diffraction by a semi-infinite metallic sheet, *Proc. R. Soc. London, Ser. A* **213** (1952), 436–458.

[44] ———, Diffraction by an imperfectly conducting half-plane at oblique incidence, *Appl. Sci. Res., Ser. B* **8** (1959), 35–61.

[45] ———, Impedance boundary conditions for imperfectly conducting surfaces, *Appl. Sci. Res., Ser. B* **8** (1960), 418–436.

[46] ———, Half-plane edge diffraction, *Radio Sci.* **10(6)** (1975), 645–650.

[47] ———, Diffraction tensors for imperfectly conducting edges, *Radio Sci.* **10(10)** (1975), 911–919.

[48] ———, Some problems involving imperfect half-planes, *in: Electromagnetic Scattering, ed. by P.L.E. Uslenghi, Academic, New York* (1978), 185–219.

[49] ———, Solution for a class of imperfect wedge problems for skew incidence, *Radio Sci.* **21** (1986), 185–191.

[50] ———, Skew incidence on a dielectric half-plane, *Electromagnetics* **9** (1989), 187–200.

[51] ———, Skew incidence on a material junction, *Radio Sci.* **26(2)** (1991), 305–311.

[52] Senior, T.B.A. and S.R. Legault, Diffraction by an anisotropic impedance half-plane at skew incidence, *Electromagnetics* **18** (1998), 207–225.

[53] Senior, T.B.A. and J.L. Volakis, *Approximate Boundary Conditions in Electromagnetics*, IEE Electromagn. Waves Ser., vol. 41, Inst. of Electr. Eng., London 1995.

[54] Spitkovskii, I.M. and A.M. Tashbaev, On the problem of effective factorization of matrix functions, *Izv. Vyssh. Uchebn. Zaved. Mat.* **33(4)** (1989), 69–76.

[55] Syed, H.H. and J.L. Volakis, On the skew incidence diffraction by an impedance wedge with arbitrary face impedances, *Electromagnetics* **15** (1995), 291–300.

[56] Uzgören, G., A. Büyükaksoy, and A.H. Serbest, Plane wave diffraction by the discontinuity formed by resistive and impedance half-planes: Oblique incidence case, *Ann. Télécommun.* **45** (1990), 410–418.

[57] Williams, W.E., Diffraction of an electromagnetic plane wave by a metallic sheet, *Proc. Roy. Soc. London, Ser. A* **257** (1960), 413–419.

Ernst Lüneburg
EML Consultants
Georg Schmid Weg 4
D-82234 Wessling, Germany
e-mail: e.lueneburg@t-online.de

Hamit A. Serbest
Çukurova University
Department of Electrical Engineering
and Electronics
01330 Balcali-Adana, Turkey
e-mail: serbest@mail.cu.edu.tr

2000 *Mathematics Subject Classification.*
Primary 78; Secondary 47.

Received: 22 May 2001.

Operator Theory:
Advances and Applications, Vol. 142, 253–261

Aperiodic Wave Diffraction by a Half-Plane

Erhard Meister

Abstract. Since A. Sommerfeld founded the mathematical theory of diffraction or scattering theory, a lot of mathematicians have tried to attack boundary value problems for time-harmonic and aperiodic wave fields. Different methods have been developed for canonical obstacles, like half-planes, wedges, cones, etc. relying on the Wiener-Hopf technique, involving the factorization of scalar and matricial Fourier symbols, or treating functional equations in strips of the complex plane, the so-called Maliuzhinets method. In the present paper we give an overview of the recent achievements in the subject by the Darmstadt research group. Some sucess in tackling time-aperiodic boundary-transmission problems allowed us to close the gap between the two above-mentioned methods. Very general representations of the time-dependent wave-fields in acoustics, electrodynamics and elastodynamics can be obtained in the 2D case using conformal mapping techniques. The explicit factorization of parameter-dependent Fourier symbols relies strongly on the time-harmonic case, but it can be extended to further situations including, for example, the integrodifferential equations of linear viscoelastodynamics.

1. Introduction

The mathematical diffraction theory started with the habilitation thesis by A. Sommerfeld [22]. Then the theory bifurcated into two different directions for canonical objects. The first was initiated by N. Wiener and E. Hopf [25], who considered half-planes and later generalizations of half-planes. The other approach, relying on so-called Sommerfeld integrals, was developed by the Leningrad school of G. Maliuzhinets [13]. General finite scatterers for electrodynamical, elasto- and thermoelastodynamical time-harmonic wave-fields were considered, e.g., by C. Müller [19], D. Colton and R. Kress [5], V. Kupradze et al. [11], D. Jones [9]. Aperiodic scattering problems were treated by R. du Cloux [7], R. Leis [12], and others.

Fourier symbol factorization is the crucial step to solve the canonical boundary-transmission problem. There are a lot of books and papers on this subject, and we here confine ourselves to mention [10] and [23]. In close cooperation, the Lisbon and Darmstadt groups contributed to the solution of this problem and published several papers [1, 2, 3, 4, 8, 14, 20, 24] during the last 15 years. A vast list of references on this subject can be found in [20].

The application of quaternionic analysis to canonical boundary-transmission problems for electrodynamic boundary conditions is discussed in [15, 16].

The present paper is devoted to recent results for aperiodic incoming wave-fields, which were partly published in [17]. The geometry of the canonical semi-infinite domains is described in more detail in [17].

2. The field equations in elasto- and viscoelastodynamics

2.1. The basic field equations

We shall concentrate our attention on the following problem. Let $\Omega \subset \mathbf{R}^3$ be a domain with Lipschitz boundary $\Gamma = \partial\Omega$, let $\Omega_T = [0,T) \times \Omega$, $0 < T \leq \infty$, and let $\mathbf{u}(t,\mathbf{x}) \in \mathbf{R}^3$ be the unknown displacement vector field in Ω_T .

The conservation laws for momentum, mass, and energy lead to equations of second order:

$$\frac{d}{dt}\int_\Omega \rho\dot{\mathbf{u}}(t,\mathbf{x})d\mathbf{x} = \int_\Gamma \mathbf{s}(t,\mathbf{x})d\Gamma(\mathbf{x}) + \int_\Omega \mathbf{f}(t,\mathbf{x})d\mathbf{x}, \quad t \in \mathbf{R} \tag{2.1}$$

$$\frac{d}{dt}\int_\Omega \rho\mathbf{x} \wedge \dot{\mathbf{u}}(t,\mathbf{x})d\mathbf{x} = \int_\Gamma \mathbf{x} \wedge \mathbf{s}(t,\mathbf{x})d\Gamma(x) + \int_\Omega \mathbf{x} \wedge \mathbf{f}(t,\mathbf{x})d\mathbf{x}, \quad t \in \mathbf{R} \tag{2.2}$$

where $\rho(t,\mathbf{x})$ is the mass density and $\mathbf{f}(t,\mathbf{x})$ is a given force density.

The surface traction is

$$\mathbf{s}(t,\mathbf{x}) = \boldsymbol{\sigma}(t,\mathbf{x})\mathbf{n}(\mathbf{x})$$

with the stress tensor $\boldsymbol{\sigma}$ depending on

$$\boldsymbol{\varepsilon}(t,\mathbf{x}) = \frac{1}{2}[\nabla\mathbf{u}(t,\mathbf{x}) + (\nabla\mathbf{u}(t,\mathbf{x}))^T]. \tag{2.3}$$

Mechanical energy balance leads to the equation

$$\int_0^T\int_\Omega \rho|\dot{\mathbf{u}}(t,\mathbf{x})|^2 d\mathbf{x}dt + \int_0^T\int_\Omega \boldsymbol{\sigma}(t,\mathbf{x}) : \boldsymbol{\varepsilon}(t,\mathbf{x})d\mathbf{x}dt = \int_\Omega \rho|\dot{\mathbf{u}}(0+,\mathbf{x})|^2 d\mathbf{x} +$$
$$\int_0^T\int_\Omega \mathbf{f}(t,\mathbf{x})\dot{\mathbf{u}}(t,\mathbf{x})d\mathbf{x}dt + \int_0^T\int_\Gamma \mathbf{s}(t,\mathbf{x})\dot{\mathbf{u}}(0+,\mathbf{x})d\Gamma(\mathbf{x})dt \tag{2.4}$$

where $\sigma : \varepsilon$ denotes a tensor contraction. The strain-energy condition gives:

$$\int_0^T\int_\Omega \sigma(t,\mathbf{x}) : \dot{\varepsilon}(t,\mathbf{x})d\mathbf{x}dt \geq 0 \quad \forall T > 0. \tag{2.5}$$

Linearization of these equations leads to a convolution integral of the stress tensor,

$$\boldsymbol{\sigma}(t,\mathbf{x}) = d\mathcal{A} * \dot{\boldsymbol{\varepsilon}}(t,\mathbf{x}) = \int_0^t d\mathcal{A}(\tau,\mathbf{x}) : \dot{\boldsymbol{\varepsilon}}(t-\tau,\mathbf{x})d\tau \tag{2.6}$$

where $\mathcal{A}$ is the 4-order **relaxation tensor,** with

$$\int_0^T\int_0^t d\mathcal{A}(\tau,\mathbf{x}) : \mathbf{F}(t-\tau) : \mathbf{F}(t)d\tau dt \geq 0 \quad \forall\mathbf{F} \in \mathbf{C}(\mathbf{R}_+, \mathbf{R}^{3\times 3}). \tag{2.7}$$

This includes the time-memory behavior.

2.2. Isotropic, homogeneous continua

For isotropic, homogeneous continua, the equations above simplify to a second order system with respect to the displacement vector $\mathbf{u}$:

$$\rho\ddot{\mathbf{u}}(t,\mathbf{x}) - da * \Delta\dot{\mathbf{u}}(t,\mathbf{x}) - d(b + \frac{a}{3}) * \operatorname{grad}\operatorname{div}\dot{\mathbf{u}}(t,\mathbf{x}) = \mathbf{f}(t,\mathbf{x}) \qquad (2.8)$$

The symbol $*$ denotes finite time convolution. Then the stress field is given by

$$\mathbf{s}(t,\mathbf{x}) = [d(b - \frac{2a}{3}) * (\operatorname{div}\,\dot{\mathbf{u}}(t,\mathbf{x}))\mathbf{I} + da * [\dot{\mathbf{u}}(t,\mathbf{x}) + (\dot{\mathbf{u}}(t,\mathbf{x}))^T]]\mathbf{n}(\mathbf{x}) \qquad (2.9)$$

where $\mathbf{I}$ is the identity 3×3 matrix and $\mathbf{n}(\mathbf{x})$ is the normal vector on the boundary Γ.

In the pure elastodynamic case the equation for displacement vectors simplifies and $\mathbf{u}(\mathbf{x}, t)$ may be split due to a Helmholtz decomposition into a gradient field and a divergenceless field (see [18]).

3. Initial boundary value problems

3.1. The general problem

Now, we turn to initial boundary value problems. We start with the definition.

Definition 3.1. Let the visco-elastic stress-strain relation under mechanical deformation be

$$\boldsymbol{\varepsilon}(t,\mathbf{x}) = d\mathcal{C} * \boldsymbol{\sigma}(t,\mathbf{x}), \quad (t,\mathbf{x}) \in \Omega_T \qquad (3.1)$$

with a four **compliance tensor** $\mathcal{C}$:

$$d\mathcal{A} * \mathcal{C}(t,\mathbf{x}) = \mathcal{A} * d\mathcal{C}(t,\mathbf{x}) = t\mathcal{I} \qquad (3.2)$$

where $d\mathcal{A}$ is the relaxation tensor and $\mathcal{I}$ is the 4 identity tensor.

Then the main problem is the following.

Problem 3.2. For given fields $(\mathbf{f}, \mathbf{s})$, material density $\rho > 0$, $\mathcal{A}(t,\mathbf{x})$ or $a(t,\mathbf{x}), b(t,\mathbf{x})$ of positive type, and initial data $\mathbf{u}_0, \mathbf{u}_1$, one has to find the displacement field $\mathbf{u}(t,\mathbf{x})$ such that

$$\rho\ddot{\mathbf{u}}(t,\mathbf{x}) = \operatorname{div}\boldsymbol{\sigma}(t,\mathbf{x}) + \mathbf{f}(t,\mathbf{x}) \;\; \forall (t,\mathbf{x}) \in \Omega_T \qquad (3.3)$$

with the initial conditions

$$\mathbf{u}(+0,\mathbf{x}) = \mathbf{u}_0, \quad \dot{\mathbf{u}}(+0,\mathbf{x}) = \mathbf{u}_1 \qquad (3.4)$$

and the traction boundary condition

$$[d\mathcal{A} * \dot{\varepsilon}(t,\mathbf{x})]\mathbf{n}(\mathbf{x}) = \mathbf{s}(t,\mathbf{x}) \;\text{ on }\; \Gamma_T. \qquad (3.5)$$

There could be an additional Dirichlet boundary condition on Γ_D.

3.2. Cauchy problems in linear visco-elasticity

The full space problem is the following.

Problem $\mathcal{P}_C$:

$$\begin{aligned} \partial_t^2 \mathbf{u}(t,\mathbf{x}) + da * \mathcal{A}\partial_t \mathbf{u}(t,\mathbf{x}) &= \mathbf{f}(t,\mathbf{x}), \\ \mathbf{u}(+0,\mathbf{x}) &= \mathbf{u}_0(\mathbf{x}), \\ \partial_t \mathbf{u}(+0,\mathbf{x}) &= \mathbf{u}_1(\mathbf{x}). \end{aligned} \tag{3.6}$$

J. Mark [14] considered this problem and proved the existence of distributional, weak, and strong solutions in different spaces, particulary, in exponentially time-weighted spaces.

Laplace transformation of Problem $\mathcal{P}_C$ leads to

$$(\mu + A)\tilde{\mathbf{u}}(\lambda) = \frac{1}{\lambda \tilde{da}(\lambda)} \tilde{\mathbf{f}}^*(\lambda), \ \ \text{Re}\lambda > \lambda_0 > 0 \tag{3.7}$$

with A being a pure space operator. The right-hand side is given by

$$\tilde{\mathbf{f}}^*(\lambda) := \tilde{\mathbf{f}}(\lambda) + \tilde{\Theta}(\lambda)\mathbf{u}_1 + \lambda\tilde{\Theta}(\lambda)\mathbf{u}_0 + \tilde{da}(\lambda)A\mathbf{u}_0 \tag{3.8}$$

where $\Theta(t)$ is the Heaviside step function and $\tilde{\Theta}(\lambda)$ is its $\mathcal{L}$-transform.

Let X, Y be basic solution (Banach) spaces and $L(X,Y)$ be the space of linear bounded mappings $X \to Y$. Further we consider the space of tempered Schwartz distributions $U : \mathbf{R} \to L(X,Y)$ supported on $\overline{\mathbf{R}}_+$ which is denoted by $\mathcal{S}'_+(L(X,Y))$.

The fundamental resolvent is an operator-valued tempered distribution U in $\mathcal{S}'_+(L(X,Y))$ which has a family of $\mathcal{L}$-transformed holomorphic operators in the region $\text{Re}\lambda > 0$,

$$\lambda\tilde{U}(\lambda) + \tilde{da}(\lambda)A\tilde{U}(\lambda) = \tilde{\Theta}(\lambda)I, \tag{3.9}$$

satisfying

$$\|\tilde{U}(\lambda)\|_{L(X,Y)} \leq M\lambda^{-1}. \tag{3.10}$$

Da Prato and Ianelli [6] gave a formula for this resolvent:

$$U(t) = D_t^2 G(t) \tag{3.11}$$

where

$$G(t) = \frac{1}{2\pi i}\int_{Re\lambda=\tau>0} e^{\lambda t}\tilde{U}(\lambda)\frac{d\lambda}{\lambda^2} \tag{3.12}$$

Lemma 3.3. *For given creep function a, force field $\mathbf{f}$, and initial displacements $\mathbf{u}_0 \in X, \mathbf{u}_1 \in Y$, the unique displacement $\mathbf{u}(t)$ is given by*

$$\begin{aligned} \mathbf{u}(t) &= U * \int_0^t \mathbf{f}^*(\tau)d\tau \\ &= U * \int_0^t \mathbf{f}(\tau)d\tau + U * \Theta(t) \otimes \mathbf{u}_1 + \Theta(t) \otimes \mathbf{u}_0 \end{aligned} \tag{3.13}$$

where U depends on a and has to be determined as follows.

The Laplace transformed solution of the Cauchy problem corresponds to the parameter-dependent Helmholtz equation

$$[\Delta + k^2(\lambda)]\tilde{\mathbf{u}}(\lambda, \mathbf{x}) = -\frac{1}{\lambda \tilde{da}(\lambda)}\tilde{\mathbf{f}}^*(\mathbf{x}), \quad \mathrm{Re}\lambda > 0 \tag{3.14}$$

where

$$k(\lambda) = \left[-\frac{\lambda}{\tilde{da}(\lambda)}\right]^{1/2}, \quad \mathrm{Im}\, k(\lambda) > 0. \tag{3.15}$$

There exist some fundamentally different cases, depending on both $a(t)$ and its $\mathcal{L}$-transform $\tilde{a}(\lambda)$: the viscous fluid, the viscous solid, the rigid fluid, the rigid solid. The detailed description of these cases is given in [14].

The explicit fundamental displacement distribution is

$$U(t, \mathbf{x}) = \mathcal{L}^{-1}[\tilde{U}(\lambda, \mathbf{x})](t)\Theta(t) \tag{3.16}$$

where

$$\tilde{U}(t, \mathbf{x}) = \frac{\exp[ik(\lambda)|\mathbf{x}|]}{4\pi \tilde{da}(\lambda)|\mathbf{x}|}. \tag{3.17}$$

Comments. Scalar initial-boundary value problems may be reduced to parameter-dependent elliptic problems. The vectorial initial-boundary value problems play an essential role here. Let us consider them more carefully.

After a Helmholtz decomposition into longitudinal (L) and transversal (T) displacements we have

$$\mathbf{u}(t, \mathbf{x}) = \mathbf{u}_L(t, \mathbf{x}) + \mathbf{u}_T(t, \mathbf{x}), \quad \mathbf{u}_L,\ \mathbf{u}_T \in \mathcal{S}'_+(\mathbf{R}^4)^3 \tag{3.18}$$

where

$$\mathrm{curl}\, \mathbf{u}_L(t, \mathbf{x}) = 0 \tag{3.19}$$

and

$$\mathrm{div}\, \mathbf{u}_T(t, \mathbf{x}) = 0, \tag{3.20}$$

which leads to

$$\mathbf{u}(t, \mathbf{x}) = \mathrm{grad}\, p_0(t, \mathbf{x}) + \mathrm{curl}\, \mathbf{p}(t, \mathbf{x}) \tag{3.21}$$

with the additional condition

$$\mathrm{div}\, \mathbf{p}(t, \mathbf{x}) = 0.$$

Two field equations result from this decomposition:

$$\rho D_t^2 \mathbf{u}_L(t, \mathbf{x}) - d(b + \frac{3}{4}a) * \Delta D_t \mathbf{u}_L(t, \mathbf{x}) = \mathrm{grad}\, q_0 \tag{3.22}$$

$$\rho D_t^2 \mathbf{u}_T(t, \mathbf{x}) - da * \Delta D_t \mathbf{u}_T(t, \mathbf{x}) = \mathrm{curl}\, \mathbf{q}(t, \mathbf{x}) \tag{3.23}$$

with given scalar q_0 and vectorial $\mathbf{q}(t, \mathbf{x})$ parts.

Definition 3.4. The fundamental resolvent 3×3 matrix $\mathcal{U} \in \mathcal{S}'_+(\mathbf{R}^4)^{3\times 3}$ satisfies:

$$\rho D_t^2 \mathcal{U}(t,\mathbf{x}) - da * \Delta D_t \mathcal{U}(t,\mathbf{x}) - d(b + \frac{1}{3}a) * \operatorname{grad}\operatorname{div} D_t \mathcal{U}(t,\mathbf{x}) \qquad (3.24)$$
$$= D_t\delta(t) \otimes \delta(\mathbf{x}) I_3$$

Application of $\mathcal{L}$ and $\mathcal{F}$ transforms leads to a matrix equation with respect to $\hat{\tilde{\mathcal{U}}}$,

$$\mathcal{B}(\lambda,\xi)\hat{\tilde{\mathcal{U}}}(\lambda,\xi) = I_3 \quad \forall(\lambda,\xi) \in \mathbf{C}_+ \times \mathbf{R}^3 \qquad (3.25)$$

with an explicitly known matrix function $\mathcal{B}(\lambda,\xi)$. The full Green tensor is given by

$$\mathcal{U}(t,\mathbf{x}) = \mathcal{L}^{-1}[\tilde{\mathcal{U}}(\lambda,\mathbf{x})](t)\Theta(t) \qquad (3.26)$$

with its $\mathcal{L}$-transform

$$\tilde{\mathcal{U}}(\lambda,\mathbf{x}) \qquad (3.27)$$
$$= \frac{1}{4\pi\tilde{da}(\lambda)}\left[\frac{1}{k_T^2(\lambda)}\nabla\otimes\nabla\left(\frac{e^{ik_T(\lambda)|\mathbf{x}|}}{|\mathbf{x}|} - \frac{e^{ik_L(\lambda)|\mathbf{x}|}}{|\mathbf{x}|}\right) - \frac{e^{ik_T(\lambda)|\mathbf{x}|}}{|\mathbf{x}|}I_3\right].$$

This allows us to get the explicit solution of the vectorial Cauchy problem

$$\mathbf{u}(t,\mathbf{x}) = U * \int_0^t \mathbf{f}(\tau,\mathbf{x})d\tau + U * [\Theta(t)\otimes\mathbf{u}_1(\mathbf{x})] + \Theta(t)\otimes\mathbf{u}_0. \qquad (3.28)$$

3.3. Half-space initial boundary problems

In the elastodynamic time-harmonic case the problem is reduced to Wiener-Hopf equations for half-plane-like cracks [18].

Applying the unilateral $\mathcal{L}$-transform to the longitudinal and transversal displacements, one gets a **pseudo-Lamé system** with the characteristic square roots

$$\eta_q(\lambda,\boldsymbol{\xi}') := \sqrt{|\boldsymbol{\xi}'|^2 - k_q^2(\lambda)}, \quad q = L \text{ or } T. \qquad (3.29)$$

The general forms of the $\mathcal{F}$-$\mathcal{L}$-transformed solutions for $\mathbf{R}^3_\pm$ are given by

$$\tilde{\mathbf{u}}_H^\pm(\lambda,\mathbf{x}) = \mathcal{G}^\pm(\lambda,\mathbf{x})\tilde{\varphi}^\pm(\lambda,\mathbf{x}') \qquad (3.30)$$

with the resolvent operators

$$\mathcal{G}^\pm(\lambda,\mathbf{x}) := \Theta(\pm x_3)\Phi^\pm(\lambda,\xi',x_3)\mathcal{F}_{\mathbf{x}'\to\xi'}\tilde{\varphi}^\pm(\lambda,\mathbf{x}') \qquad (3.31)$$

whose symbol matrices with different η_T and η_L are

$$\Phi^\pm(\lambda,\xi',x_3) := \begin{pmatrix} e^{\mp\eta_T x_3} & 0 & \mp\frac{i\xi_1}{\eta_L}e^{\mp\eta_L x_3} \\ 0 & e^{\mp\eta_T x_3} & \mp\frac{i\xi_2}{\eta_L}e^{\mp\eta_L x_3} \\ \mp\frac{i\xi_1}{\eta_T}e^{\mp\eta_T x_3} & \mp\frac{i\xi_2}{\eta_T}e^{\mp\eta_T x_3} & e^{\mp\eta_L x_3} \end{pmatrix}. \qquad (3.32)$$

The $\mathcal{F}$-transformed boundary matrices $\Phi_{(p)}$, corresponding to different boundary conditions (p), belong to a **matrix algebra** (see [18]):

$$\Phi_{(p)} = \begin{pmatrix} a_{11}\mathbf{R}_1 + b\mathbf{R}_2 & a_{12}\frac{\boldsymbol{\xi}'}{|\boldsymbol{\xi}'|} \\ a_{21}\frac{\boldsymbol{\xi}'}{|\boldsymbol{\xi}'|} & a_{22} \end{pmatrix} \tag{3.33}$$

with the rational projection matrices

$$\mathbf{R}_1 = \frac{1}{|\boldsymbol{\xi}'|^2}\begin{pmatrix} \xi_1^2 & \xi_1\xi_2 \\ \xi_1\xi_2 & \xi_2^2 \end{pmatrix} \tag{3.34}$$

$$\mathbf{R}_2 = \frac{1}{|\boldsymbol{\xi}'|^2}\begin{pmatrix} \xi_2^2 & -\xi_1\xi_2 \\ -\xi_1\xi_2 & \xi_1^2 \end{pmatrix} \tag{3.35}$$

and coefficients depending on the boundary conditions (p).

3.4. Initial boundary-transmission problems for Sommerfeld cracks

Now we consider the homogeneous boundary value problem.

Problem 3.5. For given density $\rho > 0$ and relaxation functions a, b, one has to find the displacement $\mathbf{u}_H(t, \mathbf{x}) \in H^1_\tau(\mathbf{R}_+ \times \Omega)$, where $H^1_\tau(\mathbf{R}_+ \times \Omega)$ is a time-weighted Sobolev space of the first order on Ω, such that

$$L(\partial_t, \partial_x)\mathbf{u}_H(t, \mathbf{x}) = 0 \quad \forall (t, \mathbf{x}) \in \mathbf{R}_+ \times \Omega \tag{3.36}$$

with the homogeneous initial conditions

$$\mathbf{u}_H(+0, \mathbf{x}) = \partial_t \mathbf{u}_H(+0, \mathbf{x}) = 0 \text{ in } \Omega \tag{3.37}$$

and the boundary conditions

$$\gamma^{\pm}_{(p)}\mathbf{u}_H(t, \mathbf{x}') = g^{\pm}_{(p)}(t, \mathbf{x}'), \ (t, \mathbf{x}') \in \mathbf{R}_+ \times \Sigma \tag{3.38}$$

where Σ is the crack surface on $\mathbf{R}'^2$.

Lemma 3.6. *For given $\rho > 0$ and relaxation functions a, b, the function $\tilde{\mathbf{u}}_H(\lambda, \mathbf{x})$ is the general solution to*

$$L(\lambda, \xi)\tilde{\mathbf{u}}_H(\lambda, \mathbf{x}) = 0$$

if and only if there are no jumps across the half-plane $\Sigma' \subset \mathbf{R}'^2$ complementary to the crack Σ.

The main result can be formulated as the following theorem

Theorem 3.7. *Let $\Phi_{(p)}$ be the symbol matrices corresponding with the four most important boundary conditions on the crack surfaces. The related Wiener-Hopf operators $W_{(p)}$ are linear and continuous. The corresponding boundary-transmission problems are well posed and uniquely solvable if and only if the matrices $W_{(p)}$ possess a canonical factorization.*

4. Conclusions

The viscoelastodynamic situation completely carries over from the pure elastodynamic boundary value problems for cracks in the case of time-harmonic waves as they were treated by the author and F.-O. Speck [18]. Some progress in attacking time-aperiodic boundary-transmission problems allowed us to close the gap between the Wiener-Hopf and Maliuzhinets methods. The explicit factorization of parameter-dependent Fourier symbols relies strongly on the time-harmonic case, but it can be extended to further cases including, for example, the integrodifferential equations of linear viscoelastodynamics.

Here we did not discuss the wedge problem. A treatment of this problem can be found in the habilitation thesis of K. Rottbrand [21].

References

[1] M.A. Bastos and A.F. Dos Santos, Generalized factorization for a class of 2×2 matrix functions with non-rational entries. *Appl.Anal.*, 46(1–2):101–127, 1992.

[2] M.A. Bastos and A.F. Dos Santos, Generalized factorization for a class of 2×2 matrix functions with rationally-independent entries. *Complex Variables Theory Appl.*, 22(3–4):153–174, 1993.

[3] M.C. Câmara, A.B. Lebre, and F.-O. Speck, Meromorphic factorization, partial index estimates and elastodynamic diffraction problems. *Math. Nachr.*, 157:291–317, 1992.

[4] M.C. Câmara, A.B. Lebre, and F.-O. Speck, Generalized factorization for a class of Jones form matrix functions. *Proc. R. Soc. Edinburgh*, 123A:401–422, 1993.

[5] D. Colton and R. Kress, *Inverse Acoustic and Electromagnetic Scattering Theory*, volume 93 of *Applied Mathematical Sciences.* Springer-Verlag, Heidelberg 1992.

[6] G. DaPrato and M. Iannelli, Linear abstract integro-differential equations of hyperbolic type in Hilbert space. *Rend.Sem.Mat.Univ.Padova*, 62:191–206, 1980.

[7] R. du Cloux, Symmetry properties of elastodynamic wave fields and their application to space-time scatteringtheory. Technical Report 1986–14, TU Delft, 1986.

[8] C. Erbe, On Sommerfeld's half-plane problem for the equations of linear thermoelasticity. *Math. Meth. in the Appl. Sci.*, 18:1215–1237, 1995.

[9] D.S. Jones, *The Theory of Electromagnetism.* International Series of Monographs on Pure and Applied Mathematics, Pergamon, Oxford 1964.

[10] D.S. Jones, Factorization of a Wiener-Hopf matrix. *IMA J. Appl. Math.*, (32):211–220, 1984.

[11] V.D. Kupradze, editor, *Three-Dimensional Problems of the Mathematical Theory of Elasticity and Thermoelasticity.* North-Holland, Amsterdam, 1979 (in Russian 1976).

[12] R. Leis, *Initial Boundary Value Problems in Mathematical Physics.* B.G. Teubner, Stuttgart, J.Wiley, Chichester, 1986.

[13] G.D. Maliuzhinets, Inversion formula for the Sommerfeld integral. *Sov. Phys. Dokl.*, (3):52–56, 1958.

[14] J. Mark, *Initial-Boundary Value Problems in Linear Viscoelasticity using Wiener-Hopf methods.* PhD thesis, TU Darmstadt, 2000.

[15] E. Meister and L. Meister, Quaternion boundary value problems for canonical objects in continuum physics. In *Advances in Applied Clifford Algebras*, 11(2):231-246, 2001.

[16] E. Meister and L. Meister, Some initial boundary problems in electrodynamics for canonical domains in quaternions. *Mathematica Bohemica*, 126(4):429-442, 2001 (to be published).

[17] E. Meister, A. Passow, and K. Rottbrand, New results on wave diffraction by canonical obstacles. *Oper. Theory: Adv. and Appl.*, 110:235–256, 1999.

[18] E. Meister and F.-O. Speck, The explicit solution of elastodynamical diffraction problems by symbol factorization. *Zeitschrift für Analysis und ihre Anwendungen*, 8(4):307–328, 1989.

[19] C. Müller, *Foundations of the Mathematical Theory of Electromagnetic Waves.* Springer-Verlag, Berlin, Heidelberg, New York, 1969.

[20] A. Passow, *Sommerfeld Halbebenenproblem mit elektromagnetisch anisotropen Leontovich-Randbedingungen.* PhD thesis, TU Darmstadt, 1998.

[21] K. Rottbrand, *Diffraction of Plane Waves and Exact Time-Space Solutions for Some Linear Models in Classical Physics.* Habilitationsschrift, TU Darmstadt, 2000.

[22] A. Sommerfeld, Mathematische Theorie der Diffraction. *Math. Ann.*, 47:317–374, 1896.

[23] F.-O. Speck, *General Wiener-Hopf Factorization Methods.* Pitman, London, 1985.

[24] F.S. Teixeira, Generalized factorization for a class of symbols in $[PC(\mathbf{R})]^{2\times 2}$. *Appl. Anal.*, 36:95–117, 1990.

[25] N. Wiener and E. Hopf, Über eine Klasse singulärer Integralgleichungen. *Semester-Ber. Preuss. Akad. Wiss. Berlin, Phys.-Math. Kl.*, (30/32):696–706, 1931.

Erhard Meister
Fachbereich Mathematik, Arbeitsgruppe 12
Technische Universität Darmstadt
Schlossgartenstrasse 7
D-64289 Darmstadt, Germany
e-mail: `meister@mathematik.tu-darmstadt.de`

2000 *Mathematics Subject Classification.*
Primary 74J20, secondary 35C05, 35L05, 45E10, 78A45.

Received: 29 June 2001.

Operator Theory:
Advances and Applications, Vol. 142, 263–285

An Axiomatic Approach to the Limit Operators Method

Vladimir S. Rabinovich and Steffen Roch

Abstract. We propose an axiomatic approach to the limit operators method. This approach will be applied to operators in a C^*-algebra which is generated by operators of right convolution on a homogeneous non-commutative group $\mathbb{X}$ and by operators of multiplication by functions in $L_\infty(\mathbb{X})$. In terms of limit operators, we derive necessary and sufficient conditions for these operators to be semi-Fredholm or Fredholm. As another application, we obtain necessary and sufficient conditions for the semi-Fredholmness and Fredholmness of pseudodifferential operators.

1. Introduction

The first appearance of limit operators is in Favard's paper [2] where they are used to verify the existence of almost-periodic solutions of ordinary differential equations with almost-periodic coefficients. Later, Muhamadiev [10, 11] applied limit operators to the question of solvability of elliptic partial differential equations in $\mathbb{R}^n$. The limit operators method has been further developed in the papers [6, 7, 8, 13, 14, 15] for the study of the Fredholm property of wide classes of pseudodifferential operators and convolution operators on $\mathbb{R}^n$ and $\mathbb{Z}^n$. Note also the paper [1], where the applicability of the limit operators method to the computation of the essential spectrum of singular integral operators on Carleson curves acting in general weighted L^2-spaces has been illustrated. Observe that in all of these papers, the method of limit operators is applied to a concrete class of operators acting on a concrete Banach space.

In contrast to this, it is the first goal of the present paper to propose an axiomatic scheme for the application of the method of limit operators which contains many of the afore-mentioned applications as special cases. Then, by means of this scheme, we examine the Fredholm property of operators in a C^*-algebra which is generated by operators of convolution acting on $L^2(\mathbb{X})$ over a homogeneous non-commutative group $\mathbb{X}$ and by the operators of multiplication by functions in $L^\infty(\mathbb{X})$. Our aim is to give necessary and sufficient conditions for operators in this algebra to be semi-Fredholm or Fredholm.

A well-known and important example of a homogeneous group is the Heisenberg group. Singular integral operators and pseudodifferential operators on the

Heisenberg group have been intensively studied by many authors (see, for example, the monographs [17, 12, 21, 22], which contain extensive bibliographies). Let us also mention the papers [23, 24, 4, 5] which are devoted to the analysis of double convolutions on a class of step two nilpotent Lie groups.

The Fredholm property of operators in certain algebras generated by convolution operators and operators of multiplications by bounded functions on non-commutative locally compact groups was studied in [18, 19] by means of Simonenko's local principle (see [16]).

As another illustration of the abstract scheme, we apply it to the problem of Fredholmness and semi-Fredholmness of a class of pseudodifferential operators with double symbols, which is connected with convolutions on the Heisenberg group (see, for instance, [21, 22]).

2. Abstract scheme for the method of limit operators

Let H be a Hilbert space and $L(H)$ the C^*-algebra of all bounded linear operators acting on H. Throughout what follows, suppose that we are given

(A1) operators $P, \hat{P} \in L(H)$ with $P\hat{P} = \hat{P}P = P$;

(A2) a countable set $\{U_\alpha\}_{\alpha\in\Lambda}$ of unitary operators on H such that, with $P_\alpha := U_\alpha P U_\alpha^*$ and $\hat{P}_\alpha := U_\alpha \hat{P} U_\alpha^*$,

$$\sum_{\alpha\in\Lambda} \|P_\alpha u\|^2 = \|u\|^2 \quad \text{and} \quad \sum_{\alpha\in\Lambda} \|\hat{P}_\alpha u\|^2 \leq C\|u\|^2 \tag{2.1}$$

for all $u \in H$ with a constant C independent of u;

(A3) a sequence $(W_k)_{k\in\mathbb{N}}$ of unitary operators on H and an associated sequence $(D_k)_{k\in\mathbb{N}}$ of mappings from Λ into itself such that $W_k U_\alpha = U_{D_k(\alpha)} W_k$ for all $\alpha \in \Lambda$ and $k \in \mathbb{N}$, and such that the operators $\hat{P}^{(k)} := W_k \hat{P} W_k^*$ converge strongly to the identity operator on H. We also set $P^{(k)} := W_k P W_k^*$ and $P_{k,\,\alpha} := W_k P_\alpha W_k^*$ as well as $\hat{P}_{k,\,\alpha} := W_k \hat{P}_\alpha W_k^*$;

(A4) a bounded sequence $(Q_r)_{r\in\mathbb{N}}$ of operators in $L(H)$ such that
- there is a distinguished set $\mathfrak{B}$ of sequences in Λ which contains all sequences (β_m) for which there exist a $k \in \mathbb{N}$ and a sequence (r_m) in $\mathbb{N}$ tending to infinity such that
$$P_{k,\,\beta_m} Q_{r_m} \neq 0 \quad \text{for all } m \in \mathbb{N}; \tag{2.2}$$
- every subsequence of a sequence in $\mathfrak{B}$ belongs to $\mathfrak{B}$;
- the set $\mathfrak{B}$ is invariant with respect to each of the mappings D_k, i.e., if $(\beta_m) \in \mathfrak{B}$, then $(D_k\beta_m) \in \mathfrak{B}$ for every k;
- for each $r \in \mathbb{N}$ and each sequence $(\beta_m) \in \mathfrak{B}$,
$$\text{s-lim}_{m\to\infty} U_{\beta_m}^* Q_r U_{\beta_m} = I. \tag{2.3}$$

Since both the U_α and the W_k are unitary operators, one also has

$$\sum_{\alpha\in\Lambda} \|P_{k,\alpha}u\|^2 = \|u\|^2 \quad \text{and} \quad \sum_{\alpha\in\Lambda} \|\hat{P}_{k,\alpha}u\|^2 \le C\|u\|^2$$

for all $u \in H$ and $k \in \mathbb{N}$ and

$$P_{k,\alpha}\hat{P}_{k,\alpha} = \hat{P}_{k,\alpha}P_{k,\alpha} = P_{k,\alpha}$$

for all $\alpha \in \Lambda$ and $k \in \mathbb{N}$.

Definition 2.1. We say that the operator A_β is the *limit operator of* $A \in L(H)$ with respect to the sequence $\beta = (\beta_j) \in \mathfrak{B}$ if, for every $k \in \mathbb{N}$,

$$\lim_{j\to\infty} \|(U^*_{\beta_j} A U_{\beta_j} - A_\beta)\hat{P}^{(k)}\| = \lim_{j\to\infty} \|(\hat{P}^{(k)})^*(U^*_{\beta_j} A U_{\beta_j} - A_\beta)\| = 0.$$

The set of all limit operators of A with respect to sequences in $\mathfrak{B}$ will be denoted by $\lim_{\mathfrak{B}}(A)$.

The following proposition describes some elementary properties of limit operators.

Proposition 2.2. *Let $\beta \in \mathfrak{B}$, and let A, $B \in L(H)$ be operators for which the limit operators A_β and B_β exist. Then*

(a) $\|A_\beta\| \le \|A\|$;

(b) *$(A+B)_\beta$ exists and $(A+B)_\beta = A_\beta + B_\beta$;*

(c) *$(A^*)_\beta$ exists and $(A^*)_\beta = (A_\beta)^*$;*

(d) *if C, $C_n \in L(H)$ are operators with $\|C - C_n\| \to 0$, and if the limit operators $(C_n)_\beta$ exist for all sufficiently large n, then the limit operator C_β exists and $\|C_\beta - (C_n)_\beta\| \to 0$.*

Proof. We prove assertion (a) only. The other assertions can be checked as in [15]. Let $\varepsilon > 0$. For every $k \in \mathbb{N}$, choose $m(k)$ such that

$$\|(U^*_{\beta_{m(k)}} A U_{\beta_{m(k)}} - A_\beta)\hat{P}^{(k)}\| < \varepsilon.$$

Then, for every $u \in H$ and $k \in \mathbb{N}$,

$$\|A_\beta \hat{P}^{(k)}u\| \le \|U^*_{\beta_{m(k)}} A U_{\beta_{m(k)}} \hat{P}^{(k)}u\| + \varepsilon\|u\| \le \|A\|\,\|\hat{P}^{(k)}u\| + \varepsilon\|u\|.$$

Letting k go to infinity we get $\|A_\beta u\| \le (\|A\| + \varepsilon)\|u\|$, whence the assertion. □

Definition 2.3. Let $\mathcal{A}_0(H)$ denote the set of all operators $A \in L(H)$ with the following properties:

(a) $\lim_{k\to\infty} \|[P_{k,\alpha}, A]\| = 0$ and $\lim_{k\to\infty} \|[P_{k,\alpha}, A^*]\| = 0$ uniformly with respect to $\alpha \in \Lambda$,

(b) every sequence in $\mathfrak{B}$ possesses a subsequence β for which the limit operator A_β exists,

(c) there is a $k_0 \in \mathbb{N}$ such that $P_{k,\alpha}A = P_{k,\alpha}A\hat{P}_{k,\alpha}$ for all $k \ge k_0$,

and let $\mathcal{A}(H)$ denote the closure of $\mathcal{A}_0(H)$ in $L(H)$.

It is easy to check that $\mathcal{A}_0(H)$ and $\mathcal{A}(H)$ are linear spaces. Moreover, every operator A in $\mathcal{A}(H)$ also satisfies conditions (a) and (b) (the latter follows from Proposition 2.2), and if A and B are operators which satisfy (a), then their product also satisfies this condition. On the other hand, condition (c) (which is the abstract analogue of the band property) is not stable with respect to norm limits and products of operators.

In what follows we will need the notion of the lower norm

$$\nu(A) := \inf_{\|f\|=1} \|Af\|$$

of an operator $A \in L(H)$. It is well known that the operator A is invertible from the left if and only if $\nu(A) > 0$ and invertible from the right if and only if $\nu(A^*) > 0$. Thus, A is invertible if and only if both $\nu(A) > 0$ and $\nu(A^*) > 0$.

For every non-zero (but not necessarily closed) subspace L of H we also consider the lower norm of the restriction $A|_L$ of A onto L. If, in particular, L is the range of a non-zero operator $P \in L(H)$, then we call

$$\nu(A|_{P(H)}) = \inf_{\|Pf\|=1} \|APf\|$$

the lower norm of A relative to P.

Let us also recall that an operator $A \in L(H)$ is a Φ_+-operator if A has a closed range and a finite-dimensional kernel, whereas A is called a Φ_--operator if its adjoint A^* is a Φ_+-operator. Thus, if A is both a Φ_+ and a Φ_--operator, then A is Fredholm. $\Phi_\pm$-operators are also called semi-Fredholm operators. One can show that A is a Φ_+-operator if and only if there exists an operator $P \in L(H)$ such that $I - P$ is compact and $\nu(A|_{P(H)}) > 0$ ([9], Ch. I, Lemma 2.1).

Theorem 2.4. *Let $A \in \mathcal{A}(H)$. Then*

$$\liminf_{r\to\infty} \nu(A|_{Q_r(H)}) = \inf_{A_\beta \in \lim_{\mathfrak{B}}(A)} \nu(A_\beta). \tag{2.4}$$

We split the proof into three propositions.

Proposition 2.5. *Let $A \in \mathcal{A}_0(H)$. Then, for every $A_\beta \in \lim_{\mathfrak{B}}(A)$ and every $r \in \mathbb{N}$,*

$$\nu(A|_{Q_r(H)}) \le \nu(A_\beta).$$

Proof. Let $\beta = (\beta_m) \in \mathfrak{B}$ be such that the limit operator A_β exists. Then, for all $u \in H$ and $r \in \mathbb{N}$,

$$\|AQ_r u\| \ge \nu(A|_{Q_r(H)}) \, \|Q_r u\|.$$

Consequently, for all $k \in \mathbb{N}$,

$$\|U^*_{\beta_m} A U_{\beta_m} U^*_{\beta_m} Q_r U_{\beta_m} \hat{P}^{(k)} u\| \ge \nu(A|_{Q_r(H)}) \, \|U^*_{\beta_m} Q_r U_{\beta_m} \hat{P}^{(k)} u\|.$$

Letting m go to infinity, we obtain

$$\|A_\beta \hat{P}^{(k)} u\| \ge \nu(A|_{Q_r(H)}) \, \|\hat{P}^{(k)} u\|,$$

and passing to the limit $k \to \infty$ yields the assertion. □

Proposition 2.6. *Let $A \in \mathcal{A}_0(H)$. Then*

$$\inf_k \inf_{(\beta_j)\in\mathfrak{B}} \liminf_{j\to\infty} \nu(A|_{P_{k,\,\beta_j}(H)}) \leq \liminf_{r\to\infty} \nu(A|_{Q_r(H)}).$$

Proof. Abbreviate $\mu_A := \liminf_{r\to\infty} \nu(A|_{Q_r(H)})$ and let $\varepsilon > 0$. Choose $k \in \mathbb{N}$ such that

$$\|[P_{k,\,\alpha},\, A]\| \leq \varepsilon \quad \text{for all } \alpha \in \Lambda$$

and such that

$$P_{k,\,\alpha}A = P_{k,\,\alpha}A\hat{P}_{k,\,\alpha}.$$

Further, let (r_m) be a sequence tending to infinity such that

$$\nu(A|_{Q_{r_m}(H)}) \leq \mu_A + \varepsilon \quad \text{for every } m,$$

and choose vectors $w_m = Q_{r_m}v_m$ with $\|w_m\| = 1$ such that

$$\|Aw_m\| \leq \mu_A + 2\varepsilon \quad \text{for every } m. \tag{2.5}$$

Axioms (A1) and (A2) and the Hölder inequality imply that

$$\begin{aligned}
&\sum_{\alpha\in\Lambda} \|AP_{k,\,\alpha}w_m\|^2 \\
&\quad = \sum_{\alpha\in\Lambda} \|AP_{k,\,\alpha}\hat{P}_{k,\,\alpha}w_m\|^2 \\
&\quad \leq \sum_{\alpha\in\Lambda} \Big(\|P_{k,\,\alpha}A\hat{P}_{k,\,\alpha}w_m\| + \|[P_{k,\,\alpha},\, A]\hat{P}_{k,\,\alpha}w_m\|\Big)^2 \\
&\quad = \sum_{\alpha\in\Lambda} \|P_{k,\,\alpha}Aw_m\|^2 + 2\varepsilon C\left(\sum_{\alpha\in\Lambda} \|P_{k,\,\alpha}Aw_m\|^2\right)^{1/2} \|w_m\| + (\varepsilon C)^2\|w_m\|^2 \\
&\quad = (\|Aw_m\| + \varepsilon C\|w_m\|)^2 \leq (\mu_A + (2+C)\varepsilon)^2\|w_m\|^2.
\end{aligned}$$

Taking into account the identity

$$1 = \|w_m\|^2 = \sum_{\alpha\in\Lambda} \|P_{k,\,\alpha}w_m\|^2,$$

we conclude from the last estimate that

$$\frac{\sum_{\alpha\in\Lambda} \|AP_{k,\,\alpha}Aw_m\|^2}{\sum_{\alpha\in\Lambda} \|P_{k,\,\alpha}w_m\|^2} \leq (\mu_A + (2+C)\varepsilon)^2.$$

In particular, for every $m \in \mathbb{N}$, there is an α_m such that $P_{k,\,\alpha_m}w_m \neq 0$ and

$$\frac{\|AP_{k,\,\alpha_m}Q_{r_m}v_m\|}{\|P_{k,\,\alpha_m}Q_{r_m}v_m\|} \leq \mu_A + (2+C)\varepsilon. \tag{2.6}$$

Observe that the condition $P_{k,\alpha_m}Q_{r_m}v_m \neq 0$ guarantees that the sequence (α_m) belongs to $\mathfrak{B}$. Thus, for every $\varepsilon > 0$, there exist a $k \in \mathbb{N}$ and a sequence $(\alpha_m) \in \mathfrak{B}$ such that

$$\nu(A|_{P_{k,\,\alpha_m}(H)}) \leq \mu_A + (2+C)\varepsilon.$$

This finally shows that

$$\inf_k \inf_{(\alpha_m)\in\mathfrak{B}} \liminf_{m\to\infty} \nu(A|_{P_{k,\alpha_m}(H)}) \le \mu_A + (2+C)\varepsilon.$$

Since $\varepsilon > 0$ is arbitrary, we arrive at the assertion. □

Proposition 2.7. *Let $A \in \mathcal{A}_0(H)$. Then, for all $k \in \mathbb{N}$ and $\beta = (\beta_m) \in \mathfrak{B}$,*

$$\inf_{A_\delta \in \lim_{\mathfrak{B}}(A)} \nu(A_\delta) \le \liminf_{m\to\infty} \nu(A|_{P_{k,\beta_m}(H)}).$$

Proof. Set $\nu_A := \liminf_{m\to\infty} \nu(A|_{P_{k,\beta_m}(H)})$ and let $\varepsilon > 0$. Choose a subsequence $\gamma = (\gamma_j) \in \mathfrak{B}$ of β such that

$$\nu_A = \lim_{j\to\infty} \nu(A|_{P_{k,\gamma_j}(H)})$$

and vectors $f_j \in H$ with $\|P_{k,\gamma_j} f_j\| = 1$ such that

$$\nu_A = \lim_{j\to\infty} \|AP_{k,\gamma_j} f_j\|.$$

Thus, $\|AP_{k,\gamma_j} f_j\| < \nu_A + \varepsilon$ for all sufficiently large j.

Let $\delta \in \mathfrak{B}$ be a subsequence of $(D_k\gamma_j)_{j\in\mathbb{N}}$, say $\delta = (\delta_m)_{m\in\mathbb{N}}$ with $\delta_m = D_k\gamma_{j_m}$, such that the limit operator A_δ exists. Then

$$\begin{aligned}
&\|A_\delta P^{(k)} U^*_{\delta_m} f_{j_m}\| \\
&\quad \le \|(U^*_{\delta_m} A U_{\delta_m} - A_\delta)\hat{P}^{(k)} P^{(k)} U^*_{\delta_m} f_{j_m}\| + \|U^*_{\delta_m} A U_{\delta_m} P^{(k)} U^*_{\delta_m} f_{j_m}\| \\
&\quad \le \|(U^*_{\delta_m} A U_{\delta_m} - A_\delta)\hat{P}^{(k)}\| \, \|P^{(k)} U^*_{\delta_m} f_{j_m}\| + \|A U_{\delta_m} P^{(k)} U^*_{\delta_m} f_{j_m}\|.
\end{aligned}$$

Since

$$U_{\delta_m} P^{(k)} U^*_{\delta_m} = U_{D_k\gamma_{j_m}} W_k P W_k^* U^*_{D_k\gamma_{j_m}} = W_k U_{\gamma_{j_m}} P U^*_{\gamma_{j_m}} W_k^* = P_{k,\gamma_{j_m}},$$

we conclude that

$$\|A_\delta P^{(k)} U^*_{\delta_m} f_{j_m}\| \le \varepsilon + \|AP_{k,\gamma_{j_m}} f_{j_m}\| \le \nu_A + 2\varepsilon$$

for all sufficiently large m. Hence,

$$\frac{\|A_\delta P^{(k)} U^*_{\delta_m} f_{j_m}\|}{\|P^{(k)} U^*_{\delta_m} f_{j_m}\|} = \frac{\|A_\delta P^{(k)} U^*_{\delta_m} f_{j_m}\|}{\|P_{k,\gamma_{j_m}} f_{j_m}\|} \le \nu_A + 2\varepsilon,$$

whence $\nu(A_\delta) \le \nu_A + 2\varepsilon$ and the assertion. □

We are now in a position to prove Theorem 2.4.

Proof. First let $A \in \mathcal{A}_0(H)$. From Proposition 2.5 we infer that

$$\liminf_{r\to\infty} \nu(A|_{Q_r(H)}) \le \inf_{A_\beta \in \lim_{\mathfrak{B}}(A)} \nu(A_\beta).$$

In Proposition 2.6 we proved that

$$\inf_k \inf_{(\beta_j)\in\mathfrak{B}} \liminf_{j\to\infty} \nu(A|_{P_{k,\beta_j}(H)}) \le \liminf_{r\to\infty} \nu(A|_{Q_r(H)}).$$

Finally, we conclude from Proposition 2.7 that

$$\inf_{A_\delta \in \lim_{\mathfrak{B}}(A)} \nu(A_\delta) \leq \inf_k \inf_{(\beta_j)\in\mathfrak{B}} \liminf_{j\to\infty} \nu(A|_{P_{k,\beta_j}(H)}).$$

These three estimates yield the assertion in case $A \in \mathcal{A}_0(H)$.

Let now $A \in \mathcal{A}(H)$, and abbreviate the left-hand side of (2.4) by S_l and the right-hand side by S_r. If $\varepsilon > 0$ is arbitrarily given, then there is an r_0 such that

$$\nu(A|_{Q_r(H)}) \geq S_l - \varepsilon \quad \text{for all } r \geq r_0.$$

Thus,

$$\|AQ_r u\| \geq (S_l - \varepsilon)\|Q_r u\| \quad \text{for all } r \geq r_0 \text{ and all } u \in H.$$

Let $A' \in \mathcal{A}_0(H)$ be such that $\|A - A'\| < \varepsilon$. Then

$$\|A'Q_r u\| \geq \|AQ_r u\| - \|A - A'\| \, \|Q_r u\| \geq (S_l - 2\varepsilon)\|Q_r u\|$$

for all $r \geq r_0$ and $u \in H$. Consequently, $\liminf_{r\to\infty} \nu(A'|_{Q_r(H)}) \geq S_l - 2\varepsilon$. By what has already been proved, this implies that

$$\inf_{A'_\beta \in \lim_{\mathfrak{B}}(A')} \nu(A'_\beta) \geq S_l - 2\varepsilon$$

or, equivalently,

$$\|A'_\beta u\| \geq (S_l - 2\varepsilon)\|u\| \tag{2.7}$$

for all limit operators A'_β of A' and for all $u \in H$. If the limit operator of A with respect to the sequence β exists, then the limit operator of A' with respect to β exists, too, and by Proposition 2.2 one has $\|A'_\beta - A_\beta\| \leq \|A' - A\| \leq \varepsilon$. Thus, (2.7) implies

$$\|A_\beta u\| \geq \|A'_\beta u\| - \|A'_\beta - A_\beta\| \, \|u\| \geq (S_l - 3\varepsilon)\|u\|$$

for all $u \in H$, whence $\nu(A_\beta) \geq S_l - 3\varepsilon$. Consequently, $S_r \geq S_l$, and the reverse inequality can be proved in the same way. □

3. Fredholmness of convolution operators on homogeneous groups

3.1. Some notations

Following [17], Chapter XIII, Section 5, we start with recalling some facts concerning homogeneous groups which are needed in what follows.

Homogeneous groups arise from $\mathbb{R}^m$ by equipping this space with a Lie group structure and with a family of dilations that act as group automorphisms on this space. To be precise, to make $\mathbb{R}^m$ to a homogeneous group $\mathbb{X}$, we assume that there is a pair of mappings

$$\mathbb{R}^m \to \mathbb{R}^m : (x, y) \mapsto x \cdot y \quad \text{and} \quad \mathbb{R}^m \to \mathbb{R}^m : x \mapsto x^{-1}$$

which are smooth and which provide $\mathbb{R}^m$ with a Lie group structure such that $0 \in \mathbb{R}^m$ is the identity element of the Lie group. Further we suppose that there is an m-tuple of positive integers $a_1 \leq \cdots \leq a_m$ (with the monotonicity being no essential restriction) such that the dilations

$$x = (x_1, \ldots, x_m) \mapsto D_\delta x := (\delta^{a_1} x_1, \ldots, \delta^{a_m} x_m)$$

are group automorphisms for every $\delta > 0$, i.e., that

$$D_\delta(x \cdot y) = D_\delta x \cdot D_\delta y \quad \text{for all } x,\ y \in \mathbb{R}^m.$$

As follows from these properties, the group operation is necessarily of the form

$$x \cdot y = x + y + Q(x, y),$$

where $Q : \mathbb{R}^m \times \mathbb{R}^m$ satisfies

$$Q(0,0) = Q(x,0) = Q(0,x) = 0.$$

Moreover, $Q = (Q_1, \dots, Q_m)$, where every Q_k is a polynomial in $2m$ real variables which is homogeneous of degree a_k. Thus, Q contains no pure monomials in x or y.

The Euclidean measure dx on $\mathbb{R}^m$ is both left- and right-invariant with respect to the group multiplication. Thus, it is a Haar measure on $\mathbb{X}$. Note also that $d(D_\delta x) = \delta^a dx$, where $a := a_1 + \cdots + a_m$.

A nontrivial example of a homogeneous non-commutative group is the Heisenberg group, which can be identified with $\mathbb{C}^n \times \mathbb{R}$ with the group operation

$$(w, s) \cdot (z, t) = (w + z, s + t + 2\,\mathrm{Im}\,\langle w, z\rangle)$$

where $\langle w, z\rangle := \sum_{j=1}^n z_j \bar{w}_j$.

The norm function ρ on $\mathbb{R}^m$ is defined as

$$\rho(x) := \max\{|x_j|^{1/a_j} : 1 \le j \le m\}.$$

Note that $\rho(x) \ge 0$ and $\rho(x) = 0$ if and only if $x = 0$. Also, $\rho(D_\delta x) = \delta\rho(x)$, and there exists a constant $c > 0$ such that

$$\rho(x \cdot y) \le c(\rho(x) + \rho(y)) \quad \text{and} \quad \rho(x^{-1}) \le c\rho(x).$$

Set $\rho(x, y) := \rho(x^{-1} \cdot y)$. The collection $\{B(x,\varepsilon)\}_{\varepsilon > 0}$ of all balls

$$B(x,\varepsilon) := \{y \in \mathbb{X} : \rho(x,y) < \varepsilon\}$$

forms an open neighborhood base of the point $x \in \mathbb{X}$. Since ρ is left-invariant, one also hase $B(x,\varepsilon) = x \cdot B(0,\varepsilon)$, and because the measure is left-invariant,

$$|B(x,\varepsilon)| = |B(0,\varepsilon)| = \varepsilon^a |B(0,1)|.$$

3.2. Convolution operators on the homogeneous group

Let $\mathbb{X}$ be a homogeneous group. A function f is said to be uniformly continuous on $\mathbb{X}$ if, for each $\varepsilon > 0$, there exists an $\eta > 0$ such that $|f(x) - f(y)| < \varepsilon$ whenever $\rho(x,y) < \eta$. The class of all bounded uniformly continuous functions on $\mathbb{X}$ will be denoted by $BUC(\mathbb{X})$. Clearly, $BUC(\mathbb{X})$ is a C^*-subalgebra of $L^\infty(\mathbb{X})$.

Let $k \in L^1(\mathbb{X})$. Then we define the operator $C_{k,r}$ of right convolution by k by

$$(C_{k,r}u)(x) := \int_{\mathbb{R}^m} k(x^{-1} \cdot y) u(y) dy = \int_{\mathbb{R}^m} k(z) u(x \cdot z) dz, \quad x \in \mathbb{R}^m.$$

It is well known that $C_{k,r}$ is bounded on $L^2(\mathbb{R}^m)$ and invariant with respect to the left shift, i.e.,

$$U_{l,g} C_{k,r} = C_{k,r} U_{l,g} \quad \text{where} \quad (U_{l,g} f)(x) := f(g \cdot x) \quad \text{for } g \in \mathbb{X}.$$

We denote by $V_r(\mathbb{X})$ the set of all operators $C_{k,r}$ of right convolution by a function $k \in L^1(\mathbb{R}^m)$.

Let $\mathbb{Y}$ be a discrete subgroup of the group $\mathbb{X}$ which acts freely on $\mathbb{X}$ such that $\mathbb{X}/\mathbb{Y}$ is a compact manifold. Let M be a fundamental domain of $\mathbb{X}$ with respect to the action of $\mathbb{Y}$ on $\mathbb{X}$ by left shift. In particular, M is a bounded domain in $\mathbb{X}$ such that

$$\mathbb{X} = \bigcup_{\alpha \in \mathbb{Y}} \alpha \cdot \overline{M}.$$

Let M' be an open neighborhood of $\overline{M}$ such that the family $\{\alpha M'\}_{\alpha \in \mathbb{Y}}$ provides a covering of $\mathbb{X}$ of finite multiplicity. Let $f : \mathbb{X} \to [0, 1]$ be a continuous function with $f(x) = 1$ if $x \in \overline{M}$ and $f(x) = 0$ outside M', and let φ be the non-negative function which satisfies

$$\varphi^2(x) := \frac{f(x)}{\sum_{\beta \in \mathbb{Y}} f(\beta \cdot x)}.$$

For $\alpha \in \mathbb{Y}$, set $\varphi_\alpha(x) := \varphi(\alpha \cdot x)$. Evidently, $0 \le \varphi_\alpha(x) \le 1$ and

$$\sum_{\alpha \in \mathbb{Y}} \varphi_\alpha^2(x) = 1, \quad x \in \mathbb{X}. \tag{3.1}$$

In that sense, the family $\{\varphi_\alpha^2\}_{\alpha \in \mathbb{Y}}$ forms a partition of unity on $\mathbb{X}$. For $\delta > 0$, set $\varphi_{\delta,\alpha}(x) := \varphi_\alpha(D_\delta x)$.

Proposition 3.1. *Let $K \in V_r(\mathbb{X})$. Then $\lim_{\delta \to 0} \|[\varphi_{\delta,\alpha}, K]\| = 0$ uniformly with respect to $\alpha \in \mathbb{Y}$.*

Proof. Let

$$\gamma_1(\delta, \alpha) := \sup_{x \in \mathbb{X}} \int_{\mathbb{X}} |k(x^{-1} \cdot y)| \, |\varphi_{\delta,\alpha}(x) - \varphi_{\delta,\alpha}(y)| \, dy,$$

$$\gamma_2(\delta, \alpha) := \sup_{y \in \mathbb{X}} \int_{\mathbb{X}} |k(x^{-1} \cdot y)| \, |\varphi_{\delta,\alpha}(x) - \varphi_{\delta,\alpha}(y)| \, dx.$$

Then

$$\|[\varphi_{\delta,\alpha}, K]\| \le \max\{\gamma_1(\delta, \alpha), \gamma_2(\delta, \alpha)\}.$$

Let us suppose for a moment that $k(x) = 0$ if $\rho(x, 0) \ge R$. Then, for $j = 1, 2$,

$$\begin{aligned} \gamma_j(\delta, \alpha) &\le \sup_{\rho(x,y) \le R} |\varphi_{\delta,\alpha}(x) - \varphi_{\delta,\alpha}(y)| \int_{\mathbb{X}} |k(x)| \, dx \\ &\le \sup_{\rho(x,y) \le \delta R} |\varphi(x) - \varphi(y)| \int_{\mathbb{X}} |k(x)| \, dx \end{aligned}$$

where we used that $\rho(D_\delta x, D_\delta y) = \delta \rho(x, y)$. The function φ is uniformly continuous on $\mathbb{X}$. Thus, for each $\varepsilon > 0$, we find a $\delta > 0$ such that $\gamma_j(\delta, \alpha) \le \varepsilon$ for $j = 1, 2$ and for all $\alpha \in \mathbb{Y}$.

Since the set of all functions with compact support is dense in $L^1(\mathbb{X})$, we can use a standard approximation argument to get the assertion of the proposition for arbitrary kernel functions $k \in L^1(\mathbb{X})$. $\square$

Now we can specify the axioms (A1)–(A4) of the abstract scheme for the limit operators method as follows.

(A1) We let P be the operator of multiplication by φ and P' be the operator of multiplication by the characteristic function of a set $\hat{M}$, which contains the closure if M' in its interior and for which $\{\alpha\hat{M}\}_{\alpha\in\mathbb{Y}}$ pvovides a covering of $\mathbb{X}$ of finite multiplicity.

(A2) We choose $\Lambda := \mathbb{Y}$ and let U_α, $\alpha \in \Lambda$, be the operator of left shift by α,

$$(U_\alpha u)(x) := (U_{l,\,\alpha} u)(x) = u(\alpha \cdot x).$$

Observe that then P_α is the operator of multiplication by φ_α. Hence, the first condition in (2.1) follows from (3.1)

$$\sum \|P_\alpha u\|^2 = \sum \langle \varphi_\alpha u,\, \varphi_\alpha u\rangle = \sum \langle \varphi_\alpha^2 u,\, u\rangle = \langle u,\, u\rangle = \|u\|^2,$$

and the second one follows similarly due to the finite multiplicity of the covering $\{\alpha M'\}$.

(A3) We choose a sequence $(\delta_k)_{k\in\mathbb{N}}$ of positive numbers with $\delta_k \to 0$ as $k \to \infty$ and such that $D_{\delta_k^{-1}}\mathbb{Y} \subseteq \mathbb{Y}$, and we define W_k

$$(W_k u)(x) := \delta_k^{a/2} u(D_{\delta_k} x)$$

with $a = a_1 + \cdots + a_m$. Then

$$W_k U_\alpha = U_{D_{\delta_k^{-1}}\alpha} W_k \quad \text{for all } k \in \mathbb{N} \text{ and } \alpha \in \mathbb{Y},$$

and the operators $\hat{P}^{(k)}$ converge strongly to the identity.

(A4) For $r \in \mathbb{N}$, let Q_r be the operator of multiplication by the characteristic function of $\{x \in \mathbb{X} : \rho(x,\, 0) > r\}$, and let $\mathfrak{B}$ be the set of all sequences in $\mathbb{Y}$ which tend to infinity. Then conditions (2.2) and (2.3) are fulfilled.

Definition 3.2. We denote by $\mathcal{B}(BUC(\mathbb{X}),\, V_r(\mathbb{X}))$ the smallest C^*-subalgebra of $L^2(\mathbb{X})$ which contains all compact operators and all operators of the form

$$\sum_{i=1}^{p} \prod_{j=1}^{q} a_{ij} K_{ij} b_{ij} I \tag{3.2}$$

where $p,\, q \in \mathbb{N}$, $a_{ij}I$ and $b_{ij}I$ are operators of multiplication by functions a_{ij} and b_{ij} in $BUC(\mathbb{X})$, and the operators K_{ij} belong to $V_r(\mathbb{X})$. Let further $\dot{\mathcal{B}}(BUC(\mathbb{X}),\, V_r(\mathbb{X}))$ refer to the smallest unital C^*-subalgebra of $L^2(\mathbb{X})$ which contains all operators (3.2). Finally, $\dot{\mathcal{B}}_0(BUC(\mathbb{X}),\, V_r(\mathbb{X}))$ refers to the smallest (not necessarily closed) algebra which contains the identity operator, all operators of the form

$$\varphi_\alpha C \varphi_\alpha I \quad \text{where } C \text{ is compact and } \alpha \in \mathbb{Z}^n, \tag{3.3}$$

and all operators of the form (3.2) where the kernel functions of the K_{ij} are compactly supported.

Proposition 3.3. *Let $A \in \dot{\mathcal{B}}(BUC(\mathbb{X}), V_r(\mathbb{X}))$. Then*

$$\lim_{k\to\infty} \|[A, P_{k,\alpha}]\| = 0 \quad \textit{uniformly with respect to } \alpha \in \mathbb{Y}. \tag{3.4}$$

Proof. For operators of the form (3.2), the proof follows immediately from Proposition 3.1. If A is compact (in particular, if A is of the form (3.3)), then the proof follows easily from the *-strong convergence of the operators $P_{k,\alpha}$ to $\varphi(\alpha)I$ as $k \to \infty$ and from the *-strong convergence of the operators $\hat{P}^{(m)}$ to the identity operator as $m \to \infty$. Indeed, given $\varepsilon > 0$, there is a m such that

$$\|\hat{P}^{(m)}A - A\| < \varepsilon \quad \text{and} \quad \|A\hat{P}^{(m)} - A\| < \varepsilon.$$

Thus,

$$\|[A, P_{k,\alpha}]\| \le \|A\hat{P}^{(m)}(P_{k,\alpha} - \varphi(\alpha)I)\| + \|(P_{k,\alpha} - \varphi(\alpha)I)\hat{P}^{(m)}A\| + 4\,\varepsilon.$$

Since $P_{k,\alpha} = \varphi_{k,\alpha}I$, and since the functions $\varphi_{k,\alpha}$ converge uniformly with respect to α to $\varphi(\alpha)$ on each compact subset of $\mathbb{X}$ as $k \to \infty$, one can find a k_0 such that, for $k \ge k_0$,

$$\|\hat{P}^{(m)}(P_{k,\alpha} - \varphi(\alpha)I)\| < \varepsilon \quad \text{and} \quad \|(P_{k,\alpha} - \varphi(\alpha)I)\hat{P}^{(m)}\| < \varepsilon.$$

Thus, $\|[A, P_{k,\alpha}]\| \le 2\,\|A\|\,\varepsilon + 4\,\varepsilon$ for $k \ge k_0$, whence the assertion for compact operators. Finally, since the set of all operators A which satisfy (3.4) is a C^*-subalgebra of $L(L^2(\mathbb{X}))$, we get the assertion in the general case. □

Proposition 3.4. *Let $A \in \dot{\mathcal{B}}_0(BUC(\mathbb{X}), V_r(\mathbb{X}))$. Then*

$$P_{k,\alpha}A = P_{k,\alpha}A\hat{P}_{k,\alpha}$$

for all sufficiently large k.

Proof. Given an open set $N \subseteq \mathbb{X}$, let χ^N refer to the characteristic function of N and, for $k \in \mathbb{N}$ and $\alpha \in \mathbb{Y}$, define $\chi^N_{k,\alpha}$ in analogy to $\varphi_{k,\alpha}$. Thus, $\hat{P}_{k,\alpha}$ is the operator of multiplication by $\chi^{\hat{M}}_{k,\alpha}$. We consider the set $\mathcal{C}$ of all operators B on $L^2(\mathbb{X})$ with the following property. If N_1, N_2 are open sets with $\overline{N_1} \subset N_2$ then, for every $\alpha \in \mathbb{Y}$ and every sufficiently large k,

$$\chi^{N_1}_{k,\alpha}B = \chi^{N_1}_{k,\alpha}B\chi^{N_2}_{k,\alpha}I.$$

It is evident that $\mathcal{C}$ contains all operators of the form (3.3), all operators aI of multiplication by bounded functions on $\mathbb{X}$ and all operators K of right convolution by a compactly supported measurable function on $\mathbb{X}$. Moreover, $\mathcal{C}$ also contains the adjoints of these operators. Thus, the assertion will follow once we have shown that $\mathcal{C}$ is an algebra.

Let A, B in $\mathcal{C}$. Then, clearly, $A + B \in \mathcal{C}$. To prove that $AB \in \mathcal{C}$, let N_1, N_2 be open sets with $\overline{N_1} \subset N_2$, and choose an open set N' such that

$$\overline{N_1} \subset N' \subset \overline{N'} \subset N_2.$$

Then

$$\chi^{N_1}_{k,\alpha}AB = \chi^{N_1}_{k,\alpha}A\chi^{N'}_{k,\alpha}B\chi^{N_2}_{k,\alpha}I = \chi^{N_1}_{k,\alpha}AB\chi^{N_2}_{k,\alpha}I,$$

which implies that $AB \in \mathcal{C}$. Consequently, $\dot{\mathcal{B}}_0(BUC(\mathbb{X}), V_r(\mathbb{X})) \subseteq \mathcal{C}$, whence the assertion. □

Proposition 3.5. *Let $A \in \dot{\mathcal{B}}(BUC(\mathbb{X}), V_r(\mathbb{X}))$. Then every sequence in $\mathfrak{B}$ has a subsequence β such that the limit operator A_β exists.*

Proof. We start with verifying the assertion for the operators in the algebra $\mathcal{B}_0 := \dot{\mathcal{B}}_0(BUC(\mathbb{X}), V_r(\mathbb{X}))$. Let $\mathcal{D}$ denote the set of all operators $A \in \mathcal{B}_0$ having the property that every sequence in $\mathfrak{B}$ has a subsequence $\tilde{\beta} = (\tilde{\beta}_k)$ such that the operators $U_{\tilde{\beta}_k}^{-1} A U_{\tilde{\beta}_k}$ converge *-strongly as $k \to \infty$.

We claim that $\mathcal{D}$ contains the generating operators of $\mathcal{B}_0$. Since $\mathcal{D}$ is an algebra (which can be easily shown), this implies that $\mathcal{B}_0 \subseteq \mathcal{D}$.

If C is compact, then the limit operator of C with respect to every sequence $\beta \in \mathfrak{B}$ exists and is equal to zero. If aI is the operator of multiplication by the function $a \in BUC(\mathbb{X})$, then $U_{\beta_k}^{-1} a U_{\beta_k}$ is the operator of multiplication by the function $x \mapsto a(\beta_k \cdot x)$. The functions in $BUC(\mathbb{X})$ are bounded and uniformly continuous by definition. Hence, by the Arzelà-Ascoli theorem, the sequence β possesses a subsequence $\tilde{\beta}$ such that the functions $x \mapsto a(\tilde{\beta}_k \cdot x)$ tend uniformly on compact subsets of $\mathbb{X}$ to a certain bounded function $a_{\tilde{\beta}}$ as $k \to \infty$. Consequently, the operators $U_{\tilde{\beta}_k}^{-1} a U_{\tilde{\beta}_k}$ converge strongly to $a_{\tilde{\beta}} I$, and the strong convergence of the adjoint sequence follows analogously.

If $A = K$ is a convolution operator, then there is nothing to prove because A commutes with the U_k.

The remaining part of the proof makes use of the elementary fact that, if $A_n \to A$ and $B_n^* \to B^*$ strongly, and if K is compact, then $\|A_n K B_n - AKB\| \to 0$.

Every operator in $\mathcal{B}_0$ can be written as a sum of operators of the form ACB where C is the operator of convolution by a compactly supported function, and where $A, B \in \mathcal{B}_0$. Let $\beta \in \mathfrak{B}$ and $m \in \mathbb{N}$. We choose a subsequence $\tilde{\beta}$ of β such that

$$U_{\tilde{\beta}_k}^{-1} A U_{\tilde{\beta}_k} \to A_{\tilde{\beta}} \quad \text{and} \quad U_{\tilde{\beta}_k}^{-1} B U_{\tilde{\beta}_k} \to B_{\tilde{\beta}}$$

*-strongly as $k \to \infty$. It follows from the proof of the previous proposition that, if N is sufficiently large,

$$\begin{aligned} U_{\tilde{\beta}_k}^{-1} ACB U_{\tilde{\beta}_k} \hat{P}^{(m)} &= U_{\tilde{\beta}_k}^{-1} A U_{\tilde{\beta}_k} C U_{\tilde{\beta}_k}^{-1} B U_{\tilde{\beta}_k} \hat{P}^{(m)} \\ &= U_{\tilde{\beta}_k}^{-1} A U_{\tilde{\beta}_k} C \hat{P}^{(N)} U_{\tilde{\beta}_k}^{-1} B U_{\tilde{\beta}_k} \hat{P}^{(m)}. \end{aligned}$$

Since the operator $C\hat{P}^{(N)}$ is compact, it follows from the fact just mentioned that

$$\|(U_{\tilde{\beta}_k}^{-1} ACB U_{\tilde{\beta}_k} - A_{\tilde{\beta}} C B_{\tilde{\beta}}) \hat{P}^{(m)}\| \to 0$$

as $k \to \infty$. The dual condition follows analogously. Thus, $A_{\tilde{\beta}} C B_{\tilde{\beta}}$ is a limit operator of ACB. This yields the assertion for operators in $\mathcal{B}_0$ and, employing Proposition 2.2 (d), also for operators in $\dot{\mathcal{B}}(BUC(\mathbb{X}), V_r(\mathbb{X}))$. □

For $A \in \dot{\mathcal{B}}(BUC(\mathbb{X}), V_r(\mathbb{X}))$, we denote by $\lim_\infty(A)$ the set of all limit operators of A which are defined with respect to sequences in $\mathfrak{B}$. Thus, the following theorem is a corollary of the general Theorem 2.4.

Theorem 3.6. *Let* $A \in \dot{\mathcal{B}}(BUC(\mathbb{X}), V_r(\mathbb{X}))$. *Then*

$$\liminf_{r\to\infty} \nu(A|_{Q_r(L^2(\mathbb{X}))}) = \inf\{\nu(A_\beta) : A_\beta \in \lim_\infty(A)\}.$$

Theorem 3.7. *Let* $A \in \dot{\mathcal{B}}(BUC(\mathbb{X}), V_r(\mathbb{X}))$. *Then*

(a) *A is a Φ_+-operator if and only if* $\inf\{\nu(A_\beta) : A_\beta \in \lim_\infty(A)\} > 0$;

(b) *A is a Φ_--operator if and only if* $\inf\{\nu(A_\beta^*) : A_\beta \in \lim_\infty(A)\} > 0$;

(c) *A is a Fredholm operator if and only if all operators $A_\beta \in \lim_\infty(A)$ are uniformly invertible, i.e., if and only if* $\sup\{\|A_\beta^{-1}\| : A_\beta \in \lim_\infty(A)\} > 0$.

Proof. We will show assertion (a) only. The proof of (b) proceeds similarly, and (c) is a consequence of (a) and (b).

Let $\inf\{\nu(A_\beta) : A_\beta \in \lim_\infty(A)\} > 0$. Then there exist $r \in \mathbb{N}$ and $C > 0$ such that

$$\|\langle Q_r A^* A Q_r f, Q_r f\rangle\| \geq C\|Q_r f\|^2 \quad \text{for every } f \in L^2(\mathbb{X}).$$

Thus, the operator $Q_r A^* A Q_r$ is invertible from the left on $L^2(Q_r\mathbb{X})$, i.e., there is an operator B such that

$$BQ_r A^* A Q_r = Q_r. \tag{3.5}$$

The operator B belongs to the C^*-subalgebra $\dot{\mathcal{B}}(BUC(\mathbb{X}), V_r(\mathbb{X}), Q_r)$ of $L^2(\mathbb{X})$ which is generated by the operators in $\dot{\mathcal{B}}(BUC(\mathbb{X}), V_r(\mathbb{X}))$ and by Q_r.

Let J_0 refer to the closed ideal of $\dot{\mathcal{B}}(BUC(\mathbb{X}), V_r(\mathbb{X}), Q_r)$ which is generated by the operators $I - Q_r$, $r \in \mathbb{N}$, and let J_1 stand for the smallest closed ideal of that algebra which contains all operators in $V_r(\mathbb{X})$ and all compact operators. It is evident from the definition of the algebra $\dot{\mathcal{B}}(BUC(\mathbb{X}), V_r(\mathbb{X}), Q_r)$ that for every operator G in this algebra, there is a (uniquely determined) complex number γ_G such that $G - \gamma_G I \in J_1$. Clearly, the mapping $G \mapsto \gamma_G$ is a continuous algebra homomorphism. Since $\gamma_{Q_r} = 1$, it follows from (3.5) that $\gamma_A \neq 0$.

The equality (3.5) further implies that there is an operator R' in the algebra $\dot{\mathcal{B}}(BUC(\mathbb{X}), V_r(\mathbb{X}), Q_r)$ such that $R'A - I \in J_0$. If we set $R := \gamma_A R' - AR' + I$, then

$$RA - \gamma_A I = \gamma_A R'A - AR'A + A - \gamma_A I = (\gamma_A I - A)(R'A - I).$$

Since $T_0 := R'A - I \in J_0$ and $T_1 := \gamma_A I - A \in J_1$, the operator $RA - \gamma_A I = T_1 T_0$ is compact. Hence, and because of $\gamma_A \neq 0$, A is a Φ_+-operator.

Conversely, let A be a Φ_+-operator. Then, as we have already remarked, the a priori estimate

$$C\|u\| \leq \|Au\| + \|Ku\|$$

holds with a certain compact operator K and a constant $C > 0$ ([9], Chapter I, Lemma 2.1). This estimate gives

$$\|AQ_r u\| \geq C\|Q_r u\| - \|KQ_r u\|.$$

Since Q_r converges *-strongly to 0 as $r \to \infty$, we have $\|KQ_{r_0}u\| \leq \frac{C}{2}\|Q_{r_0}u\|$ for a certain r_0. Hence, $\|AQ_{r_0}u\| \geq \frac{C}{2}\|Q_{r_0}u\|$. This implies the assertion via Theorem 3.6. □

3.3. Convolution operators on discrete subgroups of the homogeneous group

Let $l^2(\mathbb{Y})$ be the space of all complex valued functions u on the discrete group $\mathbb{Y}$ of $\mathbb{X}$ for which

$$\|u\|^2_{l^2(\mathbb{Y})} := \sum_{x\in\mathbb{Y}} |u(x)|^2 < \infty,$$

and write $l^\infty(\mathbb{Y})$ for the space of all bounded complex valued functions on $\mathbb{Y}$, provided with the norm

$$\|a\|_{l^\infty(\mathbb{Y})} := \sup_{x\in\mathbb{Y}} |a(x)|.$$

By aI we will denote the operator of multiplication by $a \in l^\infty(\mathbb{Y})$ thought of as acting on $l^2(\mathbb{Y})$. Further, given $g \in \mathbb{Y}$, we let $U_{g,l}$ and $U_{g,r}$ stand for the unitary operators of left and right shift acting at $u \in l^2(\mathbb{Y})$ by

$$(U_{g,l}u)(x) := u(g\cdot x) \quad \text{and} \quad (U_{g,r}u)(x) := u(x\cdot g), \quad x \in \mathbb{Y}.$$

Finally, for every function ψ on $\mathbb{X}$, we denote its restriction to $\mathbb{Y}$ by $\hat{\psi}$.

Definition 3.8. Let $\mathcal{B}(l^\infty(\mathbb{Y}), \{U_{g,r}\}_{g\in\mathbb{Y}})$ denote the closure in $L(l^2(\mathbb{Y}))$ of the set of all operators of the form

$$A_\Gamma := \sum_{g\in\Gamma} a_g U_{g,r} \quad \text{with } a_g \in l^\infty(\mathbb{Y}) \tag{3.6}$$

where Γ is a finite subset of $\mathbb{Y}$.

It turns out that $\mathcal{B}(l^\infty(\mathbb{Y}), \{U_{g,r}\}_{g\in\mathbb{Y}})$ is a C^*-subalgebra of $L(l^2(\mathbb{Y}))$.

Proposition 3.9. *Let $A \in \mathcal{B}(l^\infty(\mathbb{Y}), \{U_{g,r}\}_{g\in\mathbb{Y}})$ and $\varphi \in BUC(\mathbb{X})$. Then*

$$\lim_{\delta\to 0} \|[\hat{\varphi}_{\delta,\,g}, A]\|_{L(l^2(\mathbb{Y}))} = 0 \tag{3.7}$$

uniformly with respect to $g \in \mathbb{Y}$, where $\varphi_g(x) := \varphi(g\cdot x)$ and $\varphi_{\delta,\,g}(x) := \varphi_g(D_\delta x)$.

Proof. A simple calculation shows that

$$\begin{aligned}
\|[\hat{\varphi}_{\delta,\,g}, U_{z,r}]\| &= \|U_{z,r}^{-1}\hat{\varphi}_{\delta,\,g}U_{z,r} - \hat{\varphi}_{\delta,\,g}I\| \\
&\leq \sup_{y\in\mathbb{Y}} |\varphi_{\delta,\,g}(y\cdot z^{-1}) - \varphi_{\delta,\,g}(y)| \\
&= \sup_{y\in\mathbb{Y}} |\varphi(g\cdot D_\delta(y\cdot z^{-1})) - \varphi(g\cdot D_\delta y)| \\
&= \sup_{y\in\mathbb{Y}} |\varphi(g\cdot D_\delta y\cdot D_\delta z^{-1}) - \varphi(g\cdot D_\delta y)| \\
&\leq \sup_{x\in\mathbb{X}} |\varphi(x\cdot D_\delta z^{-1}) - \varphi(x)|.
\end{aligned}$$

Since φ is in BUC, the right-hand side becomes as small as desired if δ is sufficiently small. This verifies condition (3.7) for the shift operator. But then this condition holds for all operators of the form (3.6), and passage to the closure yields the proof of the proposition in the general case. □

To apply the abstract scheme proposed in Section 2, we will use the sequence of unitary operators $\{U_{g,l}\}_{g\in\mathbb{Y}}$ in place of the U_α. Further, we let P, $\hat{P}$ and W_k be as in the previous section, and we let Q_r be the operator of multiplication by the function $\hat{\chi}_r$ where χ_r is the characteristic function of $\{y \in \mathbb{Y} : \rho(y, 0) > r\}$.

We claim that every operator of the form (3.6) satisfies the conditions (a)–(c) in the definition of $\mathcal{A}_0$. Condition (a) is a consequence of Proposition 3.9, and condition (c) is obviously satisfied. To check (b), let A_Γ be an operator of the form (3.6), and let $h = (h_k)$ be a sequence in $\mathbb{Y}$ which tends to infinity. Then, for all $y \in \mathbb{Y}$,

$$\left(U_{h_{k,l}}^{-1} A_\Gamma U_{h_{k,l}} u\right)(y) = \sum_{g\in\Gamma} a_g(h_k^{-1} \cdot x)(U_{g,r}u)(y).$$

As follows from the Bolzano-Weierstrass theorem and the Cantor diagonalization procedure, there exists a subsequence $\tilde{h} = (h_{k_m})$ of h such that the pointwise limit

$$a_g(h_{k_m} \cdot y) \to (a_g)_{\tilde{h}}(y)$$

exists for each $g \in \Gamma$. This implies that, with $(A_\Gamma)_{\tilde{h}} := \sum_{g\in\Gamma}(a_g)_{\tilde{h}} U_{g,r}$,

$$\lim_{m\to\infty} \|(U_{h_{k_m},l}^{-1} A_\Gamma U_{h_{k_m},l} - (A_\Gamma)_{\tilde{h}})\hat{P}_r\| = 0 \quad \text{for all } r$$

and

$$\lim_{m\to\infty} \|\hat{P}_r^*(U_{h_{k_m},l}^{-1} A_\Gamma U_{h_{k_m},l} - (A_\Gamma)_{\tilde{h}})\| = 0 \quad \text{for all } r.$$

Thus, A_Γ belongs to $\mathcal{A}_0(l^2(\mathbb{Y}))$. Taking into account Proposition 2.2 (d), one concludes from this result that even

$$\mathcal{B}(l^\infty(\mathbb{Y}), \{U_{g,r}\}_{g\in\mathbb{Y}}) \subseteq \mathcal{A}(l^2(\mathbb{Y})).$$

The conditions (2.1), (2.2) are evidently satisfied in the present setting. So we obtain the following as a corollary of Theorem 2.4.

Theorem 3.10. *Let $A \in \mathcal{B}(l^\infty(\mathbb{Y}), \{U_{g,r}\}_{g\in\mathbb{Y}})$. Then the following assertions are equivalent:*

(a) $\liminf_{r\to\infty} \nu(A|_{Q_r(l^2(\mathbb{Y}))}) > 0$;

(b) $\inf\{\nu(A_\beta) : A_\beta \in \lim_\infty(A)\} > 0$.

The operators $I - Q_r$ are compact. So Theorem 3.10 has the following corollary, where $\lim_\infty(A)$ stands for the set of all limit operators of A with respect to the family $\{U_{g,l}\}_{g\in\mathbb{Y}}$.

Corollary 3.11. *Let $A \in \mathcal{B}(l^\infty(\mathbb{Y}), \{U_{g,r}\}_{g\in\mathbb{Y}})$. Then*

(a) *A is a Φ_+-operator if and only if* $\inf\{\nu(A_\beta) : A_\beta \in \lim_\infty(A)\} > 0$;

(b) *A is a Φ_--operator if and only if* $\inf\{\nu(A_\beta^*) : A_\beta \in \lim_\infty(A)\} > 0$;

(c) *A is a Fredholm operator if and only if all limit operators $A_\beta \in \lim_\infty(A)$ are uniformly invertible.*

4. Pseudodifferential operators

Let $a \in C^\infty(\mathbb{R}^n \times \mathbb{R}^n)$ be a function all derivatives of which are bounded, i.e.,

$$|\partial_x^\beta \partial_\xi^\alpha a(x, \xi)| \le C_{\alpha\beta}$$

for all multiindices α, $\beta \in \mathbb{N}$. We say that A is a pseudodifferential operator in the class $OPS^0_{0,0}$ with symbol a if

$$(Au)(x) = \int_{\mathbb{R}^n} \int_{\mathbb{R}^n} a(x, \xi) e^{i\langle x-y, \xi\rangle} u(y)\, dy\, d\xi$$

for all $u \in C_0^\infty(\mathbb{R}^n)$. In that case, A is denoted by $Op(a)$. If a depends on x only, then $Op(a)$ is the operator of multiplication by a which we denote by aI, and if a depends on ξ only, then $Op(a)$ is a convolution operator which we denote by $C(a)$. The Calderón-Vaillancourt theorem ([20]) states that operators in $OPS^0_{0,0}$ are bounded on $L^2(\mathbb{R}^n)$ and that

$$\|Op(a)\| \le C \sum_{|\alpha|+|\beta|\le m} \sup_{(x,\xi)\in\mathbb{R}^n\times\mathbb{R}^n} |\partial_x^\beta \partial_\xi^\alpha a(x, \xi)| \tag{4.1}$$

with a constant C independent of A.

Similarly, if $a \in C^\infty(\mathbb{R}^n \times \mathbb{R}^n \times \mathbb{R}^n)$ is a function such that

$$|\partial_x^\beta \partial_y^\gamma \partial_\xi^\alpha a(x, y, \xi)| \le C_{\alpha\beta\gamma}$$

for all multiindices α, β, $\gamma \in \mathbb{N}^n$, then the operator $A = Op_d(a)$ is called a pseudodifferential operator in the class $OPS^0_{0,0,0}$ with double symbol a if

$$(Au)(x) = (Op_d(a)u)(x) = \int_{\mathbb{R}^n} \int_{\mathbb{R}^n} a(x, y, \xi) e^{i\langle x-y, \xi\rangle} u(y)\, dy\, d\xi$$

for all $u \in C_0^\infty(\mathbb{R}^n)$. Every pseudodifferential operator $A = Op_d(a) \in OPS^0_{0,0,0}$ with double symbol can also be considered as a pseudodifferential operator in the class $OPS^0_{0,0}$ with symbol

$$(x, \xi) \mapsto (2\pi)^{-n} \int_{\mathbb{R}^n} \int_{\mathbb{R}^n} a(x, x+y, \xi+\eta) e^{-i\langle y, \eta\rangle}\, dy\, d\eta,$$

where the integral is understood as an oscillatory integral.

Let $C_b^\infty(\mathbb{R}^n)$ stand for the class of all C^∞-functions on $\mathbb{R}^n$ which are bounded together with all their derivatives.

Proposition 4.1. *Let $A \in OPS^0_{0,0}$ and $\varphi \in C_b^\infty(\mathbb{R}^n)$. For $r > 0$ and $\alpha \in \mathbb{Z}^n$, set $\varphi_\alpha(y) := \varphi(y - \alpha)$ and $\varphi_{r,\alpha}(y) := \varphi_\alpha(y/r)$. Then*

$$\lim_{r\to\infty} \|[A, \varphi_{r,\alpha} I]\| = \lim_{r\to\infty} \|[A, C(\varphi_{r,\alpha})]\| = 0 \tag{4.2}$$

uniformly with respect to $\alpha \in \mathbb{R}^n$.

Proof. This result follows easily from the composition rule for pseudodifferential operators and from estimate (4.1). □

Let $f \in C_0^\infty(\mathbb{R}^n)$ be an even function (i.e., $f(x) = f(-x)$ for all x) which is 1 on the cube $Q := \{x \in \mathbb{R}^n : |x_i| \le 2/3 \text{ for } i = 1, \dots, n\}$ and 0 outside $\frac{9}{8}Q$ and which takes values in $[0, 1]$ only. For $k \in \mathbb{N}$, define $f_k(x) := f(x/k)$ and, for every $\alpha \in \mathbb{Z}^n$, let φ_α be the non-negative function which satisfies

$$\varphi_\alpha^2(x) := \frac{f(x-\alpha)}{\sum_{\beta\in\mathbb{Z}^n} f(x-\beta)}.$$

It is evident that

$$\sum_{\alpha\in\mathbb{Z}^n} \varphi_\alpha^2(x) = 1 \quad \text{and} \quad 0 \le \varphi_\alpha(x) \le 1 \quad \text{for all } x \in \mathbb{R}^n.$$

Proposition 4.2. *Given $a \in C_b^\infty(\mathbb{R}^n \times \mathbb{R}^n)$ and $k \in \mathbb{N}$, define $a^{(k)}$ on $\mathbb{R}^n \times \mathbb{R}^n \times \mathbb{R}^n$ by $a^{(k)}(x, y, \xi) := a(x, \xi) f_k(x-y)$. Then*

$$\lim_{k\to\infty} \|Op(a) - Op_d(a^{(k)})\| = 0.$$

Proof. It is easy to check that $a^{(k)}$ lies in $C_b^\infty(\mathbb{R}^n \times \mathbb{R}^n \times \mathbb{R}^n)$. Thus,

$$Op(a) - Op_d(a^{(k)}) = Op_d(b^{(k)})$$

where $b^{(k)}(x, y, \xi) = a(x, \xi)\psi_k(x-y)$ and $\psi_k := 1 - f_k$. The symbol of $B_k := Op_d(b^{(k)})$ is given by

$$\begin{aligned}
&\sigma_{B_k}(x, \xi)\\
&\quad = (2\pi)^{-n} \int\!\!\int a(x, \xi+\eta)\psi_k(y) e^{-i\langle y, \eta\rangle} dy\, d\eta\\
&\quad = (2\pi)^{-n} \int\!\!\int e^{-i\langle y, \eta\rangle} \langle y\rangle^{-2l_1} \langle D_\eta\rangle^{2l_1} \langle \eta\rangle^{-2l_2} a(x, \xi+\eta) \langle D_y\rangle^{2l_2} \psi_k(y) dy\, d\eta
\end{aligned}$$

for every choice of $l_1, l_2 \in \mathbb{N}$. Let $2l_1 > n+1$ and $2l_2 > n$. Then

$$|\sigma_{B_k}(x, \xi)| \le \frac{C}{k} \sum_{|\gamma|\le 2l_1} \sup_{x,\,\xi} |\partial_\xi^\gamma a(x, \xi)|$$

where C does not depend on a and k. In the same way we obtain the estimates

$$|\partial_\xi^\alpha \partial_x^\beta \sigma_{B_k}(x, \xi)| \le \frac{C}{k} \sum_{|\gamma|\le 2l_1+|\alpha|} \sup_{x,\,\xi} |\partial_\xi^\gamma \partial_x^\beta a(x, \xi)|.$$

The Calderón-Vaillancourt theorem implies that $\lim_{k\to\infty} \|B_k\| = 0$. □

Corollary 4.3. *Let $\phi, \psi \in C_b^\infty(\mathbb{R}^n)$ be such that* dist (supp ϕ, supp ψ) $> k$. *Then*

$$\phi\, Op_d(a^{(k)})\, \psi I = 0.$$

To reify the abstract scheme, we set, for $\alpha \in \mathbb{Z}^n$,

$$(U_\alpha u)(x) := u(x-\alpha), \quad x \in \mathbb{R}^n,$$

and we choose $P := \varphi_0 I$ and $\hat{P} := \phi I$ where $\phi \in C_0^\infty(\mathbb{R}^n)$ is a function which is 1 on the ball $\{x \in \mathbb{R}^n : |x| \le 1\}$. Further, for $k \in \mathbb{N}$, we let

$$(W_k u)(x) := k^{-n/2} u(x/k), \quad x \in \mathbb{R}^n,$$

and we define P_α, $P_{k,\alpha}$, $\hat{P}_\alpha$, $\hat{P}_{k,\alpha}$ and $\hat{P}^{(k)}$ in accordance with the axioms. Finally, for $r \in \mathbb{N}$, let $Q_r := \chi_r I$, where χ_r is the characteristic function of $\{x \in \mathbb{R}^n : |x| \ge r\}$. Then it is easy to check that conditions (A1)–(A4) are fulfilled with $\Lambda := \mathbb{Z}^n$ and $\mathfrak{B}$ being the set of all sequences in $\mathbb{Z}^n$ which tend to infinity.

Definition 4.4. We denote by $\mathcal{P}(L^2(\mathbb{R}^n))$ the closure in $L(L^2(\mathbb{R}^n))$ of the set of all pseudodifferential operators of the form

$$A_{MN} = \sum_{i=1}^{M} \prod_{j=1}^{N} a_{jk} C(b_{jk}) c_{jk} I,$$

where a_{jk}, b_{jk}, $c_{jk} \in C_b^\infty(\mathbb{R}^n)$.

Clearly, $\mathcal{P}(L^2(\mathbb{R}^n))$ is a C^*-subalgebra of $L(L^2(\mathbb{R}^n))$.

It is evident that A_{MN} is a pseudodifferential operator with double symbol in $OPS^0_{0,0,0}$ and that

$$a_{MN}(x, y, \xi) := \sum_{i=1}^{M} \prod_{j=1}^{N} a_{jk}(x) b_{jk}(\xi) c_{jk}(y).$$

Thus, by Proposition 4.2, the operator $A_{MN} = Op_d(a_{MN})$ is the limit as $R \to \infty$ in the norm of $L(L^2(\mathbb{R}^n))$ of the sequence of the operators $A_{MN}^{(R)}$,

$$A_{MN}^{(R)} := \sum_{i=1}^{M} \prod_{j=1}^{N} a_{jk} Op_d(b_{jk}^{(R)}) c_{jk})$$

where $b_{jk}^{(R)}(x, y, \xi) := b_{jk}(\xi) f_R(x-y)$. Note that the operators $Op_d(b_{jk}^{(R)})$ commute with the operators U_α.

It follows from Propositions 4.1 and 4.2, that the conditions (a) and (c) of the definition of the class $\mathcal{A}_0(L^2(\mathbb{R}^n))$ are satisfied for the operators $A_{MN}^{(R)}$. Thus, we are left with condition (b), i.e., with the existence of sufficiently many limit operators with respect to the set $\{U_\alpha\}_{\alpha \in \mathbb{Z}^n}$.

First, let $a \in C_b^\infty(\mathbb{R}^n)$, and let $\beta = (\beta_k)$ be a sequence in $\mathfrak{B}$. By the Arzelà-Ascoli theorem, the sequence of the functions $x \mapsto a(x + \beta_m)$ has a subsequence, say the functions $x \mapsto a(x + \beta_{m_k})$, $k \in \mathbb{N}$, which converges to a function $a^{(\beta)}$ uniformly on every compact $K \subset \mathbb{R}^n$ together with all their derivatives. It is easy to see that $a^{(\beta)} \in C_b^\infty(\mathbb{R}^n)$. Thus,

$$\lim_{k\to\infty} \|\hat{P}^{(m)}(U_{-\beta_{m_k}} a U_{\beta_{m_k}} - a^{(\beta)} I)\| = \lim_{k\to\infty} \|(U_{-\beta_{m_k}} a U_{\beta_{m_k}} - a^{(\beta)} I)\hat{P}^{(m)}\| = 0.$$

Let us now examine the existence of limit operators for the operator $A_{11}^{(R)} := a Op_d(b^{(R)}) cI$. Let $\beta = (\beta_m) \in \mathfrak{B}$ be a sequence such that the functions $a_m : x \mapsto$

$a(x+\beta_m)$ and $c_m : x \mapsto c(x+\beta_m)$ converge to certain functions $a^{(\beta)}$ and $c^{(\beta)}$ in the C^∞-topology, respectively (otherwise we pass to a subsequence of β). Then, for arbitrary $k \in \mathbb{N}$ and large enough $N > k$, one has

$$\begin{aligned}
\lim_{m\to\infty} U_{-\beta_m} A_{11}^{(R)} U_{\beta_m} \hat{P}^{(k)} &= \lim_{k\to\infty} U_{-\beta_m} a Op_d(b^{(R)}) c U_{\beta_m} \hat{P}^{(k)} \\
&= \lim_{k\to\infty} a_m Op_d(b^{(R)}) c_m \hat{P}^{(k)} \\
&= \lim_{k\to\infty} a_m \hat{P}^{(N)} Op_d(b^{(R)}) c_m \hat{P}^{(k)} \\
&= a^{(\beta)} \hat{P}^{(N)} Op_d(b^{(R)}) c^{(\beta)} \hat{P}^{(k)} \\
&= a^{(\beta)} Op_d(b^{(R)}) c^{(\beta)} \hat{P}^{(k)}
\end{aligned}$$

(all limits are taken with respect to the norm convergence). In the same way one gets

$$\lim_{m\to\infty} \hat{P}^{(k)} U_{-\beta_m} A_{11}^{(R)} U_{\beta_m} = \hat{P}^{(k)} a^{(\beta)} Op_d(b^{(R)}) c^{(\beta)} I$$

for every $k \in \mathbb{N}$. Thus, every sequence $\beta \in \mathfrak{B}$ possesses a subsequence for which the limit operator of $A_{11}^{(R)}$ exists. The same reasoning gives this result also for arbitrary operators of the form $A_{MN}^{(R)}$.

Due to Propositions 2.2 (d) and 4.2, this yields finally that, for every operator $A \in \mathcal{P}(L^2(\mathbb{R}^n))$ and every sequence $\beta \in \mathfrak{B}$, there is a subsequence $\tilde{\beta}$ of β such that the limit operator $A_{\tilde{\beta}}$ of A exists. Thus, Theorem 2.4 implies the following.

Theorem 4.5. *Let $A \in \mathcal{P}(L^2(\mathbb{R}^n))$. Then*

$$\liminf_{r\to\infty} \nu(A|_{Q_r(L^2(\mathbb{R}^n))}) > 0 \tag{4.3}$$

if and only if

$$\inf\{\nu(A_\alpha) : A_\alpha \in \lim_{\mathfrak{B}}(A)\} > 0. \tag{4.4}$$

For $\psi \in C_b^\infty(\mathbb{R}^n)$ and $r > 0$, let $\psi_r(x) := \psi(x/r)$ for $x \in \mathbb{R}^n$. We denote by $\mathcal{D}(L^2(\mathbb{R}^n))$ the subset of $L(L^2(\mathbb{R}^n))$ which consists of all operators A with

$$\lim_{r\to\infty} \|[A, \psi_r I]\| = 0 \quad \text{for all } \psi \in C_b^\infty(\mathbb{R}^n).$$

It is easy to see that $\mathcal{D}(L^2(\mathbb{R}^n))$ is a C^*-subalgebra of $L(L^2(\mathbb{R}^n))$, which contains $\mathcal{P}(L^2(\mathbb{R}^n))$ due to Proposition 4.1.

Let $\rho \in C^\infty(\mathbb{R}^n)$ be a function with $\rho(x) = 1$ if $|x| \ge 2$ and $\rho(x) = 0$ if $|x| \le 1$, and set $\rho_r(x) := \rho(x/r)$ for $r > 0$. Let further $\mathcal{J}$ consist of all operators $T \in \mathcal{D}(L^2(\mathbb{R}^n))$ with

$$\lim_{r\to\infty} \|\rho_r T\| = \lim_{r\to\infty} \|T\rho_r I\| = 0,$$

which is a closed ideal of $\mathcal{D}(L^2(\mathbb{R}^n))$.

Proposition 4.6. *Condition* (4.3) *is satisfied if and only if there exist operators $L \in \mathcal{D}(L^2(\mathbb{R}^n))$ and $T \in \mathcal{J}$ such that $LA = I + T$.*

Proof. Let (4.3) be satisfied. Then there exist $\delta > 0$ and $r_0 > 0$ such that

$$\langle \chi_{r_0} A^* A \chi_{r_0} u, \chi_{r_0} u \rangle \geq \delta \|\chi_{r_0} u\|^2$$

where χ_{r_0} is the characteristic function of the set $\{x \in \mathbb{R}^n : |x| > r_0\}$. This inequality implies the existence of an operator $L \in \mathcal{D}(L^2(\mathbb{R}^n))$ such that $LA\chi_{r_0} I = \chi_{r_0} I$, which can be rewritten as

$$LA = I - LA(I - \chi_{r_0} I) + (I - \chi_{r_0} I).$$

Since $I - \chi_{r_0} I \in \mathcal{J}$, we obtain

$$T := -LA(I - \chi_{r_0} I) + (I - \chi_{r_0} I) \in \mathcal{J},$$

which is the assertion. Conversely, let $LA = I + T$ with $L \in \mathcal{D}(L^2(\mathbb{R}^n))$ and $T \in \mathcal{J}$. Then $LA\rho_r I = \rho_r I + T\rho_r I$. Choose r such that $\|T\rho_r I\| < 1$, and let r_0 be such that $\chi_{r_0}\rho_r = \chi_{r_0}$. Then $LA\chi_{r_0} I = (I + T\rho_r I)\chi_{r_0}$ and, thus,

$$(I + T\rho_r I)^{-1} LA\chi_{r_0} I = \chi_{r_0} I.$$

This identity implies estimate (4.3). □

We conclude this paper with a 'dual' application of the abstract scheme and its consequences. In this case, the unitary operators, which are now denoted by U'_α instead of U_α, are given by $(U'_\alpha u)(x) := e^{i\langle \alpha, x \rangle} u(x)$, and we choose $W_k := D_{1/k}$. Further, we let

$$P := C(\varphi_0), \quad \hat{P} := C(\phi), \quad Q_r := C(\chi_r)$$

where φ_0, ϕ and χ_r are as before. Moreover, $\mathfrak{B}$ again refers to the set of all sequences in $\mathbb{Z}^n$ which tend to infinity. Finally, we write $\lim'_{\mathfrak{B}}(A)$ for the set of all limit operators of A with respect to the unitary operators U'_α.

One can check in the same way as before that all assumptions of the axiomatic approach are satisfied. Thus, as a corollary of Theorem 2.4, we get the following.

Theorem 4.7. *Let $A \in \mathcal{P}(L^2(\mathbb{R}^n))$. Then*

$$\liminf_{r \to \infty} \nu(A|_{Q_r(L^2(\mathbb{R}^n))}) > 0 \tag{4.5}$$

if and only if

$$\inf\{\nu(A_\beta) : A_\beta \in \lim'_{\mathfrak{B}}(A)\} > 0. \tag{4.6}$$

Denote by $\mathcal{D}'(L^2(\mathbb{R}^n))$ the set of all operators $A \in L(L^2(\mathbb{R}^n))$ which satisfy

$$\lim_{r \to \infty} \|[A, C(\psi_r)]\| = 0 \quad \text{for every } \psi \in C_0^\infty(\mathbb{R}^n).$$

Clearly, $\mathcal{D}'(L^2(\mathbb{R}^n))$ is a C^*-subalgebra of $L(L^2(\mathbb{R}^n))$, which contains the algebra $\mathcal{P}(L^2(\mathbb{R}^n))$. Further, the set $\mathcal{J}'$ of all operators $A \in \mathcal{D}'(L^2(\mathbb{R}^n))$ with

$$\lim_{r \to \infty} \|C(\phi_r)A\| = \lim_{r \to \infty} \|AC(\phi_r)\| = 0$$

is a closed ideal of $\mathcal{D}'(L^2(\mathbb{R}^n))$.

Proposition 4.8. *Condition* (4.5) *holds if and only if there exist operators $L' \in \mathcal{D}'(L^2(\mathbb{R}^n))$ and $T' \in \mathcal{J}'$ such that $L'A = I + T'$.*

The proof is similar to the proof of Proposition 4.6.

The preceding two theorems have remarkable consequences for the semi-Fredholmness and Fredholmness of operators in $\mathcal{P}(L^2(\mathbb{R}^n))$.

Theorem 4.9. *Let $A \in \mathcal{P}(L^2(\mathbb{R}^n))$. Then A is a Φ_+-operator if and only if*

$$\inf\{\nu(A_\beta) : A_\beta \in \lim_{\mathfrak{B}}(A)\} > 0 \quad \text{and} \quad \inf\{\nu(A_\beta) : A_\beta \in \lim'_{\mathfrak{B}}(A)\} > 0 \tag{4.7}$$

where $\lim_{\mathfrak{B}}(A)$ is the set of all limit operators of A which are defined by means of the unitary operators U_α, whereas $\lim'_{\mathfrak{B}}(A)$ refers to the set of all limit operators of A with respect to the U'_α.

Proof. Let condition (4.7) be satisfied. Then there are operators $L \in \mathcal{D}(L^2(\mathbb{R}^n))$ and $L' \in \mathcal{D}'(L^2(\mathbb{R}^n))$ as well as operators $T \in \mathcal{J}$ and $T' \in \mathcal{J}'$ such that

$$LA = I + T \quad \text{and} \quad L'A = I + T'.$$

With the operator $B := LAL' - L - L'$ one finds $BA - I = TT'$. We claim that the operator TT' is compact. Indeed, let ϕ_r be defined as above. Then

$$\lim_{r\to\infty} \|TT'\,\phi_r I\| = \lim_{r\to\infty} \|TT'\,C(\phi_r)\| = 0.$$

Hence, the operator TT' can be approximated in the norm by the compact operators $TT'(I - \phi_r I)(I - C(\phi_r))$ as closely as desired, which proves our claim. Thus, $BA - I$ is a compact operator, which implies its Φ_+-property.

Conversely, let A be a Φ_+-operator. Then the a priori estimate

$$\delta\|u\| \le \|Au\| + \|Ku\|, \quad u \in L(L^2(\mathbb{R}^n)), \tag{4.8}$$

holds with a positive constant δ and a compact operator K. If $(U_\gamma)_{\gamma\in\mathbb{Z}^n}$ is one of the sequences of unitary operators considered in the theorem, then it follows from (4.8) that

$$\delta\|u\| \le \|U_\gamma^{-1} A U_\gamma u\| + \|U_\gamma^{-1} K U_\gamma u\|. \tag{4.9}$$

Since the U_γ converge weakly to zero as $\gamma \to \infty$, the operators $U_\gamma^{-1} K U_\gamma$ converge strongly to 0. Thus, letting in (4.9) γ go to infinity yields condition (4.7). □

Our final result is a corollary to Theorem 4.9.

Theorem 4.10. *Let $A \in \mathcal{P}(L^2(\mathbb{R}^n))$. Then*

(a) *A is a Φ_--operator if and only if*

$$\inf\{\nu(A^*_\beta) : A_\beta \in \lim_{\mathfrak{B}}(A)\} > 0 \quad \text{and} \quad \inf\{\nu(A^*_\beta) : A_\beta \in \lim'_{\mathfrak{B}}(A)\} > 0;$$

(b) *A is a Fredholm operator if and only if all operators in $\lim_{\mathfrak{B}}(A) \cup \lim'_{\mathfrak{B}}(A)$ are uniformly invertible, i.e., if*

$$\sup\{\|A_\beta^{-1} : A_\beta \in \lim_{\mathfrak{B}}(A) \cup \lim'_{\mathfrak{B}}(A)\} < \infty.$$

Acknowledgements

This work has been supported by the German Research Foundation (DFG) under Grant Nr. 436 RUS 17/24/01.

References

[1] Böttcher, A., Karlovich, Yu., Rabinovich, V., The method of limit operators for one-dimensional singular integral operators with slowly oscillating data, *J. Oper. Theory* 43 (2000), 171–198.

[2] Favard, J., Sur les équations différentiales a coéfficients presque périodiques, *Acta Math.* **51** (1927), 31–81.

[3] Gohberg, I., Krupnik, N., *One-dimensional Linear Singular Integral Operators*, Birkhäuser Verlag, Basel, Boston, Stuttgart 1992.

[4] Kisil', V.V., The spectrum of the algebra generated by two-sided convolutions on the Heisenberg group and by operators of multiplications by continuous functions, *Dokl. Akad. Nauk Ross.* **337** (1994), No. 4, 439–441 (Russian, Engl. transl. *Russ. Acad. Sci. Dokl., Math.* **50** (1995), No. 1, 93–96).

[5] Kisil', V.V., On the algebra of pseudodifferential operators which is generated by convolutions on the Heisenberg group, *Sib. Mat. Zh.* **34** (1993), No. 6, 75–85 (Russian, Engl. transl. *Sib. Math. J.* **34** (1993), No. 6, 1066–1075).

[6] Lange, B.V., Rabinovich, V.S., On the Noether property of multi-dimensional discrete convolution operators, *Mat. Zametki* **37** (1985), No. 3, 407–421 (Russian).

[7] Lange, B.V., Rabinovich, V.S., On Noetherian multidimensional convolution operators with bounded measurable coefficients, *Izv. Vyssh. Uchebn. Zaved., Mat.* **277** (1985), No. 6, 22–30 (Russian, Engl. transl. *Sov. Math.* **29** (1985), No. 6, 26–35).

[8] Lange, B.V., Rabinovich, V.S., Pseudo-differential operators on $\mathbb{R}^n$ and limit operators, *Mat. Sbornik* **129** (1986), No. 2, 175–185 (Russian, Engl. transl. *Math. USSR Sbornik* **577** (1987), No. 1, 183–194).

[9] Mikhlin, S.G., Prössdorf, S., *Singular Integral Operators*, Springer-Verlag, Heidelberg 1986.

[10] Muhamadiev, E.M., On the invertibility of differential operators in partial derivatives of elliptic type, *Dokl. Akad. Nauk SSSR* **205** (1972), no. 6, 1292–1295 (Russian).

[11] Muhamadiev, E.M., On the normal solvability and the Noether property of elliptic operators on $\mathbb{R}^n$, Zap. Nauchn. Sem. LOMI **110** (1981), 120–140 (Russian).

[12] Nagel, A., Stein, E.M., *Lectures on Pseudo-Differential Operators: Regularity Theorems and Applications to Non-Elliptic Problems*, Math. Notes, Princeton Univ. Press, 1979.

[13] Rabinovich, V.S., Fredholmness of pseudo-differential operators on $\mathbb{R}^n$ in the scale of $L_{p,q}$-spaces, *Sib. Mat. Zh.* **29** (1988), No. 4, 635–646 (Russian, Engl. transl. *Sib. Math. J.* **29** (1988), No. 4, 635–646).

[14] Rabinovich, V.S., Criterion for local invertibility of pseudo-differential operators with operator symbols and some applications, *In: Proceedings of the St. Petersburg Mathematical Society*, vol. **V**, AMS Transl. Series 2, **193** (1998), 239–260.

[15] Rabinovich, V.S., Roch, S., Silbermann, B., Fredholm theory and finite section method for band-dominated operators, *Integral Equations Oper. Theory* **30** (1998), 452–495.

[16] Simonenko, I.B., A new general method of investigating linear operator equations of the singular integral equations type, I, *Izv. Akad. Nauk SSSR, Ser. Mat.* **3** (1965), 567–586 (Russian).

[17] Stein, E.M., *Harmonic Analysis: Real-Variable Methods, Orthogonality, and Oscillatory Integrals*, Princeton Univ. Press, Princeton, New Jersey, 1993.

[18] Steinberg, B.Ya., On convolution operators on locally compact groups, *Funkts. Anal. Prilozh.* **15** (1981), No. 3, 95–96 (Russian, Engl. transl. *Funct. Anal. Appl.* **15** (1981), No. 3, 233–234).

[19] Steinberg, B.Ya., Boundedness and compactness of convolution operators on locally compact groups, *Mat. Zametki* **38** (1985), No. 2, 278–292 (Russian).

[20] Taylor, M.E., *Pseudodifferential Operators*, Princeton Univ. Press, Princeton, New Jersey, 1981.

[21] Taylor, M.E., *Noncommutative Microlocal Analysis, I*, Memoirs AMS **313** (1984).

[22] Taylor, M.E., *Noncommutative Harmonic Analysis*, Math. Surveys and Monographs **22**, AMS 1986.

[23] Vasilevski, N.L., Trujillo, R., Group convolution operators on standard CR-manifolds, I. Structural properties, *Integral Equations Oper. Theory* **19** (1994), 66–104.

[24] Vasilevski, N.L., Convolution operators on standard CR-manifolds, II. Algebras of convolution operators on the Heisenberg group, *Integral Equations Oper. Theory* **19** (1994), 327–358.

Vladimir S. Rabinovich
Instituto Politechnico National
ESIME-Zacatenco
Ed. 1, 2-do piso, Av. IPN
Mexico, D.F., 07738, Mexico
e-mail: rabinov@maya.esimez.ipn.mx

Steffen Roch
Fachbereich Mathematik
Technische Universität Darmstadt
Schloßgartenstraße 7
D-64289 Darmstadt, Germany
e-mail: roch@mathematik.tu-darmstadt.de

2000 *Mathematics Subject Classification.*
Primary 47A53; Secondary 35S05, 47G10.

Received: 23 May 2001.

[illegible]

Operator Theory:
Advances and Applications, Vol. 142, 287–310

Algebras of Approximation Sequences: Structure of Fractal Algebras

Steffen Roch

Abstract. A basic idea to study the stability of an approximation sequence is to translate the stability problem into an invertibility problem in a suitably defined C^*-algebra, which offers the applicability of C^*-tools to problems in numerical analysis. This approach associates with every class $\mathcal{L}$ of operators and with every discretization procedure D for the operators in $\mathcal{L}$ a concrete C^*-algebra $\mathfrak{A}(\mathcal{L}, D)$. The surprising result of an analysis of these algebras is that, for many apparently quite different choices of $\mathcal{L}$ and D, they show a common structure, which might be formally summarized in the so-called standard model.

The goal of this paper is to analyse how this formal structure arises from intrinsic properties of the algebras $\mathfrak{A}$. The basic of these properties are the fractality of the algebra and a certain maximality condition for an ideal in $\mathfrak{A}$ which is intimately related to the Fredholm theory in $\mathfrak{A}$. Here fractality of $\mathfrak{A}$ means that every sequence in $\mathfrak{A}$ can be completely reconstructed from each of its subsequences modulo a sequence tending to zero in the norm, whereas the Fredholm property of a sequence (A_n) is closely related to the asymptotic behavior of the smallest singular values of the matrices A_n.

1. Introduction

Nowadays, it is a standard approach to study the stability of (a class of) approximation sequences for operators on a Hilbert space by putting all sequences under consideration into a C^*-algebra, $\mathcal{A}$, and by studying its irreducible representations. Thereby, the primary interest is not in determining all irreducible representations of $\mathcal{A}$, but to single out a family of irreducible representations which is *sufficient* in the sense that an element $a \in \mathcal{A}$ is invertible if and only if $W(a)$ is invertible for all representations W in that family, and which is as small as possible.

Here are more precise statements: Let $\mathcal{H}$ be a Hilbert space and, for every n, let $\mathcal{H}_n$ be an n-dimensional subspace of $\mathcal{H}$. We suppose that the orthogonal projections P_n from $\mathcal{H}$ onto $\mathcal{H}_n$ converge strongly to the identity operator on $\mathcal{H}$. By an *approximation sequence* for an operator $A \in L(\mathcal{H})$ we mean a sequence of operators $A_n : \mathcal{H}_n \to \mathcal{H}_n$ which converges strongly to A as $n \to \infty$. The sequence (A_n) is called *stable* if the operators A_n are invertible for n large enough and if the norms of their inverses are uniformly bounded. The stability of a sequence can be translated into algebraic language as follows. We let $\mathcal{F}$ stand for the set of all bounded sequences (A_n) of operators $A_n : \mathcal{H}_n \to \mathcal{H}_n$ and provide this set with element-wisely defined operations and with the supremum norm. This makes $\mathcal{F}$ to a C^*-algebra and the set $\mathcal{G}$ of all sequences $(G_n) \in \mathcal{F}$ with $\lim \|G_n P_n\| = 0$ to a closed ideal of $\mathcal{F}$. The Banach-Steinhaus theorem implies the uniform boundedness of every approximation sequence (A_n), whence $(A_n) \in \mathcal{F}$, and a simple Neumann series argument shows that a sequence $(A_n) \in \mathcal{F}$ is stable if and only if its coset $(A_n) + \mathcal{G}$ is invertible in the quotient algebra $\mathcal{F}/\mathcal{G}$. Thus, to examine a class of concrete approximation sequences, one considers the smallest closed subalgebra $\mathcal{A}$ of $\mathcal{F}$ which contains all sequences under consideration, and what one is looking for is a family (W_t) of irreducible representations of $\mathcal{A}$ such that $\mathcal{G} \cap \mathcal{A}$ is in the kernel of W_t for every t and such that a sequence $(A_n) \in \mathcal{A}$ is stable (i.e. invertible modulo $\mathcal{G} \cap \mathcal{A}$) if and only if all operators $W_t(A_n)$ are invertible.

This program has been carried out for several concrete classes of operators and discretization methods. Here are two examples which have played a pioneering role in the development of that approach.

Finite section method for Toeplitz operators

Given an $L^\infty(\mathbb{T})$ function a with kth Fourier coefficient a_k, let $T(a)$ denote the *Toeplitz operator generated by* a, i.e., the operator that is defined by its matrix representation $(a_{i-j})_{i,j=1}^\infty$ with respect to the standard basis of l^2. Toeplitz operators with bounded generating function are bounded. Further write P_n for the projection operator on l^2 that maps the sequence $(x_i)_{i=1}^\infty$ to $(x_1, x_2, \ldots, x_n, 0, 0, \ldots)$, and consider the sequence $(P_n T(a) P_n)_{n \geq 1}$ of the finite sections of $T(a)$. The operators $P_n T(a) P_n$ can be identified with finite Toeplitz matrices acting on $\operatorname{Im} P_n \cong \mathbb{C}^n$.

Suppose we are interested in the case where a is a piecewise continuous function. The associated algebra $\mathcal{F}$ consists of sequences of operators $A_n : \operatorname{Im} P_n \to \operatorname{Im} P_n$, and we let $\mathcal{A}$ stand for the smallest closed subalgebra of $\mathcal{F}$ which contains all sequences $(P_n T(a) P_n)_{n \geq 1}$ with a a piecewise continuous function on $\mathbb{T}$. Further, write R_n for the reflection operators $(x_i) \mapsto (x_n, x_{n-1}, \ldots, x_1, 0, 0, \ldots)$ on l^2. It has been shown in [2] that $\mathcal{A}$ contains all sequences of the form $(P_n K P_n + R_n L R_n + G_n)$ with compact operators K, L on l^2 and with $(G_n) \in \mathcal{G}$, that the set $\mathcal{J}$ of all these sequences forms a closed ideal of $\mathcal{A}$ and that, for every sequence $(A_n) \in \mathcal{A}$, the strong limits

$$W(A_n) := \text{s-lim}\, A_n \quad \text{and} \quad \widetilde{W} := \text{s-lim}\, R_n A_n R_n. \tag{1.1}$$

exist. The main result from [2] reads as follows.

Theorem 1.1.

(a) *A coset $(A_n) + \mathcal{J} \in \mathcal{A}/\mathcal{J}$ is invertible if and only if the operators $W(A_n)$ and $\widetilde{W}(A_n)$ are Fredholm.*

(b) *A sequence $(A_n) \in \mathcal{A}$ is stable if and only if the operators $W(A_n)$ and $\widetilde{W}(A_n)$ are invertible.*

Observe that, indeed, the mappings W and $\widetilde{W}$ are irreducible representations of the algebra $\mathcal{A}$. This is a consequence of the identities

$$W(P_nKP_n) = K \quad \text{and} \quad \widetilde{W}(R_nLR_n) = L,$$

which hold for arbitrary compact operators K and L and which show that the ideal $K(l^2)$ of all compact operators on l^2 lies in the range of both W and $\widetilde{W}$. □

Finite section method for operators in the Toeplitz algebra

In this example we consider the same approximation method for a larger class of operators. Let $\mathcal{T}$ stand for the smallest closed subalgebra of $L(l^2)$ which contains all Toeplitz operators $T(a)$ with piecewise continuous generating function a. Again we discretize operators A in $\mathcal{T}$ by the finite section method. Accordingly, let $\mathcal{B}$ refer to the smallest closed subalgebra of $\mathcal{F}$ which contains all sequences (P_nAP_n) with $A \in \mathcal{T}$. It is not hard to prove that the strong limits $W(A_n)$ and $\widetilde{W}(A_n)$ (defined as in (1.1)) exist for every sequence $(A_n) \in \mathcal{B}$. Thus, the invertibility of the operators $W(A_n)$ and $\widetilde{W}(A_n)$ is also a necessary condition for the stability of the sequence (A_n). But it turns out that these conditions are no longer sufficient (see [11] for a quite simple argument). A detailed analysis (performed by Verbitsky, Rathsfeld, Böttcher, Silbermann, the author, and others) yields the following stability result for sequences in $\mathcal{B}$ where, besides the invertibility of $W(A_n)$ and $\widetilde{W}(A_n)$, certain *local stability conditions* occur. Let S_n refer to the subspace of the Hilbert space $L^2([0,\, 1])$ that is spanned by the functions

$$\varphi_{k,n}(x) := \begin{cases} \sqrt{n} & \text{if} \quad x \in (k/n,\, (k+1)/n) \\ 0 & \text{if} \quad x \in (0,\, k/n) \cup ((k+1)/n,\, 1) \end{cases}$$

with $k = 0, \ldots, n-1$, write E_n for the operators

$$E_n : \operatorname{Im} P_n \to S_n, \quad (x_1, \ldots, x_n, 0, \ldots) \mapsto \sum_{k=1}^{n} x_k \varphi_{k,n},$$

and define $E_{-n} : S_n \to \operatorname{Im} P_n$ by $E_{-n} := (E_n)^{-1}$. Finally, for $t \in \mathbb{T}$, let Y_t stand for the operator

$$Y_t : l^2 \to l^2, \quad (x_k)_{k=1}^{\infty} \mapsto (t^{-k} x_k)_{k=1}^{\infty}.$$

One can show that, for every sequence $(A_n) \in \mathcal{B}$ and for every $t \in \mathbb{T}$, the strong limit

$$W_t(A_n) := \text{s-lim}_{n\to\infty} E_n Y_{t^{-1}} A_n Y_t E_{-n}$$

exists and that it defines a bounded linear operator $W_t(A_n)$ on $L^2([0,\, 1])$. Moreover, for every $t \in \mathbb{T}$ and every compact operator $K \in K(L^2([0,\, 1]))$, the sequence $(P_n Y_t E_{-n} L_n K E_n Y_{t^{-1}} P_n)$ belongs to $\mathcal{B}$, and the sets

$$\mathcal{J}_t := \{(P_n Y_t E_{-n} L_n K E_n Y_{t^{-1}} P_n + G_n) : K \in K(L^2([0,\, 1])), (G_n) \in \mathcal{G}\}$$

form closed ideals of $\mathcal{B}$.

Theorem 1.2. *A sequence $(A_n) \in \mathcal{B}$ is stable if and only if the operators $W(A_n)$, $\widetilde{W}(A_n)$ and $W_t(A_n)$ are invertible for every $t \in \mathbb{T}$.*

For a proof see, e.g., [5], Theorem 4.1. Again, the mappings W, $\widetilde{W}$ and W_t are irreducible representations, which follows for W and $\widetilde{W}$ as above and for W_t from

$$W_s(P_n Y_t E_{-n} L_n K E_n Y_{t^{-1}} P_n) = \begin{cases} K & \text{if} \quad s = t \\ 0 & \text{if} \quad s \neq t \end{cases}$$

whence $K(L^2([0, 1])) \subset \operatorname{Im} W_t$. □

The standard model

A similar analysis as in these two examples has been performed for numerous approximation methods including, e.g., Galerkin, collocation and qualocation methods with spline ansatz functions and quadrature methods for singular integral operators and Mellin operators. For a more detailed overview see the monographs [5, 7]. The surprising and to some extent miraculous result of this analysis is that (almost) all of the investigated algebras (which arise from very different approximation procedures for very different classes of operators) exhibit one and the same structure. The attempt to formalize this structure leads to a standard model for algebras arising from approximation sequences. A C^*-subalgebra $\mathcal{A}$ of $\mathcal{F}$ is called a *standard algebra* if it satisfies the following axioms.

1. There is a (possibly infinite) set T, and for every $t \in T$, there is a Hilbert space H^t and a sequence (E_n^t) of partial isometries $E_n^t : H^t \to \mathcal{H}$ such that
   - the initial projections $(E_n^t)^* E_n^t$ converge strongly to the identity on H^t,
   - the range projections $E_n^t (E_n^t)^*$ coincide with P_n, and
   - the separation condition holds:
   $$(E_n^s)^* E_n^t \to 0 \quad \text{weakly for all } s \neq t. \tag{1.2}$$
2. For every $t \in T$ and every $(A_n) \in \mathcal{A}$, there exists the strong limit
   $$\text{s-lim}\, (E_n^t)^* A_n E_n^t =: W_t(A_n). \tag{1.3}$$
3. $\mathcal{A}$ is unital and contains all sequences $(E_n^t K (E_n^t)^*)$ with K compact on H^t as well as all sequences $(G_n) \in \mathcal{G}$. The closed linear span of all of these sequences forms a closed ideal $\mathcal{J}$ of $\mathcal{A}$.
4. A sequence $(A_n) \in \mathcal{A}$ is stable if and only if all associated operators $W_t(A_n)$ are invertible on H^t.

The algebras which result from applying the finite section method to Toeplitz operators or to operators in the Toeplitz algebra are obvious examples for standard algebras.

It is evident that these axioms provide only a formal description of the structure of the occurring algebras, which says nothing about the reasons which force this structure. It is the goal of the present paper to show how this structure arises from more intrinsic properties of the algebra. The main of these properties are the fractality of the algebra and the fact that a certain ideal, which is intimately related with the Fredholm theory in the algebra, is large in some sense. The notions of fractality and Fredholmness for approximation sequences were introduced in [10, 11, 12] and will be briefly recalled in Section 2. It will turn out that these properties are naturally satisfied for many algebras generated by commonly used approximation processes, although not for all. For a first discussion of a class of non-fractal approximation methods see [8].

2. Fractality and Fredholmness of approximation sequences

Fractal algebras

This class has been introduced and studied in [10, 12]. Here are the definition as well as a few properties of fractal algebras which we will need in the sequel.

Let $\mathcal{F}$ be as above. Given a strongly monotonically increasing sequence $\eta : \mathbb{N} \to \mathbb{N}$, let $\mathcal{F}_\eta$ refer to the C^*-algebra of all bounded sequences (A_n) with $A_n \in \mathbb{C}^{\eta(n)\times\eta(n)}$, and write $\mathcal{G}_\eta$ for the ideal of all sequences $(A_n) \in \mathcal{F}_\eta$ which tend to zero in the norm. Further, let R_η stand for the restriction mapping $R_\eta : \mathcal{F} \to \mathcal{F}_\eta$, $(A_n) \mapsto (A_{\eta(n)})$. This mapping is a *-homomorphism from $\mathcal{F}$ onto $\mathcal{F}_\eta$ which moreover maps $\mathcal{G}$ onto $\mathcal{G}_\eta$. Given a C^*-subalgebra $\mathcal{A}$ of $\mathcal{F}$, let $\mathcal{A}_\eta$ denote the image of $\mathcal{A}$ under R_η. Clearly, $\mathcal{A}_\eta$ is a C^*-algebra again.

Definition 2.1. Let $\mathcal{A}$ be a C^*-subalgebra of the algebra $\mathcal{F}$.

(a) A *-homomorphism $W : \mathcal{A} \to \mathcal{B}$ of $\mathcal{A}$ into a C^*-algebra $\mathcal{B}$ is *fractal* if, for every strongly monotonically increasing sequence η, there is a *-homomorphism $W_\eta : \mathcal{A}_\eta \to \mathcal{B}$ such that $W = W_\eta R_\eta$.

(b) The algebra $\mathcal{A}$ is *fractal* if the canonical homomorphism $\pi : \mathcal{A} \to \mathcal{A}/(\mathcal{A} \cap \mathcal{G})$ is fractal.

(c) A sequence $(A_n) \in \mathcal{F}$ is *fractal* if the smallest C^*-subalgebra of $\mathcal{F}$ which contains (A_n) is fractal.

Thus, given a subsequence $(a_{\eta(n)})$ of a sequence (a_n) which belongs to a fractal algebra $\mathcal{A}$, it is possible to reconstruct the original sequence (a_n) from this subsequence modulo sequences tending to zero in the norm. This assumption is very natural for sequences arising from discretization procedures. For example, employing the following theorem, one easily gets the fractality of the algebra of the finite section method for operators in the Toeplitz algebra. On the other hand, the algebra $\mathcal{F}$ of all bounded sequences fails to be fractal.

Theorem 2.2. *Let $\mathcal{A}$ be a unital C^*-subalgebra of $\mathcal{F}$ which contains the ideal $\mathcal{G}$. The algebra $\mathcal{A}$ is fractal if and only if there exists a family $\{W_t\}_{t\in T}$ of unital and fractal *-homomorphisms from $\mathcal{A}$ into unital C^*-algebras $\mathcal{B}_t$ such that the following equivalence holds for every sequence $(A_n) \in \mathcal{A}$: The coset $(A_n) + \mathcal{G}$ is invertible in $\mathcal{A}/\mathcal{G}$ if and only if $W_t(a_n)$ is invertible in $\mathcal{B}_t$ for every $t \in T$.*

The representations W, $\widetilde{W}$ and W_t occurring in Theorems 1.1 and 1.2 are given by strong limits, which immediately implies their fractality. Hence, by Theorem 2.2, the algebras of the finite section methods for Toeplitz operators and operators in the Toeplitz algebra are fractal. Another useful criterion for fractality is the following.

Theorem 2.3. *A C^*-subalgebra $\mathcal{A}$ of $\mathcal{F}$ is fractal if and only if the following implication holds for every sequence $(A_n) \in \mathcal{A}$ and every strongly monotonically increasing sequence η:*

$$R_\eta(A_n) \in \mathcal{G}_\eta \Rightarrow (A_n) \in \mathcal{A} \cap \mathcal{G}.$$

Corollary 2.4.

(a) *Every C^*-subalgebra of a fractal algebra $\mathcal{A} \subseteq \mathcal{F}$ is fractal.*

(b) *A C^*-subalgebra $\mathcal{A}$ of $\mathcal{F}$ is fractal if and only if each of its elements is fractal.*

The main feature of fractality is that it makes certain limiting processes uniform. For example, the following result for the norms of the approximation operators ([10, 12]) holds.

Proposition 2.5.

(a) *For every sequence $(A_n) \in \mathcal{F}$,*

$$\limsup_{n\to\infty} \|A_n\| = \|(A_n) + \mathcal{G}\|.$$

(b) *If $(A_n) \in \mathcal{F}$ is a fractal sequence, then $\lim_{n\to\infty} \|A_n\|$ exists.*

Similar effects can be observed for the limiting sets of the eigenvalues, pseudo-eigenvalues, singular values or Raileigh quotients of the operators A_n.

Fredholm sequences

Let $\mathcal{J}(\mathcal{F})$ stand for the smallest closed subset of $\mathcal{F}$ which contains all sequences (K_n) for which $\sup \dim \operatorname{Im} K_n$ is finite. The set $\mathcal{J}(\mathcal{F})$ is a closed two-sided ideal of $\mathcal{F}$ which contains the ideal $\mathcal{G}$ of all sequences tending to zero in the norm. A sequence $(A_n) \in \mathcal{F}$ is called a *Fredholm sequence* if it is invertible modulo the ideal $\mathcal{J}(\mathcal{F})$. If (A_n) is a Fredholm sequence then there is a number k such that $\liminf_{n\to\infty} \sigma_{k+1}^{(n)} > 0$ (Theorem 2 in [11]). The smallest number k with this property is called the *α-number* of the sequence (A_n) and will be denoted by $\alpha(A_n)$. This number plays the same role in the Fredholm theory of approximation sequences as the number $\dim \operatorname{Ker} A$ plays in the common Fredholm theory for operators A on a Hilbert space.

Sequences of rank one

Let $\mathcal{A}$ be a unital and fractal C^*-subalgebra of $\mathcal{F}$ which contains the ideal $\mathcal{G}$. A sequence (K_n) in $\mathcal{A}$ is said to be of *rank one* if every matrix K_n has range dimension less than or equal to one. Clearly, if $(K_n) \in \mathcal{A}$ is a rank one sequence then, for every sequence $(A_n) \in \mathcal{A}$, there is a bounded sequence (α_n) of complex numbers such that

$$K_n A_n K_n = \alpha_n K_n \quad \text{for all } n. \tag{2.1}$$

For the boundedness of (α_n) observe that we can choose $\alpha_n = 0$ if $K_n = 0$ and that $|\alpha_n| \le \|K_n\| \|A_n\|$ in case $K_n \ne 0$. The smallest closed ideal of $\mathcal{A}$ which contains all sequences of rank one will be denoted by $\mathcal{K}(\mathcal{A})$.

Further, we call a sequence $(K_n) \in \mathcal{A}$ of *central rank one* if, for every sequence $(A_n) \in \mathcal{A}$, there is a *convergent* sequence (α_n) such that (2.1) holds. The smallest closed two-sided ideal of $\mathcal{A}$ which contains all sequences of central rank one will be denoted by $\mathcal{J}_{cen}(\mathcal{A})$. Finally, a central rank one sequence (K_n) of $\mathcal{A}$ is said to be of *essential rank one* if it does not belong to the ideal $\mathcal{G}$ and if $K_n \neq 0$ for all n. We denote the smallest closed ideal of $\mathcal{A}$ which contains all sequences of essential rank one by $\mathcal{J}_{ess}(\mathcal{A})$. In case $\mathcal{A}$ does not possess essential rank one sequences, we simply define $\mathcal{J}_{ess}(\mathcal{A}) := \mathcal{G}$.

It is easy to see that these definitions imply the inclusions

$$\mathcal{G} \subseteq \mathcal{J}_{ess}(\mathcal{A}) \subseteq \mathcal{J}_{cen}(\mathcal{A}) \subseteq \mathcal{K}(\mathcal{A}). \tag{2.2}$$

Proposition 2.6. *If $\mathcal{A}$ is a unital and fractal C^*-subalgebra of $\mathcal{F}$ which contains the ideal $\mathcal{G}$, then*

$$\mathcal{J}_{ess}(\mathcal{A}) = \mathcal{J}_{cen}(\mathcal{A}) = \mathcal{K}(\mathcal{A}).$$

Proof. The inclusions $\subseteq$ hold by definition. Let (K_n) be a rank one sequence in $\mathcal{A}$. If $(K_n) \in \mathcal{G}$, then $(K_n) \in \mathcal{J}_{ess}(\mathcal{A})$ due to (2.2). Thus, let $(K_n) \notin \mathcal{G}$. Then, by Proposition 2.5, the limit $\lim \|K_n\| =: \alpha$ exists and is not zero. We will first show that then (K_n) is a central rank one sequence.

Assume $\{K_n\}$ is not of central rank one. Then there are a sequence $\{A_n\} \in \mathcal{A}$ and a non-convergent bounded sequence (α_n) such that

$$K_n A_n K_n = \alpha_n K_n \quad \text{for all } n.$$

Choose two partial limits $\beta \neq \gamma$ of the sequence (α_n) as well as two subsequences μ and η of the positive integers such that

$$\alpha_{\mu(n)} \to \beta \quad \text{and} \quad \alpha_{\eta(n)} \to \gamma \quad \text{as } n \to \infty.$$

Then both sequences $\{\alpha_{\mu(n)} K_{\mu(n)} - \beta K_{\mu(n)}\}$ and $\{\alpha_{\eta(n)} K_{\eta(n)} - \gamma K_{\eta(n)}\}$ tend to zero. Hence, by Theorem 2.3, both sequences $\{\alpha_n K_n - \beta K_n\}$ and $\{\alpha_n K_n - \gamma K_n\}$ are zero sequences. But then, also their difference $(\beta - \gamma)\{K_n\}$ goes to zero. Since $\|K_n\| \to \alpha \neq 0$, this implies $\beta = \gamma$ in contradiction to the choice of β and γ. Hence, (K_n) is a central rank one sequence.

Further, $\alpha \neq 0$ implies that $K_n \neq 0$ for all sufficiently large n. Thus, there is a sequence $(G_n) \in \mathcal{G}$ (with a finite number of non-vanishing entries) such that $(K_n + G_n)$ is an essential rank one sequence. Since $(G_n) \in \mathcal{J}_{ess}(\mathcal{A})$, this shows that $(K_n) \in \mathcal{J}_{ess}(\mathcal{A})$. □

The algebra $\mathcal{A}$ is called *Fredholm inverse closed* in $\mathcal{F}$ if $\mathcal{J}_{ess}(\mathcal{A}) = \mathcal{A} \cap \mathcal{J}(\mathcal{F})$.

Fredholm theory in fractal and Fredholm inverse closed algebras

For every essential rank one sequence (K_n), let $\text{Id}\,(K_n)$ refer to the smallest closed ideal of $\mathcal{A}$ which contains the sequence (K_n). It is easy to see that $\mathcal{G} \subset \text{Id}\,(K_n)$. In [11] it is shown that, if (K_n) and (L_n) are sequences of essential rank one in $\mathcal{A}$, then either $\text{Id}\,(K_n) = \text{Id}\,(L_n)$ or $\text{Id}\,(K_n) \cap \text{Id}\,(L_n) = \mathcal{G}$. We call (K_n) and (L_n) equivalent in the first case and denote by S the set of all equivalence classes with respect to this relation. With every $s \in S$, there is associated a unique irreducible representation W_s of $\mathcal{A}$ into the algebra $L(H_s)$ for some Hilbert space H_s such that the ideal $\mathrm{Id}\,(K_n)$ is mapped onto the ideal $K(H_s)$ of the compact operators on H_s and such that the kernel of the mapping $W_s : \mathrm{Id}\,(K_n) \to K(H_s)$ is $\mathcal{G}$. The main result of [11] reads as follows.

Theorem 2.7. *Let $\mathcal{A}$ be a unital, fractal and Fredholm inverse closed C^*-subalgebra of $\mathcal{F}$ which contains the ideal $\mathcal{G}$.*

(a) *If $(A_n) \in \mathcal{A}$ is a Fredholm sequence, then the operators $W_s(A_n)$ are Fredholm operators for every $s \in S$, and $\alpha(A_n) = \sum_{s\in S} \dim \mathrm{Ker}\, W_s(A_n)$.*

(b) *If $(A_n) \in \mathcal{A}$ is Fredholm and $k = \alpha(A_n) > 0$, then $\lim_{n\to\infty} \sigma_k^{(n)} = 0$.*

(c) *Assume that family $(W_s)_{s\in S}$ is sufficient for the stability of sequences in $\mathcal{A}$. If, for a sequence $(A_n) \in \mathcal{A}$, all operators $W_s(A_n)$ are Fredholm and if only a finite number of them is not invertible, then (A_n) is a Fredholm sequence.*

3. Structure of certain fractal homomorphisms

The assumption made in the introduction that the orthogonal projections $P_n : \mathcal{H} \to \mathcal{H}_n$ converge strongly to the identity operator ensures that $\mathcal{H}$ is a separable Hilbert space. Further, to avoid trivialities, we will also assume that $\mathcal{H}$ has infinite dimension. Thus, without essential loss of generality, we can identify $\mathcal{H}$ with $l^2 = l^2(\mathbb{Z}^+)$ in what follows and $\mathcal{H}_n$ with the subspace of $\mathcal{H}$ consisting of all sequences of the form $(x_1, x_2, \ldots, x_n, 0, 0, \ldots)$. In particular, also the algebra $\mathcal{F}$ is defined with respect to this special choice. The orthogonal projection from $\mathcal{H}$ onto $\mathcal{H}_n$ will be denoted by P_n, and the sequence $(0, \ldots, 0, 1, 0, \ldots)$ with the 1 standing at the ith place will be abbreviated to e_i. Thus, the vectors $(e_i)_{i\geq 1}$ form an orthonormal basis of l^2.

Let us emphasize once more that the Fredholm theory proposed in [10] and briefly sketched in the previous section is based on a decomposition of the ideal of the centrally compact sequences into ideals $\mathrm{Id}\,(K_n)$ which are generated by an essential rank one sequence and that the quotients $\mathrm{Id}\,(K_n)/\mathcal{G}$ are *-isomorphic to the ideal $K(H)$ of all compact operators on some Hilbert space H. It is the goal of this section to give a precise description of all sequences in $\mathrm{Id}\,(K_n)$ in case H is a *separable* Hilbert space. The main result of this section is the following.

Theorem 3.1. *Let $\mathcal{A}$ be a unital and fractal C^*-subalgebra of $\mathcal{F}$ which contains the ideal $\mathcal{G}$, let (K_n) be an essential rank one sequence in $\mathcal{A}$, and suppose that there is a *-homomorphism W from* $\mathrm{Id}\,(K_n)$ *onto the ideal $K(H)$ of all compact operators on a separable Hilbert space H with* $\mathrm{Ker}\, W = \mathcal{G}$. *Then there exist partial isometries $E_n : H \to l^2$ with $E_nE_n^* = P_n$ such that*

$$\mathrm{Id}\,(K_n) = \{(E_nKE_n^* + G_n) : K \in K(H),\, (G_n) \in \mathcal{G}\}. \tag{3.1}$$

Before going into the details of the proof, let us consider an example which shows that the assertion of the theorem becomes false if the ideal $\mathrm{Id}\,(K_n)$ of $\mathcal{A}$ is replaced by an ideal which is not generated by a rank one sequence.

Example. Let the operators P_n and R_n and the further notations be as in the example of the finite section method for Toeplitz matrices, and consider the sets of sequences

$$\begin{aligned}\mathcal{A} &:= \{(\alpha P_n + P_nKP_n + R_nKR_n + G_n) : K \in K(l^2),\, (G_n) \in \mathcal{G}\},\\ \mathcal{J} &:= \{(P_nKP_n + R_nKR_n + G_n) : K \in K(l^2),\, (G_n) \in \mathcal{G}\}.\end{aligned}$$

It is not hard to check that $\mathcal{A}$ is a *-subalgebra of $\mathcal{F}$ and $\mathcal{J}$ a closed ideal of $\mathcal{A}$. For example, if K and L are compact, and if we set $Q_n := I - P_n$ for brevity, then

$$\begin{aligned}&(P_nKP_n + R_nKR_n)(P_nLP_n + R_nLR_n)\\ &= P_nKLP_n + R_nKLR_n - P_nKQ_nLP_n - R_nKQ_nLR_n + P_nKR_nLR_n\\ &\qquad + R_nKR_nLP_n\\ &= P_nKLP_n + R_nKLR_n + G_n\end{aligned}$$

with $(G_n) \in \mathcal{G}$. Since $\mathcal{A}$ is a subalgebra of the algebra related with the finite section method for Toeplitz operators, and since

$$W(\alpha P_n + P_nKP_n + R_nKR_n + G_n) = \alpha I + K = \widetilde{W}(\alpha P_n + P_nKP_n + R_nKR_n + G_n)$$

we conclude from Theorem 1.1 that a sequence $(A_n) \in \mathcal{A}$ is stable if and only if the operator $W(A_n)$ is invertible. Via Theorem 2.2, this implies the fractality of the algebra $\mathcal{A}$, and it is also clear that the homomorphism W maps the ideal $\mathcal{J}$ onto $K(l^2)$ and that the kernel of that homomorphism is $\mathcal{G}$.

But $\mathcal{J}$ does not contain sequences of essential rank one. Indeed, every such sequence would be of the form $(P_nKP_n + R_nKR_n + G_n)$ with $K \neq 0$. Choose $x \in l^2$ such that $\|Kx\| = 1$ and set

$$y_n := (P_nKP_n + R_nKR_n + G_n)P_nx \quad \text{and} \quad z_n := (P_nKP_n + R_nKR_n + G_n)R_nx.$$

Then $y_n \to Kx$ and $R_nz_n \to Kx$, whence

$$\|y_n\| \to \|Kx\| = 1 \quad \text{and} \quad \|z_n\| = \|R_nz_n\| \to 1. \tag{3.2}$$

Moreover, $z_n \rightharpoonup 0$ weakly, which implies

$$\langle y_n,\, z_n\rangle = \langle y_n - Kx,\, z_n\rangle + \langle Kx,\, z_n\rangle \to 0. \tag{3.3}$$

Now suppose that there are numbers α_n such that $\alpha_n y_n = z_n$. Then $|\alpha_n|\,\|y_n\| = \|z_n\|$ whence, due to (3.2), $|\alpha_n| \to 1$. On the other hand, from (3.3) we get

$$\overline{\alpha_n}\,\|y_n\|^2 = \overline{\alpha_n}\,\langle y_n,\, y_n\rangle = \langle y_n,\, z_n\rangle \to 0$$

which yields $\alpha_n \to 0$. Contradiction. □

We prepare the proof of Theorem 3.1 by the following propositions.

Proposition 3.2. *Let the algebra $\mathcal{A}$ be as in Theorem 3.1. Then, for every essential rank one sequence $(K_n) \in \mathcal{A}$, there exists an essential rank one sequence $(\Pi_n) \in \mathcal{A}$ of projections such that* $\mathrm{Id}\,(K_n) = \mathrm{Id}\,(\Pi_n)$.

Proof. By definition, there is a convergent sequence (μ_n) such that $K_nK_n^*K_n = \mu_nK_n$, whence $(K_n^*K_n)^2 = \mu_nK_n^*K_n$. Since $\|K_n^*K_n\| = \|K_n\|^2 \neq 0$ belongs to the spectrum of the non-negative matrix $K_n^*K_n$, the functional calculus for self-adjoint matrices yields $\|K_n^*K_n\|^2 = \mu_n\|K_n^*K_n\|$ and, thus, $\mu_n = \|K_n^*K_n\|$. Since (K_n) is not in $\mathcal{G}$, the limit of the sequence (μ_n) cannot be zero. Hence, the sequence $(\mu_n^{-1}P_n)$ belongs to $\mathcal{A}$, too. The matrices

$$\Pi_n := \mu_n^{-1}P_nK_n^*K_n = \|K_n^*K_n\|^{-1}K_n^*K_n$$

are projections of rank one as one easily checks. Further, (Π_n) is a central rank one sequence in $\mathcal{A}$. Indeed, given a sequence $(A_n) \in \mathcal{A}$, there is a convergent sequence (α_n) such that $K_n(A_nK_n^*)K_n = \alpha_nK_n$, whence $\Pi_nA_n\Pi_n = \alpha_n\Pi_n$. Thus, (Π_n) is an essential rank one sequence in $\mathcal{A}$. Finally, the equalities

$$(\Pi_n) = (\mu_n^{-1}P_n)(K_n)^*(K_n) \quad \text{and} \quad (K_n) = (\mu_n^{-1}K_nK_n^*K_n) = (\mu_n^{-2}P_n)(K_n)(\Pi_n)$$

show that $\mathrm{Id}\,(\Pi_n) \subseteq \mathrm{Id}\,(K_n)$ and $\mathrm{Id}\,(K_n) \subseteq \mathrm{Id}\,(\Pi_n)$, respectively. □

Proposition 3.3. *Let the coset $p \in \mathcal{F}/\mathcal{G}$ be a projection. Then there exists a sequence of projections $(P_n) \in \mathcal{F}$ such that $(P_n) + \mathcal{G} = p$.*

The proof rests on an application of the Gelfand-Naimark theorem for commutative C^*-algebras and can be found, e.g., in [13]. In the present context, this result can be essentially completed. For, recall that an element k of a C^*-algebra $\mathcal{B}$ is *of rank one* if, for every $b \in \mathcal{B}$, there is a $\mu \in \mathbb{C}$ such that $kbk = \mu k$. Further we agree upon calling two projections p_1, $p_2 \in \mathcal{B}$ *orthogonal* if $p_1p_2 = p_2p_1 = 0$.

Proposition 3.4. *Let $\mathcal{A}$ and (K_n) be as in Theorem 3.1, let $(P_n^{(1)})$ be a sequence of projections in $\mathrm{Id}\,(K_n)$, and set $p^{(1)} := (P_n^{(1)}) + \mathcal{G}$. Further suppose that $p^{(2)} \in \mathrm{Id}\,(K_n)/\mathcal{G}$ is a projection of rank one such that the cosets $p^{(1)}$ and $p^{(2)}$ are orthogonal. Then there is a sequence $(P_n^{(2)})$ in $\mathrm{Id}\,(K_n)$ such that $(P_n^{(2)}) + \mathcal{G} = p^{(2)}$ and such that $P_n^{(2)}$ is a projection which is orthogonal to $P_n^{(1)}$ for all sufficiently large n.*

For a proof see Propositions 10 and 11 in [11]. It follows the proof of Theorem 3.1.

Proof. We will prove the theorem in case H is an infinite-dimensional Hilbert space, in which case we can assume that $H = l^2$. The proof is subdivided into several steps.

Step 1. Let (K_n) be an essential rank one sequence in $\mathcal{A}$, and let W be a *-homomorphism W from $\mathrm{Id}\,(K_n)$ onto $K(l^2)$ with $\mathrm{Ker}\, W = \mathcal{G}$. Due to Proposition 3.2, we can choose an essential rank one sequence $(\Pi_n) \in \mathcal{A}$ of projections such that $\mathrm{Id}\,(K_n) = \mathrm{Id}\,(\Pi_n)$. Observe further that W is an irreducible representation of $\mathrm{Id}\,(\Pi_n)$ since the range of W contains the ideal of all compact operators. The ideal $\mathrm{Id}\,(\Pi_n)$ will be abbreviated to $\mathcal{J}^{(0)}$ in what follows.

Our first claim is that $W(\Pi_n)$ is a non-zero rank one projection. Clearly, $W(\Pi_n)$ is a projection which cannot be zero since, otherwise, $W(\mathcal{J}^{(0)}) = \{0\}$ in contrast to the assumptions. We will check that the range dimension of $W(\Pi_n)$ is at most one. Given a compact operator T on l^2, let (T_n) be a sequence in $\mathcal{J}^{(0)}$ such that $W(T_n) = T$, and let (μ_n) be a convergent sequence with limit μ such that $\Pi_n T_n \Pi_n = \mu_n \Pi_n$. Since $\mathcal{G}$ is in the kernel of W, we find

$$W(\Pi_n)TW(\Pi_n) = W(\Pi_n T_n \Pi_n) = W(\mu\Pi_n) + W((\mu_n - \mu)\Pi_n) = \mu W(\Pi_n).$$

Hence, the operator $W(\Pi_n)$ has rank one.

Next we will show that, by a unitary transformation, we can always reach that $W(\Pi_n) = P_1$. Such transformations will be used several times in what follows, so we will go a little bit more into the details here.

Let $f_1 \in \operatorname{Im} W(\Pi_n)$ be a vector of length one. If $\operatorname{Im} W(\Pi_n) = \operatorname{span} e_1$, then we choose $f_1 := e_1$. Otherwise, we have $f_1 \notin \operatorname{span} e_1$, and the vectors f_1 and e_1 are linearly independent. Decompose l^2 into the orthogonal sum

$$l^2 = \operatorname{span}\{e_1, f_1\} \oplus H_1$$

and choose vectors $\hat{e}_1$ and $\hat{f}_1$ in $\operatorname{span}\{e_1, f_1\}$ such that both $\{e_1, \hat{e}_1\}$ and $\{f_1, \hat{f}_1\}$ become orthogonal bases of $\operatorname{span}\{e_1, f_1\}$. Thus, every element $h \in l^2$ can be written uniquely as

$$h = \alpha f_1 + \beta \hat{f}_1 + h_1 \quad \text{with} \quad \alpha, \beta \in \mathbb{C},\ h_1 \in H_1.$$

Define a mapping $J_1 : l^2 \to l^2$ by

$$J_1 h = J_1(\alpha f_1 + \beta \hat{f}_1 + h_1) := \alpha e_1 + \beta \hat{e}_1 + h_1.$$

Clearly, W is a unitary operator on l^2 which maps f_1 onto e_1, and

$$W_1 : \mathcal{J}^{(0)} \to K(l^2), \quad (A_n) \mapsto J_1 W(A_n) J_1^{-1}$$

is an irreducible representation of $\mathcal{J}^{(0)}$ with kernel $\mathcal{G}$ and range $K(l^2)$ which is unitarily equivalent to W. Thus, we can suppose without loss of generality that already $W(\Pi_n) = P_1$.

An analogous transformation can be performed for every projection Π_n (recall that the Π_n are non-zero rank one projections by their definition). Thus, there exist unitary $n \times n$ matrices $E_n^{(1)} : L(\operatorname{Im} P_n) \to L(\operatorname{Im} P_n)$ such that

$$(E_n^{(1)})^* \Pi_n E_n^{(1)} = P_1 \quad \text{for all } n.$$

So, the outcome of the first step is that the ideal $\mathcal{J}^{(0)}$ of $\mathcal{A}$ is *-isomorphic to

$$\mathcal{J}^{(1)} := \{((E_n^{(1)})^* A_n E_n^{(1)}) : (A_n) \in \mathcal{J}^{(0)}\},$$

that the isomorphism can be chosen so that it maps the sequence (Π_n) to the constant sequence (P_1), and that the mapping

$$W^{(1)} : \mathcal{J}^{(1)} \to K(l^2), \quad (A_n) \to W(E_n^{(1)} A_n (E_n^{(1)})^*)$$

is an irreducible representation of $\mathcal{J}^{(1)}$ which maps the sequence (P_1) to the operator P_1 and which has kernel $\mathcal{G}$. The quotient homomorphism

$$\mathcal{J}^{(1)}/\mathcal{G} \to K(l^2), \quad (A_n) + \mathcal{G} \to W^{(1)}(A_n)$$

will be denoted by $W_\pi^{(1)}$.

Step 2. Let us fix some notations. The orthogonal projection from l^2 onto $\mathbb{C}e_k$ will be denoted by S_k (i.e., $S_k = P_k - P_{k-1}$ for $k \geq 2$), and we let $F_{ij} \in L(l^2)$ stand for the operator which satisfies

$$S_i F_{ij} = F_{ij} S_j = F_{ij}$$

(i.e., the representation of F_{ij} with respect to the basis $(e_k)_{k\geq 1}$ of l^2 has a 1 at the ijth entry whereas all other entries are zero).

What we will show in this step is that the ideal $\mathcal{J}^{(0)}$ is *-isomorphic to an ideal $\mathcal{J}^{(2)}$ which contains the sequences $(P_n K P_n)$ for $K = S_1, S_2, F_{12}, F_{21}$ and which, thus, contains a copy of $\mathbb{C}^{2\times 2}$.

Since S_2 is a rank one projection in $K(l^2)$, there is a uniquely determined rank one coset S_2^π in $\mathcal{J}^{(1)}/\mathcal{G}$ such that $W_\pi^{(1)}(S_2^\pi) = S_2$. From Proposition 3.4 (which we here apply in the special case $P_n^{(1)} = P_1$) we infer that there is a sequence $(S_n^{(2)}) \in \mathcal{J}^{(1)}$ which owns the following properties:

- $(S_n^{(2)}) + \mathcal{G} = S_2^\pi$, whence $W^{(1)}(S_n^{(2)}) = S_2$,
- $S_1^{(2)} = 0$, and $S_n^{(2)}$ is a projection with one-dimensional range for every $n \geq 2$,
- P_1 and $S_n^{(2)}$ are orthogonal for all n.

The orthogonality of P_1 and $S_n^{(2)}$ ensures that the ranges of these projections are mutually orthogonal subspaces of l^2. Thus, as in step 1, we can find unitary matrices $\hat{E}_n^{(2)} : \operatorname{Im} P_n \to \operatorname{Im} P_n$ such that

$$(\hat{E}_n^{(2)})^* P_1 \hat{E}_n^{(2)} = P_1 \quad \text{and} \quad (\hat{E}_n^{(2)})^* S_n^{(2)} \hat{E}_n^{(2)} = S_2. \tag{3.4}$$

The $\hat{E}_n^{(2)}$ are chosen so that they act on the orthogonal complement of the linear span of $\operatorname{Im} S_n^{(2)}$ and e_2 as the identity.

On defining

$$\widetilde{E}_n^{(2)} := \begin{cases} E_n^{(1)} & \text{if } n = 1 \\ E_n^{(1)} \hat{E}_n^{(2)} & \text{if } n \geq 2 \end{cases}$$

we see that the ideal $\mathcal{J}^{(0)}$ is *-isomorphic to

$$\widetilde{\mathcal{J}}^{(2)} := \{((\widetilde{E}_n^{(2)})^* A_n \widetilde{E}_n^{(2)}) : (A_n) \in \mathcal{J}^{(0)}\}$$

with the isomorphism mapping the sequence (Π_n) to the constant sequence (P_1), that $(P_n S_2 P_n) \in \widetilde{\mathcal{J}}^{(2)}$, and that

$$\widetilde{W}^{(2)} : \widetilde{\mathcal{J}}^{(2)} \to K(l^2), \quad (A_n) \mapsto W(\widetilde{E}_n^{(2)} A_n (\widetilde{E}_n^{(2)})^*)$$

is an irreducible representation of $\widetilde{\mathcal{J}}^{(2)}$ which maps (S_1) to S_1 and $(P_n S_2 P_n)$ to S_2.

Further, since $\widetilde{W}^{(2)}$ maps $\widetilde{\mathcal{J}}^{(2)}$ onto the compact operators, there is a sequence $(\widetilde{F}_n^{12})$ in $\widetilde{\mathcal{J}}^{(2)}$ such that $\widetilde{W}^{(2)}(\widetilde{F}_n^{12}) = F_{12}$. Then, clearly, $\widetilde{W}^{(2)}(P_n S_1 \widetilde{F}_n^{12} S_2 P_n) = F_{12}$, there are complex numbers α_n such that

$$P_n S_1 \widetilde{F}_n^{12} S_2 P_n = \alpha_n P_n F_{12} P_n \quad \text{for all } n,$$

and the sequence $(\alpha_n P_n F_{12} P_n)$ belongs to $\widetilde{\mathcal{J}}^{(2)}$. The fractality of $\mathcal{A}$ (and, hence, of $\widetilde{\mathcal{J}}^{(2)}$) implies the existence of the limit

$$\lim_{n\to\infty} \|\alpha_n P_n F_{12} P_n\| = \lim_{n\to\infty} |\alpha_n|,$$

and from

$$\lim_{n\to\infty} \|\alpha_n P_n F_{12} P_n\| = \|(\alpha_n P_n F_{12} P_n) + \mathcal{G}\| = \|\widetilde{W}^{(2)}(\alpha_n P_n F_{12} P_n)\| = \|F_{12}\| = 1$$

we conclude that $\lim |\alpha_n| = 1$. Write α_n as

$$\alpha_n = r_n e^{i\varphi_n} \quad \text{with } r_n \geq 0 \text{ and } \varphi_n \in [0,\, 2\pi).$$

Since

$$\alpha_n - e^{i\varphi_n} = (r_n - 1)e^{i\varphi_n} \to 0,$$

the sequence $((\alpha_n - e^{i\varphi_n})P_n F_{12} P_n)$ belongs to $\mathcal{G}$, whence $(e^{i\varphi_n} P_n F_{12} P_n) \in \widetilde{\mathcal{J}}^{(2)}$. For $n \geq 2$, one has

$$e^{i\varphi_n} P_n F_{12} P_n = e^{i\varphi_n} F_{12} = \begin{pmatrix} 0 & e^{i\varphi_n} \\ 0 & 0 \end{pmatrix}.$$

Thus, defining unitary diagonal matrices

$$D_n^{(2)} := \operatorname{diag}(1,\, e^{-i\varphi_n},\, 1,\, \ldots,\, 1) \in \mathbb{C}^{n\times n}$$

for $n > 1$, we find

$$(D_n^{(2)})^* P_n S_1 P_n D_n^{(2)} = P_n S_1 P_n, \quad (D_n^{(2)})^* P_n S_2 P_n D_n^{(2)} = P_n S_2 P_n$$

and

$$(D_n^{(2)})^* P_n e^{i\varphi_n} F_{12} P_n D_n^{(2)} = P_n F_{12} P_n$$

for all $n > 1$. If we finally set

$$E_n^{(2)} := \begin{cases} \widetilde{E}_n^{(2)} & \text{if } n = 1 \\ \widetilde{E}_n^{(2)} D_n^{(2)} & \text{if } n > 1, \end{cases}$$

then we can summarize the result of step 2 as follows. The ideal $\mathcal{J}^{(0)}$ of $\mathcal{A}$ is $*$-isomorphic to

$$\mathcal{J}^{(2)} := \{((E_n^{(2)})^* A_n E_n^{(2)}) : (A_n) \in \mathcal{J}^{(0)}\},$$

the isomorphism can be chosen so that it maps the sequence (Π_n) to the constant sequence (P_1), and the mapping

$$W^{(2)} : \mathcal{J}^{(2)} \to K(l^2), \quad (A_n) \to W(E_n^{(2)} A_n (E_n^{(2)})^*)$$

is an irreducible representation of $\mathcal{J}^{(2)}$ which has kernel $\mathcal{G}$. Besides the sequence (P_1), the algebra $\mathcal{J}^{(2)}$ also contains the sequences

$$(P_n S_2 P_n), \quad (P_n F_{12} P_n) \quad \text{and} \quad (P_n F_{21} P_n) = (P_n F_{12} P_n)^*,$$

and the representation $W^{(2)}$ maps these sequences to the operators P_1, S_2, F_{12} and F_{21}, respectively.

Step 3. Now we proceed per induction. Suppose that, for $k \geq 1$, we are given unitary operators $E_n^{(k-1)}$ with the related algebra

$$\mathcal{J}^{(k-1)} := \{((E_n^{(k-1)})^* A_n E_n^{(k-1)}) : (A_n) \in \mathrm{Id}\,(\Pi_n)\}$$

and with the corresponding irreducible representation

$$W^{(k-1)} : \mathcal{J}^{(k-1)} \to K(l^2), \quad (A_n) \to W(E_n^{(k-1)} A_n (E_n^{(k-1)})^*)$$

having $\mathcal{G}$ as its kernel, and suppose furthermore that the algebra $\mathcal{J}^{(k-1)}$ contains all sequences $(P_n K P_n)$ with $K \in \mathbb{C}^{(k-1)\times(k-1)} = L(\mathrm{Im}\, P_{k-1})$ and that $W^{(k-1)}(P_n K P_n) = K$ for all these sequences.

As in step 2, there is a sequence $(S_n^{(k)}) \in \mathcal{J}^{(k-1)}$ such that

- $W^{(k-1)}(S_n^{(k)}) = S_k$,
- $S_n^{(k)} = 0$ for $n < k$ and $S_n^{(k)}$ is a projection with one-dimensional range for every $n \geq k$,
- P_{k-1} and $S_n^{(2)}$ are orthogonal for all n.

Further, we can choose unitary matrices $\hat{E}_n^{(k)} : \mathrm{Im}\, P_n \to \mathrm{Im}\, P_n$ so that

- $(\hat{E}_n^{(k)})^* P_n K P_n \hat{E}_n^{(k)} = P_n K P_n$ for every $K \in \mathbb{C}^{(k-1)\times(k-1)}$,
- $(\hat{E}_n^{(k)})^* S_n^{(k)} \hat{E}_n^{(k)} = P_n S_k P_n$.

Consider the matrices

$$\widetilde{E}_n^{(k)} := \begin{cases} E_n^{(k-1)} & \text{if } n < k \\ E_n^{(k-1)} \hat{E}_n^{(k)} & \text{if } n \geq k, \end{cases}$$

and introduce the related algebra

$$\widetilde{\mathcal{J}}^{(k)} := \{((\widetilde{E}_n^{(k)})^* A_n \widetilde{E}_n^{(k)}) : (A_n) \in \mathcal{J}^{(0)}\}$$

with corresponding irreducible representation

$$\widetilde{W}^{(k)} : \widetilde{\mathcal{J}}^{(k)} \to K(l^2), \quad (A_n) \mapsto W(\widetilde{E}_n^{(k)} A_n (\widetilde{E}_n^{(k)})^*).$$

In $\widetilde{\mathcal{J}}^{(k)}$, we find a sequence $(\widetilde{F}_n^{1k})$ such that $\widetilde{W}^{(k)}(\widetilde{F}_n^{1k}) = F_{1k}$. The same reasoning as in step 2 reveals the existence of unimodular numbers such that the sequence $(e^{i\varphi_n} P_n F_{1k} P_n)$ lies in $\widetilde{\mathcal{J}}^{(k)}$. For $n \geq k$, set

$$D_n^{(k)} := \mathrm{diag}\,(1, \ldots, e^{-i\varphi_n}, 1, \ldots, 1) \in \mathbb{C}^{n\times n}$$

with $e^{-i\varphi_n}$ standing at the kth place, and define

$$E_n^{(k)} := \begin{cases} \widetilde{E}_n^{(k)} & \text{if } n < k \\ \widetilde{E}_n^{(k)} D_n^{(k)} & \text{if } n \geq k. \end{cases}$$

With these sequences, we relate the algebra

$$\mathcal{J}^{(k)} := \{((E_n^{(k)})^* A_n E_n^{(k)}) : (A_n) \in \mathcal{J}^{(0)}\}$$

and its irreducible representation

$$W^{(k)} : \mathcal{J}^{(k)} \to K(l^2), \quad (A_n) \to W(E_n^{(k)} A_n (E_n^{(k)})^*),$$

which has kernel $\mathcal{G}$. We claim that $\mathcal{J}^{(k)}$ contains all sequences (P_nKP_n) with $K \in \mathbb{C}^{k\times k}$ and that $W^{(k)}(P_nKP_n) = K$ for these sequences.

Indeed, all sequences $(P_nF_{ij}P_n)$ with $i, j \leq k-1$ lie already in $\mathcal{J}^{(k-1)}$. Since the unitary operators $E_n^{(k)}$ act on $\operatorname{Im} P_{k-1}$ as the identity, these sequences also belong to $\mathcal{J}^{(k)}$. Further, by construction, we have included the sequences $(P_nS_kP_n)$ and $(P_nF_{1k}P_n)$ and, hence, also the sequence $(P_nF_{k1}P_n) = (P_nF_{1k}P_n)^*$ into $\mathcal{J}^{(k)}$. Consequently, for $l = 2, \ldots, k-1$, the sequences

$$(P_nF_{lk}P_n) = (P_nF_{l1}P_n)(P_nF_{1k}P_n) \quad \text{and} \quad (P_nF_{kl}P_n) = (P_nF_{lk}P_n)^*$$

lie in $\mathcal{J}^{(k)}$, too, which proves our claim.

Final step. For every n, let E_n be the unitary matrix $E_n^{(n)}$, and consider the algebra

$$\mathcal{J}^\infty := \{((E_n)^*A_nE_n) : (A_n) \in \mathcal{J}^{(0)}\},$$

which has

$$W^\infty : \mathcal{J}^\infty \to K(l^2), \quad (A_n) \to W(E_nA_n(E_n)^*)$$

as its irreducible representation.

Let k be a fixed positive integer and $K \in L(\operatorname{Im} P_k)$. We know from the previous step that there is a sequence (K_n) in the ideal $\operatorname{Id}(\Pi_n)$ of $\mathcal{A}$ such that

$$((E_n^{(k)})^*K_nE_n^{(k)}) = P_nKP_n \quad \text{or, equivalently,} \quad (K_n) = (E_n^{(k)}P_nKP_n(E_n^{(k)})^*).$$

Due to the definitions of $E_n^{(k)}$ and E_n,

$$E_n^{(k)}P_k = E_n^{(n)}P_k = E_nP_k \quad \text{for all } n \geq k$$

and, consequently, $K_n = E_nP_nKP_n(E_n)^*$ for all $n \geq k$. This shows that, for every $K \in \mathbb{C}^{k\times k}$, the sequence $(E_nP_nKP_n(E_n)^*)$ belongs to $\mathcal{J}^{(0)}$ and that

$$W(E_nP_nKP_n(E_n)^*) = K.$$

A simple approximation argument shows that then, for every compact operator K on l^2,

$$(E_nP_nKP_n(E_n)^*) \in \mathcal{J}^{(0)} \quad \text{and} \quad W(E_nP_nKP_n(E_n)^*) = K.$$

It is elementary to check that

$$\mathcal{J} := \{((E_nP_nKP_n(E_n)^* + G_n) : K \in K(l^2), (G_n) \in \mathcal{G}\}$$

is a closed subalgebra of $\mathcal{J}^{(0)}$ and that W maps $\mathcal{J}$ onto $K(l^2)$ in such a way that the only sequences in $\mathcal{J}$ which are mapped to 0 are the sequences in $\mathcal{G}$. Consequently, $\mathcal{J} = \mathcal{J}^{(0)}$.

To get the assertion in the form stated in the theorem, we extend E_n and $(E_n)^*$ to partial isometries on l^2 by letting them act on $\operatorname{Im}(I - P_n)$ as the zero operator. □

In case the Hilbert space H has finite dimension r, the induction process in the preceding proof terminates after r steps, and one can choose $E_n := E_n^{(r)}$ for all n. Observe also that Theorem 3.1 remains valid if we only assume the fractality of the ideal $\operatorname{Id}(K_n)$ instead of the fractality of the algebra $\mathcal{A}$.

4. The first half of the standard model

Now we will see how the results of the preceding section imply the first three axioms of the standard model.

Let $E_n : H \to l^2$ be partial isometries with range projections P_n and such that the initial projections $P'_n := E_n^* E_n$ converge strongly to the identity operator on H. Further set

$$\mathcal{J}(E_n) := \{(E_n K E_n^*) + (G_n) \text{ with } K \in K(H),\, (G_n) \in \mathcal{G}\}. \tag{4.1}$$

Theorem 4.1. *Let $\mathcal{A}$ be a C^*-subalgebra of $\mathcal{F}$ which contains $\mathcal{J}(E_n)$. Then the following assertions are equivalent.*

(a) *$\mathcal{J}(E_n)$ is an ideal of $\mathcal{A}$.*

(b) *For any sequence $(A_n) \in \mathcal{A}$, there exists the strong limit* s-lim $E_n^* A_n E_n$.

Proof. (b) $\Rightarrow$ (a): Given a sequence (A_n) in $\mathcal{A}$, let $W(A_n)$ denote the strong limit s-lim $E_n^* A_n E_n$. Then, for arbitrary operators $K \in K(H)$ and sequences $(G_n) \in \mathcal{G}$,

$$\begin{aligned} A_n(E_n K E_n^* + G_n) &= E_n E_n^* A_n E_n K E_n^* + A_n G_n \\ &= E_n W(A_n) K E_n^* + E_n(E_n^* A_n E_n - W(A_n)) K E_n^* + A_n G_n. \end{aligned} \tag{4.2}$$

The operator $W(A_n)K$ is compact, and the strong convergence of $(E_n^* A_n E_n)$ to $W(A_n)$ together with the compactness of K imply the norm convergence of $(E_n^* A_n E_n - W(A_n))$ to 0. Hence, the right-hand side of (4.2) belongs to $\mathcal{J}(E_n)$, and $\mathcal{J}(E_n)$ is a left ideal of $\mathcal{A}$. The other side of the ideal property can be checked analogously.

(a) $\Rightarrow$ (b): For the reverse implication, let $(A_n) \in \mathcal{A}$ and $K \in K(H)$. Since $\mathcal{J}(E_n)$ is an ideal of $\mathcal{A}$, there are an operator $R \in K(H)$ as well as a sequence $(G_n) \in \mathcal{G}$ such that

$$A_n E_n K E_n^* = E_n R E_n^* + G_n.$$

Multiplication by E_n^* from the left and by E_n from the right yields

$$E_n^* A_n E_n K = P'_n R P'_n + E_n^* G_n E_n = R + G'_n$$

with a sequence (G'_n) tending to zero in the norm (recall that the projections P'_n converge strongly to the identity operator on H by hypothesis). Thus, the sequence $(E_n^* A_n E_n K)$ is norm convergent for every compact operator $K \in K(H)$.

Let $x \in H$, and write P_x for the (compact) orthogonal projection from H onto its one-dimensional subspace $\mathbb{C}x$. Then the sequence $(E_n^* A_n E_n P_x)$ is norm convergent as we have just seen, and from

$$E_n^* A_n E_n x = E_n^* A_n E_n P_x x$$

we conclude that the sequence $(E_n^* A_n E_n x)$ converges in the norm of H for every x, i.e. $(U_n^* A_n U_n)$ is a strongly convergent sequence. $\square$

The restriction of the mapping $\mathcal{A} \to L(H)$, $(A_n) \mapsto \text{s-lim}\, E_n^* A_n E_n$ to $\mathcal{J}(E_n)$ is evidently a homomorphism from $\mathcal{J}(E_n)$ onto $K(H)$ which has $\mathcal{G}$ as its kernel.

Now we consider the case where two (or more) ideals of the form (4.1) occur and ask for specifications of these ideals and of the related strong limits which are involved by the separation property.

Theorem 4.2. *Let H_1 and H_2 be Hilbert spaces and let E_n and F_n be partial isometries from H_1 and H_2 into l^2 with range projections P_n and such that their initial projections converge strongly to the identity operators on H_1 and H_2, respectively. Further, let $\mathcal{A}$ be a C^*-subalgebra of $\mathcal{F}$ which contains the sets*

$$\mathcal{J}(E_n) := \{(E_n K E_n^* + G_n) \text{ with } K \in K(H_1) \text{ and } (G_n) \in \mathcal{G}\},$$

$$\mathcal{J}(F_n) := \{(F_n K F_n^* + G_n) \text{ with } K \in K(H_2) \text{ and } (G_n) \in \mathcal{G}\}$$

as its ideals, and let W_1 denote the homomorphism $(A_n) \mapsto \text{s-lim}\, E_n^* A_n E_n$. Then the following assertions are equivalent.

(a) *The separation property $W_1(\mathcal{J}(F_n)) = \{0\}$ holds.*

(b) *The sequence $(F_n^* E_n)$ converges weakly to zero.*

Proof. The separation property (a) is equivalent to

$$\text{s-lim}\, E_n^* F_n K F_n^* E_n = 0 \quad \text{for all} \quad K \in K(H_2).$$

Thus, for every $K \in K(H_2)$ and $x \in H_1$, one has

$$\lim \langle E_n^* F_n K^* K F_n^* E_n x,\, x\rangle = \lim \langle K F_n^* E_n x,\, K F_n^* E_n x\rangle = 0$$

and, consequently, $\|K V_n^* U_n x\| \to 0$. Given $y \in H_2$, let P_y refer to the orthogonal projection from H_2 onto its subspace $\mathbb{C}y$. The operator P_y is compact, hence, $\|P_y F_n^* E_n x\| \to 0$ and

$$\langle F_n^* E_n x,\, y\rangle = \langle F_n^* E_n x,\, P_y y\rangle = \langle P_y F_n^* E_n x,\, y\rangle \to 0$$

for every $x \in H_1$ and $y \in H_2$. Thus, the sequence $(F_n^* E_n)$ converges weakly to zero.

For the reverse implication note that the weak convergence of $(F_n^* E_n)$ to zero involves the strong convergence of $(K F_n^* E_n)$ to zero for every compact operator K and, thus, the strong convergence to zero of the sequence $(F_n^* E_n K F_n^* E_n)$. But this is the separation property. □

In algebras arising from concrete approximation methods, one usually has the existence of the strong limit s-lim A_n for every sequence in the algebra, i.e., one can choose $E_n := P_n$ in the preceding theorem. In this case, the separation condition implies that the F_n converge weakly to 0.

5. The second half of the standard model

Up to now, we have associated with the ideal $\mathcal{J}_{ess}(\mathcal{A})$ of the fractal algebra $\mathcal{A}$ a family $\{W_t\}$ of homomorphisms which are given in terms of strong limits. Now we will discuss the fourth axiom of the standard model, i.e., we ask whether the invertibility of all operators $W_t(A_n)$ is sufficient for the stability of the sequence $(A_n) \in \mathcal{A}$. Since $W_t(A_n)$ depends on the coset $(A_n) + \mathcal{G}$ only, one can consider the quotient mappings

$$\mathcal{A}/\mathcal{G} \to L(H_t), \quad (A_n) + \mathcal{G} \mapsto W_t(A_n). \tag{5.1}$$

These mappings are irreducible representations of $\mathcal{A}/\mathcal{G}$; the irreducibility is a consequence of the fact that the ideal $K(H_t)$ of the compact operators lies in the range of W_t. It is also evident from the construction of the W_t that the restrictions of these representations to $\mathcal{J}_{ess}(\mathcal{A})/\mathcal{G}$ are irreducible representations of this ideal. Moreover, these restrictions already exhaust the spectrum of the C^*-algebra $\mathcal{J}_{ess}(\mathcal{A})/\mathcal{G}$.

These observations suggest to consider a more general situation. Let $\mathfrak{A}$ be a C^*-algebra with identity element e and with spectrum (= set of all equivalence classes of irreducible representations) $\operatorname{Spec}\mathfrak{A}$, and let $\mathfrak{J}$ be a closed two-sided ideal of $\mathfrak{A}$. Then every irreducible representation W of $\mathfrak{J}$ determines uniquely an irreducible representation of $\mathfrak{A}$ which we denote by W again. Now our problem reads as follows: If $a \in \mathfrak{A}$ and $W(a)$ is invertible for all $W \in \operatorname{Spec}\mathfrak{J}$, is then a invertible? Let us call a subset M of $\operatorname{Spec}\mathfrak{A}$ a *sufficient family of irreducible representations* if the equivalence

$$a \text{ invertible} \iff W(a) \text{ invertible for all } W \in M$$

holds for every $a \in \mathfrak{A}$, and let us call M a *weakly sufficient family* if

$$a \text{ invertible} \iff W(a) \text{ invertible for all } W \in M \text{ and } \sup_{W \in M} \|W(a)^{-1}\| < \infty$$

for all $a \in \mathfrak{A}$. These definitions make sense since both the invertibility and the norm of $W(a)$ depend on the equivalence class of the representation W only.

Since equivalent representations have the same kernel, there is a natural surjective mapping

$$\Pi : \operatorname{Spec}\mathfrak{A} \to \operatorname{Prim}\mathfrak{A}, \quad W \mapsto \ker W$$

where $\operatorname{Prim}\mathfrak{A}$ stands for the space of the primitive ideals of $\mathfrak{A}$. This space carries a natural topology, the so-called hull-kernel or Jacobson topology. For definitions see, e.g., [4, 6].

Weakly sufficient families of representations

We start with criteria for the weak sufficiency of a family M of irreducible representations.

Theorem 5.1. *Let $\mathfrak{A}$ be a unital C^*-algebra. The following conditions are equivalent for a subset M of* $\operatorname{Spec}\mathfrak{A}$.

(a) *The family M is weakly sufficient.*

(b) *For every $a \in \mathfrak{A}$, $W(a) = 0$ for every $W \in M$ implies that $a = 0$.*

(c) *For every $a \in \mathfrak{A}$, $\|a\| = \sup_{W \in M} \|W(a)\|$.*

(d) $\Pi(M)$ *is dense in* $\operatorname{Prim}\mathfrak{A}$.

Proof. (a) $\Rightarrow$ (b): Let $a \in \mathfrak{A}$ be an element such that $W(a) = 0$ for all $W \in M$, and let b be an arbitrary invertible element of $\mathfrak{A}$. Then, evidently, $W(b)$ is invertible for all $W \in M$, and the norms $\|W(b)^{-1}\|$ are uniformly bounded by $\|b^{-1}\|$. Then also the operators $W(a+b)$ are invertible for all $W \in M$, and the norms of their inverses are uniformly bounded, too. By (a), the element $a+b$ is invertible. Hence, a lies in the radical of $\mathfrak{A}$, which consists of the zero element only.

(b) $\Rightarrow$ (c): Let $\mathfrak{F}$ stand for the set of all bounded functions on M which take at $W \in M$ a value in $W(\mathfrak{A})$. Provided with pointwisely defined operations and the supremum norm, the set $\mathfrak{F}$ becomes a C^*-algebra, and the mapping

$$\mathcal{S} : \mathfrak{A} \to \mathfrak{F}, \quad a \mapsto (W \mapsto W(a))$$

is a *-homomorphism. By (b), this homomorphism has kernel $\{0\}$ and, hence, is an isometry. Thus, (c) follows.

(c) $\Rightarrow$ (a): Let $a \in \mathfrak{A}$ be an element for which the operators $W(a)$ are invertible for all $W \in M$ and the norms $\|W(a)^{-1}\|$ are uniformly bounded. Then $\mathcal{S}(a)$ is invertible in $\mathfrak{F}$ and, due to the inverse closedness of C^*-algebras, also in $\mathcal{S}(\mathfrak{A})$. But $\mathcal{S}$ is an isometry due to (c) and, consequently, an isomorphism between $\mathfrak{A}$ and $\mathcal{S}(\mathfrak{A})$. Hence, a is invertible in $\mathfrak{A}$.

(b) $\Leftrightarrow$ (d): Let $M' := \{\ker W : W \in M\} \subseteq \operatorname{Prim}\mathfrak{A}$. The condition (b) implies that $\cap_{J \in M'} J = \{0\}$. Hence, the kernel of M' is $\{0\}$ and, consequently, the hull of M' is all of $\operatorname{Prim}\mathfrak{A}$. Thus, M' is dense in the primitive ideal space of $\mathfrak{A}$.

Let, conversely, M' be dense in $\operatorname{Prim}\mathfrak{A}$. Then the kernel of M' is contained in every primitive ideal of $\mathfrak{A}$ and, hence, in the kernel of $\operatorname{Prim}\mathfrak{A}$. Since the intersection of all primitive ideals of a C^*-algebra is the zero ideal, we have $\cap_{J \in M'} J = \{0\}$, which implies (b). □

Essential ideals

Now we are going to single out a class of ideals $\mathfrak{J}$ of $\mathfrak{A}$ having the property that $\operatorname{Spec}\mathfrak{J}$ is a weakly sufficient family of representations for $\mathfrak{A}$. Recall that an ideal $\mathfrak{J}$ of $\mathfrak{A}$ is called *essential* if the intersection of $\mathfrak{J}$ with every non-zero ideal of $\mathfrak{A}$ is non-zero. Further we need the notion of the *annulator* of an ideal, which is defined by

$$\operatorname{Ann}\mathfrak{J} := \{a \in \mathfrak{A} : a\mathfrak{J} = \mathfrak{J}a = \{0\}\}.$$

Lemma 5.2. *Let $\mathfrak{A}$ be a C^*-algebra with identity e and $\mathfrak{J}$ a closed two-sided ideal of $\mathfrak{A}$. Then*

$$\operatorname{Ann}\mathfrak{J} = \{a \in \mathfrak{A} : W(a) = 0 \text{ for all } W \in \operatorname{Spec}\mathfrak{A}\}.$$

Here we use again the same letter W for an irreducible representation of $\mathfrak{J}$ and its extension to $\mathfrak{A}$.

Proof. Let $W(a) = 0$ for all $W \in \operatorname{Spec}\mathfrak{A}$. Suppose there is a $j \in \mathfrak{J}$ such that $aj \neq 0$. Then there is an irreducible representation W of $\mathfrak{J}$ such that $W(aj) \neq 0$. For this representation, one cannot have $W(a) = 0$.

Conversely, let $aj = 0$ for all $j \in \mathfrak{J}$, and let $W : \mathfrak{J} \to L(H)$ be in $\operatorname{Spec}\mathfrak{J}$. By [4], Proposition 2.10.4, the set $W(\mathfrak{J})$ is dense in $W(\mathfrak{A})$ with respect to the strong operator topology on H. Since W maps the identity element e to the identity operator I, there is a net $(j_t) \subset \mathfrak{J}$ such that

$$W(j_t) \to I \quad \text{strongly as } t \to \infty.$$

Then

$$0 = W(aj_t) = W(a)W(j_t) \to W(a) \quad \text{strongly as } t \to \infty,$$

whence $W(a) = 0$. □

Theorem 5.3. *Let $\mathfrak{A}$ be a unital C^*-algebra. The following conditions are equivalent for a closed two-sided ideal $\mathfrak{J}$ of $\mathfrak{A}$.*

(a) $\mathfrak{J}$ *is an essential ideal of* $\mathfrak{A}$.

(b) $\operatorname{Ann}\mathfrak{J} = \{0\}$.

(c) $\operatorname{Spec}\mathfrak{J}$ *is weakly sufficient for* $\mathfrak{A}$.

(d) *For every $a \in \mathfrak{A}$, $W(a) = 0$ for every $W \in \operatorname{Spec}\mathfrak{J}$ implies that $a = 0$.*

(e) *For every $a \in \mathfrak{A}$, $\|a\| = \sup_{W \in \operatorname{Spec}\mathfrak{J}} \|W(a)\|$.*

(f) $\operatorname{Prim}\mathfrak{J}$ *is dense in* $\operatorname{Prim}\mathfrak{A}$.

Assertion (f) has to be interpreted as follows. Set

$$\operatorname{Spec}^{\mathfrak{J}}\mathfrak{A} := \{W \in \operatorname{Spec}\mathfrak{A} : W|_{\mathfrak{J}} \neq 0\}, \quad \operatorname{Prim}^{\mathfrak{J}}\mathfrak{A} := \{I \in \operatorname{Prim}\mathfrak{A} : \mathfrak{J} \not\subseteq I\}.$$

It is well known that the mappings $W \mapsto W|_{\mathfrak{J}}$ and $I \mapsto I \cap \mathfrak{J}$ are homemorphisms from $\operatorname{Spec}^{\mathfrak{J}}\mathfrak{A}$ onto $\operatorname{Spec}\mathfrak{J}$ and from $\operatorname{Prim}^{\mathfrak{J}}\mathfrak{A}$ onto $\operatorname{Prim}\mathfrak{J}$, respectively (see [4], Sections 2.11 and 3.2), and we have identified the corresponding sets in the formulation of (f). In this sense, statement (f) is equivalent to saying that

$\Pi(\operatorname{Spec}^{\mathfrak{J}}\mathfrak{A})$ (which evidently coincides with $\operatorname{Prim}^{\mathfrak{J}}\mathfrak{A}$) is dense in $\operatorname{Prim}\mathfrak{A}$.

Proof. The simple proof of (a) $\Leftrightarrow$ (b) can be found, e.g., in [9], Lemma 2.36. The equivalence between (b) and (d) follows from the preceding lemma, and the equivalence of (c)–(f) is a consequence of Theorem 5.1. □

Recall that a C^*-algebra $\mathfrak{A}$ is called *postliminal* if $K(H) \subseteq W(\mathfrak{A})$ for every irreducible representation (H, W) of $\mathfrak{A}$. One feature of postliminal algebras is that the mapping Π is a bijection from $\operatorname{Spec}\mathfrak{A}$ onto $\operatorname{Prim}\mathfrak{A}$.

It turns out that every postliminal C^*-algebra contains some canonical essential ideals. Following Pedersen, we call an element a of a C^*-algebra $\mathfrak{A}$ *abelian* if $a \geq 0$ and if the algebra $a\mathfrak{A}a$ is commutative. By $\mathcal{J}_{abel}(\mathfrak{A})$ we denote the smallest closed two-sided ideal of $\mathfrak{A}$ which contains all abelian elements of $\mathfrak{A}$.

Theorem 5.4. *If $\mathfrak{A}$ is postliminal, then $\mathcal{J}_{abel}(\mathfrak{A})$ is an essential ideal of $\mathfrak{A}$.*

Proof. In [6], Theorem 6.2.11, it is shown that every postliminal algebra possesses an essential ideal with continuous trace. Further, by Theorem 6.1.11 in [6], every algebra with continuous trace is generated by its abelian elements and, by Proposition 6.1.7 of [6], $\mathcal{J}_{abel}(\mathfrak{A})$ is the largest ideal of $\mathfrak{A}$ which is generated by its abelian elements. Hence, this ideal contains an essential ideal, and so it is essential, too. □

Further, let $\mathcal{J}_{com}(\mathfrak{A})$ stand for the set of all elements $a \in \mathfrak{A}$ such that $W(a)$ is a compact operator for every irreducible representation W of $\mathfrak{A}$. Similarly to the preceding theorem, one has the following.

Theorem 5.5. *If $\mathfrak{A}$ is postliminal, then $\mathcal{J}_{com}(\mathfrak{A})$ is an essential ideal of $\mathfrak{A}$.*

Proof. From Theorem 5.4 we know that $\mathcal{J}_{abel}(\mathfrak{A})$ is an essential ideal of $\mathfrak{A}$. Since $\mathcal{J}_{abel}(\mathfrak{A})$ is a liminal ideal (Corollary 6.1.6 in [6]) and $\mathcal{J}_{com}(\mathfrak{A})$ is the largest liminal ideal of $\mathfrak{A}$ (Lemma 6.2.2 in [6]), we conclude that $\mathcal{J}_{abel}(\mathfrak{A}) \subseteq \mathcal{J}_{com}(\mathfrak{A})$ and, hence, $\mathcal{J}_{com}(\mathfrak{A})$ is essential, too. □

Essential dual ideals

In general, the irreducible representations of an essential ideal yield a *weakly* sufficient family only. To force the sufficiency of that family, one needs stronger properties of the ideal. One version which fits to our purposes is as follows. A C^*-algebra is called *elementary* if it is *-isomorphic to the ideal $K(H)$ of the compact operators on some Hilbert space H. A C^*-algebra is *dual* if it is the restricted direct sum of elementary algebras. This is not the original definition of a dual algebra, but it is an equivalent one (for this and other equivalent descriptions of dual algebras see [1]).

Theorem 5.6. *Let $\mathfrak{A}$ be a unital C^*-algebra and $\mathfrak{J}$ be an essential dual ideal of $\mathfrak{A}$. Then* Spec $\mathfrak{J}$ *is a sufficient family of representations for $\mathfrak{A}$.*

For the proof, we need the following result.

Theorem 5.7. *Let $\mathfrak{A}$ be a unital C^*-algebra, and let $M \subseteq$ Spec $\mathfrak{A}$ be a family of representations. This family is sufficient if and only if, for every $a \in \mathfrak{A}$, there is a $W_0 \in M$ such that $\|a\| = \|W_0(a)\|$.*

Proof. Let M be a sufficient family, but contrary to what we want assume that there is an $a \in \mathfrak{A}$ such that

$$\|W_0(a)\| < \sup_{W \in M} \|W(a)\| \quad \text{for all } W_0 \in M. \tag{5.2}$$

Since $\|W(a)\| = \|W((a^*a)^{1/2})\|$ by the C^*-axiom, we can without loss of generality suppose that the element a in (5.2) is self-adjoint and non-negative. The norm of the self-adjoint element a coincides with its spectral radius $\rho(a)$. Thus, (5.2) can be rewritten as

$$\rho(W_0(a)) < \sup_{W \in M} \rho(W(a)) \quad \text{for all} \quad W_0 \in M. \tag{5.3}$$

Denote the supremum on the right-hand side of (5.3) by C and set $c = a - Ce$. The elements $W(c) = W(a) - CW(e)$ are invertible for all $W \in M$ since $\rho(W(a)) < C$, and the hypothesis yields the invertibility of $c = a - Ce$. Then, clearly, $ba - ce$ is invertible for all c belonging to some neighborhood U of C. On the other hand, since $\sup_{W \in M} \rho(W(a)) = C$, for every neighborhood U of C there is an $W_U \in M$ such that $m_U := \rho(W_U(a)) \in U$. The element $W_U(a) - m_U W_U(e)$ is not invertible, because the spectral radius of a non-negative element belongs to the spectrum of this element. Hence, $a - m_U e$ is not invertible. This contradiction proves the only-if-part of the assertion.

To verify its if-part, let $a \in \mathfrak{A}$ be a non-invertible element. We have to show the existence of a $W_0 \in M$ such that $W_0(a)$ is not invertible.

If a is not invertible, then at least one of the self-adjoint elements aa^* or a^*a is not invertible, for definiteness say a^*a. Since a^*a is non-negative, a simple application of the Gelfand-Naimark theorem for commutative C^*-algebras shows that

$$\| \, \|a^*a\|e - a^*a\| = \|a^*a\|. \tag{5.4}$$

Set $b := \|a^*a\|e - a^*a$. By hypothesis, there is a $W_0 \in M$ such that $\|W_0(b)\| = \|b\|$, which together with (5.4) implies

$$\| \, \|a^*a\|W_0(e) - W_0(a^*a)\| = \|W_0(\|a^*a\|e - a^*a)\| = \| \, \|a^*a\|e - a^*a\| = \|a^*a\|.$$

Since $\|W_0(a^*a)\| \leq \|a^*a\|$, we can apply the Gelfand-Naimark theorem once more to obtain the non-invertibility of $W_0(a^*a)$ and, thus, of $W_0(a)$. □

Now we can prove Theorem 5.6.

Proof. By the preceding theorem, what we have to prove is that, for every $a \in \mathfrak{A}$, there is a $W_0 \in \operatorname{Spec} \mathfrak{A}$ such that $\|a\| = \|W_0(a)\|$. For $a = 0$, this is trivially satisfied. Let $\|a\| > 0$. Since $\mathfrak{J}$ is an essential ideal, the set

$$M_0 := \{W \in \operatorname{Spec} \mathfrak{J} : \|W(a)\| \geq \|a\|/2\}$$

is not empty due to Theorem 5.3(e). Further, this set is finite since $\mathfrak{J}$ is a dual ideal. Thus, there is a $W_0 \in M_0$ such that

$$\|W_0(a)\| \geq \|W(a)\| \quad \text{for all } W \in M_0.$$

Since

$$\sup\{\|W(a)\| : W \in M_0\} = \sup\{\|W(a)\| : W \in \operatorname{Spec} \mathfrak{J}\} = \|a\|$$

by Theorem 5.3(e) again, we have $\|W_0(a)\| = \|a\|$. □

Applications to algebras of approximation sequences

Now we are in a position to establish an equivalent characterization of standard algebras.

Theorem 5.8. *Let $\mathcal{A}$ be a unital and fractal C^*-subalgebra of $\mathcal{F}$ which contains the ideal $\mathcal{G}y$ of the zero sequences. If $\mathcal{J}_{ess}(\mathcal{A})$ is separable and $\mathcal{J}_{ess}(\mathcal{A})/\mathcal{G}$ is an essential ideal in $\mathcal{A}/\mathcal{G}$ (for example, if $\mathfrak{A}$ is postliminal and $\mathcal{J}_{ess}(\mathcal{A})/\mathcal{G} = \mathcal{J}_{abel}(\mathcal{A}/\mathcal{G})$), then $\mathcal{A}$ is a standard algebra.*

Proof. The ideal $\mathcal{J}_{ess}(\mathcal{A})/\mathcal{G}$ is generated by the ideals $\mathrm{Id}\,(K_n)/\mathcal{G}$ where (K_n) runs through the essential rank one sequences of $\mathcal{A}$. Since each of the ideals $\mathrm{Id}\,(K_n)/\mathcal{G}$ is *-isomorphic to $K(H)$ for some Hilbert space H (see [11]), $\mathcal{J}_{ess}(\mathcal{A})/\mathcal{G}$ is a dual ideal in $\mathcal{A}/\mathcal{G}$. By hypothesis, this ideal is also essential. Hence, by Theorem 5.6, the set $T := \mathrm{Spec}\,\mathcal{J}_{ess}(\mathcal{A})/\mathcal{G}$ provides us with a sufficient family of representations for the invertibility in $\mathcal{A}/\mathcal{G}$ and, thus, for the stability of sequences in $\mathcal{A}$.

Further, it is easy to see that the spectrum T of $\mathcal{J}_{ess}(\mathcal{A})/\mathcal{G}$ is in one-to-one correspondence with the set S of equivalence classes of ideals $\mathrm{Id}\,(K_n)/\mathcal{G}$ introduced before Theorem 2.7. Hence, with each of these ideals, there is associated a representation $W_t \in T$ with

$$W_t(\mathrm{Id}\,(K_n)/\mathcal{G}) = K(H_t).$$

The separability assumption ensures that every ideal $\mathrm{Id}\,(K_n)/\mathcal{G}$ is separable. Consequently, the associated Hilbert space H_t is separable, too, and from Theorem 3.1 we conclude that

$$\mathrm{Id}\,(K_n) = \{(E_n^t K (E_n^t)^*) + G_n) : K \in K(H_t),\, (G_n) \in \mathcal{G}\}$$

with certain partial isometries $E_n^t : H_t \to l^2$ satisfying $E_n^t (E_n^t)^* = P_n$. Now the first three conditions in the definition of a standard algebra follow from Theorems 4.1 and 4.2. □

References

[1] Berglund, M.C.F., *Ideal C^*-algebras*, Duke Math. J. **40**(1973), 241–257.

[2] Böttcher, A., Silbermann, B., The finite section method for Toeplitz operators on the quarter plane with piecewise continuous symbols, *Math. Nachr.* **110**(1983), 279–291.

[3] Böttcher, A., Silbermann, B., *Introduction to Large Truncated Toeplitz Matrices*, Springer-Verlag, New York, Berlin, Heidelberg 1999.

[4] Dixmier, J., *C^*-Algebras*, North Holland Publ. Comp., Amsterdam, New York, Oxford 1982.

[5] Hagen, R., Roch, S., Silbermann, B., *Spectral Theory of Approximation Methods for Convolution Equations*, Birkhäuser Verlag, Basel, Boston, Berlin 1995.

[6] Pedersen, G.K., *C^*-Algebras and Their Automorphism Groups*, Academic Press, London, New York, San Francisco 1979.

[7] Prössdorf, S., Silbermann, B., *Numerical Analysis for Integral and Related Operator Equations*, Akademie-Verlag, Berlin, 1991, and Birkhäuser Verlag, Basel, Boston, Stuttgart 1991.

[8] Rabinovich, V.S., Roch, S., Silbermann, B., Algebras of approximation sequences: Finite sections of band-dominated operators, *Acta Appl. Math.* **65**(2001), 315–332.

[9] Raeburn, I., Williams D.P., *Morita Equivalence and Continuous-Trace C^*-Algebras*, Mathematical Surveys and Monographs Vol. 60, American Mathematical Society 1998.

[10] Roch, S., Algebras of approximation sequences: Fractality, in Operator Theory: Adv. and Appl. **121**, Birkhäuser Verlag, Basel 2001, 471–497.

[11] Roch, S., Algebras of approximation sequences: Fredholmness, Preprint 2048(1999) TU Darmstadt, to appear in *J. Oper. Theory.*

[12] Roch, S., Silbermann, B., C^*-algebra techniques in numerical analysis, *J. Operator Theory* **35**(1996), 2, 241–280.

[13] Roch, S., Silbermann, B., Index calculus for approximation methods and singular value decomposition, *J. Math. Anal. Appl.* **225**(1998), 401–426.

Steffen Roch
Technische Universität Darmstadt
Fachbereich Mathematik
Schlossgartenstrasse 7
D-64289 Darmstadt, FRG
e-mail: `roch@mathematik.tu-darmstadt.de`

2000 *Mathematics Subject Classification.*
Primary 46N40; Secondary 65R20.

Received: 20 December 2000.

Operator Theory:
Advances and Applications, Vol. 142, 311–344

Noncanonical Factorizations of Almost Periodic Multivariable Matrix Functions

Leiba Rodman, Ilya M. Spitkovsky and Hugo J. Woerdeman

Abstract. Noncanonical factorizations of almost periodic matrix-valued functions of several real variables are studied. In particular, results are proved concerning hereditary properties of such factorizations, symmetric factorization of matrix functions that possess certain symmetries, behavior of factorization indices under small perturbations, connections between left and right indices, and relations between factorization and Fredholmness properties of the associated Toeplitz operators. The last section is devoted to uses of factorization for normalization of bases, an important problem in wavelets and other applications. Conjectures and open problems are stated.

1. Introduction

Factorizations have been used for a long time to study Toeplitz and singular integral operators, and closely related Wiener-Hopf equations and Riemann boundary value problems. Initiated in the ground breaking papers by Krein [30] and Gohberg-Krein [17], Wiener-Hopf type factorization properties of the symbol function, which may be scalar, matrix, or operator-valued, played a key role in these studies. Since the literature on this subject is voluminous, we mention here only a few (by no means all) relevant books. Convolution equations and their discrete analogs are studied in [14] using Wiener-Hopf factorizations of matrix-valued functions defined on the unit circle. Generalized factorizations are used in [18] (scalar symbols) and [8], [33] (matrix symbols) to study singular integral equations on rather general classes of curves. For rational-valued symbols, a state space approach leads to explicit factorization and description of the corresponding Toeplitz operators and solutions of systems of Wiener-Hopf equations; see [16]. The book [15] is a graduate level textbook where an exposition of basic theory of Toeplitz operators and factorization of their symbols is found.

Invertibility properties (one-sided and two-sided) of certain integro-differential operators of Wiener-Hopf type were described in [13], [9] in terms of factorization of *scalar* almost periodic functions. In recent years, in connection with systems of convolution type and Wiener-Hopf equations on finite intervals, factorization of almost periodic *matrix* functions became a subject of intensive research, see [24], [25], [26], [27], [6], [40]. It turns out that the factorization problem for 2×2 triangular almost periodic matrices is closely related with the almost periodic version of the corona problem, see [4], [5], [41]. General relations between Fredholm properties of Toeplitz operators with almost periodic matrix symbols and factorization of the symbols were discussed in [22], [28], [3]. A thorough exposition of many of these developments is found in the new book [7]. Still another area of application of almost periodic factorization was found in positive and contractive extension problems in the context of the abstract band method: see [49], [42], [44], and [1], [2], where matrix periodic functions of two and more variables were considered.

Motivated by these developments, a study of factorization of matrix-valued almost periodic functions of several real variables and its applications was initiated in [43], [46], [47]. The focus in these papers is on *canonical* factorization and its various applications: Toeplitz operators [43], contraction extension problems and distance formulas [46], abstract band method and positive extensions [47].

In the present paper, the main emphasis is on *noncanonical* factorizations of almost periodic matrix functions of k real variables. Here, as well as in our previous papers [42], [43], [46], [45], [47], we consider functions with Fourier spectrum in a given additive subgroup of $\mathbb{R}^k$. By doing so, we are able to treat simultaneously many particular cases, for example, periodic functions, or functions of mixed type (e.g., periodic in one variable and non-periodic in other variables).

We now introduce one of several variations of the concept of noncanonical almost periodic factorization that we study in this paper. Denote by (AP^k) the algebra of complex-valued almost periodic functions of k real variables $t = (t_1, \ldots, t_k) \in \mathbb{R}^k$, i.e., the closed subalgebra of $L^\infty(\mathbb{R}^k)$ (with respect to the standard Lebesgue measure) generated by all the functions $e_\lambda(t) = e^{i(t_1\lambda_1+\cdots+t_k\lambda_k)}$, where $\lambda = (\lambda_1, \ldots, \lambda_k) \in \mathbb{R}^k$. Given a half-space S of $\mathbb{R}^k$ (see the next section for the definition of a half-space), and an additive subgroup $\Lambda \subseteq \mathbb{R}^k$, a *left* $(AP_S)_\Lambda$ *factorization* of an $n \times n$ matrix G with entries in the algebra (AP^k) is a representation of the form

$$G = G_+ \begin{bmatrix} e_{\lambda_1} & 0 & \cdots & 0 \\ 0 & e_{\lambda_2} & \cdots & 0 \\ \vdots & \vdots & \ddots & \vdots \\ 0 & 0 & \cdots & e_{\lambda_n} \end{bmatrix} G_-, \tag{1.1}$$

where $\lambda_j \in \Lambda$ are ordered so that $\lambda_i - \lambda_j \in S$ for $i < j$, and the $n \times n$ matrices G_+ and G_- have their entries in (AP^k), are invertible in $(AP^k)^{n\times n}$, and the Fourier spectrum (defined in the next section) of $G_-^{\pm 1}$ (resp., of $G_+^{\pm 1}$) lies in the intersection $S \cap \Lambda$ (resp., in $(-S) \cap \Lambda$). The classical Wiener-Hopf factorization of continuous periodic functions of one variable is a particular case when $k = 1$, $S = [0, \infty)$, and $\Lambda = \mathbb{Z}$.

The paper is organized as follows. In Section 2 we introduce definitions, notation, and present preliminary results on algebras of almost periodic functions. The main new result there is Theorem 2.5 that asserts that, under suitable hypotheses, existence of a noncanonical almost periodic factorization of a matrix function G is hereditary, in the sense that if G has a left $(AP_S)_\Lambda$ factorization, then G also has a left $(AP_S)_{\Lambda_0}$ factorization, where Λ_0 is the subgroup generated by the Fourier spectrum of G. In Section 3, the degree of nonuniqueness of a noncanonical factorization is described, and subsequently used to prove results on symmetric factorizations of $n \times n$ matrix functions G with entries in (AP^k), provided G has certain symmetries, such as $G(t) = (G(t))^*$, or $G(t) = \overline{G(t)^{-1}}$. Section 4 is devoted to connections between Fredholmness properties of Toeplitz operators with almost periodic multivariable matrix symbols and noncanonical factorizations of the symbols. These problems for canonical factorizations, on one hand, and one-sided invertibility, on the other hand, were resolved in [43]. In Section 5, we describe possible changes of the left factorization indices (λ_j in (1.1)) under small perturbations of G, in particular, we determine when they are stable. The relation between these indices and their right counterpart (corresponding to the factorization (1.1) with interchanged G_+ and G_-) is also discussed in this section. Applications of almost periodic factorization to problems of normalizing bases are outlined in the last section.

To indicate possible future directions of research on noncanonical almost periodic factorization, we state several open problems and conjectures. Problem 2.5 raises the question whether noncanonical almost periodic factorizations always have the hereditary property. Conjecture 4.8 proposes necessary and sufficient conditions for Fredholmness of Toeplitz operators with almost periodic multivariable matrix symbols, in terms of noncanonical factorization of their symbols. Conjecture 5.4 deals with the behavior of the parameters λ_j under small perturbations of the factorization (1.1).

Notation: $\mathbb{N}$, $\mathbb{Z}$, $\mathbb{Q}$, $\mathbb{R}$, $\mathbb{C}$ stand for the sets of the positive integers, the integers, the rational numbers, the real numbers, and the complex numbers, respectively.

2. Definitions and preliminaries

Recall from the introduction that (AP^k) is the closed subalgebra of $L^\infty(\mathbb{R}^k)$ generated by all the functions $e_\lambda(t) = e^{i\langle\lambda,t\rangle}$, where

$$\langle\lambda, t\rangle = \sum_{j=1}^{k} \lambda_j t_j$$

is the standard inner product in $\mathbb{R}^k$. The norm in (AP^k) (inherited from $L^\infty(\mathbb{R}^k)$) will be denoted by $\|\cdot\|_\infty$.

The next proposition is standard (see Section 1.1 in [38]).

Proposition 2.1. *(AP^k) is a commutative unital C^*-algebra, and therefore can be identified with the algebra $C(\mathcal{B})$ of complex-valued continuous functions on a compact Hausdorff topological space $\mathcal{B}$, called the* Bohr compactification *of $\mathbb{R}^k$. Moreover, $\mathbb{R}^k$ is dense in $\mathcal{B}$.*

Recall that for any $f \in (AP^k)$ its *Fourier series* is defined by the formal sum

$$\sum_{\lambda} f_\lambda e^{i\langle \lambda, t\rangle}, \tag{2.1}$$

where

$$f_\lambda = \lim_{T\to\infty} \frac{1}{(2T)^k} \int_{[-T,T]^k} e^{-i\langle\lambda,t\rangle} f(t)dt, \ \lambda \in \mathbb{R}^k, \tag{2.2}$$

and the sum in (2.1) is taken over the set $\sigma(f) = \{\lambda \in \mathbb{R}^k : \ f_\lambda \neq 0\}$, called the *Fourier spectrum* of f. The Fourier spectrum of every $f \in (AP^k)$ is at most a countable set. The *mean* $M\{f\}$ of $f \in (AP^k)$ is defined by $M\{f\} = f_0 = \lim_{T\to\infty} \frac{1}{(2T)^k} \int_{[-T,T]^k} f(t)dt$. The *Wiener algebra* (APW^k) is defined as the set of all $f \in (AP^k)$ such that the Fourier series of f converges absolutely. The Wiener algebra is a Banach $*$-algebra with respect to the *Wiener norm* $\|f\|_W = \sum_{\lambda\in\mathbb{R}^k} |f_\lambda|$ (the multiplication in (APW^k) is pointwise). Note that (APW^k) is dense in (AP^k). For the general theory of almost periodic functions of one and several variables we refer the reader to the books [10], [31], [32] and to Chapter 1 in [38].

Let Δ be a non-empty subset of $\mathbb{R}^k$. Denote

$$\begin{aligned}(AP^k)_\Delta &= \{f \in (AP^k) : \ \sigma(f) \subseteq \Delta\},\\ (APW^k)_\Delta &= \{f \in (APW^k) : \ \sigma(f) \subseteq \Delta\}.\end{aligned}$$

If Δ is an additive subgroup of $\mathbb{R}^k$, then $(AP^k)_\Delta$ (resp. $(APW^k)_\Delta$) is a unital subalgebra of (AP^k) (resp. (APW^k)). The algebra $(AP^k)_\Delta$ is identified with $C(\mathcal{B}_\Delta)$, where the compact Hausdorff topological space $\mathcal{B}_\Delta$ is the *Bohr compactification* of Δ.

Proposition 2.2. *Let Δ be an additive subgroup of $\mathbb{R}^k$. Then:*

(a) *$f \in (AP^k)_\Delta$ is invertible in $(AP^k)_\Delta$ if and only if f is invertible in (AP^k) if and only if there exists $\epsilon > 0$ such that $|f(t)| \geq \epsilon$ for every $t \in \mathbb{R}^k$;*

(b) *$f \in (APW^k)_\Delta$ is invertible in $(APW^k)_\Delta$ if and only if f is invertible in (APW^k) if and only if there exists $\epsilon > 0$ such that $|f(t)| \geq \epsilon$ for every $t \in \mathbb{R}^k$.*

For the proof see [43].

Introduce an inner product on (AP^k) by the formula

$$\langle f, g\rangle = M\{fg^*\}, \ f, g \in (AP^k). \tag{2.3}$$

The completion of (AP^k) with respect to this inner product is called the *Besikovitch space* and is denoted by (B^k). Thus (B^k) is a Hilbert space. For a nonempty set $\Lambda \subseteq \mathbb{R}^k$, define the projection

$$\Pi_\Lambda\left(\sum_{\lambda\in\sigma(f)} f_\lambda e^{i\langle\lambda,t\rangle}\right) = \sum_{\lambda\in\sigma(f)\cap\Lambda} f_\lambda e^{i\langle\lambda,t\rangle},$$

where $f \in (APW^k)$. The projection Π_Λ extends by continuity to the orthogonal projection (also denoted Π_Λ) on (B^k). We denote by $(B^k)_\Lambda$ the range of Π_Λ, or, equivalently, the completion of $(AP^k)_\Lambda$ with respect to the inner product (2.3). The vector-valued Besikovitch space $(B^k)^{n\times 1}$ consists of $n \times 1$ columns with components in (B^k), with the standard Hilbert space structure. Similarly, $(B^k)^{n\times 1}_\Lambda$ is the Hilbert space of $n \times 1$ columns with components in $(B^k)_\Lambda$. In the periodic case ($\Lambda = \mathbb{Z}^k$) we may identify $(B^k)_\Lambda$ with $L_2(\mathbb{T}^k)$.

We now introduce the notion of canonical factorization with respect to a half-space. A subset S of $\mathbb{R}^k$ is called a *half-space* if it has the following properties:

(i) $\mathbb{R}^k = S \cup (-S)$;

(ii) $S \cap (-S) = \{0\}$;

(iii) if $x, y \in S$ then $x + y \in S$;

(iv) if $x \in S$ and α is a nonnegative real number, then $\alpha x \in S$.

A standard example of a half-space is given by

$$E_k = \{(x_1,\dots,x_k)^T \in \mathbb{R}^k \setminus \{0\} : x_1 = x_2 = \cdots = x_{j-1} = 0, x_j \neq 0 \Rightarrow x_j > 0\} \cup \{0\}. \quad (2.4)$$

(The vectors in $\mathbb{R}^k$ are understood as column vectors; the superscript ^{T} denotes the transpose.) Clearly, when $k = 1$ the only half-spaces are $[0, \infty)(= E_1)$ and $(-\infty, 0]$. In general, we have the following statement:

Proposition 2.3. *A set $S \subset \mathbb{R}^k$ is a half-space if and only if there exists a real invertible $k \times k$ matrix A such that*

$$S = AE_k \stackrel{\text{def}}{=} \{Ax \ : \ x \in E_k\}. \quad (2.5)$$

Moreover, for a given half-space S the matrix A satisfying (2.5) *is unique up to multiplication on the right by a lower triangular real matrix with positive entries on the diagonal.*

For a proof see [43], for example.

A half-space S induces naturally a total order $\geq_S$ on $\mathbb{R}^k$ by the rule: $\lambda \geq_S \mu$ if and only if $\lambda - \mu \in S$. We write $\lambda >_S \mu$ if $\lambda \geq_S \mu$ and $\lambda \neq \mu$.

Let $G \in (AP^k)^{n\times n}$. A representation

$$G(t) = G_+(t)G_-(t), \quad t \in \mathbb{R}^k, \quad (2.6)$$

where

$$G_+^{\pm 1} \in (AP^k)^{n\times n}_S, \quad G_-^{\pm 1} \in (AP^k)^{n\times n}_{-S} \quad (2.7)$$

is called a *left AP_S canonical factorization* of G. We say that (2.6) is a left APW_S canonical factorization of G if $G_\pm$ satisfy the stronger than (2.7) conditions $G_+^{\pm 1} \in (APW^k)^{n\times n}_S$, $G_-^{\pm 1} \in (APW^k)^{n\times n}_{-S}$. If Λ is an additive subgroup of $\mathbb{R}^k$, then a representation (2.6) is called a left canonical $(AP_S)_\Lambda$ factorization if $G_+^{\pm 1} \in (AP^k)^{n\times n}_{S\cap\Lambda}$, $G_-^{\pm 1} \in (AP^k)^{n\times n}_{(-S)\cap\Lambda}$, and a left canonical $(APW_S)_\Lambda$ factorization if $G_+^{\pm 1} \in (APW^k)^{n\times n}_{S\cap\Lambda}$, $G_-^{\pm 1} \in (APW^k)^{n\times n}_{(-S)\cap\Lambda}$. Of course, G must belong to $(APW^k)^{n\times n}$ (respectively, $(AP^k)^{n\times n}_\Lambda$, or $(APW^k)^{n\times n}_\Lambda$) in order to potentially admit a left canonical APW_S (respectively, $(AP_S)_\Lambda$, or $(APW_S)_\Lambda$) factorization.

A more general factorization that involves a diagonal term, is introduced now. Let $G \in (AP^k)^{n\times n}$. A *left AP_S factorization* of G is a representation of the form

$$G = G_+ \operatorname{diag}(e_{\lambda_1}, \dots, e_{\lambda_n}) G_-, \tag{2.8}$$

where $G_+^{\pm 1} \in (AP^k)_S^{n\times n}$, $G_-^{\pm 1} \in (AP^k)_{-S}^{n\times n}$, and $\lambda_j \in \mathbb{R}^k$ are ordered so that

$$\lambda_1 \geq_S \lambda_2 \geq_S \cdots \geq_S \lambda_n.$$

In (2.8) and elsewhere, we denote by $\operatorname{diag}(a_1, \dots, a_n)$ the $n \times n$ diagonal matrix with elements $a_1, \dots, a_n$ (in this order) on the main diagonal. We shall see in the next section that the λ_j are uniquely defined; we call them the *left factorization indices.* The variations APW_S, $(AP_S)_\Lambda$, $(APW_S)_\Lambda$ are defined similarly to the corresponding variations of the left canonical factorization (we require that $\lambda_j \in \Lambda$ as well for the $(AP_S)_\Lambda$ and $(APW_S)_\Lambda$ factorizations); in fact, the left $(AP_S)_\Lambda$ factorization was defined in the introduction.

Right AP_S factorization, as well as the APW_S, $(AP_S)_\Lambda$ and $(APW_S)_\Lambda$ variations, of $G \in (AP^k)^{n\times n}$ is defined analogously, using equation

$$G = G_- \operatorname{diag}(e_{\lambda_1}, \dots, e_{\lambda_n}) G_+, \tag{2.9}$$

in place of (2.8), with the same properties of G_-, G_+ and λ_j. Clearly, (2.9) is a right AP_S factorization of G if and only if (2.9) is a left AP_{-S} factorization of G if and only if

$$(G(t))^* = (G_+(t))^* \operatorname{diag}(e_{-\lambda_1}, \dots, e_{-\lambda_n})(G_-(t))^*$$

is a right AP_S factorization of $(G(t))^*$. This observation allows us to treat for most problems left factorizations only, with the understanding that the corresponding results for right factorizations hold true as well.

In the scalar case, in contrast to the matrix case, factorizability is equivalent to invertibility, for functions in the Wiener algebra. Of course, in the scalar case left and right factorizations coincide because of commutativity.

Proposition 2.4. *Let Λ be a subgroup of $\mathbb{R}^k$. A function $G \in (APW^k)_\Lambda^{1\times 1}$ admits a left (equivalently, right) $(APW_S)_\Lambda$ factorization if and only if*

$$|G(t)| \geq \epsilon > 0 \text{ for all } t \in \mathbb{R}^k, \tag{2.10}$$

where ϵ is independent of t.

Proof. Use the following representation of $G \in (APW^k)_\Lambda^{1\times 1}$ satisfying (2.10):

$$G(t) = |G(t)| e_c(t) e^{u(t)}, \quad t \in \mathbb{R}^k, \tag{2.11}$$

where $u \in (APW^k)_\Lambda^{1\times 1}$, and $c \in \Lambda$; the element c is called the *mean motion* of G (see [43] for the discussion of this notion in case of several variables and [39] for the proof of representation (2.11)). It turns out that $\log|G(t)| \in (APW^k)_\Lambda^{1\times 1}$ by Proposition 2.3 of [43]. Now standard arguments complete the proof, by writing

$$\log|G(t)| + u(t) = v_S(t) + v_{-S}(t),$$

where $v_S \in (APW^k)_{\Lambda\cap S}^{1\times 1}$, $v_{-S} \in (APW^k)_{\Lambda\cap(-S)}^{1\times 1}$. ☐

Problem 2.5. *Let $G \in (AP^k)_\Lambda^{n\times n}$, where Λ is a subgroup of $\mathbb{R}^k$. If G admits a left AP_S factorization, does it follow that G admits a left $(AP_S)_\Lambda$ factorization? Analogous problem for $G \in (APW^k)^{n\times n}$ such that G admits a left APW_S factorization.*

By Theorem 4.1 in Section 4, Problem 2.5 has an affirmative answer for $G \in (APW^k)^{n\times n}$ in case the left APW_S factorization is canonical. Proposition 2.4 shows that Problem 2.5 has an affirmative answer also in the scalar case.

In connection with Problem 2.5, we introduce the following concept. A pair of subgroups $\Lambda_0 \subseteq \Lambda$ is said to have the *hereditary factorization property* with respect to S if each time a function $G \in (APW^k)_{\Lambda_0}^{n\times n}$ admits a left $(APW_S)_\Lambda$ factorization, it admits also a left $(APW_S)_{\Lambda_0}$ factorization. Under additional hypotheses, we prove this property:

Theorem 2.6. *Assume that Λ, a subgroup of $\mathbb{R}^k$, has a minimal (in the total order induced by S) element μ_0 with the property that $\mu_0 >_S 0$, and let $\mathbb{Z}\mu_0$ be the cyclic subgroup generated by μ_0. Assume in addition that there exists a subgroup $\Lambda_0 \subseteq \Lambda$ such that Λ is a direct sum $\mathbb{Z}\mu_0 \dot{+} \Lambda_0$. Then the pair $\mathbb{Z}\mu_0 \subseteq \Lambda$ has the hereditary factorization property with respect to S.*

Notice that the additional hypothesis means, by definition, that the short exact sequence

$$\{0\} \to \mathbb{Z}\mu_0 \to \Lambda \to \Lambda/\mathbb{Z}\mu_0 \to \{0\} \tag{2.12}$$

splits. For example, (2.12) always splits if Λ is finitely generated.

Proof. We will use the following properties of Λ.

(a) If $\lambda_1, \lambda_2 \in \Lambda \cap S$ and $\lambda_1 + \lambda_2 \in \mathbb{Z}\mu_0$, then $\lambda_1, \lambda_2 \in \mathbb{Z}\mu_0$.

(b) If $\lambda_1, \lambda_2 \in \Lambda \cap (-S)$ and $\lambda_1 + \lambda_2 \in \mathbb{Z}\mu_0$, then $\lambda_1, \lambda_2 \in \mathbb{Z}\mu_0$.

The proofs of (a) and (b) are easily obtained using the minimal property of μ_0.

In view of (a) and (b) we have that

$$\begin{aligned} F_1 \in (APW^k)_{\Lambda\cap S}^{p\times q}, \quad F_2 \in (APW^k)_{\Lambda\cap S}^{q\times r}, \quad F_1F_2 \in (APW^k)_{\mathbb{Z}\mu_0}^{p\times r} \\ \Longrightarrow \quad F_1 \in (APW^k)_{\mathbb{Z}\mu_0}^{p\times q}, \quad F_2 \in (APW^k)_{\mathbb{Z}\mu_0}^{q\times r}; \end{aligned} \tag{2.13}$$

the analogous statement with S replaced by $-S$ is valid as well.

Define the projection Q as follows:

$$Q(e_\lambda) = e_{s\mu_0}, \quad \lambda \in \Lambda, \quad s \in \mathbb{Z},$$

where s is taken from the unique decomposition

$$\lambda = s\mu_0 + \lambda_0, \quad \lambda_0 \in \Lambda_0.$$

We extend Q by linearity and continuity to a (bounded) projection, also denoted by Q,

$$Q : (APW^k)_\Lambda \to (APW^k)_{\mathbb{Z}\mu_0}.$$

By applying Q entrywise, we further extend the definition of Q to matrix valued functions.

Now assume that $G \in (APW^k)^{n\times n}_{\mathbb{Z}\mu_0}$ admits a left $(APW_S)_\Lambda$ factorization

$$G = G_+\mathrm{diag}(e_{\lambda_1},\cdots,e_{\lambda_n})G_-,\ G_+^{\pm 1} \in (APW^k)^{n\times n}_{S\cap\Lambda},\ G_-^{\pm 1} \in (APW^k)^{n\times n}_{-S\cap\Lambda}, \lambda_j \in \Lambda,$$

where

$$\lambda_1 \geq_S \lambda_2 \geq_S \cdots \lambda_k \geq_S 0 \geq_S \lambda_{k+1} \geq_S \cdots \geq_S \lambda_n.$$

Letting

$$D_1 = \mathrm{diag}(e_{\lambda_1},\cdots,e_{\lambda_k}),\quad D_2 = \mathrm{diag}(e_{\lambda_{k+1}},\cdots,e_{\lambda_n}),$$

we rewrite the factorization in the form

$$\begin{bmatrix} I & 0 \\ 0 & D_2^{-1} \end{bmatrix} G_+^{-1}G = \begin{bmatrix} D_1 & 0 \\ 0 & I \end{bmatrix} G_-.$$

Apply the projection Q to this equality. In view of (2.13), we have $Q(G_\pm^{-1}) = (Q(G_\pm))^{-1}$, $Q(D_j^{-1}) = (Q(D_j))^{-1}$ for $j = 1,2$, and therefore

$$\begin{bmatrix} I & 0 \\ 0 & Q(D_2)^{-1} \end{bmatrix} Q(G_+)^{-1}G = \begin{bmatrix} Q(D_1) & 0 \\ 0 & I \end{bmatrix} Q(G_-).$$

Thus,

$$G = Q(G_+)\begin{bmatrix} Q(D_1) & 0 \\ 0 & Q(D_2) \end{bmatrix} Q(G_-)$$

is a left $(APW_S)_{\mathbb{Z}\mu_0}$ factorization of G. □

3. Symmetric factorizations

We start with a basic result describing possible non-uniqueness of left AP_S factorization (2.8); notice that the indices are unique.

Theorem 3.1. *If $G \in (AP^k)^{n\times n}$ admits a left factorization of one of the types AP_S, APW^S, $(AP_S)_\Lambda$, $(APW_S)_\Lambda$,*

$$G = G_+\,\mathrm{diag}(e_{\lambda_1},\ldots,e_{\lambda_n})\,G_-, \tag{3.1}$$

then the left factorization indices are unique, and every other left AP_W (or left APW_S, $(AP_S)_\Lambda$, or $(APW_S)_\Lambda$, as appropriate) factorization of G is given by the formula

$$G = \widetilde{G}_+\,\mathrm{diag}(e_{\lambda_1},\ldots,e_{\lambda_n})\,\widetilde{G}_-, \tag{3.2}$$

where $\widetilde{G}_-(t) = H_-(t)G_-(t)$,

$$\widetilde{G}_+(t) = G_+(t)\,\mathrm{diag}(e_{\lambda_1},\ldots,e_{\lambda_n})\,(H_-(t))^{-1}\,\mathrm{diag}(e_{-\lambda_1},\ldots,e_{-\lambda_n}),$$

and

$$H_-(t) = [h_{j,k}]_{j,k=1}^n \in (AP^k)^{n\times n}_{-S}$$

(or $\in (APW^k)^{n\times n}_{-S}, \in (AP^k)^{n\times n}_{(-S)\cap\Lambda}$, or $\in (APW^k)^{n\times n}_{(-S)\cap\Lambda}$, as appropriate)

is invertible in $(AP^k)^{n\times n}_{-S}$ (or in $(APW^k)^{n\times n}_{-S}$, $(AP^k)^{n\times n}_{(-S)\cap\Lambda}$, or $(APW^k)^{n\times n}_{(-S)\cap\Lambda}$ as the case may be), and satisfies the following conditions:

(i) $h_{j,k}(t) = 0$ *if* $\lambda_k >_S \lambda_j$;

(ii) $h_{j,k}(t)$ *is a constant if* $\lambda_k = \lambda_j$;

(iii) *the Fourier spectrum of* $h_{j,k}$ *belongs to the interval* $[\lambda_k - \lambda_j, 0]$ *(in the total order induced by* S*) if* $\lambda_k <_S \lambda_j$.

The proof is standard, and can be obtained by following the proofs of Theorems VIII.1.1 and VIII.1.2 of [14], for example.

Note that the function H_- from Theorem 3.1 has a block triangular structure: If

$$\operatorname{diag}(e_{\lambda_1}, \dots, e_{\lambda_n}) = \operatorname{diag}(e_{\mu_1} I_{m_1}, \dots, e_{\mu_p} I_{m_p}), \tag{3.3}$$

where $\mu_1 >_S \cdots >_S \mu_p$, is the diagonal term in the left factorization (3.1), then

$$H_- = \begin{bmatrix} Q_{1,1} & Q_{1,2} & \cdots & Q_{1,p} \\ 0 & Q_{2,2} & \cdots & Q_{2,p} \\ \vdots & \vdots & \ddots & \vdots \\ 0 & 0 & \cdots & Q_{p,p} \end{bmatrix}, \tag{3.4}$$

where $Q_{j,j}$ is a constant $m_j \times m_j$ invertible matrix, and $\sigma(Q_{j,k}) \subseteq [\mu_k - \mu_j, 0]$ for $1 \le j < k \le p$. It follows that H_-^{-1} has the same block triangular structure.

Theorem 3.1, when applied to Hermitian-valued matrix functions, leads to the following far reaching generalization of a result proved in [37] for factorizations of Hermitian matrix functions on the unit circle. We denote by $P_m = [\delta_{i+j,m+1}]_{i,j=1}^m$ the $m \times m$ permutation matrix with 1's on the southwest-northeast diagonal.

Theorem 3.2. *Let* $G \in (AP^k)^{n\times n}$ *be Hermitian valued:* $G(t) = (G(t))^*$ *for every* $t \in \mathbb{R}^k$. *Assume that* G *admits a left factorization of one of the types* AP_S, APW_S, $(AP_S)_\Lambda$, $(APW_S)_\Lambda$. *Then* $G(t)$ *admits a factorization*

$$G(t) = G_+(t) \begin{bmatrix} 0 & 0 & \cdots & 0 & e_{\mu_1} P_{m_1} \\ 0 & 0 & \cdots & e_{\mu_2} P_{m_2} & 0 \\ \vdots & \vdots & \vdots & \vdots & \vdots \\ 0 & 0 & \cdots J \cdots & 0 & 0 \\ \vdots & \vdots & \vdots & \vdots & \vdots \\ 0 & e_{-\mu_2} P_{m_2} & \cdots & 0 & 0 \\ e_{-\mu_1} P_{m_1} & 0 & \cdots & 0 & 0 \end{bmatrix} (G_+(t))^*. \tag{3.5}$$

Here G_+ *and its inverse belong to* $(AP^k)_S^{n\times n}$ *(or* $(APW^k)_S^{n\times n}$, $(AP^k)_{S\cap\Lambda}^{n\times n}$, *or* $(APW^k)_{S\cap\Lambda}^{n\times n}$, *as the case may be),* J *is a constant invertible Hermitian matrix (which can be taken in the form* $J = \begin{bmatrix} I_{k_1} & 0 \\ 0 & -I_{k_2} \end{bmatrix}$*), and* $\mu_1 >_S \cdots >_S \mu_q >_S 0$; *for left factorizations of types* $(AP_S)_\Lambda$ *or* $(APW_S)_\Lambda$, *in addition,* $\mu_j \in \Lambda$.

Of course, it may happen that the blocks associated with J are absent in (3.5), or that the blocks associated with $e_{\mu_j} I_{m_j}$ are absent; the latter case corresponds to canonical left factorization of G.

The proof of Theorem 3.2, as well as other proofs in this section, follows a well-known pattern that has been exploited often in the literature.

Proof. First note that in case G admits a canonical left factorization, the result of Theorem 3.2 is known (see Theorem 5.1 in [43]).

Let

$$G = G_+ \mathrm{diag}(e_{\lambda_1}, \dots, e_{\lambda_n}) G_- \tag{3.6}$$

be a left AP_S factorization (the proof for other types of left factorization is completely analogous). Then we have

$$G = G^* = G_-^* \ \mathrm{diag}(e_{-\lambda_1}, \dots, e_{-\lambda_n})\, G_+^* = (G_-^* P_n)\, \mathrm{diag}(e_{-\lambda_n}, \dots, e_{-\lambda_1})\, (P_n G_+^*). \tag{3.7}$$

Clearly, (3.7) is also a left AP_S factorization of G. Since the left factorization indices are unique, we must have $\lambda_j = -\lambda_{n+1-j}$, $(j = 1, \dots, n)$. Also, by Theorem 3.1, $P_n G_+^* = H_- G_-$, where H_- is given by (3.4). Thus $G_- = H_-^{-1} P_n G_+^*$, and substituting in (3.6), we have

$$G = G_+ \mathrm{diag}(e_{\lambda_1}, \dots, e_{\lambda_n}) H_-^{-1} P_n G_+^* = G_+ \mathrm{diag}(e_{\lambda_1}, \dots, e_{\lambda_n}) P_n \cdot P_n H_-^{-1} P_n G_+^*.$$

Now we may write

$$\mathrm{diag}(e_{\lambda_1}, \dots, e_{\lambda_n}) P_n = \begin{bmatrix} 0 & 0 & \cdots & & 0 & e_{\mu_1} P_{m_1} \\ 0 & 0 & \cdots & & e_{\mu_2} P_{m_2} & 0 \\ \vdots & \vdots & \vdots & & \vdots & \vdots \\ 0 & 0 & \cdots & P_{m_{q+1}} \cdots & 0 & 0 \\ \vdots & \vdots & \vdots & & \vdots & \vdots \\ 0 & e_{-\mu_2} P_{m_2} & \cdots & & 0 & 0 \\ e_{-\mu_1} P_{m_1} & 0 & \cdots & & 0 & 0 \end{bmatrix}, \tag{3.8}$$

where $\mu_1 >_S \cdots >_S \mu_q >_S 0$ are the distinct elements among the λ_j's that belong to S. (We assume here that the number of the λ_j's that are equal to zero is $m_{q+1} \geq 1$; if no λ_j is equal to zero, the subsequent proof goes through with self-evident changes.) On the other hand, $P_n H_-^{-1} P_n$ is computed to be of the form

$$\begin{bmatrix} Q_{p,p} & 0 & \cdots & 0 \\ Q_{p-1,p} & Q_{p-1,p-1} & \cdots & 0 \\ \vdots & \vdots & \ddots & \vdots \\ Q_{1,p} & Q_{1,p-1} & \cdots & Q_{1,1} \end{bmatrix}, \tag{3.9}$$

where $p = 2q + 1$ and $\sigma(Q_{j,k}) \subseteq [\mu_k - \mu_j, 0]$; we define $\mu_{q+1} = 0$ and $\mu_{q+1+r} = -\mu_{q+1-r}$ for $r = 1, 2, \dots, q$. The sizes of blocks in (3.9) coincide with the sizes of corresponding blocks in (3.8).

Consider the function

$$\widetilde{G} = \begin{bmatrix} 0 & 0 & \cdots & 0 & e_{\mu_1}P_{m_1} \\ 0 & 0 & \cdots & e_{\mu_2}P_{m_2} & 0 \\ \vdots & \vdots & \vdots & \vdots & \vdots \\ 0 & 0 & \cdots\ P_{m_{q+1}}\ \cdots & 0 & 0 \\ \vdots & \vdots & \vdots & \vdots & \vdots \\ 0 & e_{-\mu_2}P_{m_2} & \cdots & 0 & 0 \\ e_{-\mu_1}P_{m_1} & 0 & \cdots & 0 & 0 \end{bmatrix} P_n H_-^{-1} P_n.$$

Clearly, $\widetilde{G} = \widetilde{G}^*$. Moreover, using (3.8) and (3.9), we have that $\widetilde{G}$ is equal to

$$\begin{bmatrix} e_{\mu_1}P_{m_1}Q_{1,p} & e_{\mu_1}P_{m_1}Q_{1,p-1} & e_{\mu_1}P_{m_1}Q_{1,p-2} & \cdots & e_{\mu_1}P_{m_1}Q_{1,2} & e_{\mu_1}P_{m_1}Q_{1,1} \\ e_{\mu_2}P_{m_2}Q_{2,p} & e_{\mu_2}P_{m_2}Q_{2,p-1} & e_{\mu_2}P_{m_2}Q_{2,p-2} & \cdots & e_{\mu_2}P_{m_2}Q_{2,2} & 0 \\ \vdots & \vdots & \vdots & \vdots & \vdots & \vdots \\ e_{-\mu_2}P_{m_2}Q_{p-1,p} & e_{-\mu_2}P_{m_2}Q_{p-1,p-1} & 0 & \cdots & 0 & 0 \\ e_{-\mu_1}P_{m_1}Q_{p,p} & 0 & 0 & \cdots & 0 & 0 \end{bmatrix}.$$

In particular,

$$P_{m_1}Q_{p,p} = (P_{m_1}Q_{1,1})^*, \quad \text{and} \quad e_{\mu_1}P_{m_1}Q_{1,p} = e_{-\mu_1}Q_{1,p}^*P_{m_1}. \tag{3.10}$$

Define the matrix function

$$X = \begin{bmatrix} I_{m_1} & 0 & 0 & \cdots & 0 & 0 \\ 0 & I_{m_2} & 0 & \cdots & 0 & 0 \\ \vdots & \vdots & \vdots & \vdots & \vdots & \vdots \\ 0 & 0 & 0 & \cdots & I_{m_2} & 0 \\ X_p & X_{p-1} & X_{p-2} & \cdots & X_2 & I_{m_1} \end{bmatrix},$$

where $X_j = -(Q_{1,1})^{-1}Q_{1,j}$ for $j = 2, 3, \ldots, p-1$, and $X_p = -\frac{1}{2}(Q_{1,1})^{-1}Q_{1,p}$. Clearly, $X^{\pm 1} \in (AP^k)_{-S}^{n\times n}$. A straightforward computation, using (3.10), shows that $X^*\widetilde{G}X$ has the block form

$$\begin{bmatrix} 0 & 0 & e_{\mu_1}P_{m_1}Q_{1,1} \\ 0 & * & 0 \\ e_{-\mu_1}P_{m_1}Q_{p,p} & 0 & 0 \end{bmatrix}. \tag{3.11}$$

In turn, multiplying (3.11) on the right and on the left by

$$\begin{bmatrix} 0 & 0 & I \\ 0 & I & 0 \\ (Q_{1,1})^{*-1} & 0 & 0 \end{bmatrix} \quad \text{and} \quad \begin{bmatrix} 0 & 0 & (Q_{1,1})^* \\ 0 & I & 0 \\ I & 0 & 0 \end{bmatrix},$$

respectively, we obtain

$$\begin{bmatrix} 0 & 0 & e_{\mu_1}P_{m_1} \\ 0 & * & 0 \\ e_{-\mu_1}P_{m_1} & 0 & 0 \end{bmatrix}.$$

Now induction on q completes the proof. □

The proof of Theorem 3.2 can be used (with obvious modifications) to prove a factorization result for matrix functions G satisfying the equality $G(t) = (G(-t))^T$:

Theorem 3.3. *Let $G \in (AP^k)^{n\times n}$ be such that $G(t) = (G(-t))^T$ for every $t \in \mathbb{R}^k$. Assume that G admits a left factorization of one of the types AP_S, APW_S, $(AP_S)_\Lambda$, $(APW_S)_\Lambda$. Then $G(t)$ admits a factorization*

$$G(t) = G_+(t) \begin{bmatrix} 0 & 0 & \cdots & 0 & e_{\mu_1}P_{m_1} \\ 0 & 0 & \cdots & e_{\mu_2}P_{m_2} & 0 \\ \vdots & \vdots & \vdots & \vdots & \vdots \\ 0 & 0 & \cdots K \cdots & 0 & 0 \\ \vdots & \vdots & \vdots & \vdots & \vdots \\ 0 & e_{-\mu_2}P_{m_2} & \cdots & 0 & 0 \\ e_{-\mu_1}P_{m_1} & 0 & \cdots & 0 & 0 \end{bmatrix} (G_+(-t))^T. \tag{3.12}$$

Here G_+ and its inverse belong to $(AP^k)^{n\times n}_S$ (or $(APW^k)^{n\times n}_S$, $(AP^k)^{n\times n}_{S\cap\Lambda}$, or $(APW^k)^{n\times n}_{S\cap\Lambda}$, as the case may be), K is a constant invertible complex symmetric matrix, and $\mu_1 >_S \cdots >_S \mu_q >_S 0$; for left factorizations of types $(AP_S)_\Lambda$ or $(APW_S)_\Lambda$, in addition, $\mu_j \in \Lambda$.

Proof. Following the method of the proof of Theorem 3.2, we only need to consider the case of left canonical factorization. If

$$G = G_+G_- \tag{3.13}$$

is a left canonical factorization, then taking transposes yields

$$G(t) = (G(-t))^T = (G_-(-t))^T (G_+(-t))^T,$$

which is again a left canonical factorization, and by Theorem 3.1 $(G_+(-t))^T = QG_-$ for some constant invertible matrix Q. Thus $G_- = Q^{-1}(G_+(-t))^T$, and substituting for G_- in (3.13) the result follows. □

For functions $G \in (AP^k)^{n\times n}$ satisfying the symmetry

$$G(t) = -(G(t))^*, \quad t \in \mathbb{R}^k \tag{3.14}$$

or the symmetry

$$G(t) = -(G(-t))^T, \quad t \in \mathbb{R}^k \tag{3.15}$$

results analogous to Theorems 3.2 and 3.3 hold true.

Namely, for G satisfying (3.14) formula (3.5) is replaced by

$$G(t) = G_+(t) \begin{bmatrix} 0 & 0 & \cdots & 0 & e_{\mu_1}P_{m_1} \\ 0 & 0 & \cdots & e_{\mu_2}P_{m_2} & 0 \\ \vdots & \vdots & \vdots & \vdots & \vdots \\ 0 & 0 & \cdots J \cdots & 0 & 0 \\ \vdots & \vdots & \vdots & \vdots & \vdots \\ 0 & -e_{-\mu_2}P_{m_2} & \cdots & 0 & 0 \\ -e_{-\mu_1}P_{m_1} & 0 & \cdots & 0 & 0 \end{bmatrix} (G_+(t))^*, \tag{3.16}$$

with constant invertible skew Hermitian matrix J, and for G satisfying (3.15) formula (3.12) is replaced by

$$G(t) = G_+(t) \begin{bmatrix} 0 & 0 & \cdots & 0 & e_{\mu_1}P_{m_1} \\ 0 & 0 & \cdots & e_{\mu_2}P_{m_2} & 0 \\ \vdots & \vdots & \vdots & \vdots & \vdots \\ 0 & 0 & \cdots K \cdots & 0 & 0 \\ \vdots & \vdots & \vdots & \vdots & \vdots \\ 0 & -e_{-\mu_2}P_{m_2} & \cdots & 0 & 0 \\ -e_{-\mu_1}P_{m_1} & 0 & \cdots & 0 & 0 \end{bmatrix} (G_+(-t))^T, \tag{3.17}$$

where $K = -K^T$ is a constant invertible matrix. The proofs are again obvious modifications of the proof of Theorem 3.2.

Consider also the symmetries

$$G(t) = G(-t)^{-1}, \quad t \in \mathbb{R}^k \tag{3.18}$$

and

$$G(t) = \overline{G(t)^{-1}}, \quad t \in \mathbb{R}^k. \tag{3.19}$$

Symmetries of type (3.18), and left factorizations of G, were studied in [11] (for periodic functions G of one variable). Symmetric factorizations of functions with the symmetry (3.18) take the following form.

Theorem 3.4. *Let $G \in (AP^k)^{n\times n}$ be such that (3.18) holds. Assume that G admits a left factorization of one of the types AP_S, APW_S, $(AP_S)_\Lambda$, $(APW_S)_\Lambda$. Then $G(t)$ admits a factorization*

$$G(t) = G_+(t)\,\mathrm{diag}(\delta_1 e_{\lambda_1}, \dots, \delta_n e_{\lambda_n}) G_+(-t)^{-1}. \tag{3.20}$$

Here G_+ and its inverse belong to $(AP^k)_S^{n\times n}$ (or $(APW^k)_S^{n\times n}$, $(AP^k)_{S\cap\Lambda}^{n\times n}$, or $(APW^k)_{S\cap\Lambda}^{n\times n}$, as the case may be), and $\delta_j = \pm 1$, $j = 1, 2, \dots, n$; for left factorizations of types $(AP_S)_\Lambda$ or $(APW_S)_\Lambda$, in addition, $\lambda_j \in \Lambda$.

Moreover, the diagonal term $\mathrm{diag}(\delta_1 e_{\lambda_1}, \dots, \delta_n e_{\lambda_n})$ in (3.20) is uniquely determined up to permutation of the diagonal entries.

Proof. First, we prove existence of the symmetric factorization (3.20).

Let

$$G = G_+ \mathrm{diag}(e_{\mu_1} I_{m_1}, \ldots, e_{\mu_p} I_{m_p}) G_-$$

be a left AP_S factorization, where $\mu_1 >_S \cdots >_S \mu_p$. Then

$$G(t) = G(-t)^{-1} = G_-(-t)^{-1} \mathrm{diag}(e_{\mu_1} I_{m_1}, \ldots, e_{\mu_p} I_{m_p}) G_+(-t)^{-1},$$

and comparison of these two left AP_S factorizations, by Theorem 3.1, leads to the equality

$$G(t) = G_+(t) \mathrm{diag}(e_{\mu_1} I_{m_1}, \ldots, e_{\mu_p} I_{m_p}) H_-(t)^{-1} G_+(-t)^{-1},$$

where H is given by (3.4). Letting

$$H_-(t)^{-1} = \begin{bmatrix} \widetilde{Q}_{1,1} & \widetilde{Q}_{1,2} & \cdots & \widetilde{Q}_{1,p} \\ 0 & \widetilde{Q}_{2,2} & \cdots & \widetilde{Q}_{2,p} \\ \vdots & \vdots & \ddots & \vdots \\ 0 & 0 & \cdots & \widetilde{Q}_{p,p} \end{bmatrix},$$

where $\widetilde{Q}_{j,j}$ is a constant $m_j \times m_j$ invertible matrix for $j = 1, 2, \cdots, p$, and $\widetilde{Q}_{j,k} \in (AP^k)^{n \times n}$ with $\sigma(\widetilde{Q}_{j,k}) \subseteq [\mu_k - \mu_j, 0]$ for $1 \le j < k \le p$, and letting

$$F(t) = \begin{bmatrix} e_{\mu_1}\widetilde{Q}_{1,1} & e_{\mu_1}\widetilde{Q}_{1,2} & \cdots & e_{\mu_1}\widetilde{Q}_{1,p} \\ 0 & e_{\mu_2}\widetilde{Q}_{2,2} & \cdots & e_{\mu_2}\widetilde{Q}_{2,p} \\ \vdots & \vdots & \ddots & \vdots \\ 0 & 0 & \cdots & e_{\mu_p}\widetilde{Q}_{p,p} \end{bmatrix}, \tag{3.21}$$

we note that $F(t) = F(-t)^{-1}$, and reduce the proof to the function F in place of G.

Now we use induction on p. If $p = 1$, then $F(t) = e_{\mu_1}\widetilde{Q}_{1,1}$, and the condition $F(t) = F(-t)^{-1}$ means that $\widetilde{Q}_{1,1}^2 = I$. Writing $\widetilde{Q}_{1,1} = S \begin{bmatrix} I_{s_1} & 0 \\ 0 & -I_{s_2} \end{bmatrix} S^{-1}$ for a suitable invertible matrix S, we obtain the desired factorization of F in the form

$$F(t) = S \begin{bmatrix} e_{\mu_1} I_{(s_1)} & 0 \\ 0 & -e_{\mu_1} I_{s_2} \end{bmatrix} S^{-1}.$$

Next, assume $p = 2$. Thus, we let

$$F(t) = \begin{bmatrix} e_{\mu_1}(t)\widetilde{Q}_{1,1} & e_{\mu_1}(t)\widetilde{Q}_{1,2}(t) \\ 0 & e_{\mu_2}(t)\widetilde{Q}_{2,2} \end{bmatrix} = F(-t)^{-1},$$

where $\widetilde{Q}_{1,1}$, $\widetilde{Q}_{2,2}$ are constant invertible matrices, $\mu_1 >_S \mu_2$, and $\sigma(\widetilde{Q}_{1,2})$ is a subset of $[\mu_2 - \mu_1, 0]$. The condition $F(t) = F(-t)^{-1}$ easily implies that

$$e_{\mu_1}(t)\widetilde{Q}_{1,2}(t) = -\widetilde{Q}_{1,1}^{-1}\widetilde{Q}_{1,2}(-t) e_{\mu_2}(t)\widetilde{Q}_{2,2}. \tag{3.22}$$

Now define the functions

$$Z_+(t) = \frac{1}{2}\widetilde{Q}_{1,1}^{-1}\widetilde{Q}_{1,2}(-t) \in (AP^k)_S^{m_1 \times m_2},$$

and

$$X(t) = \begin{bmatrix} I & Z_+(t) \\ 0 & I \end{bmatrix}.$$

Using (3.22), we verify the symmetric factorization

$$X(t)F(t)X(-t)^{-1} = \begin{bmatrix} e_{\mu_1}(t)\widetilde{Q}_{1,1} & 0 \\ 0 & e_{\mu_2}(t)\widetilde{Q}_{2,2} \end{bmatrix},$$

and since this matrix is block diagonal, the proof is reduced to the case $p = 1$, which has been taken care of.

Assume now that $F(t) = F(-t)^{-1}$ is given by (3.21), where $p > 2$, and assume that the existence of factorizations (3.20) has been already proved for all matrix functions of the form (3.21), but with a smaller positive integer in place of p. Applying the induction hypothesis to the upper left $(m_1 + \cdots + m_{p-1}) \times (m_1 + \cdots + m_{p-1})$ block of F, we may (and do) assume that $\widetilde{Q}_{i,j} = 0$ for $i < j < p$. Next, applying the induction hypothesis (with $p = 2$) to the submatrix $\begin{bmatrix} e_{\mu_1}\widetilde{Q}_{1,1} & e_{\mu_1}\widetilde{Q}_{1,p} \\ 0 & e_{\mu_p}(t)\widetilde{Q}_{p,p} \end{bmatrix}$ of F, we obtain a new matrix function, call it $\widetilde{F}$, of the form

$$\widetilde{F}(t) = F_+(t)F(t)F_+(-t)^{-1}, \quad (F_+)^{\pm 1} \in (AP^k)_S^{n\times n},$$

where

$$\widetilde{F} = \begin{bmatrix} e_{\mu_1}\widetilde{Q}_{1,1} & 0 \\ 0 & \widetilde{F}_1 \end{bmatrix},$$

and $\widetilde{F}_1$ has the form (3.21), with p replaced by $p-1$. Now the induction completes the proof.

It remains to prove the uniqueness of the diagonal term in (3.20). Thus, assume that we have two left AP_S factorizations of G as in (3.20), where we have combined together diagonal terms having the same diagonal factor e_{λ_j}:

$$\begin{aligned} G(t) &= G_+(t)\text{diag}(e_{\mu_1}K_{1,1}, \ldots, e_{\mu_p}K_{1,p})G_+(-t)^{-1} \\ &= \widetilde{G}_+\text{diag}(e_{\mu_1}K_{2,1}, \ldots, e_{\mu_p}K_{2,p})\widetilde{G}_+(-t)^{-1}, \end{aligned}$$

where $\mu_1 >_S \cdots >_S \mu_p$, and the matrices $K_{1,k}$ and $K_{2,k}$ are diagonal with ± 1 on the main diagonal and of size $m_k \times m_k$, for $k = 1, \ldots, p$. The uniqueness of the left factorization indices guarantees that the numbers $\mu_1, \ldots, \mu_p$ and the sizes $m_1, \ldots, m_k$ are the same for both factorizations

$$G(t) = G_+(t)\text{diag}(e_{\mu_1}K_{1,1}, \ldots, e_{\mu_p}K_{1,p})G_+(-t)^{-1}$$

and

$$G(t) = \widetilde{G}_+\text{diag}(e_{\mu_1}K_{2,1}, \cdots, e_{\mu_p}K_{2,p})\widetilde{G}_+(-t)^{-1}.$$

Denote

$$Z(t) = \widetilde{G}_+(t)^{-1}G_+(t) \in (AP^k)_S^{n\times n}.$$

Then we have

$$Z(t)\mathrm{diag}(e_{\mu_1}K_{1,1},\dots,e_{\mu_p}K_{1,p}) = \mathrm{diag}(e_{\mu_1}K_{2,1},\dots,e_{\mu_p}K_{2,p})Z(-t). \qquad (3.23)$$

Partition $Z = [Z_{j,k}]_{j,k=1}^p$, where $Z_{j,k}$ is $m_j \times m_k$. For $j > k$ we have

$$Z_{j,k}e_{\mu_k}K_{1,k} \in e_{\mu_k}(AP^k)_S^{m_j\times m_k},$$

$$e_{\mu_j}K_{2,j}Z_{j,k}(-t) \in e_{\mu_j}(AP^k)_{-S}^{m_j\times m_k},$$

and

$$Z_{j,k}e_{\mu_k}K_{1,k} = e_{\mu_j}K_{2,j}Z_{j,k}(-t).$$

But

$$\left(e_{\mu_k}(AP^k)_S^{m_j\times m_k}\right) \cap \left(e_{\mu_j}(AP^k)_{-S}^{m_j\times m_k}\right) = \{0\},$$

so we conclude that $Z_{j,k} = 0$ for $j > k$. Analogously, we obtain that $Z_{j,j}$ is a constant matrix (which is necessarily invertible) for $j = 1, 2, \dots, p$. Now (3.23) implies that $Z_{j,j}K_{1,j} = K_{2,j}Z_{j,j}$. In particular, for every $j = 1, 2, \dots, p$, the matrices $K_{1,j}$ and $K_{2,j}$ are similar, in other words, they have the same number of 1's and the same number of -1's. This proves the uniqueness of the diagonal term in (3.20) as claimed in the theorem. □

For the symmetry $G(t) = \overline{G(t)^{-1}}$ we have the following result.

Theorem 3.5. *Let $G \in (AP^k)^{n\times n}$ be such that* (3.19) *holds. Assume that G admits a left factorization of one of the types AP_S, APW_S, $(AP_S)_\Lambda$, $(APW_S)_\Lambda$. Then $G(t)$ admits a factorization*

$$G(t) = G_+(t)\,\mathrm{diag}(e_{\lambda_1},\dots,e_{\lambda_n})\overline{G_+(t)^{-1}}. \qquad (3.24)$$

Here G_+ and its inverse belong to $(AP^k)_S^{n\times n}$ (or $(APW^k)_S^{n\times n}$, $(AP^k)_{S\cap\Lambda}^{n\times n}$, or $(APW^k)_{S\cap\Lambda}^{n\times n}$, as the case may be); for left factorizations of types $(AP_S)_\Lambda$ or $(APW_S)_\Lambda$, in addition, $\lambda_j \in \Lambda$.

The proof is completely analogous to that of Theorem 3.4. The only difference is that instead of using the fact that a matrix $\widetilde{Q}_{1,1}$ satisfies $\widetilde{Q}_{1,1}^2 = I$ if and only it is similar to a diagonal matrix $\begin{bmatrix} I_{s_1} & 0 \\ 0 & -I_{s_2} \end{bmatrix}$, we now use the following fact: A matrix A satisfies $A\overline{A} = I$ if and only if $A = S\overline{S^{-1}}$ for some invertible matrix S (see, e.g., Lemma 4.6.9 in [20] for a simple proof).

4. Factorization and Toeplitz operators

Throughout this section, we fix a half-space S of $\mathbb{R}^k$, and fix an additive subgroup Λ of $\mathbb{R}^k$.

For $f \in (APW^k)^{n\times n}$ and $\Psi \subseteq \mathbb{R}^k$, we let

$$T(f)_\Psi : (B^k)_\Psi^{n\times 1} \to (B^k)_\Psi^{n\times 1}$$

be the Toeplitz operator defined by

$$T(f)_\Psi(g) = \Pi_\Psi(fg), \tag{4.1}$$

$g \in (B^k)_\Psi^{n\times 1}$. More precisely, we use (4.1) for $g \in (B^k)_\Psi^{n\times 1}$ having a finite Fourier spectrum, and extend by continuity to the whole Hilbert space $(B^k)_\Psi^{n\times 1}$.

As shown in the following theorem, there are close connections between invertibility of the Toeplitz operator $T(G)_{(-S)\cap\Lambda}$, for $G \in (APW^k)_\Lambda^{n\times n}$, and left canonical factorizations of G.

Theorem 4.1. *Let $G \in (APW^k)_\Lambda^{n\times n}$. Then the following statements are equivalent:*

i) *the operator $T(G)_{(-S)\cap\Lambda'}$ is invertible for some additive subgroup $\Lambda' \supseteq \Lambda$;*
ii) *the operator $T(G)_{(-S)\cap\Lambda'}$ is invertible for every additive subgroup $\Lambda' \supseteq \Lambda$;*
iii) *the operator $T(G)_{(-S\setminus\{0\})\cap\Lambda'}$ is invertible for some additive subgroup $\Lambda' \supseteq \Lambda$;*
iv) *the operator $T(G)_{(-S\setminus\{0\})\cap\Lambda'}$ is invertible for every additive subgroup $\Lambda' \supseteq \Lambda$;*
v) *G admits a left canonical AP_S factorization;*
vi) *G admits a left canonical APW_S factorization;*
vii) *G admits a left canonical $(AP_S)_\Lambda$ factorization;*
viii) *G admits a left canonical $(APW_S)_\Lambda$ factorization.*

In contrast, one-sided invertibility of $T(G)_{(-S)\cap\Lambda}$ corresponds to factorizations of more general type, as in (4.2), than the left AP_S factorizations:

Theorem 4.2. *Let $G \in (APW^k)_\Lambda^{n\times n}$. Then the following statements are equivalent:*

i) *the operator $T(G)_{(-S)\cap\Lambda'}$ is right invertible for some additive subgroup $\Lambda' \supseteq \Lambda$;*
ii) *the operator $T(G)_{(-S)\cap\Lambda'}$ is right invertible for every additive subgroup $\Lambda' \supseteq \Lambda$;*
iii) *the operator $T(G)_{(-S\setminus\{0\})\cap\Lambda'}$ is right invertible for some additive subgroup $\Lambda' \supseteq \Lambda$;*
iv) *the operator $T(G)_{(-S\setminus\{0\})\cap\Lambda'}$ is right invertible for every additive subgroup $\Lambda' \supseteq \Lambda$;*
v) *G admits a representation*

$$G = \Phi_+ U \Phi_-, \tag{4.2}$$

where $\Phi_+^{\pm 1} \in (AP^k)_S^{n\times n}$, $\Phi_-^{\pm 1} \in (AP^k)_{-S}^{n\times n}$, and where $U \in (AP^k)_S^{n\times n}$ is unitary-valued,

vi) *G admits a representation (4.2), where*

$$\Phi_+^{\pm 1} \in (APW^k)_{S\cap\Lambda}^{n\times n}, \quad \Phi_-^{\pm 1} \in (APW^k)_{(-S)\cap\Lambda}^{n\times n}, \quad \text{and} \quad U = \begin{bmatrix} V & 0 \\ 0 & I \end{bmatrix}$$

with unitary-valued $V \in (APW^k)_{(S\setminus\{0\})\cap\Lambda}^{m\times m}$, where $0 \le m \le n$.

These two theorems (with several more statements added on) were proved for almost periodic functions of several variables in [43]. For the one variable case various parts of Theorems 4.1 and 4.2 were proved in [21], [48], [42], and in full generality in [3].

Observe that in both Theorems 4.1 and 4.2, a necessary condition for G to satisfy the equivalent conditions listed in the theorem is that G is invertible in $(APW^k)^{n\times n}_\Lambda$; see [43]. Notice that representation (4.2) in vi) of Theorem 4.2 is not necessarily a left AP_S factorization, and in fact there exist examples of right invertible Toeplitz operators $T(G)_{(-S)\cap\Lambda}$ whose symbol $G \in (APW^k)^{n\times n}_\Lambda$ does not admit a left AP_S factorization (see [7], for example).

In this section we study connections between AP_S factorizations and properties of the corresponding Toeplitz operators beyond one–sided invertibility. We start with a general statement concerning normal solvability:

Proposition 4.3. *Let $G \in (APW^k)^{n\times n}_\Lambda$. If G admits a left $(APW_S)_\Lambda$ factorization, then the Toeplitz operator is normally solvable, i.e., its range is closed.*

The converse of Proposition 4.3 does not hold in general, even assuming that G is invertible in $(APW^k)^{n\times n}_\Lambda$.

A proof of Proposition 4.3 follows without difficulty from the following lemma whose verification is standard.

Lemma 4.4. *If $F \in (APW^k)^{n\times n}_\Lambda$ admits a factorization $F = F_+\Sigma F_-$, where $F_+ \in (APW^k)^{n\times n}_{S\cap\Lambda}$, $F_- \in (APW^k)^{n\times n}_{(-S)\cap\Lambda}$, then*

$$T(F)_{(-S)\cap\Lambda} = T(F_+)_{(-S)\cap\Lambda}T(\Sigma)_{(-S)\cap\Lambda}T(F_-)_{(-S)\cap\Lambda}. \tag{4.3}$$

Next, we study Fredholmness of Toeplitz operators in terms of (noncanonical) left APW_S factorizations.

Theorem 4.5. *Let $G \in (APW^k)^{n\times n}_\Lambda$. Then the Toeplitz operator $T(G)_{(-S)\cap\Lambda}$ is simultaneously Fredholm and right (resp., left) invertible if and only if G admits a factorization*

$$G = G_+\operatorname{diag}(e_{\lambda_1},\dots,e_{\lambda_n})G_-$$

where $G_+^{\pm1} \in (APW^k)^{n\times n}_{S\cap\Lambda}$, $G_-^{\pm1} \in (APW^k)^{n\times n}_{(-S)\cap\Lambda}$, $\lambda_j \in S\cap\Lambda$ (resp., $\lambda_j \in (-S)\cap\Lambda$), and there are only finitely many elements of Λ between zero and any of the λ_j in the total ordering induced by S.

Proof. It suffices to consider the case of right invertibility; left invertibility can then be covered by passing to the adjoint operator $(T(G)_{(-S)\cap\Lambda})^* = T(G^*)_{(-S)\cap\Lambda}$. The "if" part of the theorem follows without difficulties from Lemma 4.4.

For the "only if" part of the theorem, we proceed by induction on n. Let $n = 1$. By Theorem 4.2, G satisfies (2.10), and therefore by Proposition 2.4 admits a (left and right) $(APW_S)_\Lambda$ factorization. Fredholmness of $T(G)_{(-S)\cap\Lambda}$ easily implies the condition that there are only finitely many elements of Λ between zero and the factorization index of G.

Assume now $n > 1$. According to Theorem 4.2, there exists a representation

$$G = H_{+,1} \begin{bmatrix} V & 0 \\ 0 & I \end{bmatrix} H_{-,1}, \tag{4.4}$$

where

$$H_{+,1}^{\pm 1} \in (APW^k)^{n\times n}_{S\cap\Lambda}, \quad H_{-,1}^{\pm 1} \in (APW^k)^{n\times n}_{(-S)\cap\Lambda},$$

and a (possibly void) block $V \in (APW^k)^{m_1\times m_1}_{(S\setminus\{0\})\cap\Lambda}$ is a unitary-valued matrix function.

If $m_1 = 0$, that is the block V is absent, representation (4.4) is a canonical APW_S factorization of G, which is a particular case of (2.8) (corresponding to $\lambda_1 = \cdots \lambda_n = 0$), and we are done. In this (and only this) case the operator $T(G)_{(-S)\cap\Lambda}$ is invertible. If $n > m_1 > 0$, the Toeplitz operator $T(V)_{(-S)\cap\Lambda}$ is Fredholm and right invertible simultaneously with $T(G)_{(-S)\cap\Lambda}$, by (4.3). Applying the induction hypothesis, we obtain the required factorization of V, and hence by (4.4) also of G.

Let now $m_1 = n$. Consider $\widetilde{V} \in (APW^k)^{n\times n}$ with a finite Fourier spectrum $\sigma(\widetilde{V}) \subseteq \sigma(V)$ and so close to V (in the supremum norm) that the operator $T(\widetilde{V})_{(-S)\cap\Lambda}$ is still Fredholm. Let μ_0 be the smallest (in the total ordering imposed by S) element of $\sigma(\widetilde{V})$. For every $\mu \in \Lambda$ lying between 0 and μ_0 (in the same total ordering), and every constant vector $x \in \mathbb{C}^n$, we have

$$e_{-\mu}x \in \operatorname{Ker} T(\widetilde{V})_{(-S)\cap\Lambda}.$$

By Fredholmness, there are only finitely many such μ. Since $\mu_0 \in \sigma(V) \subseteq \Lambda$, this implies that $\sigma(V)$ contains a (unique) minimal element in the sense of the total order induced by S. Call this minimal element μ_1.

The matrix function $V_1 = e_{-\mu_1}V$ belongs to $(APW^k)^{n\times n}_{S\cap\Lambda}$, due to the minimality of μ_1. But then the operator $T(V_1)_{(-S)\cap\Lambda}$ is right invertible, with the right inverse being $T(V_1^*)_{(-S)\cap\Lambda}$. It is also Fredholm, because

$$e_{-\mu_1}\left(\operatorname{Ker} T(V_1)_{(-S)\cap\Lambda}\right) \subseteq \operatorname{Ker} T(V_1)_{(-S)\cap\Lambda}.$$

As the operator $C_{-\mu_1}$ of multiplication by $e_{-\mu_1}$ is an isometry on $(B^k)^{n\times 1}_{(-S)\cap\Lambda}$, the subspace $\operatorname{Ker} T(V_1)_{(-S)\cap\Lambda}$ is finite-dimensional. We may apply the above reasoning to V_1 in place of G, and show that

$$G = H_{+,1}e_{\mu_1}H_{+,2} \begin{bmatrix} e_{\mu_2}V_2 & 0 \\ 0 & I \end{bmatrix} H_{-,2}H_{-,1}, \tag{4.5}$$

where

$$H_{+,2}^{\pm 1} \in (APW^k)^{n\times n}_{S\cap\Lambda}, \quad H_{-,2}^{\pm 1} \in (APW^k)^{n\times n}_{(-S)\cap\Lambda}, \quad \mu_2 \in (S\setminus\{0\})\cap\Lambda,$$

and $V_2 \in (APW^k)^{m_2\times m_2}_{(S\setminus\{0\})\cap\Lambda}$ is unitary. Here m_2 is an integer between 0 and n. If $m_2 = 0$, then (4.5) is, up to a change in notation, a desired factorization of G. If $0 < m_2 < n$, then induction on n, and application of the induction hypothesis to $e_{-\mu_2}V_2$, conclude the proof. If $m_2 = n$, we continue the process.

Having gone through p such steps without decreasing the size of V_j, we arrive at the representation

$$G = e_{\mu_1+\cdots+\mu_p} H_{+,1} \cdots H_{+,p} V_p H_{-,p} \cdots H_{-,1},$$

where

$$H_{+,j}^{\pm 1} \in (APW^k)_{S\cap\Lambda}^{n\times n}, \ H_{-,j}^{\pm 1} \in (APW^k)_{(-S)\cap\Lambda}^{n\times n}, \ \mu_j \in (S \setminus \{0\}) \cap \Lambda, \ j = 1, \dots, p,$$

and $V_2p \in (APW^k)_{(S\setminus\{0\})\cap\Lambda}^{n\times n}$ is unitary. But then all vector functions of the form $e_{-(\mu_1+\cdots+\mu_j)} H_{-,1}^{-1} \cdots H_{-,p}^{-1} x$, $x \in \mathbb{C}^n$, $j = 0, \dots, p-1$, lie in the kernel of $T(G)_{(-S)\cap\Lambda}$. Since there are pn linearly independent of them and the above mentioned kernel is finite-dimensional, at some step the size of V_j will actually decrease so that the induction hypothesis will become applicable for sure. □

Corollary 4.6. *Denote by $\mathcal{F}$ the set of all functions $G \in (APW^k)_\Lambda^{n\times n}$, Λ a subgroup of $\mathbb{R}^k$, with the property that G admits a left $(APW_S)_\Lambda$ factorization*

$$G = G_+ \mathrm{diag}(e_{\lambda_1}, \dots, e_{\lambda_n}) G_-$$

such that there are only finitely many elements of Λ between λ_1 and λ_n in the total ordering induced by S. Then for every $G \in \mathcal{F}$ there exists $\epsilon > 0$ such that

$$F \in (APW^k)_\Lambda^{n\times n}, \quad \|F - G\|_\infty < \epsilon \quad \Longrightarrow \quad F \in \mathcal{F}.$$

The proof follows from Theorem 4.5, the openness of the set of Fredholm one-sided invertible operators, and the following proposition:

Proposition 4.7. *For $f_1, f_2 \in (APW^k)_\Lambda^{n\times n}$ we have*

$$\|T(f_1)_{(-S)\cap\Lambda} - T(f_2)_{(-S)\cap\Lambda}\| \leq \sup_{t\in\mathbb{R}^k} \|f_1(t) - f_2(t)\|. \tag{4.6}$$

For the proof of Proposition 4.7 notice that the norm of the multiplication operator $g \mapsto fg$, $g \in (B^k)_\Lambda^{n\times 1}$ coincides with $\|f\|_\infty$, where $f \in (AP^k)_\Lambda^{m\times n}$ (or more generally $f \in (L^\infty(\mathcal{B}_\Lambda))^{m\times n}$; here $\mathcal{B}_\Lambda$ is the Bohr compactification of Λ).

Theorem 4.5 verifies in particular cases the following conjecture.

Conjecture 4.8. Let $G \in (APW^k)_\Lambda^{n\times n}$, where Λ is a subgroup of $\mathbb{R}^k$. Then the Toeplitz operator $T(G)_{(-S)\cap\Lambda}$ is Fredholm if and only if G admits a factorization

$$G = G_+ \mathrm{diag}(e_{\lambda_1}, \dots, e_{\lambda_n}) G_-, \tag{4.7}$$

where $G_+^{\pm 1} \in (APW^k)_{S\cap\Lambda}^{n\times n}$, $G_-^{\pm 1} \in (APW^k)_{(-S)\cap\Lambda}^{n\times n}$, $\lambda_j \in \Lambda$ and there are only finitely many elements of Λ between zero and any of the λ_j in the total ordering in Λ induced by S.

Notice that the "if" part of Conjecture 4.8 is easily seen. In fact, assuming (4.7) holds, let

$$\kappa_j = \#\{\mu \in \Lambda \ : \ 0 <_S \mu \leq_S \lambda_j\} \quad \text{or} \quad \kappa_j = \#\{\mu \in \Lambda \ : \ 0 >_S \mu \geq_S \lambda_j\}$$

if $\lambda_j >_S 0$ or $\lambda_j <_S 0$, respectively; then

$$\dim \operatorname{Ker} T(G)_{(-S)\cap\Lambda} = \sum_{\{j:\lambda_j >_S 0\}} \kappa_j, \quad \operatorname{codim} \operatorname{Range} T(G)_{(-S)\cap\Lambda} = \sum_{\{j:\lambda_j <_S 0\}} \kappa_j.$$

Another result supporting Conjecture 4.8 is proved in [45]; it asserts that Fredholm Toeplitz operators with almost periodic matrix symbols are automatically invertible, provided a certain property is satisfied concerning relevant subgroups and half-spaces of $\mathbb{R}^k$.

We prove that Conjecture 4.8 is valid under an additional hypothesis that Λ contains a minimal $>_S$-positive element:

Theorem 4.9. *Let $G \in (APW^k)^{n\times n}_\Lambda$, where Λ is a subgroup of $\mathbb{R}^k$. Assume that there exists a minimal (with respect to the total order induced by S) element $\mu_0 \in \Lambda$ with the property that $\mu_0 >_S 0$. Then the Toeplitz operator $T(G)_{(-S)\cap\Lambda}$ is Fredholm if and only if G admits a factorization*

$$G = G_+ \operatorname{diag}(e_{\lambda_1}, \dots, e_{\lambda_n}) G_- \tag{4.8}$$

where $G_+^{\pm 1} \in (APW^k)^{n\times n}_{S\cap\Lambda}$, $G_-^{\pm 1} \in (APW^k)^{n\times n}_{(-S)\cap\Lambda}$, and λ_j are integer multiples of μ_0.

In connection with Theorem 4.9 note that μ_0 is unique. Also, for $\lambda \in \Lambda$, there are only finitely many elements of Λ between zero and λ in the total ordering in Λ induced by S if and only if $\lambda = q\mu_0$ for some integer q.

Proof. We prove the "if" part first. It is easy to see that for every $q \in \mathbb{Z}$, the operator $T(e_{q\mu_0} I)_{(-S)\cap\Lambda}$ is Fredholm. Choose $q > 0$ so that $\lambda_j + q\mu_0 \in S$ for every λ_j in (4.8). By Theorem 4.5 the operator $T(e_{q\mu_0} G)_{(-S)\cap\Lambda}$ is simultaneously Fredholm and right invertible. Therefore, the operator $T(G)_{(-S)\cap\Lambda}$ which by Lemma 4.4 is equal to $T(e_{q\mu_0} G)_{(-S)\cap\Lambda} T(e_{-q\mu_0} I)_{(-S)\cap\Lambda}$ is also Fredholm.

Now assume that $T(G)_{(-S)\cap\Lambda}$ is Fredholm. If $\operatorname{Ker} T(G)_{(-S)\cap\Lambda} = \{0\}$, then $T(G)_{(-S)\cap\Lambda}$ is left invertible, and we are done by Theorem 4.5. So we suppose $\operatorname{Ker} T(G)_{(-S)\cap\Lambda} \neq \{0\}$.

Consider the operators

$$A(q) = T(e_{-q\mu_0} G)_{(-S)\cap\Lambda}, \quad q = 0, 1, 2, \dots.$$

Clearly, all $A(q)$ are Fredholm; indeed,

$$A(q) = T(G)_{(-S)\cap\Lambda} T(e_{-q\mu_0} I)_{(-S)\cap\Lambda},$$

and both $T(G)_{(-S)\cap\Lambda}$ and $T(e_{-q\mu_0} I)_{(-S)\cap\Lambda}$ are Fredholm. Also,

$$\operatorname{Ker} A(q_1) \subseteq \operatorname{Ker} A(q_2) \quad \text{for} \quad q_1 > q_2. \tag{4.9}$$

We claim that there exists a positive integer q such that

$$\operatorname{Ker} A(q) = \{0\}. \tag{4.10}$$

Since the kernels of $A(q)$ are all finite-dimensional, to prove the claim, in view of (4.9) we only need to verify that

$$\operatorname{Ker} A(q_0) \neq \operatorname{Ker} T(G)_{(-S)\cap\Lambda} \tag{4.11}$$

for some q_0. Let

$$f \in \operatorname{Ker} T(G)_{(-S)\cap\Lambda}, \quad f \neq 0.$$

Then $Gf \in (B^k)^{n\times 1}_{\Lambda\cap(S\setminus\{0\})}$. Write $Gf = h_0 + h_1$, where

$$h_0 \in (B^k)^{n\times 1}_{(\mathbb{Z}\mu_0)\cap(S\setminus\{0\})}, \quad h_1 \in (B^k)^{n\times 1}_{(\Lambda\setminus(\mathbb{Z}\mu_0))\cap(S\setminus\{0\})},$$

where we denote by $\mathbb{Z}\mu_0$ the subgroup generated by μ_0. We must have $h_0 \neq 0$, for otherwise the kernel of $T(G)_{(-S)\cap\Lambda}$ would contain an infinite linearly independent set $e_{-m\mu_0}f$, $m = 0, 1, \dots$ (indeed, in view of the minimality of μ_0, $\nu_1 + \nu_0 \in S\setminus\{0\}$ for every $\nu_1 \in (\Lambda\setminus(\mathbb{Z}\mu_0))\cap S$ and every $\nu_0 \in \mathbb{Z}\mu_0$). Let $q_0\mu_0$ be the minimal element in $\mathbb{Z}\mu_0$ for which the orthogonal projection of h_0 onto $(B^k)^{n\times 1}_{q_0\mu_0}$ is nonzero. Then $f \notin \operatorname{Ker} A(q_0)$, and (4.11) is proved.

Once (4.10) is established, we apply Theorem 4.5 to $A(q)$ to complete the proof. □

5. Factorization indices under small perturbations; left vs. right indices

As in Section 4, we fix a half-space S of $\mathbb{R}^k$, and fix an additive subgroup Λ of $\mathbb{R}^k$. We also denote by $\Lambda^n_\oplus$ the set of all ordered n-tuples $\lambda = (\lambda_1, \dots, \lambda_n)$, $\lambda_j \in \Lambda$, such that $\lambda_1 \geq_S \lambda_2 \geq_S \cdots \geq_S \lambda_n$.

Observe first of all that, in general, the existence of a noncanonical AP_S factorization (right or left) is not a stable property. The respective examples (even for $k = 1$, $n = 2$, and among APW matrix functions) are available in [24]. Therefore, we restrict our attention to (AP^k) matrix functions which are a priori AP_S factorable.

We will say that $\lambda' = (\lambda'_1, \dots, \lambda'_n) \in (\mathbb{R}^k)^n_\oplus$ is obtained from $\lambda = (\lambda_1, \dots, \lambda_n) \in \Lambda^n_\oplus$ by an *elementary operation* if there exist a pair of indices $i, k \in \{1, \dots, n\}$ and $\nu \in \Lambda \cap S$ such that $\lambda_i - \lambda_k \geq_S \nu$ and $\lambda'_1, \dots, \lambda'_n$ is a permutation of $\theta_1, \dots, \theta_n$, where

$$\theta_j = \begin{cases} \lambda_i - \nu, & j = i \\ \lambda_k + \nu & j = k \\ \lambda_j & \text{otherwise.} \end{cases} \tag{5.1}$$

Observe that automatically $\lambda' \in \Lambda^n_\oplus$ and $i \geq k$.

We will write $(\lambda_1, \dots, \lambda_n) \succ (\mu_1, \dots, \mu_n)$, and say that $(\mu_1, \dots, \mu_n)$ is *majorized* by $(\lambda_1, \dots, \lambda_n)$, if $(\mu_1, \dots, \mu_n)$ can be obtained from $(\lambda_1, \dots, \lambda_n)$ by a finite sequence of elementary operations. Of course, $\succ$ is a partial order on $\Lambda^n_\oplus$; for $(\lambda_1, \dots, \lambda_n)$ and $(\mu_1, \dots, \mu_n)$ to be comparable it is necessary (but for $n > 2$ not sufficient) that $\lambda_1 + \cdots + \lambda_n = \mu_1 + \cdots + \mu_n$. The minimal elements with respect to this order are the n-tuples $(\lambda_1, \dots, \lambda_n)$ for which the interval $(0, \lambda_1 - \lambda_n)$ in the ordering $>_S$ is empty. For an infinitely divisible Λ, i.e., such that $\lambda \in \Lambda$ implies that $\frac{1}{q}\lambda \in \Lambda$ for some $q \in \mathbb{N}$, $q > 1$, (in particular, for $\Lambda = \mathbb{R}^k$) this happens only if $\lambda_1 = \lambda_n$, that is, the minimal elements in this case are of the form $(\lambda_0, \ldots, \lambda_0)$. On the other hand, for $\Lambda = \mathbb{Z}$ and $S = [0, \infty)$ the minimality condition is $\lambda_1 - \lambda_n \leq 1$.

For $\Lambda = \mathbb{R}$ (and $S = [0, \infty)$) the partial order $\succ$ is discussed in great detail in [34]. Its discrete version ($\Lambda = \mathbb{Z}$) was used in [17] to investigate the stability properties of factorization indices for classical Wiener-Hopf factorization. Next three theorems, as well as their proofs, are straightforward generalizations of the results of [17].

Theorem 5.1. *Let G be a left $(AP_S)_\Lambda$ (resp., $(APW_S)_\Lambda$) factorable $n \times n$ matrix function with left factorization indices $\lambda_1, \ldots, \lambda_n$, and let $(\mu_1, \ldots, \mu_n)$ be any n-tuple majorized by $(\lambda_1, \ldots, \lambda_n)$. Then for every $\epsilon > 0$ there exists a left $(AP_S)_\Lambda$ (resp., $(APW_S)_\Lambda$) factorable matrix function F such that $\|G - F\|_\infty < \epsilon$ (resp., $\|G - F\|_W < \epsilon$) and the left factorization indices of F are $\mu_1, \ldots, \mu_n$.*

Proof. It is sufficient to consider the case when $(\mu_1, \ldots, \mu_n)$ is obtained from $(\lambda_1, \ldots, \lambda_n)$ via one elementary operation. A straightforward computation shows that, for any number $\delta \neq 0$,

$$\begin{bmatrix} e_{\lambda_i} & 0 \\ \delta e_{\lambda_i - \nu} & e_{\lambda_k} \end{bmatrix} = \begin{bmatrix} e_\nu & -1 \\ \delta & 0 \end{bmatrix} \operatorname{diag}(e_{\lambda_i - \nu}, e_{\lambda_k + \nu}) \begin{bmatrix} 1 & \delta^{-1} e_{\lambda_k + \nu - \lambda_i} \\ 0 & \delta^{-1} \end{bmatrix}. \quad (5.2)$$

Denote by E_{rs} the $n \times n$ matrix with (r, s)-entry equal to 1 and all other entries equal to zero, and let

$$A_+ = I + (e_\nu - 1)E_{ii} - E_{kk} - E_{ik} + \delta E_{ki},$$

$$A_- = I + (\delta^{-1} - 1)E_{kk} + \delta^{-1} e_{\lambda_k + \nu - \lambda_i} E_{ik}.$$

Also, let Π denote the permutation matrix corresponding to the permutation from $\mu_1, \ldots, \mu_n$ to $\theta_1, \ldots, \theta_n$, where θ_j are given by (5.1). Then (5.2) implies

$$\operatorname{diag}(e_{\lambda_1}, \ldots, e_{\lambda_n}) + \delta e_{\lambda_i - \nu} E_{ki} = A_+ \Pi \operatorname{diag}(e_{\mu_1}, \ldots, e_{\mu_n}) \Pi A_-.$$

Due to the property $\lambda_i - \lambda_k \geq_S \nu \geq_S 0$, $A_\pm \in (APW^k)_{\pm S}^{n \times n}$. On top of that, $\det A_\pm = \delta^{\pm 1} = \text{const}$, and therefore $A_\pm^{-1} \in (APW^k)_{\pm S}^{n \times n}$ as well. Thus, a left $(AP_S)_\Lambda$ factorization (2.8) of G generates a left $(AP_S)_\Lambda$ factorization

$$(G_+ A_+ \Pi) \operatorname{diag}(e_{\mu_1}, \ldots, e_{\mu_n})(\Pi A_- G_-)$$

of

$$G_\delta = G_+(\operatorname{diag}(e_{\lambda_1}, \ldots, e_{\lambda_n}) + \delta e_{\lambda_i - \nu} E_{ki}) G_-.$$

Since $G_\delta - G = \delta e_{\lambda_i - \nu} G_+ E_{ki} G_-$, the norm of $G_\delta - G$ can be made arbitrarily small by an appropriate choice of δ. For the left $(APW_S)_\Lambda$ factorization the same proof works. □

Theorem 5.2. *Let G be a left $(AP_S)_\Lambda$ factorable $n \times n$ matrix function with left factorization indices $\lambda_1, \ldots, \lambda_n$. Then there exists such a neighborhood of G (in the*

$\|\cdot\|_\infty$ norm) that for every left $(AP_S)_\Lambda$ factorable $n \times n$ matrix function F from this neighborhood its left factorization indices $\mu_1, \dots, \mu_n$ satisfy the conditions

$$\mu_1 + \cdots + \mu_n = \lambda_1 + \cdots + \lambda_n, \qquad \lambda_1 \geq_S \mu_1 (\geq_S) \mu_n \geq_S \lambda_n. \tag{5.3}$$

Proof. Taking determinants of both sides in (2.8) we observe that $\lambda_1 + \cdots + \lambda_n$ is nothing but the mean motion of the function $\det G$ (see Proposition 2.4 and its proof), and the mean motion is indeed stable.

To prove the inequality $\mu_n \geq_S \lambda_n$, consider the matrix function $G_1 = e_{-\lambda_n}$. All its left factorization indices are in S, and the Toeplitz operator $T(G_1)_{(-S)\cap\Lambda}$ is therefore right invertible (trivial, but also follows from Theorem 4.2 if $G \in (APW^k)_\Lambda^{n\times n}$). The latter property is stable, so that $T(e_{-\lambda_n}F)_{(-S)\cap\Lambda}$ is right invertible for all F sufficiently close to G. Provided that F (and therefore $F_1 = e_{-\lambda_n}F$) is left $(AP_S)_\Lambda$ factorable, this means that the left factorization indices of F_1 lie in S. In other words, $\mu_j \geq_S \lambda_n$ for all $j = 1, \dots, n$.

The inequality $\lambda_1 \geq_S \mu_1$ can be proved in a similar way or by applying the already proved result for the smallest factorization index to G^* and F^*. □

Combining the last two theorems, we arrive at the following stability criterion.

Theorem 5.3. *Let G be a left $(AP_S)_\Lambda$ factorable $n \times n$ matrix function. The set $\lambda_1, \dots, \lambda_n$ of its left factorization indices is stable under small perturbations (preserving left $(AP_S)_\Lambda$ factorability) in the $\|\cdot\|_\infty$ norm if and only if $(\lambda_1, \dots, \lambda_n)$ is a minimal element with respect to the order $\succ$.*

Proof. Necessity follows from Theorem 5.1. To prove sufficiency, suppose that $(\lambda_1, \dots, \lambda_n)$ is minimal, that is, there are no elements of Λ strictly between 0 and $\lambda_1 - \lambda_n$ (with respect to the $\geq_S$ ordering). We now consider two cases ($\lambda_1 = \lambda_n$ and $\lambda_1 \neq \lambda_n$) separately.

Case 1. $\lambda_1 = \lambda_n$. The second of conditions (5.3) then implies that $\mu_j = \lambda_1 (= \lambda_n)$ for all $j = 1, \dots, n$ when F is close enough to G.

Case 2. $\lambda_1 \neq \lambda_n$. Then $\lambda_1 = \dots = \lambda_k$ and $\lambda_{k+1} = \dots = \lambda_n$ for some $k \in \{1, \dots, n-1\}$. The second of conditions (5.3) now implies that $\mu_1 = \dots = \mu_l = \lambda_1$ and $\mu_{l+1} = \dots = \mu_n = \lambda_n$ for some $l \in \{0, \dots, n\}$; it remains to show only that $l = k$. This follows from the first of conditions (5.3). □

For the case $\Lambda = \mathbb{R}$ Theorems 5.1–5.3 were established in [24].

Theorem 5.1 describes *some* possible changes of factorization indices under small perturbations. In fact, for $\Lambda = \mathbb{Z}$ or $\Lambda = \mathbb{R}$ these are the *only* possible changes. In the former case, this is again a result from [17]. In the latter case, the result is established in [23]. We believe that the result holds in general:

Conjecture 5.4. Let G be as in Theorem 5.1. If $(\mu_1, \dots, \mu_n) \in \Lambda_\oplus^n$ is such that for every $\epsilon > 0$ there exists a left $(AP_S)_\Lambda$ (resp., $(APW_S)_\Lambda$) factorable matrix function F with factorization indices $\mu_1, \dots, \mu_n$ and such that $\|G - F\|_\infty < \epsilon$ (resp., $\|G - F\|_W < \epsilon$), then $(\lambda_1, \dots, \lambda_n) \succ (\mu_1, \dots, \mu_n)$.

Theorem 5.3 confirms Conjecture 5.4 in the situation when $(\lambda_1, \dots, \lambda_n)$ is minimal with respect to $\succ$.

We will now entertain the following question: let G be an $n \times n$ matrix function admitting both left and right $(AP_S)_\Lambda$ factorization. What are possible relations between the sets $(\lambda_1, \dots, \lambda_n)$ of its left and $(\rho_1, \dots, \rho_n)$ of its right factorization indices?

As was mentioned earlier, the sum $\lambda_1 + \dots + \lambda_n$ coincides with the mean motion of $\det G$. Similarly, $\rho_1 + \dots + \rho_n$ coincides with the mean motion of $\det G$ as well. Thus,

$$\lambda_1 + \dots + \lambda_n = \rho_1 + \dots + \rho_n. \tag{5.4}$$

As it happens, (5.4) is the only required relation between the sets of left and right factorization indices. In other words, the following result holds.

Theorem 5.5. *Let $(\lambda_1, \dots, \lambda_n)$ and $(\rho_1, \dots, \rho_n)$ be two elements of $\Lambda^n_\oplus$ satisfying (5.4). Then there exist $n \times n$ matrix functions G, both left and right $(AP_S)_\Lambda$ factorable, such that $(\lambda_1, \dots, \lambda_n)$ and $(\rho_1, \dots, \rho_n)$ are, respectively, the set of their left/right factorization indices.*

Even in the classical setting $\Lambda = \mathbb{Z}$, Theorem 5.5 was established quite recently [12]; see this paper also for the history of the subject. Our proof follows the lines of [12] but some changes are necessary. Two auxiliary results are in order.

Lemma 5.6. *Let $\lambda = (\lambda_1, \dots, \lambda_n), \rho = (\rho_1, \dots, \rho_n) \in \Lambda^n_\oplus$ be such that (5.4) holds. Then there exists a common majorant for λ and ρ, that is, $\sigma = (\sigma_1, \dots, \sigma_n) \in \Lambda^n_\oplus$ for which $\sigma \succ \lambda$ and $\sigma \succ \rho$.*

Proof. We will use induction on n. The base of induction is trivial since for $n = 1$ (5.4) implies $\rho = \lambda$, and σ coinciding with them will do the job (actually, this is the only choice). Suppose now that the statement holds for $n = k$, and let $\lambda = (\lambda_1, \dots, \lambda_{k+1}), \rho = (\rho_1, \dots, \rho_{k+1}) \in \Lambda^{k+1}_\oplus$ satisfy $\lambda_1 + \dots + \lambda_{k+1} = \rho_1 + \dots + \rho_{k+1}$.

If $\lambda_{k+1} = \rho_{k+1} (\overset{\text{def}}{=} \sigma_{k+1})$, then $(\lambda_1, \dots, \lambda_k)$ and $(\rho_1, \dots, \rho_k)$ have a common majorant $(\sigma_1, \dots, \sigma_k)$ due to induction hypothesis. But then $(\sigma_1, \dots, \sigma_k, \sigma_{k+1})$ is a common majorant for λ and ρ.

If $\lambda_{k+1} >_S \rho_{k+1}$, let

$$\mu = (\mu_1, \dots, \mu_{k+1}) = (\lambda_1 + \lambda_{k+1} - \rho_{k+1}, \lambda_2, \dots, \lambda_k, \rho_{k+1}).$$

According to the previous reasoning, μ and ρ have a common majorant σ. Since $\mu \succ \lambda$, the same σ is also a common majorant for λ and ρ. The case $\lambda_{k+1} <_S \rho_{k+1}$ can be taken care of by interchanging the roles of λ and ρ. □

Lemma 5.7. *Let $\alpha = (\alpha_1, \dots, \alpha_n), \beta = (\beta_1, \dots, \beta_n) \in \Lambda^n_\oplus$, and $\beta \succ \alpha$. Then there exist $E_+, F_+ \in (APW^k)^{n\times n}_{S\cap\Lambda}$ and $E_- \in (APW^k)^{n\times n}_{(-S)\cap\Lambda}$ such that $E_+^{-1}, F_+^{-1} \in (APW^k)^{n\times n}_{S\cap\Lambda}$, $E_-^{-1} \in (APW^k)^{n\times n}_{(-S)\cap\Lambda}$ and*

$$F_+ \operatorname{diag}(e_{\beta_1}, \dots, e_{\beta_n}) = E_- \operatorname{diag}(e_{\alpha_1}, \dots, e_{\alpha_n}) E_+.$$

Proof. According to the analog of Theorem 5.1 for right factorization, there exist $n \times n$ matrix functions $M \in (APW^k)^{n\times n}_{\Lambda}$ of arbitrarily small $\|\cdot\|_W$ norm such that

$$\operatorname{diag}(e_{\beta_1},\dots,e_{\beta_n}) + M = G_- \operatorname{diag}(e_{\alpha_1},\dots,e_{\alpha_n})E_+, \tag{5.5}$$

where $G_-^{\pm1} \in (APW^k)^{n\times n}_{(-S)\cap\Lambda}$, $E_+^{\pm1} \in (APW^k)^{n\times n}_{S\cap\Lambda}$. Choose the norm of M so small that

$$\|M \operatorname{diag}(e_{-\beta_1},\dots,e_{-\beta_n})\| < 1.$$

Then $I + M\operatorname{diag}(e_{-\beta_1},\dots,e_{-\beta_n})$ has positive definite real part. The Toeplitz operator $T(M\operatorname{diag}(e_{-\beta_1},\dots,e_{-\beta_n}))_{S\cap\Lambda}$ then also has a positive definite real part and is therefore invertible. According to Theorem 4.1 (with S replaced by $-S$), the matrix function $I + M\operatorname{diag}(e_{-\beta_1},\dots,e_{-\beta_n})$ admits a right canonical $(APW_S)_\Lambda$ factorization:

$$I + M\operatorname{diag}(e_{-\beta_1},\dots,e_{-\beta_n}) = F_-F_+. \tag{5.6}$$

Combining (5.5) and(5.6):

$$F_-F_+\operatorname{diag}(e_{\beta_1},\dots,e_{\beta_n}) = G_-\operatorname{diag}(e_{\alpha_1},\dots,e_{\alpha_n})E_+.$$

It remains to set $E_- = F_-^{-1}G_-$. □

of Theorem 5.5. For given $\lambda = (\lambda_1,\dots,\lambda_n)$, $\rho = (\rho_1,\dots,\rho_n)$ choose their common majorant $\sigma = (\sigma_1,\dots,\sigma_n)$; such σ exists due to condition (5.4) and Lemma 5.6. Applying now Lemma 5.7 twice, with $\alpha = \lambda,\rho$ and $\beta = \sigma$, we end up with two representations:

$$A_+^{(1)}\operatorname{diag}(e_{\sigma_1},\dots,e_{\sigma_n}) = B_-^{(1)}\operatorname{diag}(e_{\rho_1},\dots,e_{\rho_n})B_+^{(1)}$$

and

$$A_+^{(2)}\operatorname{diag}(e_{\sigma_1},\dots,e_{\sigma_n}) = B_-^{(2)}\operatorname{diag}(e_{\lambda_1},\dots,e_{\lambda_n})B_+^{(2)}.$$

Here $(A_+^{(j)})^{\pm1}, (B_+^{(j)})^{\pm1} \in (APW^k)^{n\times n}_{S\cap\Lambda}$, $(B_-^{(j)})^{\pm1} \in (APW^k)^{n\times n}_{(-S)\cap\Lambda}$, for $j = 1,2$. But then

$$G = B_-^{(1)}\operatorname{diag}(e_{\rho_1},\dots,e_{\rho_n})B_+^{(1)}(A_+^{(2)})^T = A_+^{(1)}\operatorname{diag}(e_{\sigma_1},\dots,e_{\sigma_n})(A_+^{(2)})^T =$$
$$A_+^{(1)}(B_+^{(2)})^T\operatorname{diag}(e_{\lambda_1},\dots,e_{\lambda_n})(B_-^{(2)})^T$$

is the desired matrix function. □

6. Application of (non)canonical factorization to normalizing bases

In this section we shall explain how the factorization theory we developed may be used in finding a basis transformation that normalizes a generating set of the form $\{\phi(\cdot - \nu) : \nu \in \Lambda\}$. Here and in what follows Λ is a subgroup of $\mathbb{R}^k$. We shall see that in the case of spectral factorizations the normalization is in fact orthonormalization, and in case of other types of factorization we obtain other types of normalization.

Let $\mathcal{H}$ denote a Hilbert space or a Krein space with inner product $\langle\cdot,\cdot\rangle$. Thus, the inner product may be definite or indefinite. In addition, we have operations $A_\nu : \mathcal{H} \to \mathcal{H}$, $\nu \in \Lambda$, satisfying

$$A_\nu A_\mu = A_{\nu+\mu}\ ,\ A_\nu = A^*_{-\nu}\ ,\ A_0 = I.$$

The basic settings to think of are the following two.

Setting 6.1. Let $\langle\cdot,\cdot\rangle_{\mathbb{C}^n}$ be an indefinite or definite inner product on $\mathbb{C}^n$. Let $\mathcal{H} = L^2(\mathbb{R}^k, \mathbb{C}^n)$ be the Lebesgue space of square integrable functions acting $\mathbb{R}^k \to \mathbb{C}^n$ with inner product

$$\langle f, g\rangle = \int_{\mathbb{R}^k} \langle f(t), g(t)\rangle_{\mathbb{C}^n} dt.$$

Define A_ν, $\nu \in \Lambda$, by

$$(A_\nu f)(t) = f(t+\nu), t \in \mathbb{R}^k.$$

Setting 6.2. Let $\langle\cdot,\cdot\rangle_{\mathbb{C}^n}$ be an indefinite or definite inner product on $\mathbb{C}^n$. Let $\mathcal{H} = (B^k)^n_{\Lambda'}$, where Λ' is a subgroup of $\mathbb{R}^k$ containing Λ, with inner product defined via

$$\langle f, g\rangle = M\{\langle f, g\rangle_{\mathbb{C}^n}\},\ f, g \in (AP^k_{\Lambda'})^n,$$

and extended to $(B^k)^n_{\Lambda'}$ by continuity. Define A_ν, $\nu \in \Lambda$, by

$$A_\nu(\sum \alpha_\mu e_\mu) = \sum \alpha_\mu e_{\mu+\nu}.$$

We consider now the following problem. Let $\phi_i \in \mathcal{H}$, $i = 1,\ldots,p$, be given and consider the set

$$\{A_\nu\phi_i\ :\ \nu \in \Lambda, i = 1,\ldots,p\}. \tag{6.1}$$

In case of a Hilbert space, we should like to know how to orthonormalize this set, if possible. In the case of a Krein space, we are interested in a basis transformation that converts (6.1) to a normalized form, the exact nature of which to be determined later. We will explain the basic idea how to do this in general, and then illustrate the procedure in some instances.

First observe that

$$\langle A_\nu\phi_i, A_\mu\phi_j\rangle = \langle A_{\nu-\mu}\phi_i, \phi_j\rangle, \qquad \nu, \mu \in \Lambda;\quad i, j = 1,\ldots,p.$$

Let us denote

$$c^{(\nu)}_{ij} = \langle A_\nu\phi_j, \phi_i\rangle,\ \nu \in \Lambda,\ i, j = 1,\ldots,p,$$

and put

$$C_\nu = (c^{(\nu)}_{ij})^p_{i,j=1},\ \nu \in \Lambda.$$

Notice that $C^*_\nu = C_{-\nu}$, $\nu \in \Lambda$. In case only countably many C_ν are nonzero and

$$\sum_{\nu\in\Lambda} \|C_\nu\| < \infty,$$

we may define the $p \times p$ matrix-valued almost periodic function

$$F(t) = \sum_{\nu \in \Lambda} C_\nu e_\nu(t), \quad t \in \mathbb{R}^k. \tag{6.2}$$

Note that $F(t) = F(t)^*$, $t \in \mathbb{R}^k$.

Let $S \subset \mathbb{R}^k$ be a half-space. Let us assume that F allows a symmetric right $(APW_S)_\Lambda$ factorization

$$F(t) = F_+(t)^* D(t) F_+(t), \tag{6.3}$$

where $F_+^{\pm} \in (APW^k)^{p\times p}_{S\cap\Lambda}$ and $D \in (APW^k)^{p\times p}_{\Lambda}$ has the form of the middle factor in the right-hand side of (3.5). For example, when $F > 0$ such a factorization exists with $D \equiv I$. Write

$$F_+^{-1}(t) = \sum_{\nu \in \Lambda\cap S} G_\nu e_\nu(t),$$

and $G_\nu = (g_{ij}^{(\nu)})_{i,j=1}^p$. Define now

$$\psi_j = \sum_{\nu \in \Lambda\cap S} \sum_{i=1}^{p} g_{ij}^{(\nu)} A_{-\nu}\phi_i, \ j = 1, \ldots, p. \tag{6.4}$$

Notice that since only countably many G_ν are nonzero, and since $\sum \|G_\nu\| < \infty$, we get that $\psi_i, i = 1, \ldots, p$, are well defined elements of $\mathcal{H}$. Next, one may check that (6.3) implies that for $\nu \in \Lambda$ we have that

$$((\langle A_\nu \psi_i, \psi_j \rangle))_{i,j=1}^p = \sum_{\mu_1 \in S\cap\Lambda} \sum_{\mu_2 \in S\cap\Lambda} G_{\mu_2}^* ((\langle A_{\nu-\mu_1}\phi_i, A_{-\mu_2}\phi_j\rangle))_{i,j=1}^p G_{\mu_1} \tag{6.5}$$

$$= (F_+^{*-1} F F_+^{-1})_\nu = D_\nu,$$

where $D = \sum_{\nu\in\Lambda} D_\nu e_\nu$. We have thus obtained that the collection

$$\{A_\nu \psi_i, \ \nu \in \Lambda, i = 1, \ldots, p\}$$

is of some normalized form, the exact form of which depends on the function D. For instance, when $D \equiv I$, we obtain an orthonormalized set.

Example 6.3. Let $\langle \cdot, \cdot \rangle_{\mathbb{C}^n}$ be the usual Euclidean inner product on $\mathbb{C}^n$, and let $\mathcal{H} = L^2(\mathbb{R}^2, \mathbb{C}^n)$. Choose $\phi \in \mathcal{H}$ so that

$$\langle \phi, \phi \rangle = 2\zeta, \ \zeta > 2 \ , \langle \phi(\cdot \pm 1, \cdot), \phi \rangle = \langle \phi(\cdot, \cdot \pm 1), \phi \rangle = -1,$$

and

$$\langle \phi(\cdot + n, \cdot + m), \phi \rangle = 0, (n, m) \in \mathbb{Z}^2 \setminus \{(0,0), (\pm 1, 0), (0, \pm 1)\}.$$

Such a ϕ may easily be constructed, e.g., ϕ could be a spline. In this case

$$F = -e_{(-1,0)} - e_{(0,-1)} + 2\zeta - e_{(1,0)} - e_{(0,1)},$$

or, in other words,

$$F(t_1, t_2) = 2(\zeta - \cos t_1 - \cos t_2) > 0, (t_1, t_2) \in \mathbb{R}^2.$$

Since F is uniformly positive definite, it has a canonical factorization

$$F(t) = F_+(t)^* F_+(t),$$

where $F_+^{\pm 1} \in (APW^2)_{S\cap\mathbb{Z}^2}$ with $S = E_2$ the standard half-space (defined by (2.4)) in $\mathbb{R}^2$. Defining ψ by (6.4) yields an element of $L^2(\mathbb{R}^2, \mathbb{C}^n)$ with the property that $\{\psi(\cdot+\nu) : \nu \in \mathbb{Z}^2\}$ is an orthonormal set. Orthonormal sets consisting of integer translates of a single function are important in the theory of wavelets. In the papers [19], [35], [36] examples of this kind are presented, with emphasis on numerical procedures to normalize sets $\{\phi(\cdot+\nu) : \nu \in \mathbb{Z}^2\}$. Most of the procedures are based on choosing a total order on $\mathbb{Z}^2$ and computing Cholesky factorizations of finite matrices. For the above function F these calculations were performed in [35]. As a result the orthonormalized set is not of the form $\{\psi(\cdot+\nu) : \nu \in \mathbb{Z}^2\}$, that is, they are not the integer translates of a single function. They do, however, exhibit the phenomena of a so-called "limiting profile" which can be related to the function ψ that we find (see Theorem 5.3 in [36]).

Example 6.4. Let $\langle\cdot,\cdot\rangle_{\mathbb{C}^2}$ be defined by

$$\langle x, y\rangle = x_1\overline{y}_1 - x_2\overline{y}_2,$$

and let $\Lambda = \mathbb{Q}+\alpha\mathbb{Q}+\beta\mathbb{Q}$, where $\{1,\alpha,\beta\}$ are linearly independent over $\mathbb{Q}$. Consider $\mathcal{H}$ and A_ν, $\nu \in \Lambda$ as in Setting 5.2. Let

$$\phi_1(t) = \begin{bmatrix} a_1 + a_2 e_\alpha(t) \\ b_1 + b_2 e_\alpha(t) \end{bmatrix} \in (B^1_\Lambda)^2, \phi_2 = \begin{bmatrix} e_\beta(t) \\ d e_\beta(t) \end{bmatrix} \in (B^1_\Lambda)^2, \tag{6.6}$$

where a_1, a_2, b_1, b_2, c and d are complex numbers so that

$$a_1\overline{c} - b_1\overline{d} = 1, \overline{c}_2 a_2 - \overline{b}_2 d = 0.$$

Put $b = |a_1|^2 + |a_2|^2 - |b_1|^2 - |b_2|^2$ and $a = a_1\overline{a}_2 - b_1\overline{b}_2$. Then it is straightforward to check that F in (6.2) is given by

$$F(t) = \begin{bmatrix} a e_\alpha(t) + b + \overline{a} e_{-\alpha}(t) & e_\beta(t) \\ e_{-\beta}(t) & 0 \end{bmatrix}. \tag{6.7}$$

This is a Hermitian matrix-valued almost periodic trigonometric polynomial. Let $S = [0,\infty)$ and denote by m the *ceiling* $\lceil\frac{\beta}{\alpha}\rceil$ of β/α, that is, the smallest integer number bigger than β/α.

Theorem 6.5. *Let $\{1,\alpha,\beta\}$ be linearly independent over $\mathbb{Q}$, and put $m = \lceil\frac{\beta}{\alpha}\rceil$. The matrix function (6.7) always admits a right $(APW_S)_\Lambda$ factorization. The (right) factorization indices of F are equal $\pm\mu$, where*

$$\mu = \begin{cases} \beta & \text{if } a = b = 0 \\ \min\{|2l\alpha - \beta| : l \in \mathbb{N}\} & \text{if } b = 0, a \neq 0 \\ \beta - (m-1)\alpha & \text{if } 0 \neq b = 2|a|\cos\frac{\nu\pi}{m} \text{ for some } \nu \in \{1,\ldots,m-1\} \\ m\alpha - \beta & \text{if } 0 \neq b = 2|a|\cos\frac{\nu\pi}{m+1} \text{ for some } \nu \in \{1,\ldots,m\} \\ 0 & \text{in all other cases.} \end{cases}$$

Moreover, the factorization of F can be chosen in the form

$$F(t) = F_+(t)^* \begin{bmatrix} 1 & 0 \\ 0 & -1 \end{bmatrix} F_+(t), \tag{6.8}$$

for $\mu = 0$, *and*

$$F(t) = F_+(t)^* \begin{bmatrix} 0 & e_\mu(t) \\ e_{-\mu}(t) & 0 \end{bmatrix} F_+(t), \tag{6.9}$$

for $\mu \neq 0$; *in both cases* $F_+^{\pm 1} \in (APW^1)^{2\times 2}_{\Lambda\cap S}$.

We mention here that for the case when $\{1, \alpha, \beta\}$ is linearly dependent over $\mathbb{Q}$ one may also describe the $(APW_S)_\Lambda$ factorization of (6.7). We omit this result here as it does not pertain to our application. In addition, we mention that the techniques in [27] may be used to construct $F_+(t)$.

Proof. The case $a = b = 0$ is trivial. In all other cases, passing to transposed matrices, we may switch to the left $(APW_S)_\Lambda$ factorization of

$$\begin{bmatrix} ae_\alpha + b + \bar{a}e_{-\alpha} & e_{-\beta} \\ e_\beta & 0 \end{bmatrix}.$$

The latter matrix function is Hermitian simultaneously with F; its determinant equals -1 identically, thus, its eigenvalues are of the opposite sign. The required special form of its factorization, provided that it exists, follows then from [48]. To establish the existence, we may switch the rows of the matrix under consideration, that is, put it in the form

$$G_f^{(\beta)} = \begin{bmatrix} e_\beta & 0 \\ f & e_{-\beta} \end{bmatrix},$$

where $f = \bar{a}e_{-\alpha} + b + ae_\alpha$. The factorization indices do not change under such transformation.

If $a = 0$, $b \neq 0$, or $a \neq 0$, $b = 0$, then the off diagonal entry f of $G_f^{(\beta)}$ is at most a binomial, and factorability of $G_f^{(\beta)}$ (along with the formula for μ) follows from [24]. If $ab \neq 0$, then f is a trinomial with equidistant points $-\alpha, 0, \alpha$ of its Fourier spectrum. The factorability of $G_f^{(\beta)}$ in this case follows from a (more general) result of [27] on $G_f^{(\beta)}$ with commensurable distances between the points of the Fourier spectrum of f. An alternative approach of [6], however, leads to more explicit formulas, especially useful in our particular case. Namely, according to [6], the matrix function $G_f^{(\beta)}$ is factorable only simultaneously with (and has the same factorization indices as) the matrix $G_{f_1}^{(\alpha)}$, where

$$f_1 = \Pi_{(-\alpha,\alpha)}(g_2 e_{-\beta})$$

and g_2 is a solution of the equation

$$g_1 e_{\alpha+\beta} + g_2 e_\alpha f = 1; \qquad g_1, g_2 \in (APW^1)_S. \tag{6.10}$$

Such g_2 must be of the form

$$g_2 = \sum_{k=0}^{m} c_k e_{k\alpha} + he_{\alpha+\beta}, \quad h \in (APW^1)_S,$$

and the column $c = (c_0, \ldots, c_m)^T$ satisfies the equation $Dc = b$, where $b = (1, 0, \ldots, 0)^T$ and D is the $(m+1) \times (m+1)$ Toeplitz matrix (d_{i-j}) with $d_0 = \overline{a}$, $d_1 = b$, $d_2 = a$ and all other d_j equal zero.

Thus, f_1 is a binomial $c_{m-1}e_{(m-1)\alpha-\beta}+c_m e_{m\alpha-\beta}$, where $c_j = (-\overline{a})^{-j-1}\det T_j$, and T_j is a tridiagonal Hermitian Toeplitz $j \times j$ matrix with b on its main diagonal and $a, \overline{a}$ on its adjacent diagonals; $j = m-1, m$. According to [24], the factorization indices of $G_{f_1}^{(\alpha)}$ (and therefore F) are equal to 0 if $\det T_{m-1} \det T_m \neq 0$, $\pm(m\alpha - \beta)$ if $\det T_m \neq 0$, $\det T_{m-1} = 0$, and $\pm(\beta - (m-1)\alpha)$ if $\det T_{m-1} \neq 0$, $\det T_m = 0$. It remains to observe that $\det T_j = |a|^j U_j(b/|a|)$, where U_j are Chebyshev polynomials of the second kind, and use the explicit formula for the roots of the latter. □

If under the conditions of Theorem 6.5 we have that $\mu = 0$, then defining ψ_1 and ψ_2 by (6.4) we get that

$$\langle A_\nu \psi_j, \psi_i \rangle = \begin{cases} 0, & \text{if } \nu \neq 0 \text{ or } i \neq j; \\ 1, & \text{if } \nu = 0 \text{ and } i = j = 1; \\ -1, & \text{if } \nu = 0 \text{ and } i = j = 2. \end{cases}$$

In the case of $\mu \neq 0$ we get when we define ψ_1 and ψ_2 by (6.4), that

$$\langle A_\nu \psi_j, \psi_i \rangle = \begin{cases} 0, & \text{if } \nu \neq \pm\mu \text{ or } i = j; \\ 1, & \text{if } \nu = \mu \text{ and } i = 2, j = 1; \\ 1, & \text{if } \nu = -\mu \text{ and } i = 1, j = 2. \end{cases}$$

Acknowledgements

We thank C.V.M. van der Mee for attracting our attention to results obtained in [19], [35], and [36], T. Ehrhardt for sending us the preprint [11], and J. Martinez for a consultation concerning extensions of groups.

Research of the authors is partially supported by NSF grant DMS-9988579. HJW was also supported by a Faculty Research Assignment grant from the College of William and Mary.

References

[1] Bakonyi, M., Rodman, L., Spitkovsky, I.M., and Woerdeman, H.J., Positive extensions of matrix functions of two variables with support in an infinite band, *C.R. Acad. Sci. Paris Sér. I Math.* 323(8) (1996), 859–863.

[2] Bakonyi, M., Rodman, L., Spitkovsky, I.M., and Woerdeman, H.J., Positive matrix functions on the bitorus with prescribed coefficients in a band, *J. Fourier Analysis and Applications* 5 (1999), 789–812.

[3] Ball, J.A., Karlovich, Yu.I., Rodman, L., and Spitkovsky, I.M., Sarason interpolation and Toeplitz corona theorem for almost periodic matrix functions, *Integral Equations and Operator Theory* 32 (1998), 243–281.

[4] Bastos, M.A., Karlovich, Yu.I., dos Santos, F.A., and Tishin, P.M., The Corona theorem and the existence of canonical factorization of triangular AP-matrix functions, *J. Math. Anal. Appl.*, 223 (1998), 494–522.

[5] Bastos, M.A., Karlovich, Yu.I., dos Santos, F.A., and Tishin, P.M., The Corona theorem and the canonical factorization of triangular AP-matrix functions – Effective criteria and explicit formulas, *J. Math. Anal. Appl.* 223 (1998), 523–550.

[6] Bastos, M.A., Karlovich, Yu.I., Spitkovsky, I.M., and Tishin, P.M., On a new algorithm for almost periodic factorization, *Operator Theory: Advances and Applications* 103 (1998), 53–74.

[7] Böttcher, A., Karlovich, Yu.I, and Spitkovsky, I.M. *Convolution Operators and Factorization of Almost Periodic Matrix Functions*, Birkhäuser, Basel and Boston, 2002.

[8] Clancey, K.F., and Gohberg, I. *Factorization of Matrix Functions and Singular Integral Operators*, Birkhäuser, Basel and Boston, 1981.

[9] Coburn, L.A., and Douglas, R.G., Translation operators on the half-line, *Proc. Nat. Acad. Sci. U.S.A.*, 62 (1969), 1010–1013.

[10] Corduneanu, C. *Almost Periodic Functions*, J. Wiley & Sons, 1968.

[11] Ehrhardt, T., Invertibility theory for Toeplitz + Hankel operators and singular integral operators with flip, preprint.

[12] Feldman, I. and Markus, A., On some properties of factorization indices, *Integral Equations and Operator Theory*, 30 (1998), 326–337.

[13] Gohberg, I.C., and Fel'dman, I.A., Integro-difference Wiener-Hopf equations, *Acta Sci. Math. (Szeged)*, 30 (1969), 199–224. (Russian.)

[14] Gohberg, I., and Feldman, I. *Convolution Equations and Projection Methods for their Solution*, Nauka, Moscow, 1971. English translation *Amer. Math. Soc. Transl. of Math. Monographs* **41**, Providence, R.I. 1974.

[15] Gohberg, I., Goldberg, S., and Kaashoek, M.A. *Classes of Linear Operators. II*, Birkhäuser, Basel and Boston, 1993.

[16] Gohberg, I., and Kaashoek, M.A. (eds.). *Constructive Methods of Wiener-Hopf Interpolation*, Birkhäuser, Basel and Boston, 1986.

[17] Gohberg, I., and Krein, M.G., Systems of integral equations on a half-line with kernel depending upon the difference of the arguments, *Amer. Math. Soc. Transl.*, 14 (1960), 217–287.

[18] Gohberg, I., and Krupnik, N.Ya. *One-Dimensional Linear Singular Integral Equations.; Vol. I: Introduction; Vol. 2: General Theory and Applications*, Birkhäuser, Basel and Boston, 1992.

[19] Goodman, T.N.T., Micchelli, C.A., Rodriguez, G., and Seatzu, S. On the Cholesky factorization of the Gram matrix of multivariate functions, *SIAM J. Matrix Anal. Appl.* 22 (2000), 501–526.

[20] Horn, R.A., and Johnson, C.R. *Matrix Analysis*, Cambridge University Press, 1985.

[21] Karlovich, Yu.I. *Algebras of Convolution Type Operators with Discrete Groups of Shifts and Oscillating Coefficients*, Doctoral Dissertation, Mathematical Institute, Georgian Academy of Sciences, Tbilisi, 1991.

[22] Karlovich, Yu.I., On the Haseman problem, *Demonstratio Math.*, 26 (1993), 581–595.

[23] Karlovich, Yu.I., and Spitkovsky, I.M., *Factorization of Almost periodic Matrix Functions and (semi) Fredholmness of Some Convolution Type Equations*, VINITI, Moscow, 1985.

[24] Karlovich, Yu.I., and Spitkovsky, I.M., Factorization of almost periodic matrix-valued functions and the Noether theory for certain classes of equations of convolution type, *Mathematics of the USSR, Izvestiya* 34 (1990), 281–316.

[25] Karlovich, Yu.I., and Spitkovsky, I.M., (Semi)-Fredholmness of convolution operators on the spaces of Bessel potentials, *Operator Theory: Advances and Applications*, 71 (1994), 122–152.

[26] Karlovich, Yu.I., and Spitkovsky, I.M., Factorization of almost periodic matrix functions, *J. Math. Anal. Appl.*, 193 (1995), 209–232.

[27] Karlovich, Yu.I., and Spitkovsky, I.M., Almost periodic factorization: An analogue of Chebotarev's algorithm, *Contemporary Math.* 189 (1995),327–352.

[28] Karlovich, Yu.I., and Spitkovsky, I.M., Semi-Fredholm properties of certain singular integral operators, *Operator Theory: Advances and Applications*, 90 (1996), 264–287.

[29] Karlovich, Yu.I., Spitkovsky, I.M. and Walker, R.A., Almost periodic factorization of block triangular matrix functions revisited. *Linear Algebra Appl.* 293 (1999), 199–232.

[30] Krein, M.G., Integral equations on a half-line with kernel depending upon the difference of the arguments, *Amer. Math. Soc. Transl., Series 2*, 22 (1962), 163–288.

[31] Levitan, B.M. *Almost Periodic Functions*, GITTL, Moscow, 1953. (Russian.)

[32] Levitan, B.M., and Zhikov, V.V. *Almost Periodic Functions and Differential Equations*, Cambridge University Press, 1982.

[33] Litvinchuk, L.S., and Spitkovsky, I.M. *Factorization of Measurable Matrix Functions*, Birkhäuser Verlag, Basel and Boston, 1987.

[34] Marshall, A.W. and Olkin, I.*Inequalities: Theory of Majorization and its Applications*, Academic Press, New York-London, 1979.

[35] van der Mee, C.V.M., Rodriguez, G., and Seatzu, S., Spectral factorization of bi-infinite multi-index block Toeplitz matrices, *Linear Algebra Appl.*, 343/344 (2002), 355–380.

[36] van der Mee, C.V.M., Rodriguez, G., and Seatzu, S., Spectral factorization of bi-infinite block Toeplitz matrices with applications, preprint.

[37] Nikolaičuk, A.M., and Spitkovsky, I.M., On the Riemann boundary-value problem with Hermitian matrix, *Soviet Math. Doklady*, 16 (1975), 1280–1283.

[38] Pankov, A.A. *Bounded and Almost Periodic Solutions of Nonlinear Differential Operator Equations*, Kluwer, Dordrecht/Boston/London, 1990.

[39] Perov, A.I., and Kibenko, A.V.,*A theorem on the argument of an almost periodic function of several variables*, Litovskii Matematicheskii Sbornik, 7 (1967), 505–508 (Russian).

[40] Quint, D., Rodman, L., and Spitkovsky, I.M., New cases of almost periodic factorization of triangular matrix functions, *Michigan Math. J.*, 45(1) (1998), 73–102.

[41] Rodman, L., and Spitkovsky, I.M., Almost periodic factorization and corona theorem, *Indiana Univ. Math. J.*, 47 (1998), 1243–1256.

[42] Rodman, L., Spitkovsky, I.M., and Woerdeman, H.J., Carathéodory-Toeplitz and Nehari problems for matrix-valued almost periodic functions, *Trans. Amer. Math. Soc.*, 350 (1998), 2185–2227.

[43] Rodman, L., Spitkovsky, I.M., and Woerdeman, H.J., Factorization of almost periodic matrix functions of several variables and Toeplitz operators, *Operator Theory: Advances and Applications,* 122 (2001), (H. Bart, I. Gohberg, A.C.M. Ran, eds.), 385–416.

[44] Rodman, L., Spitkovsky, I.M., and Woerdeman, H. J., Multiblock problems for almost periodic matrix functions of several variables, *New York J. Math.* 7 (2001), 117–148.

[45] Rodman, L., Spitkovsky, I.M., and Woerdeman, H.J., Fredholmness and invertibility of Toeplitz operators with matrix almost periodic symbols, *Proc. Amer. Math. Soc.*, 130 (2002), 1365–1370.

[46] Rodman, L., Spitkovsky, I.M., and Woerdeman, H.J., Contractive extension problems for matrix-valued almost periodic functions of several variables, *Journal of Operator Theory,* to appear.

[47] Rodman, L., Spitkovsky, I.M., and Woerdeman, H.J., Abstract band method via factorization, positive and band extensions of multivariable almost periodic matrix functions, and spectral estimation, *Memoirs of the AMS*, to appear.

[48] Spitkovsky, I.M., On the factorization of almost periodic functions. *Math. Notes*, 45 (1989), 482–488.

[49] Spitkovsky, I.M., and Woerdeman, H.J., The Carathéodory-Toeplitz problem for almost periodic functions, *J. Functional Analysis*, 115(2) (1993), 281–293.

Leiba Rodman, Ilya M. Spitkovsky and Hugo J. Woerdeman
Department of Mathematics
College of William and Mary
Williamsburg VA 23187-8795, USA
e-mail: `lxrodm@math.wm.edu`
e-mail: `ilya@math.wm.edu`
e-mail: `hugo@math.wm.edu`

2000 *Mathematics Subject Classification.*
42A75, 26B99, 15A54, 47A68, 47A56, 47B35.

Received: 31 May 2001.

Operator Theory:
Advances and Applications, Vol. 142, 345–356

Criterion for the Fredholmness of Singular Operators with Piecewise Continuous Coefficients in Generalized Hölder Spaces with Weight

Natasha Samko

Abstract. In a previous paper we found conditions for a singular integral operator with piecewise continuous coefficients to be Fredholm in a weighted generalized Hölder space $H_0^\omega(\Gamma, \rho)$ together with a formula for the index. The conditions were given in terms of Boyd-type indices of the space $H_0^\omega(\Gamma, \rho)$. In this paper we prove that those conditions are also necessary for a singular integral operator to be Fredholm.

1. Introduction

Let

$$N = aP_+ + bP_- \tag{1.1}$$

be the well-known singular operator with piecewise continuous coefficients, where $P_\pm = \frac{1}{2}(I \pm S)$ are the corresponding projectors generated by the singular integral operator

$$S\varphi = \frac{1}{\pi i} \int_\Gamma \frac{\varphi(\tau)\, d\tau}{\tau - t}$$

over a closed smooth curve Γ. We consider this operator in the weighted generalized Hölder space $H_0^\omega(\Gamma, \rho)$ (that is, in the weighted space of continuous functions with prescribed behavior of their continuity modulus).

In the paper [SA3] we gave conditions for the operator N to be Fredholm in the spaces $H_0^\omega(\Gamma, \rho)$, together with a formula for its index. These sufficient conditions were formulated in terms of the jumps of the coefficients and the so-called indices of the space $H_0^\omega(\Gamma, \rho)$ (similar, in a sense, to the Boyd indices).

In this paper we prove that those conditions are also necessary for the operator N to be Fredholm.

2. The results on Fredholmness

To formulate the main result on the sufficiency of the conditions obtained in [SA3], we need some definitions and notation.

1. Basic definitions

A non-negative function $\varphi(x)$ on $[0,\ell]$, $0<\ell\le\infty$, is called *almost increasing (a.i.)* or *almost decreasing (a.d.)* if

$$\varphi(x)\le C\varphi(y) \quad \text{for all} \quad x\le y \quad (x\ge y, \text{ respectively})$$

with some $C\ge 1$.

Everywhere below we use the notation

$$W=W(0,\ell)=\{\varphi(x)\in C([0,\ell]):\ \varphi(0)=0,\ \varphi(x)>0 \ \text{ for } \ x>0,\ \varphi(x) \text{ is } a.i.\}\ .$$

The generalized Hölder space $H^\omega(\Gamma)$ on the curve Γ (closed or open) is defined as the space of continuous functions $f(x)$ such that

$$\max_{\substack{t_1\in\Gamma,t_2\in\Gamma\\ |t_1-t_2|\le h}}|f(t_1)-f(t_2)|\le c\omega(h).$$

We deal with the space

$$H_0^\omega(\Gamma,\rho) \ = \ \left\{u(t):\rho(t)u(t)\in H^\omega(\Gamma)\ ,\ \ \lim_{t\to t_k}[\rho(t)u(t)]=0,\ \ k=1,\dots,n\right\} \tag{2.1}$$

where $t_k\in\Gamma$ are the points at which the non-negative weight function $\rho(t)$ vanishes. We take the weight in the form

$$\rho(t)=\prod_{k=1}^{n} r_k(|t-t_k|)\ ;\ \ t,t_k\in\Gamma,\ \ k=1,2,\dots,n\ , \tag{2.2}$$

where the factors $r_k(x)$ will be assumed to be in a certain class W_α^* introduced below. The function $\omega(h)$ is referred to as the *characteristic* of the space $H^\omega(\Gamma)$ or $H_0^\omega(\Gamma,\rho)$. We assume that

$$\omega(h)\in\Phi$$

where $\Phi=\Phi([0,\ell])$ is the class of functions known as the Bari-Stechkin class of functions $\omega(h)\in[0,\ell]$ satisfying the conditions

$$\int_0^h\frac{\omega(x)}{x}dx\le c\omega(h)$$

and

$$\int_h^\ell\frac{\omega(x)}{x^2}dx\le c\frac{\omega(h)}{h}\ .$$

Definition 2.1. The numbers

$$m_\omega=\sup_{t>1}\frac{\ln\left[\underline{\lim}_{h\to 0}\frac{\omega(th)}{\omega(h)}\right]}{\ln\ t}\ ,\quad M_\omega=\inf_{t>1}\frac{\ln\left[\overline{\lim}_{h\to 0}\frac{\omega(th)}{\omega(h)}\right]}{\ln\ t}$$

will be referred to as the *lower and upper index numbers* of a function $\omega(x)\in\Phi$. (These numbers are of the type of the Boyd indices.)

Theorem 2.2 ([SA3]). *A function $\omega(x) \in W([0,\ell])$ is in the Bari-Stechkin class Φ if and only if*

$$0 < m_\omega \leq M_\omega < 1.$$

As in [SA3], we use the following definition.

Definition 2.3. The characteristic $\omega(h) \in \Phi$ is called *equilibrated* if

$$m_\omega = M_\omega.$$

The condition $\lim_{h\to 0} \frac{\omega(th)}{\omega(h)} = t^\gamma$, $\gamma = const$, is sufficient for $\omega(x)$ to be equilibrated. For example, the functions $x^\gamma \left(\ln \frac{1}{x}\right)^\alpha$, $x^\gamma \left(\ln\, \ln \frac{1}{x}\right)^\alpha, x^{\gamma+\frac{c}{\ln x}}$, satisfy this condition. One may also take $\omega(x) = x^{\gamma(x)}$ with $\gamma(x)$ satisfying the Dini condition $|\gamma(x+h) - \gamma(x)| = o\left(|\ln\, |h|\,|^{-1}\right)$.

Definition 2.4. We say that $\varphi(x) \in V_\alpha = V_\alpha([0,\ell]), \alpha > 0$, if

1) $\varphi(x)$ is non-decreasing,
2) $x^{-\alpha}\varphi(x)$ is non-increasing,
3) in the case $\alpha > 1$, the conditions

$$x^{-1}\varphi(x) \quad \text{is } a.i. \text{ and} \quad \lim_{x\to 0+} \varphi(x) = 0$$

are also satisfied.

Definition 2.5. A function $\varphi(x)$ is said to belong to $W_\alpha = W_\alpha([0,\ell])$, $\alpha > 0$, if

1) $\varphi(x) \in W$,
2) $x^{-\alpha}\varphi(x)$ is *a.d.*,
3) in the case $\alpha > 1$, the following conditions are also satisfied:

$$x^{-1}\varphi(x) \qquad \text{is} \qquad a.i.$$

and

$$|\varphi(x) - \varphi(y)| \leq c|x-y| \left(\frac{\varphi(x)}{x} + \frac{\varphi(y)}{y}\right). \tag{2.3}$$

Remark 2.6. It may be shown that for $\varphi(x) \in V_\alpha$, condition (2.3) is automatically satisfied, so that

$$V_\alpha \subset W_\alpha \subset C([0,\ell]).$$

Definition 2.7. We say that $\varphi(x) \in W_\alpha^*$ if

$$\varphi(x) \in W_\alpha \quad \text{and} \quad x^{\epsilon-\alpha}\varphi(x) \text{ is } \quad a.i. \quad \forall\epsilon > 0\,. \tag{2.4}$$

In what follows we shall assume that the weight factors $r_k(x)$ in (2.2) belong to the class W_α^*.

We consider a curve of Lyapunov type. Because of the generalized Hölder setting of the problem, it is natural to assume that the curve Γ under the consideration is also a generalized Lyapunov curve. The curve $\Gamma = \{t = t(s), 0 \leq s \leq \ell\}$, where s is the arc length, is said to be a generalized Lyapunov curve (GLC) if

$$t'(s) \in H^\Delta([0,\ell]) \qquad \text{with some} \qquad \Delta(h) \in W,$$

where ℓ is the length of the curve. We write

$$\Gamma \in \mathrm{Lyap}^{\Delta}$$

in this case. The case when $\Delta(h)$ is a power function corresponds to a usual Lyapunov curve. Everywhere below, dealing with a GLC, we assume that

$$\Delta(h) \in W \text{ is semiadditive and } \frac{\Delta(h)}{h} \text{ is } a.d. \tag{2.5}$$

2. Formulation of the result on Fredholmess (sufficient conditions)

Let $a(t), b(t) \in PC(\Gamma)$, where Γ is a closed curve. We assume Γ to be simple, that is, to have no intersections. We suppose that $a(t)$ and $b(t)$ may have jumps at the points t_k and that

$$a(t), b(t) \in H^{\omega}(\widetilde{[t_k, t_{k+1}]}),\ k = 1, 2, \ldots, n+1\ (t_{n+1} = t_1). \tag{2.6}$$

The main result we proved in [SA3] is stated below as Theorem A. Let $\vec{\mu} = (\lambda, \alpha_1, \ldots, \alpha_n)$, where $0 < \lambda < 1$ and $\lambda < \alpha_k < \lambda + 1,\ k = 1, 2, \ldots, n$. In formula (2.7) of Theorem A we use the notion of the $\vec{\mu}$-index of a function $G(t) \in PC(\Gamma)$, which was introduced by R. Duduchava [DU] in the case of power characteristic $\omega(h) = h^{\lambda}$ and power weights $r_k(x) = x^{\alpha_k}$. We denote this $\vec{\mu}$-index as

$$\mathrm{ind}_{\vec{\mu}} G(t).$$

Theorem A. *Let a closed curve Γ satisfy assumptions (2.5) and also the assumption*

$$\Delta(h)(1 + |\ln\ h|) \leq c\omega(h)\ ,\quad 0 \leq h \leq \ell\ .$$

Let $\omega(h) \in \Phi$ be equilibrated and let $\rho(t)$ be the weight function (2.2) with $r_k(x) \in W^*_{\alpha_k}$ and

$$m_{\omega} < \alpha_k < 1 + m_{\omega},\ k = 1, 2, \ldots, n.$$

The operator N with coefficients satisfying (2.6) is Fredholm in $H_0^{\omega}(\Gamma, \rho)$ if

1) $\inf_{t \in \Gamma} |a(t)| \neq 0, \quad \inf_{t \in \Gamma} |b(t)| \neq 0$
2) $\frac{1}{2\pi} \arg \frac{G(t_k - 0)}{G(t_k + 0)} \neq m_{\omega} - \alpha_k \pmod{1}$, *$k = 1, 2, \ldots, n$, where* $G(t) = \frac{b(t)}{a(t)}$.

The index of the operator N in $H_0^{\omega}(\Gamma, \rho)$ is equal to

$$\kappa_{H_0^{\omega}(\Gamma, \rho)}(N) = \mathrm{ind}_{\vec{\mu}} \frac{a(t)}{b(t)}\ ,\quad \vec{\mu} = (m_{\omega}, \alpha_1, \ldots, \alpha_n) \tag{2.7}$$

(and the deficiency numbers of the operator N are κ and 0 if $\kappa \geq 0$ and 0 and $|\kappa|$ if $\kappa \leq 0$).

We wish to show that the conditions 1)–2) of Theorem A are also necessary for the operator N to be Fredholm in the space $H_0^{\omega}(\Gamma, \rho)$.

3. The proof of necessity

1st step. The case of continuous coefficients

We first observe that in the case of continuous coefficients it suffices to consider only the coefficient $a(t)$ due to the following well-known fact (see, for example, [SI]), in which the quasicommutativity of operators means their commutativity modulo compact terms.

Lemma 3.1. *Let X be a Banach space and let $P_\pm$ be orthogonal projections in X such that $P_+ + P_- = I$ and $P_-P_+ = P_+P_- = 0$. Let further A and B be linear operators in X which quasicommute with the projections $P_\pm$. The operator*

$$\Pi = AP_+ + BP_- \tag{3.1}$$

is Fredholm in X if and only if each of the operators $AP_+ + P_-$ and $P_+ + BP_-$ is Fredholm in X. In the case of Fredholmness, $\text{Ind}\ \Pi = \text{Ind}\ (AP_+ + P_-) + \text{Ind}(P_+ + BP_-)$.

This lemma is applicable to the case under consideration since the condition of quasicommutativity is satisfied. Indeed, in the case of Hölder continuous coefficients the commutator of the singular operator with the operator of multiplication by a Hölder function is compact in the space $H_0^\omega(\Gamma, \rho)$, which follows from our compactness result obtained in [SA1]–[SA3] and cited below as the Compactness theorem.

By Lemma 3.1, the operator $N = aP_+ + bP_-$ is Fredholm in the space $H_0^\omega(\Gamma, \rho)$ if and only if each of the operators $aP_+ + P_-$, $P_+ + bP_-$ is Fredholm in this space. Therefore, it suffices to deal only with the case when $b(t) \equiv 1$.

Let X denote any Banach space of functions defined on a curve Γ such that

i) $\overline{\varphi} \in X$ for all $\varphi \in X$,

so that the operator of complex conjugation given by

$$Q\varphi = \overline{\varphi} \tag{3.2}$$

is well defined in X.

The following lemma is very useful for necessity problems.

Lemma 3.2 ([GZ]). *If the singular integral operator $aP_+ + bP_-$ with continuous coefficients is Fredholm in X, then the operator $\overline{a}P_+ + \overline{b}P_-$ is also Fredholm in X.*

This lemma, stated in [GZ] in another setting, holds for any space X in which the operator of complex conjugation is well defined, since its proof is based on a general fact from the theory of rings: *if a and b are commuting elements of a ring K having real spectra and $a + bi$ is an invertible element, then $a - bi$ is also invertible* (this fact was proved by N. Krupnik).

Let now X denote any Banach space of functions defined on a closed curve Γ which satisfies the assumption i) and also the following conditions:

ii) the anticommutant $QS + SQ$ is compact in X;

iii) if a function $a(t)$ is a multiplier in X, then the operator of multiplication by $a(t)$ is invertible if and only if $\inf_{t\in\Gamma} |a(t| \neq 0$.

Lemma 3.3. *Let X be a Banach space of functions satisfying the conditions* i)–iii) *and let a continuous function $a(t)$ on Γ be a multiplier in X. The operator $N = aP_+ + P_-$ is Fredholm in X if and only if $a(t) \neq 0$ for all $t \in \Gamma$.*

Proof. Let the operator N be Fredholm in X. We denote by X_r the space X as a Banach space over the field of real numbers (elements of the space X_r are, generally speaking, complex valued functions, but their linear dependence or independence is considered with real coefficients). If the operator N is Fredholm in X, it is also Fredholm in X_r. In the space X_r the operator Q of complex conjugation is linear and invertible (and, consequently, Fredholm). Then the operator QNQ is also Fredholm. By assumption ii) we have $QP_+Q = P_- + T_1$ and $QP_-Q = P_+ + T_2$, where T_1 and T_2 are compact operators. Therefore,

$$QNQ = P_+ + \bar{a}P_- + T,$$

where T is compact. Consequently, besides the operator $N = aP_+ + P_-$, the operator $\overline{N} = P_+ + \bar{a}P_-$ is Fredholm as well. Then their composition

$$N\overline{N} = aP_+ + \bar{a}P_- \tag{3.3}$$

is Fredholm.

To continue the proof, we apply Lemma 3.2 to the operator (3.3) and obtain that the operator

$$\bar{a}P_+ + aP_- \tag{3.4}$$

is Fredholm. Then the composition of the operators (3.3) and (3.4) is also Fredholm. This composition is equal to

$$a\bar{a}P_+ + \bar{a}aP_- = |a(t)|^2\, I \tag{3.5}$$

where I is the identity operator. It remains to make use of condition iii) to complete the proof of the lemma. □

Corollary. *Let the characteristic $\omega(h)$ and the weight function $\rho(t)$ of the space $H_0^\omega(\Gamma, \rho)$ satisfy the assumptions of Theorem A. If the operator $N = aP_+ + bP_-$ with coefficients in $H^\omega(\Gamma)$ is Fredholm in the space $H_0^\omega(\Gamma, \rho)$, then $a(t) \neq 0$ for all $t \in \Gamma$.*

Indeed, it suffices to check the validity of conditions i)–iii). Condition i) is obvious and iii) is easily verified. Let us check ii).

Verification of condition ii) for the space $H_0^\omega(\Gamma,\rho)$.

In the following theorem we use the notion of a generalized Lyapunov curve, as introduced above.

Theorem 3.4. *Let $(Q\varphi)(t)=\overline{\varphi(t)}$. The operator $QS+SQ$ is compact in the space $H_0^\omega(\Gamma,\rho)$ with $\Gamma\in \mathrm{Lyap}^\Delta$ and $\omega\in W$ if*

$$\Delta(h)(1+\ln\,|h|)\le c\omega(h)\ . \tag{3.6}$$

Proof. It is known that the operator $QS+SQ$ has the form

$$(QS+SQ)\varphi=\int_\Gamma k(t,\tau)\varphi(\tau)\,d\tau$$

where

$$k(t,\tau)=\left[\frac{\tau'(\sigma)}{\tau-t}-\frac{\overline{\tau'(\sigma)}}{\overline{\tau}-\overline{t}}\right]\overline{\tau'(\sigma)}=2i\frac{\partial\theta}{\partial\sigma}\overline{\tau'(\sigma)}\ ,$$

with $\theta=\arg(\tau-t),\ t=t(s),\ \ \tau=t(\sigma)$; see for example, [L], p. 36. Here

$$\overline{\tau'(\sigma)}\in H^\Delta(\Gamma), \tag{3.7}$$

so that we have to study only the first factor $\frac{\partial\theta}{\partial\sigma}$. We make use of the known formula (see the same book [L], p. 37)

$$\frac{\partial\theta}{\partial\sigma}=\frac{\partial Y}{\partial\sigma}\frac{X}{X^2+Y^2}-\frac{\partial X}{\partial\sigma}\frac{Y}{X^2+Y^2}, \tag{3.8}$$

where $X=X(\sigma,s)=\frac{x(\sigma)-x(s)}{\sigma-s}$, $\quad Y=Y(\sigma,s)=\frac{y(\sigma)-y(s)}{\sigma-s}$, and x and y are the cartesian coordinates of the point $t\in\Gamma$: $t=x(s)+iy(s),\ \ \tau=x(\sigma)+iy(\sigma)$.

To prove the Hölder properties of the fractions $\frac{X}{X^2+Y^2}$ and $\frac{Y}{X^2+Y^2}$, we observe that

$$X=\int_0^1 x'[s+u(\sigma-s)]du\in H^\Delta(\Gamma\times\Gamma)$$

and similarly $Y\in H^\Delta(\Gamma\times\Gamma)$. Then $X^2+Y^2\in H^\Delta(\Gamma\times\Gamma)$. We observe that $X^2+Y^2\neq 0$ on the whole curve Γ. Indeed, for $\sigma=s$ it is obvious that

$$X^2(s,s)+Y^2(s,s)=|t'(s)|^2=1\neq 0$$

whence $X^2(\sigma,s)+Y^2(\sigma,s)\neq 0$ for σ close to s.

Let further $s\neq\sigma$. Then the equality $X^2+Y^2=0$ is only possible if $x(s)=x(\sigma)$ and $y(s)=y(\sigma)$. Since the curve has no intersections, these two equations may only be valid in the case where the curve is closed and we hit the initial and final values of the arc length, that is, $\sigma=0$ and $s=\ell$ or $\sigma=\ell$ and $s=0$. But in this case again $X^2+Y^2=1$.

Since $X^2+Y^2\neq 0$, we have $\frac{1}{X^2+Y^2}\in H^\Delta(\Gamma\times\Gamma)$ and hence also

$$\frac{X}{X^2+Y^2}\in H^\Delta(\Gamma\times\Gamma)\qquad\text{and}\qquad\frac{Y}{X^2+Y^2}\in H^\Delta(\Gamma\times\Gamma).$$

To clarify the behavior of the derivatives $\frac{\partial X}{\partial \sigma}$ and $\frac{\partial Y}{\partial \sigma}$ we make use of the known representation (see [L], p. 38)

$$\frac{\partial X}{\partial \sigma} = \frac{A(\sigma, s)}{\sigma - s} \qquad \text{with} \qquad A(s,s) = 0,$$

where

$$A(\sigma, s) = x'(\sigma) - \int_0^1 x'[s + u(\sigma - s)]du \in H^{\Delta}(\Gamma \times \Gamma),$$

and a similar equality for $\frac{\partial Y}{\partial \sigma}$.

Therefore, the kernel $k(t, \tau)$ of the integral operator $QS+SQ$ is representable as

$$k(t, \tau) = \frac{B(\sigma, s)}{s - \sigma} \tag{3.9}$$

where the function $B(\sigma, s)$ satisfies the conditions

$$B(\sigma, s) \in H^{\Delta}(\Gamma \times \Gamma) \qquad \text{and} \qquad B(s,s) = 0. \tag{3.10}$$

The compactness in $H_0^{\omega}(\Gamma, \rho)$ of integral operators with kernels of such a structure is a consequence of the following theorem.

Compactness theorem. *Let the kernel $k(t, \tau)$ have the form (3.9), (3.10) and let*

$$\Delta(h) \in W, \quad \Delta(h)(1 + |\ln h|) \le c\omega(h) \quad \textit{and} \quad \frac{\Delta(h)}{h} \quad \textit{be a.d.}$$

Then the operator

$$\int_{\Gamma} k(t, \tau)\varphi(\tau) d\,\tau$$

is compact in the space $H_0^{\omega}(\Gamma, \rho)$ if

$$m_{\omega} < \alpha_k < m_{\omega} + 1, \ k = 1, 2, \dots, n.$$

This theorem is from [SA1]–[SA3], where even in a more general form of it was stated and proved.

It remains to apply this theorem. By this theorem, our operator $QS + SQ$ is compact in $H^{\omega}(\Gamma, \rho)$ under condition (3.6) and under the assumption that the weight function satisfies the natural condition $m_{\omega} < \alpha_k < 1 + m_{\omega}$. □

2nd step. The case where the coefficients are discontinuous but it is assumed that $a(t_k \pm 0) \neq 0, b(t_k \pm 0) \neq 0$ at the discontinuity points of the coefficients

We have to show that if the operator N is Fredholm, then necessarily $a(t) \neq 0$ for all other points and that the required conditions on the jumps are satisfied.

a). Representation of the operator N as a composition of "model" operators with jumps and an operator with continuous coefficients. Let $t_k, k = 1, \dots, m,\ m \le n$, be the points of discontinuity of $a(t)$ and let $t_k, k = m+1, \dots, n$, be the points of discontinuity of $b(t)$ (some of these points may coincide).

Since we suppose that $a(t_k \pm 0) \neq 0, b(t_k \pm 0) \neq 0$, the numbers $\gamma_k(a) = \frac{1}{2\pi i} \arg \frac{a(t_k-0)}{a(t_k+0)}$ and $\gamma_k(b) = \frac{1}{2\pi i} \arg \frac{b(t_k-0)}{b(t_k+0)}$ are well defined and we may introduce the functions

$$\omega_a(t) = \prod_{k=1}^{m} (t - z_0)_k^{\gamma_k(a)} \quad \text{and} \quad \omega_b(t) = \prod_{k=m+1}^{n} (t - z_0)_k^{\gamma_k(b)}$$

so that the quotients

$$a_0(t) = \frac{a(t)}{\omega_a(t)} \quad \text{and} \quad b_0(t) = \frac{b(t)}{\omega_b(t)}$$

are continuous functions on Γ. Let

$$\omega(t) = \frac{\omega_b(t)}{\omega_a(t)} = \frac{\omega^+(t)}{\omega^-(t)}$$

where

$$\omega^+(t) = \prod_{k=m+1}^{n} (t - t_k)^{\gamma_k(b)} \prod_{k=1}^{m} (t - t_k)^{-\gamma_k(a)}$$

and

$$\omega^-(t) = \prod_{k=m+1}^{n} \left(\frac{t - t_k}{t - z_0}\right)^{\gamma_k(b)} \prod_{k=1}^{m} \left(\frac{t - t_k}{t - z_0}\right)^{-\gamma_k(a)}.$$

The operator N may be represented in the familiar form

$$N\varphi = \omega_a(t) A_0 (P_+ + \omega P_-)\varphi \tag{3.11}$$

where A_0 is the weighted operator with continuous coefficients given by

$$A_0 = \omega_+(t)[a_0(t)P_+ + b_0(t)P_-]\frac{1}{\omega_+}. \tag{3.12}$$

This representation is easily verified for "nice" functions $\varphi(t)$ and may be extended by continuity to some weighted space of the type $L_p(\Gamma, \rho_1)$ into which our space $H^\omega(\Gamma, \rho)$ is embedded.

We need the following fact from the theory of Fredholm operators.

Yood's Theorem. *If the operator $C = AB$ is a Φ_+-operator, then the operator B is also a Φ_+-operator. If $C = AB$ is a Φ_--operator, then A is a Φ_--operator.*

b). On the necessity of the conditions on the jumps for the "model" operator. First of all we prove the following lemma for the factor

$$\Psi = P_+ + \omega P_-$$

in the right-hand side of (3.11).

Lemma 3.5. *Let the characteristic $\omega(h)$ and the weight $\rho(t)$ of the space $H_0^\omega(\Gamma, \rho)$ satisfy the assumptions of Theorem A. The operator $\Psi = P_+ + \omega P_-$ is a Φ_+- or Φ_--operator in the space $H^\omega(\Gamma, \rho)$ if and only if*

$$\Re\gamma_k \neq m_\omega - \alpha_k \pmod 1 \quad \textit{for all} \quad k = 1, 2, \ldots, n. \tag{3.13}$$

Proof. By Theorem A, condition (3.13) is sufficient for the operator Ψ to be a Φ-operator. Let us prove the necessity. Suppose that $\Re\gamma_k = m_\omega - \alpha_k + r$ for some $r = 0, \pm 1, \pm 2, \ldots$ and for some k, say $k = 1$, but that the operator Ψ is a Φ_+- or Φ_--operator. Let first $\Re\gamma_k \neq m_\omega - \alpha_k \pmod 1$ for all other $k = 2, 3, \ldots, n$.

Let us consider the approximating operators

$$\Psi_{\pm\varepsilon} = P_+ + \omega_{\pm\varepsilon} P_-, \quad \varepsilon > 0,$$

where $\omega_{\pm\varepsilon} = (t - z_0)_1^{\pm\varepsilon}\omega(t)$. This new function has the new approximating numbers

$$\gamma_1^{\pm\varepsilon} = \gamma_1 \pm \varepsilon.$$

We choose ε small enough, so that

$$\Re\gamma_1 \pm \varepsilon - (m_\omega - \alpha_1) \quad \text{is not an integer.}$$

Then, by the sufficiency part of Theorem A, the operators Ψ_ε and $\Psi_{-\varepsilon}$ are Fredholm operators in the space $H_0^\omega(\Gamma, \rho)$. The calculation of the index in this situation shows that the index of the operator $P_+ + (t - z_0)^\beta P_-$ with $\Re\beta \neq m_\omega - \alpha_1 \pmod 1$ in the space $H_0^\omega(\Gamma, \rho)$ is equal to

$$\text{Ind}\,(P_+ + (t - z_0)^\beta P_-) = [\Re\beta + \alpha_1 - m_\omega]. \tag{3.14}$$

Therefore,

$$\text{Ind}\,\Psi_\varepsilon - \text{Ind}\,\Psi_{-\varepsilon} = [\Re\gamma + \varepsilon + \alpha_1 - m_\omega] - [\Re\gamma - \varepsilon + \alpha_1 - m_\omega] = 1. \tag{3.15}$$

But, on the other hand,

$$\|\Psi_{\pm\varepsilon} - \Psi\|_{H_0^\omega(\Gamma,\rho)\to H_0^\omega(\Gamma,\rho)} \leq \|\omega_\varepsilon - \omega\|_{H^\omega(\Gamma;t_1,\ldots,t_n)}$$

where $H^\omega(\Gamma; t_1, \ldots, t_n)$ is the space of piecewise generalized Hölder functions with jumps at the points $t_1, \ldots, t_n$. Then

$$\|\Psi_{\pm\varepsilon} - \Psi\|_{H_0^\omega(\Gamma,\rho)\to H_0^\omega(\Gamma,\rho)} \leq c\|(t - z_0)_1^{\pm\varepsilon} - 1\|_{H^\omega(\Gamma;t_1)} \leq c_1\varepsilon.$$

The last inequality is proved in the following way. Since $\omega(h) \geq ch$, we obtain $\|f\|_{H^\omega(\Gamma;t_1)} \leq c\|f\|_{C^1(\Gamma;t_1)}$ where $C^1(\Gamma; t_1)$ is the space of functions differentiable everywhere on Γ except for the point t_1, at which they may have a jump. Obviously,

$$\|(t - z_0)_1^\varepsilon - 1\|_{C^1(\Gamma;t_1)} = \varepsilon\|(t - z_0)_1^{\varepsilon-1}\|_{C(\Gamma;t_1)} \leq \mathit{const}\ \varepsilon.$$

Therefore, by the stability theorem for $\Phi_\pm$-operators,

$$|\text{Ind}\,\Psi_{\pm\varepsilon} - \text{Ind}\,\Psi| \to 0$$

as $\varepsilon \to 0$, which contradicts equality (3.15).

This proves the lemma for the case $k = 1$. If it happens that condition (3.13) is violated for several $k = n_1, \ldots, n_m$, the arguments are similar: the operators $\Psi_{\pm\varepsilon}$ must then be introduced with the functions

$$\omega_{\pm\varepsilon}(t) = \prod_{i=1}^{m}(t - z_0)_{n_i}^{\pm\varepsilon}\omega(t).$$

At this point the proof of the lemma is complete. □

c). Proof of the necessity for the operator N itself. Now we return to the operator N. Suppose that the operator N is a Fredholm operator in the space $H_0^\omega(\Gamma,\rho)$. Then from (3.11) and Yood's theorem it follows that the operator $P_+ + \omega P_-$ is a Φ_+-operator. Thus, by Lemma 3.5, conditions (3.13) are satisfied. Consequently, by our Theorem A, this operator is a Fredholm operator.

We shall use the following well-known fact.

Lemma 3.6. *Let A and B be linear operators in a Banach space X and let $C = AB$. If C and one of the operators A and B are Fredholm operators, then the remaining operator is Fredholm as well.*

By this lemma, we obtain from (3.11) that the operator $\omega_a(t)A_0$ is Fredholm. The operator of multiplication by the piecewise continuous function $\omega_a(t)$ is Fredholm, because we already know that conditions (3.13) are satisfied. Then by the same Lemma 3.6, the operator A_0 is Fredholm. Thus, by the corollary to Lemma 3.3, $a_0(t) \neq 0$ and $b_0(t) \neq 0$, which implies that $a(t) \neq 0$ and $b(t) \neq 0$.

3rd step. It remains to get rid of the assumptions $a(t_k \pm 0) \neq 0, b(t_k \pm 0) \neq 0$. Finally, it remains to prove that the temporarily assumed conditions $a(t_k \pm 0) \neq 0, b(t_k \pm 0) \neq 0$ are also necessary for our operator N to be Fredholm.

Suppose that some of the numbers $a(t_k \pm 0)$ and $b(t_k \pm 0)$ are equal to zero and the operator A is Fredholm in $H_0^\omega(\Gamma,\rho)$. Then there exists a complex number ε with an arbitrarily small modulus and a point t_0 close to t_k such that $a(t_k \pm 0) + \varepsilon \neq 0$ and $b(t_k \pm 0) + \varepsilon \neq 0$, but $a(t_0) + \varepsilon = 0$ (or $b(t_0) + \varepsilon = 0$). Let

$$A_\varepsilon = (a+\varepsilon)P_+ + (b+\varepsilon)P_- \quad .$$

Evidently,

$$\|A_\varepsilon - A\| = \|\varepsilon I\| = \varepsilon.$$

Therefore, by the stability theorem for Fredholm operators, we obtain that the operator A_ε is Fredholm for sufficiently small ε. This contradicts the preceding part.

Acknowledgements

This paper appeared as a result of Prof. F.-O. Speck's question after the presentation of Theorem A at the seminar of "Centro de Matemática Aplicada" of Instituto Superior Técnico, Lisbon, on October 23, 1998, whether it is possible to prove that the conditions of Theorem A are also necessary. The author expresses her thanks to Prof. Speck for that question.

References

[DU] Duduchava, R., Singular integral equations in weighted Hölder spaces (Russian), *Matem. Issledov* **5** (1970), 104–124.

[GZ] Gohberg, I., Zambitskii, M.K., To the theory of linear operators in spaces with two norms (Russian), *Ukrain. Mat. Jurnal* **81** (1966), 11–23.

[L] Litvinchuk, G.S., *Boundary Value Problems and Singular Integral Equations with Shift (Russian)*, Nauka, Moscow 1977.

[SA1] Samko, N., The singular operator and operators with a weak singularity in weighted generalized Hölder spaces (Russian), *Deposited in VINITI, 22.05.90,* **2798-901** (1990), 1–38.

[SA2] Samko, N., *Singular Integral Operators with Discontinuous Coefficients in Generalized Hölder Spaces,* PhD Thesis (Russian), Rostov University, 1991.

[SA3] Samko, N.G., Singular integral operators in weighted spaces with generalized Hölder condition, *Proc. A. Razmadze Math. Inst.* **120** (1999), 107–134.

[SI] Simonenko, I.B., Some general questions in the theory of the Riemann boundary value problem (Russian), *Izv. Akad. Nauk SSSR, ser. Mat.* **32** (1968), 1138–1146; transl. in Math. USSR Izvestija, **2** (1968), no 5, 1091–1099.

Natasha Samko
Department of Mathematics and Mechanics
Rostov State University
Rostov-na-Donu, Zorge 5, 344104
Russia
e-mail: `ssamko@ualg.pt`

2000 *Mathematics Subject Classification.*
Primary 47G10, Secondary 47B38 .

Received: 10 February 2001.

Operator Theory:
Advances and Applications, Vol. 142, 357–368

On Inversion of Fractional Spherical Potentials by Spherical Hypersingular Operators

Stefan Samko

Abstract. A new proof of the inversion formula for spherical Riesz type fractional potentials in the case $0 < \Re\alpha < 2$ is presented and a constructive reduction of the case $\Re\alpha > 2$ to the case $0 < \Re\alpha < 2$ is given.

1. Introduction

Let $S^{n-1} = \{x \in R^n : |x| = 1\}$ be the unit sphere in R^n. The fractional spherical potentials, named also Riesz spherical potentials, are known in the form

$$(K^\alpha\varphi)(x) = \frac{1}{\gamma_{n-1}(\alpha)} \int_{S^{n-1}} \frac{\varphi(\sigma)d\sigma}{|x-\sigma|^{n-1-\alpha}}, \quad \alpha > 0, \ \alpha \neq n-1, n+1, n+3, \dots \tag{1.1}$$

where $|x| = 1$ and the normalizing constant $\gamma_{n-1}(\lambda)$ is defined by the formula

$$\gamma_{n-1}(\alpha) = \frac{2^\alpha \pi^{\frac{n-1}{2}} \Gamma\left(\frac{\alpha}{2}\right)}{\Gamma\left(\frac{n-1-\alpha}{2}\right)} . \tag{1.2}$$

The inversion of this potential type operator was given more than two decades ago in [PS1], [PS2] in the case $0 < \alpha < 2$ in terms of spherical hypersingular integrals. We here take up the story again and present another proof, which is more transparent and effective and also covers the case $0 < \Re\alpha < 2$. As for the case $\Re\alpha > 2$, $\alpha - n \neq 2k + 1$, $k = 0, 1, 2, \dots$, we show that this case may be constructively reduced to the case $0 < \Re\alpha < 2$.

The main statements are given in Theorems 3.5 and 4.1 below.

2. Preliminaries

a) Spherical multiplier of the Riesz potential operator

We use the basics of the theory of spherical harmonics and Fourier-Laplace expansions into series of spherical harmonics. We refer, for instance, to the books [SW], [Mu], or [Sa], Section 2 of Chapter 1. In particular, we use the notion of spherical Laplace-Fourier multipliers.

It is known that in the case $\alpha \neq n-1+2k, k=0,1,2,\ldots$, the Laplace-Fourier multiplier of the operator K^α is equal to

$$\{k_m^\alpha\}_{m=0}^\infty = \left\{ \frac{\Gamma\left(m+\frac{n-1-\alpha}{2}\right)}{\Gamma\left(m+\frac{n-1+\alpha}{2}\right)} \right\}_{m=0}^\infty . \tag{2.1}$$

b) Spherical hypersingular integrals of order $0 < \Re\alpha < 2$

The spherical hypersingular integral is introduced ([PS]) as

$$(D^\alpha f)(x) = \frac{1}{\gamma_{n-1}(-\alpha)} \int_{S^{n-1}} \frac{f(\sigma)-f(x)}{|x-\sigma|^{n-1-\alpha}}\, d\sigma, \qquad x \in S^{n-1}, \tag{2.2}$$

which converges absolutely for $0 < \Re\alpha < 1$ in the case of "nice" functions $f(\sigma)$. It converges also for $1 \leq \Re\alpha < 2$ in the case of sufficiently "nice" functions $f(\sigma)$ provided the integral is interpreted as the limit of *truncated* hypersingular operators

$$(D_\varepsilon^\alpha f)(x) = \frac{1}{\gamma_{n-1}(-\alpha)} \int_{S_\varepsilon^{n-1}(x)} \frac{f(\sigma)-f(x)}{|x-\sigma|^{n-1-\alpha}}\, d\sigma, \qquad x \in S^{n-1}, \tag{2.3}$$

where $S_\varepsilon^{n-1}(x) = \{\sigma \in S^{n-1} : |\sigma - x| > \epsilon\}$. For completeness we prove this convergence in the lemma below.

The integral (2.3) is known ([PS1]) to be representable in the form

$$(D_\varepsilon^\alpha f)(x) = \frac{|S^{n-2}|}{\gamma_{n-1}(-\alpha)2^{(n+\alpha-1)/2}} \int_{-1}^{1-\frac{\varepsilon}{2}} \frac{M_f(x,t)-f(x)}{(1-t)^{\frac{n-1+\alpha}{2}}}(1-t^2)^{\frac{n-3}{2}}\, dt \tag{2.4}$$

where $M_f(x,t)$ are the means

$$M_f(x,t) = \frac{1}{|S^{n-2}|(1-t^2)^{\frac{n-2}{2}}} \int_{\substack{|\sigma|=1 \\ \sigma\cdot x = t}} f(\sigma)\, ds_\sigma. \tag{2.5}$$

The representation (2.4) follows from (2.3) by the formula

$$\int_{S^{n-1}} f(\sigma)\varphi(x\cdot\sigma)\, d\sigma = |S^{n-2}| \int_{-1}^{1} M_f(x,t)(1-t^2)^{\frac{n-3}{2}}\varphi(t)\, dt, \tag{2.6}$$

which can be named a Cavalieri type principle; for its proof see [Sa2] or [Sa], Lemma 4.13.

When writing that $f(\sigma) \in C^N(S^{n-1})$, we mean that $f\left(\frac{x}{|x|}\right) \in C^N(R^n\backslash\{0\})$.

Lemma 2.1. *Let $f(\sigma) \in C^{2N}(S^{n-1})$, $N > \frac{3}{4}n$. Then the limit*

$$\lim_{\varepsilon\to 0} D_\varepsilon^\alpha f(x)$$

exists in $C(S^{n-1})$ for every α with $0 < \Re\alpha < 2$.

Proof. We expand $f(x)$ in a Fourier-Laplace series of spherical harmonics, which converges by the known properties of series of spherical harmonics. We have

$$(D_\varepsilon^\alpha f)(x) = \frac{1}{\gamma_{n-1}(-\alpha)} \sum_{m\mu}^{\infty} \int_{2(1-x\cdot\sigma)>\varepsilon^2} \frac{Y_{m\mu}(\sigma)-Y_{m\mu}(x)}{[2(1-x\cdot\sigma)]^{\frac{n-1+\alpha}{2}}}\, d\sigma\, . \tag{2.7}$$

We make use of the Funk-Hekke formula

$$\int_{S^{n-1}} f(x\cdot\sigma)Y_m(\sigma)\,d\sigma = \lambda Y_m(x), \tag{2.8}$$

where

$$\lambda = |S^{n-2}|\int_{-1}^{1} f(t)P_m(t)(1-t^2)^{\frac{n-3}{2}}\,dt \tag{2.9}$$

and $P_m(t)$ is the m-th Legendre polynomial. As is known, $P_m(t)$ can be expressed in terms of the Gegenbauer polynomials via

$$P_m(t) = \binom{m+n-3}{m}^{-1} C^{\frac{n-2}{2}}(t), \qquad n\geq 3 \tag{2.10}$$

(for more details on the Funk-Hekke formula, see, for example, [Sa], Section 2 of Chapter 1). Formula (2.8) is valid for any function $f(t)$ such that $f(t)(1-t^2)^{\frac{n-3}{2}}$ is in $L_1([-1,1])$. By means of this formula we obtain

$$(D_\varepsilon^\alpha f)(x) = \frac{1}{2^{\frac{n-1+\alpha}{2}}\gamma_{n-1}(-\alpha)}\sum_{m\mu}^{\infty} f_{m\mu}b_{m,\varepsilon}(\alpha)Y_{m\mu}(x) \tag{2.11}$$

where

$$b_{m,\varepsilon}(\alpha) = \int_{-1}^{1-\frac{\varepsilon^2}{2}} (1+t)^{\frac{n-3}{2}}(1-t)^{-1-\frac{\alpha}{2}}[P_m(t)-1]\,dt. \tag{2.12}$$

To estimate this coefficient, we observe that

$$|P_m'(t)| \leq \frac{m(m+n-2)}{n-1}\,, \tag{2.13}$$

which follows from (2.10) and the properties

$$\frac{d}{dt}C_m^\lambda(t) = 2\lambda C_{m-1}^{\lambda+1}(t), \qquad |C_m^\lambda(t)| \leq \frac{\Gamma(m+2\lambda)}{m!\Gamma(2\lambda)}$$

of Gegenbauer polynomials. From (2.13) the estimate

$$|P_m(t)-1| \leq \frac{m(m+n-2)}{n-1}(1-t)$$

follows. Hence

$$|b_{m,\epsilon}(\alpha)| \leq cm^2$$

with c not depending on m and ε. Therefore, to demonstrate the uniform convergence of the series (2.11), it suffices to prove the uniform convergence of the series $\sum_{m=0}^{\infty} m^2|f_{m\mu}Y_{m\mu}(x)|$. But this follows from known theorems on the convergence of Fourier-Laplace series of smooth functions (see [Ne], p.232) with the inequality $|Y_{m\mu}(x)| \leq cm^{\frac{n-2}{2}}$ taken into account. □

c) The multiplier of the spherical hypersingular operator

From the expansion (2.11) we obtain that the operator (2.2) has the spherical Fourier multiplier λ_m (that is, $D^\alpha Y_m = \lambda_m Y_m$) given by

$$\lambda_m = \frac{|S^{n-2}|}{2^{\frac{n-1+\alpha}{2}} \gamma_{n-1}(\alpha)} b_{m,0}(\alpha) \, . \tag{2.14}$$

To calculate $b_{m,0}(\alpha)$, we note that $b_{m,0}(\alpha)$ is an analytic function of the parameter α in the half-plane $\Re\alpha < 2$. Taking $\Re\alpha < 0$, we easily calculate $b_{m,0}(\alpha)$ by means of formula 7.311.3 of [GR], which yields the formula

$$\lambda_m = \frac{\Gamma\left(m + \frac{n-1+\alpha}{2}\right)}{\Gamma\left(m + \frac{n-1-\alpha}{2}\right)} - \frac{\Gamma\left(\frac{n-1+\alpha}{2}\right)}{\Gamma\left(\frac{n-1-\alpha}{2}\right)}. \tag{2.15}$$

As a consequence of (2.1) and (2.15), the operator $(K^\alpha)^{-1}$ inverse to the spherical potential operator K^α is expected to be

$$(K^\alpha)^{-1} = c_\alpha I + D^\alpha, \qquad c_\alpha = \frac{\Gamma\left(\frac{n-1+\alpha}{2}\right)}{\Gamma\left(\frac{n-1-\alpha}{2}\right)} \, , \tag{2.16}$$

where I is the identity operator.

d) Stereographic projection

The change of variables in R^n defined by

$$\xi = s(x) = \{s_1(x), \cdots , s_n(x)\} \tag{2.17}$$

with $s_k(x) = \frac{2x_k}{1+|x|^2}$, $k = 1, \cdots , n-1$, and $s_n(x) = \frac{|x|^2-1}{|x|^2+1}$, $x = (x_1, \cdots , x_n) \in R^n$, $|x| = (x_1^2 + \cdots + x_n^2)^{\frac{1}{2}}$, generates the mapping known as the *stereographic projection.* It maps the subspace $\dot{R}^{n-1}$ one-to-one onto the unit sphere $S^{n-1} \subset R^n$, $\dot{R}^{n-1}$ being the completion of R^{n-1} by a single infinite point. It is known, see [Mi], p. 35–36, that the following relations hold:

$$|\xi - \sigma| = \frac{2|x-y|}{(1+|x|^2)^{\frac{1}{2}}(1+|y|^2)^{\frac{1}{2}}}, \qquad \xi = s(x), \quad \sigma = s(y), \tag{2.18}$$

$$d\sigma = \frac{2^{n-1} dy}{(1+|y|^2)^{n-1}}, \qquad \xi = s(x), \quad \sigma = s(y). \tag{2.19}$$

This immediately yields the relation

$$\int_{S^{n-1}} \frac{f(\sigma)\, d\sigma}{|\xi - \sigma|^{n-1-\alpha}} = 2^\alpha (1+|x|^2)^{\frac{n-1-\alpha}{2}} \int_{R^{n-1}} \frac{f[s(y)]\, dy}{|x-y|^{n-1-\alpha}(1+|y|^2)^{\frac{n-1-\alpha}{2}}} \, , \tag{2.20}$$

$s(y)$ being the stereographic projection of R^{n-1} onto S^{n-1}.

3. Justification of the inversion in $L_p(S^{n-1})$ and identity approximation on the sphere

We wish to show that the operator $c_\alpha + D^\alpha$ interpreted as $c_\alpha I + \lim_{\substack{\epsilon\to 0\\(L_p)}} D_\epsilon^\alpha$ is a left inverse to the operator K^α in the spaces $L_p(S^{n-1})$:

$$(c_\alpha I + \lim_{\substack{\epsilon\to 0\\(L_p)}} D_\epsilon^\alpha)K^\alpha\varphi = \varphi, \qquad \varphi \in L_p(S^{n-1}) . \tag{3.1}$$

Direct calculations yield

$$(D_\epsilon^\alpha K^\alpha\varphi)(x) = \int_{S^{n-1}} \mathcal{L}_\epsilon(x\cdot\sigma)\varphi(\sigma)d\sigma \tag{3.2}$$

where the kernel $\mathcal{L}_\epsilon(x\cdot\sigma)$ is defined by

$$\mathcal{L}_\varepsilon(x\cdot\sigma) = \frac{1}{\gamma}\int_{S_\varepsilon^{n-1}(x)} \frac{|\tau-\sigma|^{1-n+\alpha} - |x-\sigma|^{1-n+\alpha}}{|\tau-x|^{n-1+\alpha}}\, d\tau \tag{3.3}$$

and $\gamma = \gamma_{n-1}(\alpha)\gamma_{n-1}(-\alpha) = -d_{n-1,1}(\alpha)\gamma_{n-1}(\alpha)$. We know beforehand that the kernel in (3.3) may depend only on the inner product $x\cdot\sigma$ because $D_\epsilon^\alpha K^\alpha$ is a composition of two spherical convolution operators.

As we shall see, the kernel $\mathcal{L}_\epsilon(x\cdot\sigma)$ involves an identity approximation kernel, so we dwell on this notion. We say that the spherical convolution operator

$$(A_\epsilon\varphi)(x) = \int_{S^{n-1}} \mathcal{A}_\epsilon(x\cdot\sigma)\varphi(\sigma)d\sigma \tag{3.4}$$

is a spherical identity approximation operator in $L_p(S^{n-1})$ if

$$\lim_{\substack{\epsilon\to 0\\(L_p)}} \|A_\epsilon\varphi - \varphi\|_{L_p(S^{n-1})} = 0 .$$

The following theorem ([BBP], p.210) provides sufficient conditions for a kernel $\mathcal{A}_\epsilon(x\cdot\sigma)$ to be an identity approximation kernel.

Theorem 3.1. *Let $\mathcal{A}_\epsilon(t)$ satisfy the conditions*

i) $\lim_{\epsilon\to 0}\int_{S^{n-1}} \mathcal{A}_\epsilon(x\cdot\sigma)d\sigma = 1$;

ii) $\int_{S^{n-1}} |\mathcal{A}_\epsilon(x\cdot\sigma)|d\sigma \le M < \infty$, $0<\epsilon<\epsilon_0$, *with M not depending on ϵ;*

iii) $\lim_{\epsilon\to 0}\int_{x\cdot\sigma<1-t_0} |\mathcal{A}_\epsilon(x\cdot\sigma)|d\sigma = 0$ *for each $t_0\in(0,2)$.*

Then the operator (3.4) is an identity approximation in $L_p(S^{n-1})$, $1\le p<\infty$ (and in $C(S^{n-1})$ as well).

The main job to be done now is to single out the identity approximation term in the kernel (3.3), which will be accomplished in Lemma 3.3 below. Lemma 3.2 is crucial for this purpose.

Let $u \in R^{n-1}$ be the vector connected with $x, \sigma \in S^{n-1}$ by the relation

$$s(u) \;=\; \omega(\sigma), \tag{3.5}$$

where s is the stereographical projection (2.17) and $\omega(\sigma) = \omega_x(\sigma)$ is any rotation of the sphere such that

$$\omega(x) = -e_n = (0, 0, \ldots, 0, -1). \tag{3.6}$$

We also use the notation

$$\langle u \rangle = \sqrt{1 + |u|^2}\ .$$

Lemma 3.2. *Let $x, \sigma \in S^{n-1}$. Then*

$$\int_{S^{n-1}} f(|x-\tau|, |\sigma-\tau|)\, d\tau = 2^{n-1} \int_{R^{n-1}} f\left(\frac{2|v|}{\langle v \rangle}, \frac{2|v|}{\langle v \rangle}\right) \frac{dv}{\langle v \rangle^{2(n-1)}} \tag{3.7}$$

under the assumption that the integrals converge. Here u is the vector defined in (3.5), so that

$$|u| = \frac{|x-\sigma|}{|x+\sigma|},\ \langle u \rangle = \frac{2}{|x+\sigma|},\ \frac{\langle u \rangle}{|u|} = \frac{2}{|x-\sigma|}\ . \tag{3.8}$$

Proof. Making the rotation change of variables $\eta = \omega(\tau)$ with the rotation (3.6), we transform the left-hand side of (3.7) to

$$\int_{S^{n-1}} f(|\eta + e_n|, |\xi - \eta|) d\eta.$$

After that, we may apply the stereographical projection (2.17) by putting

$$\eta = s(v), \quad v \in R^{n-1},$$

(so that $-\eta_n = s(0)$). Making use of the relations (2.18) and (2.19), we arrive at formula (3.7).

To check the relations (3.8), we denote $\xi = \omega(\sigma)$ and have

$$x \cdot \sigma = \omega^{-1}(e_n) \cdot \omega^{-1}(\xi) = e_n \cdot \xi = -\xi_n.$$

Since (3.5) implies $s_n(u) = \xi_n$, we obtain $\xi_n = \frac{|u|^2-1}{|u|^2+1}$. Therefore,

$$|u|^2 = \frac{1+\xi_n}{1-\xi_n} = \frac{1 - x\cdot\sigma}{1 + x\cdot\sigma} = \frac{|x-\sigma|^2}{|x+\sigma|^2}, \tag{3.9}$$

from which the first of the relations (3.8) follows, the two others being its consequences. □

Remark. The representation (3.7) may be rewritten in the form

$$\int_{S^{n-1}} f(|x-\tau|, |\sigma-\tau|)\, d\tau = 2^{n-1} \int_{R^{n-1}} f\left(\frac{2|y|}{\langle y \rangle}, \frac{2|re_1 - y|}{\sqrt{1+r^2}\langle y \rangle}\right) \frac{dy}{\langle y \rangle^{2(n-1)}} \tag{3.10}$$

where $r = |u| = \frac{|x-\sigma|}{|x+\sigma|} = \frac{1-x\cdot\sigma}{1-x\cdot\sigma}$, which obviously reveals the dependence of this integral on the inner product $x \cdot \sigma$ only.

To get (3.10) from (3.7), it suffices to make another change of variables, namely, the rotation $v = rot\ y$ in R^{n-1} which rotates e_1 to $\frac{u}{|u|}$: $rot\ e_1 = \frac{u}{|u|}$. Then $|u-v| = ||u|e_1 - y|$, which yields (3.10).

In the following lemma on the representation of the kernel $\mathcal{L}_\epsilon(x\cdot\sigma)$ introduced in (3.2), we use the kernel

$$\mathcal{K}_{1,\alpha}(r) = \frac{1}{\gamma r^{n-1}} \int_{|y|\langle r} \left(|y-e_1|^{\alpha-n+1} - |y|^{\alpha-n+1}\right) dy \tag{3.11}$$

with the integration over the ball in R^{n-1}. We remark that this kernel is a direct analog of a similar kernel familiar in the theory of spatial hypersingular integrals (see [Sa1]; [SKM], formula (26.30); [Sa], formula (3.46)).

Lemma 3.3. *The kernel $\mathcal{L}_\epsilon(x\cdot\sigma)$ may be represented as*

$$\mathcal{L}_\epsilon(x\cdot\sigma) \ = \ \frac{c(\epsilon)}{|x-\sigma|^{n-1-\alpha}} \ + \ \mathcal{K}_\epsilon(x\cdot\sigma), \tag{3.12}$$

where

$$c(\varepsilon) = \frac{|S^{n-2}|}{\gamma 2^\alpha} \int_\delta^\infty \frac{1-(1+r^2)^{\frac{\alpha+1-n}{2}}}{r^{1+\alpha}}\, dr, \qquad \delta = \frac{\varepsilon}{\sqrt{4-\varepsilon^2}}, \tag{3.13}$$

and

$$\mathcal{K}_\varepsilon(x\cdot\sigma) = \frac{2^{-\alpha}}{|x+\sigma|^{n-1-\alpha}}\ \frac{1}{\delta^{n-1}}\mathcal{K}_{1,\alpha}\left(\frac{|x-\sigma|}{\delta|x+\sigma|}\right) = \frac{2^{1-n}\langle u\rangle^{n-1-\alpha}}{\delta^{n-1}}\mathcal{K}_{1,\alpha}\left(\frac{|u|}{\delta}\right), \tag{3.14}$$

u being the vector defined in (3.5).

Proof. Applying formula (3.7) to the spherical integral (3.3), we arrive at the representation

$$\mathcal{L}_\varepsilon(x\cdot\sigma) = \frac{2^{1-n}\langle u\rangle^{n-1-\alpha}}{\gamma|u|^{n-1-\alpha}} \int_{R_\delta^{n-1}} \frac{|u|^{n-1-\alpha} - \left(\frac{|u-v|}{\langle v\rangle}\right)^{n-1-\alpha}}{\langle v\rangle^{n-1+\alpha}|u-v|^{n-1-\alpha}}\, dv$$

where $R_\delta^{n-1} = \{v \in R^{n-1} : |v| > \delta\}$. Hence

$$\mathcal{L}_\varepsilon(x\cdot\sigma) = \frac{2^{1-n}}{\gamma}\left(\frac{\langle u\rangle}{|u|}\right)^{n-1-\alpha}\left(\int_{|v|>\delta} \frac{|u|^{n-1-\alpha} - |u-v|^{n-1-\alpha}}{|v|^{n-1+\alpha}|u-v|^{n-1-\alpha}}\, dv\right. \tag{3.15}$$

$$\left. + \int_{|v|>\delta} \frac{1-\langle v\rangle^{\alpha-n+1}}{|v|^{n-1+\alpha}}\, dv\right) =: \frac{\langle u\rangle}{|u|^{n-1-\alpha}}\left(J_\delta(u) + c_\varepsilon\right)$$

with

$$c_\epsilon \ = \ 2^{1-n} \int_{|v|>\delta} \frac{\langle v\rangle^{n-1-\alpha} - 1}{|v|^{n-1-\alpha}\langle v\rangle^{n-1-\alpha}}\, dv.$$

Passing to polar coordinates, we transform c_ϵ to $c_\epsilon = 2^{\alpha-(n-1)}c(\epsilon)$ with $c(\epsilon)$ given by (3.13). It remains to transform the term J_δ in (3.15). To this end, we make the same rotation change of variables as in the proof of the remark after Lemma 3.2 and get

$$J_\delta(u) = \frac{2^{1-n}}{\gamma}|u|^{n-1-\alpha}\int_{|y|>\frac{\delta}{|u|}} \frac{|y-e_1|^{\alpha-n+1}-1}{|y|^{n-1+\alpha}}\,dy.$$

Making the inversion change $y = \frac{t}{|t|^2}$, $dy = \frac{dt}{|t|^{2(n-1)}}$, and using the relation $\left|\frac{y}{|y|^2}-e_1\right| = \frac{|y-e_1|}{|y|}$, we obtain

$$J_\delta(u) = \frac{2^{1-n}}{\gamma|u|^\alpha}\int_{|y|<\frac{|u|}{\delta}} \left(|y-e_1|^{\alpha-n+1}-|y|^{\alpha-n+1}\right)\,dy,$$

which transforms (3.15) into (3.12)–(3.14). □

Lemma 3.4. *Let* $0 < \Re\alpha < 2$. *The kernel* $\mathcal{K}_\epsilon(x\cdot\sigma)$ *satisfies conditions* i)-iii) *of Theorem 3.1 and is therefore an identity approximation kernel.*

Proof. To check condition ii) of Theorem 3.1, we have, according to (3.14), to estimate the integral

$$\int_{S^{n-1}} |\mathcal{K}_\varepsilon(x\cdot\sigma)|\,d\sigma = \frac{2^{1-n}}{\delta^{n-1}}\int_{S^{n-1}} \langle u\rangle^{n-1-\alpha}\left|\mathcal{K}_\varepsilon\left(\frac{|u|}{\delta}\right)\right|\,d\sigma$$

where $\delta = \frac{\varepsilon}{\sqrt{4-\varepsilon^2}}$. We transform this integral to one over R^{n-1} by means of the change of variables (3.5). Using relation (2.19) for the stereographical projection, we arrive at

$$\int_{S^{n-1}}|\mathcal{K}_\varepsilon(x\cdot\sigma)|\,d\sigma = \frac{1}{\delta^{n-1}}\int_{R^{n-1}}\left|\mathcal{K}_{1,\alpha}\left(\frac{|x|}{\delta}\right)\right|\frac{du}{\langle u\rangle^{n-1+\alpha}} \le \int_{R^{n-1}}|\mathcal{K}_{1,\alpha}(|u|)|\,du < \infty$$

by the known fact that the kernel $\mathcal{K}_{1,\alpha}(|u|)$ is integrable as shown in [Sa1], Theorem 1 or [SKM], Lemma 26.4.

To check i), we proceed similarly:

$$\lim_{\varepsilon\to 0}\int_{S^{n-1}} \mathcal{K}_\varepsilon(x\cdot\sigma)\,d\sigma = \lim_{\delta\to 0}\int_{R^{n-1}} |\mathcal{K}_{1,\alpha}(|u|)\,\frac{du}{(1+\delta^2|u|^2)^{\frac{n-1+\alpha}{2}}} = 1$$

because $\mathcal{K}_{1,\alpha}(|u|)|$ is an averaging kernel; see [Sa1], formula (2.8) or [SKM], formula (26.42). Finally, condition iii) is verified in a similar fashion:

$$\lim_{\varepsilon\to 0}\int_{x\cdot\sigma<1-t_0} |\mathcal{K}_\varepsilon(x\cdot\sigma)|\,d\sigma = \lim_{\delta\to 0}\int\limits_{|u|>\frac{1}{\delta}\sqrt{\frac{t_0}{2-t_0}}} |\mathcal{K}_{1,\alpha}(|u|)|\,\frac{du}{(1+\delta^2|u|^2)^{\frac{n-1+\alpha}{2}}} = 0.$$

□

Theorem 3.5. *Let* $0 < \Re\alpha < 2$. *The operator*

$$T^\alpha = c_\alpha I + D^\alpha = c_\alpha I + \lim_{\substack{\epsilon\to 0\\(L_p)}} D^\alpha_\epsilon \tag{3.16}$$

with the constant c_α *defined by (2.16) is a left inverse to* $\mathcal{K}^\alpha$ *in* $L_p(S^{n-1})$, $1 \le p < \infty$ *and in* $C(S^{n-1})$.

Proof. From (3.2) and (3.12) we have

$$(T^\alpha K^\alpha \varphi)(x) = c_\alpha (K^\alpha \varphi)(x) + \lim_{\substack{\varepsilon\to 0\\(L_p)}} \int_{S^{n-1}} \left[\frac{c(\varepsilon)}{|x-\sigma|^{n-1-\alpha}} + K_\varepsilon(x\cdot\sigma)\right] \varphi(\sigma)\, d\sigma,$$

or

$$(T^\alpha K^\alpha \varphi)(x) = \left[c_\alpha + \gamma_{n-1}(\alpha) \lim_{\varepsilon\to 0} c(\varepsilon)\right] (K^\alpha\varphi)(x) \; + \; \lim_{\substack{\varepsilon\to 0\\(L_p)}} \int_{S^{n-1}} K_\varepsilon(x\cdot\sigma)\varphi(\sigma)\, d\sigma\,. \tag{3.17}$$

Evidently,

$$\begin{aligned}\gamma_{n-1}(\alpha)\lim_{\varepsilon\to 0} c(\varepsilon) &= \frac{2^{-\alpha}|S^{n-2}|}{\gamma_{n-1}(\alpha)} \int_0^\infty \frac{1-(1+r^2)^{\frac{\alpha+1-n}{2}}}{r^{1+\alpha}}\, dr \\ &= \frac{\Gamma\left(\frac{n-1+\alpha}{2}\right)}{\Gamma\left(\frac{n-1}{2}\right)\Gamma\left(-\frac{\alpha}{2}\right)} \int_0^\infty \frac{1-(1+t)^{\frac{\alpha+1-n}{2}}}{t^{1+\frac{\alpha}{2}}}\, dt\,.\end{aligned}$$

We make use of the formula

$$\int_0^\infty \frac{1-(1+t)^{-\mu}}{t^{1+\beta}} dt = \frac{\Gamma(\mu+\beta)\Gamma(1-\beta)}{\beta\Gamma(\mu)}\,, \quad 0 < \Re\beta < 1,\ \Re(\mu+\beta) > 0,$$

which is easily obtained by integration by parts, and obtain

$$\gamma_{n-1}(\alpha)\lim_{\epsilon\to 0} c_\epsilon = \frac{2\Gamma(1-\frac{\alpha}{2})\Gamma(\frac{n-1+\alpha}{2})}{\alpha\Gamma(-\frac{\alpha}{2})\Gamma(\frac{n-1-\alpha}{2})} = -c_\alpha\,.$$

Therefore, by (3.17),

$$(T^\alpha K^\alpha \varphi)(x) = \lim_{\substack{\epsilon\to 0\\(L_p)}} \int_{S^{n-1}} \mathcal{K}_\epsilon(x\cdot\sigma)\varphi(\sigma) d\sigma.$$

It remains to make use of Lemma 3.4. □

Remark. Since the spherical potential operator K^α can be transformed to a similar operator over R^{n-1}, see (2.20), we could try to use the known inversion results for the spatial case (see [Sa1], Theorem 2, or [SKM], Theorem 26.3). However, this way proves to be inappropriate because the natural truncation of the hypersingular integral in R^{n-1}, by the ball of the radius ϵ, is transformed by the stereographic projection, used in (2.20), into an artificial truncation on the sphere, which, in addition, depends on the pole of the projection. The spherical hypersingular integral truncated in such an induced way, proves to be non-commuting with rotations. Above we used the direct truncation, which is rotation invariant.

4. Inversion of the spherical Riesz potential in the case $\Re\alpha > 2$

To lower the order of the potential operator, it is natural to use the Beltrami-Laplace operator

$$\delta f = \mathcal{R}\Delta P f, \quad f = f(\sigma), \quad \sigma \in S^{n-1},$$

where $Pf = f\left(\frac{x}{|x|}\right)$ is the operator of the continuation of a function $f(\sigma)$ on S^{n-1} to a homogeneous function in R^n and $\mathcal{R}$ is the operator of restriction to S^{n-1}. In what follows the Beltrami-Laplace operator for a function $f(x)$ in the range of the operator K^α is treated in the distributional sense:

$$(\delta f, \psi) = (f, \delta\psi), \quad \psi \in C^\infty(S^{n-1}). \tag{4.1}$$

Theorem 4.1. *Let $f = K^\alpha\varphi$, $\varphi \in L_p(S^{n-1})$, $1 \le p < \infty$, where $\Re\alpha > 0$, $\Re\alpha \neq 2, 4, 6, \ldots$ and $\alpha \neq n-1+2k$, $k = 0, 1, 2, 3, \ldots$ Then*

$$\varphi(x) = c_\beta g(x) + D^\beta g(x) \tag{4.2}$$

where $\beta = \alpha - 2N$, $N = \left[\frac{\Re\alpha}{2}\right]$, so that $0 < \beta < 2$, c_β being the constant defined by (2.16) and g being the function

$$g(x) = \mathcal{P}_N(-\delta)f(x), \tag{4.3}$$

where $\mathcal{P}_N(t)$ is the polynomial

$$\mathcal{P}_N(t) = (t - t_1)\cdots(t - t_N) \tag{4.4}$$

with the roots

$$t_j = \left(j + \frac{\beta + n - 3}{2}\right)\left(j + \frac{\beta + 1 - n}{2}\right). \tag{4.5}$$

The inversion (4.2) is also valid for $\alpha = 2, 4, 6, \ldots$. In this case $\varphi(x) = g(x) = \mathcal{P}_{\frac{\alpha}{2}}(-\delta)f(x)$.

Proof. We first construct the inverse operator on nice functions $\varphi \in C^\infty(S^{n-1})$. By the property $\Gamma(x+1) = x\Gamma(x)$ of the function $\Gamma(x)$, the spherical multiplier

$$\lambda_m(\alpha) = \frac{\Gamma\left(m + \frac{n-1-\alpha}{2}\right)}{\Gamma\left(m + \frac{n-1+\alpha}{2}\right)}$$

of the operator K^α is reduced to $\lambda_m(\beta)$, $\beta = \alpha - 2N$:

$$\lambda_m(\alpha) = \frac{\lambda_m(\beta)}{\left(m + \frac{n-1+\beta}{2}\right)_N \left(m - N + \frac{n-1-\beta}{2}\right)_N}$$

where $(a)_N$ stands for the Pochhammer symbol.

Hence, the operator inverse to K^α must have the multiplier

$$\frac{1}{\lambda_m(\alpha)} = \frac{[m^2 + m(n-2) + \gamma_+(\gamma_- - 1)]\cdots[m^2 + m(n-2) + (\gamma_+ + N - 1)(\gamma_- - N)]}{\lambda_m(\beta)}$$

where $\gamma_\pm = \frac{n-1\pm\beta}{2}$. In the numerator we have a polynomial in $t_m = m^2 + m(n-2)$. The latter is the multiplier of the Beltrami-Laplace operator $-\delta$. Therefore,

$$\frac{1}{\lambda_m(\alpha)} = \frac{\mathcal{P}_N(t_m)}{\lambda_m(\beta)}, \tag{4.6}$$

where

$$\mathcal{P}_N(t) = \prod_{j=1}^{N} [t + (\gamma_+ + j - 1)(\gamma_- - j)],$$

which coincides with (4.4)–(4.5).

Thus, the inversion formula (4.2) is proved in the case of functions $\varphi \in C^\infty(S^{n-1})$. To extend it to the case where $\varphi \in L_p(S^{n-1})$, we observe that $\mathcal{P}_N(-\delta)f$ is defined in the distributional sense (4.1) on functions $f := K^\alpha \varphi$ with $\varphi \in L_p$ if $2N < \alpha$. Moreover,

$$\mathcal{P}_N(-\delta)f = K^{\alpha-2N}\varphi, \tag{4.7}$$

since $(\mathcal{P}_N(-\delta)f, \psi) = (f, K^\alpha \mathcal{P}_N(-\delta)\psi) = (f, K^{\alpha-2N}\psi)$ for $\psi \in C^\infty(S^{n-1})$. Thus, it suffices to invert the right-hand side of (4.7) by means of Theorem 3.5. □

We mention the paper [Ru] in which there was developed a certain technique, based on the Cavalieri type principle (4.24), to organize finite differences of order ℓ with respect to the "shift" $M_f(x', t)$. This led to a construction of the operator inverse to the Riesz potential operator K^α for all $0 < \alpha < \infty$, although in terms of rather complicated constructions.

Finally, as an open question we note the development of the method of approximative inverse operators for spherical potential operators. In application to spatial potential operators this method was developed in a series of papers, see [ZN], [NS1], [NS2], [Sa3], [Sa4]. This idea is especially attractive in the case of large values of $\Re\alpha$. We hope to shed light on this question in the future.

Acknowledgements

Partially supported by Centro de Matemática Aplicada at Instituto Superior Técnico, Lisbon, and by Russian Funds of Fundamental Investigations, Grant 00-01-00046a.

References

[BBP] Berens, H., Butzer, P.L., Pawelke, S., Limitierungsverfahren von Reihen mehrdimensionaler Kugelfunktionen und deren Saturationsverhalten, *Publs. Res. Inst. Math. Sci., Ser. A.* **4** (1968), 201–268.

[GR] Gradshtein, I.S., Ryzhik, I.M., *Tables of Integrals, Sums, Series and Products*, 5th Edition, Academic Press, Inc. 1994.

[Mi] Mikhlin, S.G., *Multi-dimensional Singular Integrals and Integral Equations (Russian)*, Moscow, Fizmatgiz 1962. (English transl.: Pergamon Press, Oxford-New York-Paris , 1965)

[Mu] Müller, C., *Spherical Harmonics*, Lect. Notes in Math., Springer, Berlin 1966.

[Ne] Neri, U., *Singular Integrals*, Lect. Notes in Math, Springer, Berlin 1971.

[NS1] Nogin, V.A., Samko, S.G., Method of approximating inverse operators and its applications to inversion of potential type integral operators, *Integr. Transforms and Special Funct.* **8** (1999), 89–104.

[NS2] Nogin, V.A., Samko, S.G., Some applications of potentials and approximative inverse operators in multi-dimensional fractional calculus, *Fract. Calculus and Appl. Anal.* **2** (1999), 205–228.

[PS1] Pavlov, P.M., Samko, S.G., A description of the space $L_p^\alpha(S_{n-1})$ in terms of spherical hypersingular integrals (Russian), *Dokl. Akad. Nauk SSSR, ser. Matem.* **276** (1984), 546–550 (*Transl. in Soviet Math. Dokl.*, 29 (1984), 549–553).

[PS2] Pavlov, P.M., Samko, S.G., Inversion of Riesz potentials on a sphere and a characterization of the space $L_p^\alpha(S_{n-1})$ in terms of hypersingular integrals (Russian), *Deposited in VINITI, Moscow* **3800-85** (1985), 1–17.

[Ru] Rubin, B.S., The inversion of fractional integrals on a sphere, *Israel. J. Math.* **79** (1992), 47–81.

[Sa1] Samko, S.G., Spaces of Riesz potentials (Russian), *Izv. Akad. Nauk SSSR, ser. Matem.* **40** (1976), 1143–1172 (*Transl. in Math. USSR Izvestija* **10** (1976), no 5, 1089–1117).

[Sa2] Samko, S.G., Generalized Riesz potentials and hypersingular integrals with homogeneous characteristics; their symbols and inversion (Russian), *Trudy Mat. Inst. Steklov* **156** (1980), 157–222 (*Transl. in Proc. Steklov Inst. Mat.* (1983), Issue 2, 173–243).

[Sa3] Samko, S.G., A new approach to the inversion of the of the Riesz potential operator, *Fract. Calculus and Appl. Anal.* **1** (1998), 225–245.

[Sa4] Samko, S.G., Some remarks to author's paper "A new approach to the inversion of the Riesz potential operator", *Fract. Calculus and Appl. Anal.* **2** (1999), 63–66.

[Sa] Samko, S.G., *Hypersingular Integrals and Their Applications*, Taylor & Francis, Series "Analytic Methods and Special Functions", v.5, 2002.

[SKM] Samko, S.G., Kilbas, A.A., Marichev, O.I., *Fractional Integrals and Derivatives. Theory and Applications*, Gordon & Breach Sci. Publ. 1993.

[SW] Stein, E., Weiss, G., *Introduction to Fourier Analysis on Euclidean Space*, Princeton Univ. Press 1971.

[ZN] Zavolzhenskii, M.M., Nogin, V.A., Approximating approach to inversion of the generalized Riesz potentials, *Dokl. Akad. Nauk Rossii, ser. matem.* **324** (1992), 738–741.

Stefan Samko
Universidade do Algarve
Unidade de Ciencias Exactas
Campus de Gambelas
Faro 8000, Portugal
e-mail: `ssamko@ualg.pt`

2000 *Mathematics Subject Classification.*
Primary 47G10, Secondary 45P05, 26A33.

Received: 10 February 2001.

Operator Theory:
Advances and Applications, Vol. 142, 369–381

An Axiomatic Approach to Stability of Approximation Sequences

Pedro A. Santos and Bernd Silbermann

Abstract. The paper contains an axiomatic approach to the problem of the applicability of approximation methods to operators that can be represented as a piecewise continuous function of a shift. Using this approach, a small set of conditions that involve the operator and the method itself are proved to guarantee the applicability of a given approximation method.

1. Introduction

Numerical analysis for convolution like operators has a long history. Already in the (now) classical work [5] by Gohberg and Feldman, some axiomatic approach to Galerkin methods for convolution equations was proposed. The approach was heavily based on the fact that, in the Hilbert space setting, the convolutions under consideration can be understood as continuous functions of a shift. Another key assumption was that the underlying shift operator V is connected, in a very simple way, with the sequence of orthogonal projections P_τ through the formula

$$P_\tau V P_\tau = P_\tau V. \tag{1.1}$$

In the beginning of the eighties one of the authors proposed a new Banach algebra approach to the finite section method for a class of familiar Toeplitz operators (with non-continuous generating function). This approach was based on the observation that the algebra generated by the finite sections of Toeplitz operators contains two ideals which are kernels of two homomorphisms W and $\widetilde{W}$ with special properties. This concept has found numerous applications (see for instance [1], [2], [6], [7], [10], [14]). One of its main advantages is the possibility of using localization techniques to find conditions for the stability of approximation sequences.

Interestingly, it was observed, especially for operators generated by shifts, that a variety of approximation methods (such as spline Galerkin methods, quadrature methods and so on) behave like the finite section method for Toeplitz operators, at least in the scalar case (see [10, Chapter 5]). Moreover, the approach proposed in [10] was nothing else then a slight modification of the Gohberg/Feldman approach. An algebraic approach like the one for the finite section method for Toeplitz operators that would also work for other methods and other operators would be of interest in order to answer some interesting questions, for instance, the question on how to handle the block case or the question on what can be said about the asymptotic behavior of the singular values of the approximating matrices. The main problem with this algebraic approach in the general situation is the need for the homomorphisms W and $\widetilde{W}$. Thus, the following question appeared: Is there a set of conditions that allows us to study discretizations of convolution operators from an abstract point of view and that shows what is really needed?

We here propose such an approach, which is in fact a non-trivial mixture of the algebraic approach to the finite section method for Toeplitz operators and some ideas which are based on properties similar to (1.1). Among other things, we algebraize the Gohberg/Feldman approach.

The paper is organized as follows. This first section contains some definitions and relations that will be necessary in what follows. In Section 2 we present some purely algebraic results that will be needed later. In the third section we review some ideas on functions of shifts. Section 4 contains the main results. In the last section, the finite section method for Toeplitz operators is used as an example to illustrate the proposed approach.

The set of continuous functions defined on the unit circle $\mathbb{T}$ is denoted by $C(\mathbb{T})$, and the set of piecewise continuous functions on $\mathbb{T}$, that is, functions with well-defined finite one-sided limits at all points of $\mathbb{T}$, is denoted by $PC(\mathbb{T})$. These sets are considered as subalgebras of $L^\infty(\mathbb{T})$, the algebra of essentially bounded measurable functions with the essential supremum norm $\|.\|_\infty$. We use $a(t^+)$ and $a(t^-)$ to denote the clockwise and anticlockwise one-sided limits of the function a at the point t, respectively.

As usual, the Hilbert space of one-sided infinite sequences $x := (x_j)_{j=0}^\infty$ such that

$$\|x\| := \left(\sum_{j=0}^{\infty} |x_j|^2\right)^{\frac{1}{2}}$$

is finite will be denoted by l^2. We denote by e_k, $k \in \mathbb{Z}$, the function defined on $\mathbb{T}$ by $e_k(t) = t^k$. Given a bounded measurable matrix-valued function a acting on the unit circle we define its kth Fourier coefficient by

$$a_k = \int_0^1 a(e^{2\pi i s}) e^{-2\pi i k s} \, ds. \tag{1.2}$$

The function a can then be formally written as $a = \sum_{k=-\infty}^{+\infty} a_k e_k$. The Toeplitz operator $T(a)$ acting boundedly on l^2 is defined by

$$(T(a)x)_k = \sum_{j \in \mathbb{Z}_+} a_{k-j} x_j, \quad k \in \mathbb{Z}_+, \tag{1.3}$$

where (a_k) is the sequence of the Fourier coefficients of the function $a \in L^\infty(\mathbb{T})$.

2. Algebraic results

Given a C*-algebra $\mathcal{A}$ and a set $S \subset \mathcal{A}$ we will denote by $\operatorname{alg} S$ the smallest closed C*-subalgebra of $\mathcal{A}$ that contains the elements in S. By $\operatorname{id} S$ we will designate the smallest closed two-sided ideal of $\mathcal{A}$ that contains the elements in S.

Let now $\mathcal{A}$ be a unital C*-algebra and let $v \in \mathcal{A}$ be such that $v^*v = e$ but $vv^* \neq e$. We put

$$v_n := \begin{cases} v^n & \text{if } n > 0 \\ e & \text{if } n = 0 \\ (v^*)^{-n} & \text{if } n < 0 \end{cases}. \tag{2.1}$$

To any complex polynomial $p(\lambda) = \sum_{j=-n}^{n} a_j \lambda^j$, $\lambda \in \mathbb{T}$, we will associate the element $p(v) = \sum_{j=-n}^{n} a_j v_j$ of $\mathcal{A}$. It is well known from [5] that

$$\|p(v)\| = \|p\|_\infty. \tag{2.2}$$

As any continuous function f defined on $\mathbb{T}$ is the norm-limit of a sequence of polynomials p_n, it is possible to define the operator $f(v)$ as the limit of the sequence $p_n(v)$. Thus we are given an isometry $T_v : C(\mathbb{T}) \to \mathcal{A}$, $T_v(f) = f(v)$. Note the obvious relation

$$v^* T_v(f) v = T_v(f). \tag{2.3}$$

Lemma 2.1. *Let $\mathcal{B} = \operatorname{alg}\{a \in \mathcal{A} : v^*av = a\}$. Then for any $f \in C(\mathbb{T})$ we have $f(v)a - af(v) \in \operatorname{id}\{e - vv^*\}$ for all $a \in \mathcal{B}$.*

Proof. We have that $e = vv^* + (e - vv^*)$. So for a such that $v^*av = a$,

$$\begin{aligned} va - av &= (vv^* + (e - vv^*))(va - av) = vv^*(va - av) + k \\ &= v(v^*va - v^*av) + k = v(a - a) + k = k \end{aligned} \tag{2.4}$$

with $k \in \operatorname{id}\{e - vv^*\}$. Analogously

$$\begin{aligned} v^*a - av^* &= (v^*a - av^*)(vv^* + (e - vv^*)) = (v^*a - av^*)vv^* + k \\ &= (v^*av - av^*v)v^* + k = (a - a)v^* + k = k \end{aligned} \tag{2.5}$$

with k again in $\operatorname{id}\{e - vv^*\}$. From the above it is easy to see that the result is true for any polynomial $p(v)$ due to the associative and distributive properties of the multiplication. Now, if $f \in C(\mathbb{T})$, there is a sequence of polynomials p_n such that $\|c(v) - p_n(v)\| \to 0$, and as $\operatorname{id}\{e - vv^*\}$ is closed the result follows. Using a similar reasoning we extend the result to any $a \in \mathcal{B}$. □

The above innocent looking result has far-reaching consequences. For instance, if $\mathcal{A}$ is the algebra of bounded operators in l^2, $\mathcal{L}(l^2)$, and v is the unilateral shift $(x_0, x_1, x_2, \ldots) \mapsto (0, x_0, x_1, \ldots)$, then it is easy to see that $\mathcal{B}$ is the algebra generated by all Toeplitz operators. The following (known) result then is immediate.

Corollary 2.2. *If $a \in L^\infty(\mathbb{T})$ and $f \in C(\mathbb{T})$, then the commutator $T(f)T(a) - T(a)T(f)$ is compact.*

Proof. Just note that $e - vv^*$ is a rank 1 operator, and so $\mathrm{id}\{e - vv^*\}$ consists only of compact operators. □

3. Abstract Toeplitz operators

Let H be a Hilbert space and let $V : H \to H$ be a shift, that is, an isometry such that $(V^*)^n \to 0$ as n tends to infinity.

Write $P_0 := I - VV^*$ and consider the space H' of the elements $x' = (x_0, x_1, \ldots)$ such that $x_j \in \operatorname{Im} P_0$, with norm $\|x'\| := \left(\sum_{j=0}^{\infty} \|x_j\|^2\right)^{\frac{1}{2}}$. Then the transformation $\Theta : H' \to H$ defined by

$$\Theta(x') := \sum_{j=1}^{\infty} V_j x_j \tag{3.1}$$

and having the inverse

$$\Theta^{-1}(x)_j = P_0 V_{-j} x \tag{3.2}$$

is an isometry between H' and H (see [12]). So each operator A in H can be represented by the infinite matrix $\Theta^{-1} A \Theta$ whose entries are given by $A_{ij} = P_0 V_{-i} A V_j P_0$ and composition of operators in $\mathcal{L}(H)$ corresponds to the multiplication of the infinite matrices in $\mathcal{L}(H')$. We have

$$\Theta^{-1} V \Theta = \begin{bmatrix} 0 & 0 & 0 & 0 & \ddots \\ P_0 & 0 & 0 & 0 & \ddots \\ 0 & P_0 & 0 & 0 & \ddots \\ 0 & 0 & P_0 & 0 & \ddots \\ \ddots & \ddots & \ddots & \ddots & \ddots \end{bmatrix}, \quad \Theta^{-1} V^* \Theta = \begin{bmatrix} 0 & P_0 & 0 & 0 & \ddots \\ 0 & 0 & P_0 & 0 & \ddots \\ 0 & 0 & 0 & P_0 & \ddots \\ 0 & 0 & 0 & 0 & \ddots \\ \ddots & \ddots & \ddots & \ddots & \ddots \end{bmatrix}, \tag{3.3}$$

Both these operators have the so-called V-structure, that is, their entries are of the form $A_{ij} = c_{ij} P_0$, with $c_{ij} \in \mathbb{C}$. In [12] S. Roch and one of the authors proved that the collection of all operators which have the V-structure is isometrically isomorphic to $\mathcal{L}(l^2)$, the operator V corresponding to the unilateral shift $v = (x_0, x_1, x_2, \ldots) \mapsto (0, x_0, x_1, \ldots)$.

Having in mind the decomposition of H by means of the shift V, we say that an operator $T : H \to H$ whose representation in H' is of the form $[A_{i-j}]_{i,j=1}^{\infty}$ with $A_k : \operatorname{Im} P_0 \to \operatorname{Im} P_0$ is an *abstract Toeplitz operator*. This is equivalent to the condition $V^* T V = T$.

We are interested in abstract Toeplitz operators with V-structure, that is, operators whose entries are of the form $A_{ij} = a_{i-j} P_0$, $a_k \in \mathbb{C}$. Because of the isometric isomorphism with $\mathcal{L}(l^2)$ we see that the sequence $(a_k)_{k=-\infty}^{+\infty}$ is the sequence of the Fourier coefficients of some function $a \in L^\infty(\mathbb{T})$, and so we will write T as $T_V(a)$ (see (1.3)). By the same isomorphism, to each function $a \in L^\infty(\mathbb{T})$, we can associate the corresponding operator $T_V(a) : H \to H$.

So, given a function $a \in L^\infty(\mathbb{T})$, we will define the operator $a(V) : H \to H$ as $a(V) := T_V(a)$. In case a is continuous, this definition agrees with the definition in Section 2.

For a set $S \subseteq L^\infty(\mathbb{T})$ of functions defined on $\mathbb{T}$ let $\mathcal{B}(S) := \operatorname{alg}\{a(V) : a \in S\}$. We can resume the results of this section in the following theorem.

Theorem 3.1. *The subalgebra* $\operatorname{alg}\{T(a) : a \in L^\infty(\mathbb{T})\}$ *of* $\mathcal{L}(l^2)$ *is isometrically isomorphic to the subalgebra* $\mathcal{B}(L^\infty)$ *of* $\mathcal{L}(H)$, *and the isomorphism is given by* $T(a) \mapsto a(V)$. *In particular we have* $\|a(V)\| = \|a\|_\infty$.

4. Approximation methods for the operator $a(V)$, $a \in PC(\mathbb{T})$

Let $\mathbb{I} \subset \mathbb{N}$ be an infinite index set and let $(A_n)_{n\in\mathbb{I}}$ be a sequence of operators in $\mathcal{L}(H)$. The sequence $(A_n)_{n\in\mathbb{I}}$ is said to converge uniformly to an operator $A \in \mathcal{L}(H)$ if $\|A_n - A\| \to 0$ as $n \in \mathbb{I}$ and $n \to \infty$, and it is said to converge strongly to A if $\|A_n x - Ax\| \to 0$ as $n \in \mathbb{I}$ and $n \to \infty$, for all $x \in H$.

Now suppose we have a non-unitary isometry V with $\dim\operatorname{coker}(V) < \infty$ and a sequence $(H_n)_{n\in\mathbb{I}}$ of subspaces of H such that the orthogonal projections $P^n : H \to H_n$ converge strongly to the identity operator.

We denote by $\mathcal{F}$ the C*-algebra of the sequences $(A_n)_{n\in\mathbb{I}}$ such that $A_n : H_n \to H_n$ and both sequences $(A_n P^n)$ and $(A_n^* P^n)$ converge strongly to some operators in $\mathcal{B}(L^\infty)$. The operations and the norm in $\mathcal{F}$ are given by $(A_n) + (B_n) := (A_n + B_n)$, $\alpha(A_n) := (\alpha A_n)$, $(A_n)(B_n) := (A_n B_n)$, $(A_n)^* := (A_n^*)$ and $\|(A_n)\| := \sup_{n\in\mathbb{I}} \|A_n\|$. The strong limit of (A_n) will be denoted by $W(A_n)$. Let $\mathcal{G} \subset \mathcal{F}$ denote the ideal of the sequences tending to zero in the norm. As usual we will say that a sequence $(A_n) \in \mathcal{F}$ is stable if there exists an $n_0 \in \mathbb{I}$ such that for every $n > n_0$ the operators A_n are invertible in $\operatorname{Im} P^n$ and the set $\{\|(A_n)^{-1}\| : n > n_0\}$ is bounded.

Let $\mathcal{K}$ denote the closed two-sided ideal of all compact operators and consider the subset $\mathcal{J} := \{(P^n K P^n + C_n) : K \in \mathcal{K}, (C_n) \in \mathcal{G}\}$ of $\mathcal{F}$. It is well known (see for instance [6, Chapter 1]) that the set $\mathcal{J}$ is a closed two-sided ideal of $\mathcal{F}$. The following result is also proved in [6, Chapter 1].

Proposition 4.1. *A sequence* $(A_n) \in \mathcal{F}$ *is stable if and only if the operator* $W(A_n)$ *is invertible in* $\mathcal{B}(L^\infty)$ *and the coset* $(A_n) + \mathcal{J}$ *is invertible in* $\mathcal{F}/\mathcal{J}$.

Let now D be a linear bounded mapping from $\{a(V) : a \in PC(\mathbb{T})\}$ into $\mathcal{F}$ such that $W(D(A)) = A$ and $D(A^*) = D(A)^*$ for every operator $A \in \{a(V) : a \in PC(\mathbb{T})\}$. We will call D a discretization. Suppose in addition that D has the following properties:

(H1) $D((e_1 a)(V)) - D(V)D(a(V)) \in \mathcal{J}$ for every $a \in S$;

(H2) $D((a e_{-1})(V)) - D(a(V))D(V^*) \in \mathcal{J}$ for every $a \in S$.

Lemma 4.2. *Under the assumption* (H1),

$$D(V^j) - D(V)^j \in \mathcal{G} \text{ for } j \in \mathbb{Z}_+; \tag{4.1}$$

$$I - D(V)D(V^*) \in \mathcal{J}. \tag{4.2}$$

Proof. To prove the first assertion we apply (H1) j times. We get that

$$D(V^j) - D(V)^j = (P^n K P^n + C_n)$$

for some $K \in \mathcal{K}$ and $(C_n) \in \mathcal{G}$. But as the left-hand side tends strongly to zero, we conclude that K must be equal to zero and so $D(V^j) - D(V)^j \in \mathcal{G}$. The second assertion comes immediately from (H1) by putting $a(V) = V^*$. □

Write $\mathcal{A}(S)$ for $\operatorname{alg}\{D(a(V)) : a \in S\} + \mathcal{G} \subset \mathcal{F}$. Define the ideals

$$\mathcal{J}_1 := \operatorname{id}\{I - D(V)D(V^*)\} + \mathcal{G}, \tag{4.3}$$

$$\mathcal{J}_2 := \operatorname{id}\{I - D(V^*)D(V)\} + \mathcal{G} \tag{4.4}$$

in $\mathcal{A}(C)$

Lemma 4.3. *We have $I - D(V)D(V^*) = P^n(I - VV^*)P^n + C_n$ with $C_n \in \mathcal{G}$, and the ideals $\mathcal{J}_1$ and $(\mathcal{J} \cap \mathcal{A}(S)) + \mathcal{G}$ coincide.*

Proof. We start by proving that $I - D(V)D(V^*) = P^n(I - VV^*)P^n + C_n$. The sequence $I - D(V)D(V^*)$ tends strongly to $I - VV^*$. But by (4.2) it is of the form $P^n K P^n + C_n$, which tends in the norm to K. So K must be equal to $I - VV^*$. To prove the second part, it suffices to note that the ideal $(\mathcal{J} \cap \mathcal{A}(PC)) + \mathcal{G}$ is generated by the sequences $(P^n(I - VV^*)P^n + C_n)$, as can be checked easily. □

Lemma 4.4. $\mathcal{J}_1 \cap \mathcal{J}_2 \subset \mathcal{G}$.

Proof. Let $K \in \mathcal{K}$, $(C_n) \in \mathcal{G}$, and suppose $(A_n) := (P^n K P^n + C_n)$ belongs also to $\mathcal{J}_2$. The strong limit of a sequence in $\mathcal{J}_2$ is zero, and this implies $K = 0$. So $(A_n) \in \mathcal{G}$. □

We write $\mathcal{A}^{\mathcal{J}_1}(PC)$ (respectively $\mathcal{A}^{\mathcal{J}_1}(C)$) for the quotient algebra $\mathcal{A}(PC)/\mathcal{J}_1$ ($\mathcal{A}(C)/\mathcal{J}_1$). The canonical homorphism will be denoted by Φ. Due to Proposition 4.1, in order to obtain stability conditions for $D(a(V))$ we just have to study the invertibility of $\Phi(D(a(V)))$ in $\mathcal{A}^{\mathcal{J}_1}(PC)$. Denote $\Phi(D(V^*))$ by $\widetilde{V}$. By construction, $\widetilde{V}$ is a left invertible element, with left inverse $\widetilde{V}^*$, and we also have $\|\widetilde{V}\| = 1$.

Consider $\mathcal{A}^{\mathcal{J}_1}(C)$. In this case (4.1) further implies that $\mathcal{A}^{\mathcal{J}_1}(C)$ is generated by the coset $\widetilde{V}$ and its adjoint. It is not difficult to see that for every polynomial p, $p(\lambda) = \sum_{j=-k}^{k} a_j \lambda^j$, we have $\Phi(D(p(V))) = \tilde{p}(\Phi(D(V^*)))$ with $\tilde{p}(\lambda) = \sum_{j=-k}^{k} a_j \lambda^{-j}$. This gives the equality $\Phi(D(f(V))) = \tilde{f}(\widetilde{V})$ with $\tilde{f}(\lambda) = f(1/\lambda)$ for every $f \in C(\mathbb{T})$.

So there are two possibilities. Either $\widetilde{V}$ is a unitary isometry in $\mathcal{A}^{\mathcal{J}_1}(C)$ or a non-unitary one. If $\widetilde{V}$ is unitary, the invertibility of $f(V)$, $f \in C(\mathbb{T})$, implies the invertibility of the function f, and so the invertibility of $\tilde{f}$. As $\widetilde{V}$ is unitary the invertibility of $\tilde{f}(\widetilde{V})$ comes immediately.

Proposition 4.5. *If $\widetilde{V}$ is non-unitary in $\mathcal{A}^{\mathcal{J}_1}(C)$, then there exists an isometric *-homomorphism $\widetilde{W} : \mathcal{A}^{\mathcal{J}_1}(C) \to \mathcal{B}(C)$ which takes $\Phi(D(f(V)))$ into $\tilde{f}(V)$ for every $f \in C(\mathbb{T})$.*

Proof. If $\widetilde{V}$ is non-unitary in $\mathcal{A}^{\mathcal{J}_1}(C)$, then the algebra is generated by $\widetilde{V}$ and $\widetilde{V}^*$. By [3, Section 1], $\operatorname{alg}\{\widetilde{V}, \widetilde{V}^*\}$ is isometrically isomorphic to the algebra generated by a shift (for example $\operatorname{alg}\{V, V^*\}$). Since $\Phi(D(f(V))) = \tilde{f}(\widetilde{V})$, the result then follows from Theorem 3.1. □

The main results for the continuous case may be resumed in the following Theorem (see also [6, Theorem 6.72]).

Theorem 4.6. *Let $(A_n) \in \mathcal{A}(C)$.*

1. *If $\Phi(D(V^*))$ is unitary in $\mathcal{A}^{\mathcal{J}_1}(C)$ then the sequence (A_n) is stable if and only if the operator $W(A_n)$ is invertible.*
2. *If $\Phi(D(V^*))$ is non-unitary in $\mathcal{A}^{\mathcal{J}_1}(C)$ then the sequence (A_n) is stable if and only if the operators $W(A_n)$ and $\widetilde{W}(A_n)$ are invertible.*

We now turn our attention to the algebra $\mathcal{A}^{\mathcal{J}_1}(PC)$. In what follows we assume that

(H3) The ideal $\mathcal{J}_2$ of $\mathcal{A}(C)$ coincides with $\operatorname{id}\{I - D(V^*)D(V)\} + \mathcal{G}$ in $\mathcal{A}(PC)$.

Since by Lemma 4.3 $\mathcal{J}_1$ is already an ideal in $\mathcal{A}(PC)$ the above condition says that the commutator ideal $\operatorname{com}\mathcal{A}(C)$ of $\mathcal{A}(C)$ is an ideal in $\mathcal{A}(PC)$.

Lemma 4.7. *If $f \in C(\mathbb{T})$ and $a \in PC(\mathbb{T})$ then*

$$(\Phi(D(a(V))) + \mathcal{J}_2^{\mathcal{J}_1})(\Phi(D(f(V))) + \mathcal{J}_2^{\mathcal{J}_1}) = \Phi(D(af(V))) + \mathcal{J}_2^{\mathcal{J}_1}.$$

Proof. The result is easily proved by taking into account (H1) and (H2) to see that the result holds for every polynomial $p(V) = \sum_{k=0}^{m} f_k V^k + \sum_{k=1}^{m} f_{-k}(V^*)^k$. □

Lemma 4.8. *The algebra $\mathcal{A}^{\mathcal{J}_1}(C)/\mathcal{J}_2^{\mathcal{J}_1}$ is a central subalgebra of $\mathcal{A}^{\mathcal{J}_1}(PC)/\mathcal{J}_2^{\mathcal{J}_1}$ and is isometrically isomorphic to $C(\mathbb{T})$.*

Proof. The proof of the first part is the same as that of Lemma 2.1, by taking into account (H1) and (H2) to see that $\widetilde{V}^*\Phi(D(a(V)))\widetilde{V} = \Phi(D(a(V)))$. For the second part note that $\mathcal{A}^{\mathcal{J}_1}(C)/\mathcal{J}_2^{\mathcal{J}_1}$ is $\operatorname{alg}\{\widetilde{V}, \widetilde{V}^*\}/QC(\widetilde{V})$, where $QC(\widetilde{V})$ is the quasicommutator ideal, and that this algebra is isometrically isomorphic to $C(\mathbb{T})$ (see [10, Chapter 4]). □

We say that an element $k \in \mathcal{A}$ ($\mathcal{A}$ being an arbitrarily given Banach algebra) is a rank one element in $\mathcal{A}$ if for each $b \in \mathcal{A}$ there exists a number $\mu(b) \in \mathbb{C}$ such that $kbk = \mu(b)k$.

Lemma 4.9. *The element $I - \widetilde{V}\widetilde{V}^*$ is a rank one element in $\mathcal{A}^{\mathcal{J}_1}(PC)$.*

Proof. We start by proving that $I - \widetilde{V}\widetilde{V}^*$ is a rank one element in $\mathcal{A}^{\mathcal{J}_1}(C)$. If $\widetilde{V}$ is unitary the result is obvious. So suppose it is not unitary. It is easy to verify that $K_0 := I - \widetilde{V}\widetilde{V}^*$ is idempotent. Hence by simple calculus, if $b \in C(\mathbb{T})$ is a polynomial $b(\lambda) = \sum_{j=-k}^{k} b_j \lambda^j$ we have $K_0 b(\widetilde{V}) K_0 = b_0 K_0$. By the density of the polynomials in the continuous functions, the result is true for $\mathcal{A}^{\mathcal{J}_1}(C)$. In the case $b \in S$ we have $K_0 b K_0 = K_0(bK_0)K_0$ with $bK_0 \in \operatorname{com} \mathcal{A}(C)$ by (H3) because $K_0 \in \operatorname{com} \mathcal{A}(C)$, and so the result follows. □

Theorem 4.10. *Let $(A_n) \in \mathcal{A}(PC)$ and $\mathcal{J}_0 := \mathcal{J}_1 + \mathcal{J}_2$, which, by (H3), is an ideal of $\mathcal{A}(PC)$.*

1. *If $\Phi(D(V^*))$ is unitary in $\mathcal{A}^{\mathcal{J}_1}(C)$ then the sequence (A_n) is stable if and only if the operator $W(A_n)$ is invertible and the coset $(A_n) + \mathcal{J}_0$ is invertible in $\mathcal{A}^{\mathcal{J}_0}$.*
2. *If $\Phi(D(V^*))$ is non-unitary in $\mathcal{A}^{\mathcal{J}_1}(C)$ then there exists a unital *-homomorphism $\widetilde{W} : \mathcal{A} \to \mathcal{B}(L^\infty)$ such that the sequence (A_n) is stable if and only if the operator $W(A_n)$ and another operator (denoted by $\widetilde{W}(A_n)$) are invertible, and if the coset $(A_n) + \mathcal{J}_0$ is invertible in $\mathcal{A}^{\mathcal{J}_0}$.*

Proof. Suppose $\Phi(D(V^*))$ is unitary in $\mathcal{A}^{\mathcal{J}_1}(C)$. This happens if and only if $I - D(V^*)D(V)$ is uniformly convergent to the operator 0 and $\mathcal{J}_2 = \mathcal{G}$, and the result follows by Proposition 4.1.

Suppose now $\widetilde{V} \equiv \Phi(D(V^*))$ is a non-unitary isometry in $\mathcal{A}^{\mathcal{J}_1}(C)$. We will need the representation of the algebra $\mathcal{A}^{\mathcal{J}_1}(C)$ as an algebra of bounded operators acting on a Hilbert space H^0. To simplify the notation, we will use the same symbol for an element in the original algebra and its representation as an operator. Using the Wold decomposition (see, for instance, [15]) we get an orthogonal decomposition $H^0 = H^1 \oplus H^2$ into reducing subspaces for $\widetilde{V}$, with $\widetilde{V}_{|_{H^1}}$ a shift operator and $\widetilde{V}_{|_{H^2}}$ a unitary operator. Write P_1 and P_2 for the orthogonal projections onto H^1 and H^2, and $\pi(a)$ for the representation of $\Phi(D(a(V)))$. Every element $\Phi(A_n)$ (now seen as an operator in the representation) can be written as a matrix operator

$$\begin{bmatrix} A_{11} & A_{12} \\ A_{21} & A_{22} \end{bmatrix}$$

with $A_{ij} : H^i \to H^j$, $i, j = 1, 2$. Note that if $\Phi(A_n) \in \Phi(\mathcal{A}(C))$, it follows immediately that A_{12} and A_{21} equal zero. Now suppose $a \in S$. The element $(I - \widetilde{V}^n(\widetilde{V}^*)^n)_{|_{H^1}}$ belongs to the ideal $\Phi(\mathcal{J}_2) \subset \Phi(\operatorname{com} \mathcal{A}(C))$. Hence

$$(I - \widetilde{V}^n(\widetilde{V}^*)^n)_{|_{H^1}} A_{12} \in \Phi(\operatorname{com} \mathcal{A}(C))$$

due to (H3) and so $(I - \widetilde{V}^n(\widetilde{V}^*)^n)_{|_{H^1}} A_{12} = 0$. As $I - \widetilde{V}^n(\widetilde{V}^*)^n$ converges strongly to the identity, $(I - \widetilde{V}^n(\widetilde{V}^*)^n)_{|_{H^1}} A_{12}$ converges strongly to A_{12}, and we conclude that $A_{12} = 0$. A similar reasoning leads to $A_{21} = 0$. Thus, the spaces H^1 and H^2 are reducing for all elements of $\mathcal{A}^{\mathcal{J}_1}(PC)$. The invertibility of an element $\Phi(A_n)$

in the representation of $\mathcal{A}^{\mathcal{J}_1}(PC)$ is then equivalent to the invertibility of the operators $P_1\Phi(A_n)P_1$ $(=\Phi(A_n)P_1)$ in H^1 and $P_2\pi(a)P_2$ in H^2. The invertibility in H^2 is as in part (a) of this theorem. We are left with studying the operator $P_1\Phi(A_n)P_1$. By Section 3, we can represent this operator as a matrix, with entries given by

$$B_{ij} = (I - \widetilde{V}\widetilde{V}_{-1})\widetilde{V}_{-i}\Phi(A_n)P_1\widetilde{V}_j(I - \widetilde{V}\widetilde{V}_{-1}), \quad i,j \geq 1. \tag{4.5}$$

By Lemma 4.9, the projection $I - \widetilde{V}\widetilde{V}_{-1}$ has rank one, and hence the entries are scalar multiples of $I - \widetilde{V}\widetilde{V}_{-1}$. On the other hand, from (H1) and (H2) we obtain that

$$\widetilde{V}^*\Phi(A_n)\widetilde{V}P_1 = \Phi(A_n)P_1,$$

which implies that

$$B_{ij} := b_{i-j} \quad \text{with} \quad b_k = \begin{cases} (I-\widetilde{V}\widetilde{V}_{-1})\Phi(A_n)P_1\widetilde{V}_{-k}(I-\widetilde{V}\widetilde{V}_{-1}) & \text{if} \quad k \leq 0 \\ (I-\widetilde{V}\widetilde{V}_{-1})\widetilde{V}_{-k}\Phi(A_n)P_1(I-\widetilde{V}\widetilde{V}_{-1}) & \text{if} \quad k > 0 \end{cases}. \tag{4.6}$$

The above means that $\operatorname{alg}\{P_1\Phi(A_n)P_1 : (A_n) \in \mathcal{A}\}$ is an algebra generated by Toeplitz operators with scalar valued functions. This algebra is isomorphic to a subalgebra of $\mathcal{B}(L^\infty)$. To $P_1\Phi(A_n)P_1$ or the corresponding sequence (A_n) we can then associate a scalar Toeplitz operator $\widetilde{W}(A_n)$ whose invertibility, by Theorem 3.1, is equivalent to the invertibility of $P_1\Phi(A_n)P_1$. □

We are now going to identify the scalar Toeplitz operator appearing in Theorem 4.10.

Proposition 4.11. *Let $a \in PC(\mathbb{T})$ be continuous at the point $t_0 \in \mathbb{T}$ and $(A_n) = D(a(V))$. Then the generating function $b \in L^\infty(\mathbb{T})$ of $P_1\Phi(D(a(V)))P_1$ is continuous at the point $1/t_0$ and $b(1/t_0) = a(t_0)$.*

Proof. We will apply localization techniques to identify the generating function of the Toeplitz operator $P_1\pi(a)P_1$. By Lemma 4.8 $\mathcal{A}^{\mathcal{J}_1}(C)/\mathcal{J}_2^{\mathcal{J}_1}$ is a central subalgebra of $\mathcal{A}^{\mathcal{J}_1}(PC)/\mathcal{J}_2^{\mathcal{J}_1}$. The maximal ideal space of $\mathcal{A}^{\mathcal{J}_1}(C)/\mathcal{J}_2^{\mathcal{J}_1}$ is homeomorphic to $\mathbb{T}$. The homeomorphism can be chosen so that the maximal ideal $\mathcal{M}_t$ assigned to $t \in \mathbb{T}$ is the set

$$\begin{aligned} \mathcal{M}_t &= \{f(\Phi(D(V^*))) + \mathcal{J}_2^{\mathcal{J}_1} : f \in C(\mathbb{T}) \text{ and } f(t) = 0\} && (4.7) \\ &= \{\Phi(D(\tilde{f}(V))) + \mathcal{J}_2^{\mathcal{J}_1} : \tilde{f} \in C(\mathbb{T}) \text{ and } \tilde{f}(1/t) = 0\}. && (4.8) \end{aligned}$$

The same remains true if we replace $\mathcal{A}^{\mathcal{J}_1}(C)$ by $P_1\mathcal{A}^{\mathcal{J}_1}(C)P_1$ and $\mathcal{J}_2^{\mathcal{J}_1}$ by $P_1\mathcal{J}_2^{\mathcal{J}_1}P_1$. Let $\mathcal{I}_t$ be the smallest closed ideal in $\mathcal{A}^{\mathcal{J}}(PC)/\mathcal{J}_2^{\mathcal{J}_1}$ containing $\mathcal{M}_t$, write $\Psi^{\mathcal{J}}$ for the homomorphism $\mathcal{A}^{\mathcal{J}_1}(PC) \to \mathcal{A}^{\mathcal{J}_1}(PC)/\mathcal{J}_2^{\mathcal{J}_1}$ and $\Psi_t^{\mathcal{J}_2}$ for the homomorphism $\mathcal{A}^{\mathcal{J}_1}(PC) \to \mathcal{A}_t^{\mathcal{J}} := \mathcal{A}^{\mathcal{J}_1}(PC)/\mathcal{J}_2^{\mathcal{J}_1}/\mathcal{I}_t$. Take $P_1\pi(f)P_1$ with $f \in C(\mathbb{T})$. Then we already know that the Toeplitz operator $P_1\pi(f)P_1$ has $\tilde{f}$ as its generating function. Thus

$$\begin{aligned} \mathcal{M}_t &= \{f(\Phi(D(V^*))) + \mathcal{J}_2^{\mathcal{J}_1} : f \in C(\mathbb{T}) \text{ and } f(t) = 0\} && (4.9) \\ &= \{\tilde{f}(\Phi(D(V))) + \mathcal{J}_2^{\mathcal{J}_1} : \tilde{f} \in C(\mathbb{T}) \text{ and } \tilde{f}(1/t) = 0\}. && (4.10) \end{aligned}$$

Suppose $a(t_0) = 0$. Define a continuous function $f_{t_0,\epsilon}$ such that $0 \leq f_{t_0,\epsilon} \leq 1$, $f_{t_0,\epsilon}$ is zero outside the arc $[(1/t_0)e^{-i\epsilon}, (1/t_0)e^{+i\epsilon}]$ and $f_{1/t_0,\epsilon}(t) = 1$. The coset $\Psi_{t_0}^{\mathcal{J}_2}(f_{t_0,\epsilon}(\Phi(D(V^*))))$ is the identity in the local algebra $\mathcal{A}_{1/t_0}^{\mathcal{J}}$ and we have

$$\begin{aligned}
\|\Psi_{t_0}^{\mathcal{J}_2}(\Phi(D(a(V))))\| &= \|\Psi_{t_0}^{\mathcal{J}_2}(\Phi(D(a(V))))\Psi_{t_0}^{\mathcal{J}_2}(f_{t_0,\epsilon}(\Phi(D(V^*))))\| \\
&= \|\Psi_{t_0}^{\mathcal{J}_2}(\Phi(D(a(V)))f_{t_0,\epsilon}(\Phi(D(V^*))))\| \\
&= \|\Psi_{t_0}^{\mathcal{J}_2}(\Phi(D(a(V))D(\tilde{f}_{t_0,\epsilon}(V))))\| \\
&= \|\Psi_{t_0}^{\mathcal{J}_2}(\Phi(D(a\tilde{f}_{t_0,\epsilon}(V))))\| \qquad \text{by (H1)} \\
&\leq \|a\tilde{f}_{t_0,\epsilon}\|_\infty,
\end{aligned}$$

where the norm $\|a\tilde{f}_{t_0,\epsilon}\|_\infty$ can be made as small as desired by choosing ϵ sufficiently small. This means that $\Psi^{\mathcal{J}_2}(\Phi(D(a(V)))) \in \mathcal{I}_{1/t_0}$. If a is continuous at t_0, write $a = a(t_0) + (a - a(t_0))$. We obtain then that $\Psi_{1/t_0}^{\mathcal{J}_2}(\Phi(D(a(V)))) = a(t_0)I$. On the other hand, we have $\Psi^{\mathcal{J}_2}(\Phi(D(a(V)))) = \Psi^{\mathcal{J}_2}(T_{\widetilde{V}P_1}(b))$, and localizing at $1/t_0$ we get $a(t_0) = b(1/t_0)$, which implies that b is continuous at the point $1/t_0$ and that $b(1/t_0) = a(t_0)$. □

Corollary 4.12. *If $(A_n) = D(a(V))$ then we have $\widetilde{W}(A_n) = T_{\widetilde{V}}(\tilde{a})$ for every non-unitary $\widetilde{V}$.*

Proof. We can apply the localization described in Proposition 4.11 almost everywhere on $\mathbb{T}$. So the function b (connected with $\Phi(D(a(V)))$) is seen to be equal to $\tilde{a}$ at all points except a set of measure zero. □

As the following result shows, under additional conditions it is possible to identify all local algebras and so to obtain invertibility criteria for the coset $(A_n) + \mathcal{J}_0$ in $\mathcal{A}^{\mathcal{J}_0}$.

Corollary 4.13. *Suppose D transforms positive elements into positive elements and let $(A_n) \in \mathcal{A}(PC)$. Then the sequence (A_n) is stable if and only if the operators $W(A_n)$ and $\widetilde{W}(A_n)$ are invertible. Moreover, if $a \in PC(\mathbb{T})$, then the sequence $(D(a(V)))$ is stable if and only if the operator $a(V)$ is invertible.*

Proof. Define χ_t^+ as the characteristic function of the arc $[t, te^{i\pi}]$ and $\chi_t^- := 1 - \chi_t^+$. Given $a \in PC(\mathbb{T})$ define also $a_t = a(t^+)\chi_t^+ + a(t^-)\chi_t^-$. Proceeding as in the proof of Proposition 4.11, it is not difficult to see that $\Psi_{t_0}^{\mathcal{J}_2}(\Phi(D(a(V)))) = \Psi_{t_0}^{\mathcal{J}_2}(\Phi(D(a_{t_0}(V))))$ at all points $t_0 \in \mathbb{T}$. So the local algebras are generated by the identity and $(\Phi(D(\chi_t^+(V)))) + \mathcal{J}_2^{\mathcal{J}_1}$. To get a complete description of invertibility in this algebra it is sufficient to know the local spectrum of $\Phi(D(\chi_t^+(V))) + \mathcal{J}_2^{\mathcal{J}_1}$. The element $(\Phi(D(\chi_t^+(V)))) + \mathcal{J}_2^{\mathcal{J}_1}$ is positive because $\chi_t^+(V)$ is positive, and $\|\Phi(D(\chi_t^+(V)))) + \mathcal{J}_2^{\mathcal{J}_1}\| \leq \|\chi_t^+\|_\infty = 1$. Therefore the spectrum must be contained in the segment $[0, 1]$.

On the other hand the homomorphism W is well defined from $\mathcal{A}^{\mathcal{J}_1}(PC)$ into the Calkin algebra $\mathcal{B}^{\mathcal{K}}(PC)$ and $W(\mathcal{J}_2^{\mathcal{J}_1}) = 0$. So we can define a homomorphism from $\mathcal{A}^{\mathcal{J}_0}$ into $\mathcal{B}^{\mathcal{K}}(PC)$, which will be denoted by the same symbol W. As $W(\Phi(D(\chi_t^+(V))) + \mathcal{J}_2^{\mathcal{J}_1}) = \chi_t^+(V)$, the spectrum of $(\Phi(D(\chi_t^+(V)))) + \mathcal{J}_2^{\mathcal{J}_1}$ must contain the spectrum of $\chi_t^+(V)$ in $\mathcal{B}^{\mathcal{K}}(PC)$, which is exactly the segment $[0,1]$ (see, for instance, [5]). So the Fredholmness of $W(A_n)$ implies the invertibility of the coset $(A_n) + \mathcal{J}_0$ in $\mathcal{A}^{\mathcal{J}_0}$.

Regarding the last part of the assertion, in the case where $\Phi(D(V^*))$ is unitary in $\mathcal{A}^{\mathcal{J}_1}(C)$ the result is proved. If $\Phi(D(V^*))$ is non-unitary, then note that for scalar $a \in L^\infty(\mathbb{T})$, the invertibility of $T_{\tilde{V}}(\tilde{a})$ is equivalent to the invertibility of $W(D(a(V))) = a(V) = T_V(a)$. □

5. Finite section method for Toeplitz operators as example

The simplest of all applications of the results above is the finite section method for Toeplitz operators. Let $\mathbb{I} = \mathbb{N}$, $H = l^2$, V be the unilateral shift

$$(x_0, x_1, x_2, \ldots) \mapsto (0, x_0, x_1, \ldots)$$

and $H_n = \mathbb{C}^n$, with P^n being the projection

$$(x_0, x_1, x_2, \ldots) \mapsto (x_0, x_1, \ldots, x_{n-1}, 0, 0, \ldots).$$

The discretization D is then defined by $D(a(V))) = (P^n T(a) P^n)$, and so $\mathcal{A} = \operatorname{alg}\{(P^n T(a) P^n) : a \in PC\}$. Properties (H1) and (H2) can be easily verified by simple computation. Regarding (H3), we have (see, for instance, [6, Chapter 1]) that

$$\mathcal{J}_2 = \{(R^n K R^n) : K \in \mathcal{K}\} + \mathcal{G},$$

where R^n is the mapping $(x_0, x_1, x_2, \ldots) \mapsto (x_{n-1}, x_{n-2}, \ldots, x_2, x_1, 0, 0, \ldots)$. So $\mathcal{J}_2$ coincides with the ideal corresponding to the continuous case. It can also be easily verified that

$$R^n T(a) R^n = P^n T(\tilde{a}) P^n$$

and hence the homomorphism $\widetilde{W}$ can be defined directly by

$$\widetilde{W}(A_n) = \operatorname{s-lim} R^n A_n R^n.$$

Note that this example has a very important peculiarity, namely that the projections P^n are directly related to the shift V, through the identity

$$P^n = I - V_n V_{-n}.$$

In general, that is not the case.

We remark that the results of [7] concerning fractality can be incorporated in the axiomatic approach discussed here. In this way it is possible to prove convergence of condition numbers and limiting sets of singular values, for instance. In a forthcoming paper the authors intend to explore this subject more deeply and to extend the concepts presented herein to more general types of sequences (multiindiced sequences) and more general generating functions.

Acknowledgements

Bernd Silbermann was supported by grant Praxis XXI/BCC/20247/99 from FCT.

References

[1] Böttcher, A., and Silbermann, B., *Analysis of Toeplitz Operators*, Akademie-Verlag, Berlin 1989 and Springer-Verlag, Berlin, Heidelberg, New York 1990.

[2] Böttcher, A., and Silbermann, B., *Introduction to Large Truncated Toeplitz Matrices*, Springer-Verlag, New York 1999.

[3] Coburn, L.A., The C^*-algebra generated by an isometry I, *Bull. Amer. Math. Soc.* **73** (1967), 722–726.

[4] Coburn, L.A., The C^*-algebra generated by an isometry II, *Trans. Amer. Math. Soc.* **137** (1969), 211–217.

[5] Gohberg, I.C., and Feldman, I.A., *Convolution Equations and Projection Methods for Their Solution*, Amer. Math. Soc. Transl. of Math. Monographs 41, Providence, R.I. 1974.

[6] Hagen, R., Roch, S., and Silbermann, B., *Spectral Theory of Approximation Methods for Convolution Equations*, Operator Theory: Advances and Applications, Vol 74, Birkhäuser Verlag, Basel 1995.

[7] Hagen, R., Roch, S., and Silbermann, B., *C^*-algebras and numerical analysis*, Monographs and Textbooks in Pure and Applied Mathematics, 236. Marcel Dekker, Inc., New York 2001.

[8] Kozak, A., A local method in the theory of projection methods, *Soviet Math. Dokl.* **14** (1974), 1580–1583.

[9] Nikolskii, N.K., *Treatise on the Shift Operator*, Springer-Verlag, Berlin 1985.

[10] Prössdorf, S., and Silbermann, B., *Numerical Analysis for Integral and Related Operator Equations*, Akademie-Verlag, Berlin 1991 and Birkhäuser Verlag, Basel 1991.

[11] Roch, S., and Silbermann, B., Toeplitz-like operators, quasicommutator ideals, numerical analysis, I, *Math. Nach.* **120** (1985), 141–173.

[12] Roch, S., and Silbermann, B., Toeplitz-like operators, quasicommutator ideals, numerical analysis, II, *Math. Nach.* **134** (1987), 245–255.

[13] Roch, S., and Silbermann, B., C^*-algebra techniques in numerical analysis, *J. Operator Theory* **35** (1996), 241–280.

[14] Santos, P.A., and Silbermann, B., Galerkin method for Wiener-Hopf operators with piecewise continuous symbol, *Integral Equations and Operator Theory* **38** (2000), 66–80.

[15] Rosenblum, M., and Rovnyak, J., *Hardy Classes and Operator Theory*, Oxford University Press, New York 1985.

Pedro A. Santos
Departamento de Matemática
Instituto Superior Técnico
Av. Rovisco Pais
1049-001 Lisboa, Portugal
e-mail: pasantos@math.ist.utl.pt

Bernd Silbermann
Fakultät für Mathematik
TU Chemnitz
D-09107 Chemnitz, Germany
e-mail: bernd.silbermann@mathematik.tu-chemnitz.de

2000 *Mathematics Subject Classification.*
Primary 65J10; Secondary 47B35.

Received: 7 May 2001.

MIX
Papier aus verantwortungsvollen Quellen
Paper from responsible sources
FSC® C105338

If you have any concerns about our products,
you can contact us on
ProductSafety@springernature.com

In case Publisher is established outside the EU,
the EU authorized representative is:
Springer Nature Customer Service Center GmbH
Europaplatz 3, 69115 Heidelberg, Germany

Printed by Libri Plureos GmbH
in Hamburg, Germany